Right Triangle

$c^2 = a^2 + b^2$ [Pythagorean theorem]

$\text{area} = \dfrac{1}{2}ab$

$\sin\theta = \dfrac{b}{c} = \dfrac{\text{opp}}{\text{hyp}}$

$\cos\theta = \dfrac{a}{c} = \dfrac{\text{adj}}{\text{hyp}}$

$\tan\theta = \dfrac{\sin\theta}{\cos\theta} = \dfrac{b}{a} = \dfrac{\text{opp}}{\text{adj}}$

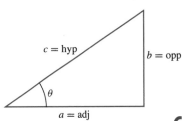

Circular Sector

$\text{area} = \dfrac{1}{2}r^2\theta$

$\text{arclength} = s = r\theta$

Circular Cylinder

$\text{volume} = \pi r^2 h$

$\text{area of curved surface} = 2\pi r h$

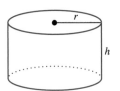

Any Triangle

$\text{area} = \dfrac{1}{2}bh = \dfrac{1}{2}ab\sin C$

$\left.\begin{array}{l} a^2 = b^2 + c^2 - 2bc\cos A \\ b^2 = a^2 + c^2 - 2ac\cos B \\ c^2 = a^2 + b^2 - 2ab\cos C \end{array}\right\}$ [Law of cosines]

$\dfrac{\sin A}{a} = \dfrac{\sin B}{b} = \dfrac{\sin C}{c}$ [Law of sines]

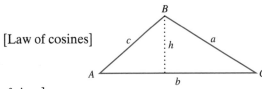

Trapezoid

$\text{area} = \dfrac{1}{2}b(h_1 + h_2)$

Circular Cone

$\text{volume} = \dfrac{1}{3}\pi r^2 h$

$\text{area of curved surface} = \pi r\sqrt{r^2 + h^2}$

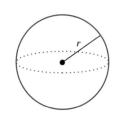

Circle

$\text{area} = \pi r^2$

$\text{circumference} = 2\pi r$

Sphere

$\text{volume} = \dfrac{4}{3}\pi r^3$

$\text{surface area} = 4\pi r^2$

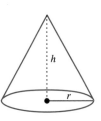

Algebraic Identities

$ax^2 + bx + c = 0 \iff x = \dfrac{-b \pm \sqrt{b^2 - 4ac}}{2a}$

[Quadratic formula]

$a^2 - b^2 = (a+b)(a-b)$

$a^3 + b^3 = (a+b)(a^2 - ab + b^2)$

$a^3 - b^3 = (a-b)(a^2 + ab + b^2)$

$(a+b)^n = a^n + na^{n-1}b + \cdots + \binom{n}{k}a^{n-k}b^k + \cdots + b^n;$

$\binom{n}{k} = \dfrac{n!}{(n-k)!\,k!}$ [Binomial theorem]

Trigonometric Identities

$\sin^2 x + \cos^2 x = 1$

$\sec^2 x = 1 + \tan^2 x$

$\sin(2x) = 2\sin x \cos x$

$\cos(2x) = \cos^2 x - \sin^2 x = 2\cos^2 x - 1 = 1 - 2\sin^2 x$

$\sin^2 x = \dfrac{1}{2}(1 - \cos(2x))$

$\cos^2 x = \dfrac{1}{2}(1 + \cos(2x))$

$\sin(x \pm y) = \sin x \cos y \pm \cos x \sin y$

$\cos(x \pm y) = \cos x \cos y \mp \sin x \sin y$

$\tan(x \pm y) = \dfrac{\tan x \pm \tan y}{1 \mp \tan x \tan y}$

$\sin x \sin y = \dfrac{1}{2}(\cos(x - y) - \cos(x + y))$

$\cos x \cos y = \dfrac{1}{2}(\cos(x - y) + \cos(x + y))$

$\sin x \cos y = \dfrac{1}{2}(\sin(x + y) + \sin(x - y))$

Derivative Formulas

$\dfrac{d}{dx} c = (c)' = 0$

$\dfrac{d}{dx} x = (x)' = 1$

$\dfrac{d}{dx} x^n = (x^n)' = nx^{n-1}$

$\dfrac{d}{dx} e^x = (e^x)' = e^x$

$\dfrac{d}{dx} \ln x = (\ln x)' = \dfrac{1}{x}$

$\dfrac{d}{dx} \sin x = (\sin x)' = \cos x$

$\dfrac{d}{dx} \cos x = (\cos x)' = -\sin x$

$\dfrac{d}{dx} \tan x = (\tan x)' = \sec^2 x$

$\dfrac{d}{dx} \sec x = (\sec x)' = \sec x \tan x$

$\dfrac{d}{dx} \arcsin x = (\arcsin x)' = \dfrac{1}{\sqrt{1 - x^2}}$

$\dfrac{d}{dx} \arctan x = (\arctan x)' = \dfrac{1}{1 + x^2}$

Derivative Rules

$(f(x) + g(x))' = f'(x) + g'(x)$ [Sum Rule]

$(f(x)g(x))' = f'(x)g(x) + f(x)g'(x)$ [Product Rule]

$\left(\dfrac{f(x)}{g(x)}\right)' = \dfrac{f'(x)g(x) - f(x)g'(x)}{(g(x))^2}$ [Quotient Rule]

$((f \circ g)(x))' = \left(f(g(x))\right)' = f'(g(x))g'(x)$ [Chain Rule]

Hyperbolic Functions

$\cosh x = \dfrac{e^x + e^{-x}}{2}$

$\sinh x = \dfrac{e^x - e^{-x}}{2}$

$\tanh x = \dfrac{\sinh x}{\cosh x} = \dfrac{e^x - e^{-x}}{e^x + e^{-x}}$

$\cosh^2 x - \sinh^2 x = 1$

$\cosh(x + y) = \cosh x \cosh y + \sinh x \sinh y$

$\sinh(x + y) = \sinh x \cosh y + \cosh x \sinh y$

$\dfrac{d}{dx} \cosh x = (\cosh x)' = \sinh x$

$\dfrac{d}{dx} \sinh x = (\sinh x)' = \cosh x$

$\dfrac{d}{dx} \tanh x = (\tanh x)' = \dfrac{1}{\cosh^2 x}$

$\cosh^{-1} x = \ln\left(x + \sqrt{x^2 + 1}\right)$

$\sinh^{-1} x = \ln\left(x + \sqrt{x^2 - 1}\right), \quad x \geq 1$

$\tanh^{-1} x = \dfrac{1}{2} \ln\left(\dfrac{1 + x}{1 - x}\right), \quad |x| < 1$

$\dfrac{d}{dx} \cosh^{-1} x = (\cosh^{-1} x)' = \dfrac{1}{\sqrt{x^2 - 1}}$

$\dfrac{d}{dx} \sinh^{-1} x = (\sinh^{-1} x)' = \dfrac{1}{\sqrt{1 + x^2}}$

$\dfrac{d}{dx} \tanh^{-1} x = (\tanh^{-1} x)' = \dfrac{1}{1 - x^2}$

Calculus

From Graphical, Numerical, and Symbolic Points of View

SECOND EDITION

Volume 2

Calculus

From Graphical, Numerical, and Symbolic Points of View

SECOND EDITION

Volume 2

Arnold Ostebee
St. Olaf College

Paul Zorn
St. Olaf College

BROOKS/COLE

THOMSON LEARNING

Australia • Canada • Mexico • Singapore • Spain
United Kingdom • United States

Math Editor: Angus McDonald
Development Editor: Leslie Lahr
Marketing Manager: Julia Conover
Production Manager: Alicia Jackson
Print/Media Buyer: Vanessa Jennings
Production Service: TechBooks
Text Designer: Lisa Adamitis
Art Director: Carol Bleistine
Copy Editor: John Joswick
Illustrator: TechBooks
Cover Designer: Lisa Adamitis
Cover Image: Surf: Maui, Hawaii © Bill
Brooks/Masterfile
Cover Printer: Lehigh Press
Compositor: TechBooks
Printer: R.R. Donnelley

*Calculus: from Graphical, Numerical, and
Symbolic Points of View, Volume 2, second edition*
ISBN 0-03-025676-3
Library of Congress Control Number:
2001090881

Asia
Thomson Learning
60 Albert Complex, #15-01
Albert Complex
Singapore 189969

Australia
Nelson Thomson Learning
102 Dodds Street
South Street
South Melbourne, Victoria 3205
Australia

Canada
Nelson Thomson Learning
1120 Birchmount Road
Toronto, Ontario M1K 5G4
Canada

Europe/Middle East/South Africa
Thomson Learning
Berkshire House
168-173 High Holborn
London WC1 V7AA
United Kingdom

Latin America
Thomson Learning
Seneca, 53
Colonia Polanco
11560 Mexico D.F.
Mexico

Printed in the United States of America
1 2 3 4 5 6 7   05 04 03 02 01

Spain
Paraninfo Thomson Learning
Calle/Magallanes, 25
28015 Madrid, Spain

**For more information about our products,
contact us at:
Thomson Learning Academic Resource Center
1-800-423-0563**

**For permission to use material from this text,
contact us by:
Phone: 1-800-730-2214
Fax: 1-800-731-2215
Web: www.thomsonrights.com**

To Kay, Kristin, and Paul
To Janet, Anne, and Libby

ABOUT THIS BOOK:
Notes for Instructors

This book—like the first edition—aims to do exactly what its title suggests: present calculus from graphical, numerical, and symbolic points of view. In this preface we elaborate briefly on what this goal means, why we chose it, and how we try to achieve it.

For more information   More detailed information on many topics related to this book can be found at the following Web sites:

```
www.stolaf.edu/people/zorn/ozcalc/
www.harcourtcollege.com/math/oz2/
```

Three volumes

We treat single-variable calculus in two volumes, each suitable for a semester-length course; Volume 3 offers a one-semester introduction to multivariable and vector calculus, culminating with Green's theorem and its three-dimensional analogues. Several "combination" volumes are also available; please see the Web sites for details.

Changes from the first edition

Why did we revise the first edition of our textbook? In short, to make the text easier to use, for teachers *and* for students. To this end, building on many suggestions from teachers and students, we've made various changes to narrative, exercises, content, emphasis, and order of presentation. Some of these changes are outlined briefly below. Much more detailed information about the second edition, including a chapter-by-chapter overview of changes and a list of frequently asked questions, can be found at the Web sites.

- **What hasn't changed**   The basic principles and strategies underlying the first edition remain unchanged. Conceptual understanding is still our main goal, and combining various viewpoints is still the main strategy for achieving it. This edition and the first share basic assumptions and operating premises: an emphasis on concepts and sense-making, complementing symbolic with graphical and numerical points of view, exercises of varied nature and difficulty, and a narrative aimed at student (rather than faculty) readers.

- **Getting more quickly to derivatives** We've somewhat compressed and redistributed the "precalculus" material of the first edition to get to the derivative idea faster. (The word *derivative* now appears in Section 1.4.) Chapter 1 now includes essentially complete coverage of the graphical approach, Chapter 2 introduces and interprets the symbolic point of view, and Chapter 3 presents the combinatorial rules for differentiation.

- **Earlier differential equations** Differential equations (DEs) appear a little earlier and reappear more often. The idea of a DE is introduced in Section 2.5, soon after the ideas of derivative and antiderivative are first met in symbolic form. Mentioning DEs early lets us say early (and economically) that polynomial, exponential, and trigonometric functions have important growth properties that help account for the importance of these function classes. DEs also provide a natural approach to scientific and engineering applications. Once introduced, DEs reappear often in following sections and in exercises.

- **Not a DE course** Although we give DEs somewhat increased emphasis in this edition, we are not jamming a DE course into an already well-filled calculus syllabus. We want to convey the *idea* of a DE and its solutions, but we make *no* effort to cover or even catalog the huge variety of DEs and solution techniques.

- **A new chapter on function approximation** Chapter 9 treats approximation of one function by another. Taylor polynomials are the main objects of study; Taylor's theorem is the main theoretical result. Separating these ideas from convergence and divergence of series should aid students' understanding. It also permits teachers who wish to do so to cover Taylor's theorem without requiring a long introduction to infinite series.

- **Helping students read** We want students to *read* the narrative—not just scan the assigned problems and search for cloned examples. To help students read more successfully, we've added many more examples, rewritten many parts of the narrative, and included more detail and brief commentary on many calculations.

- **Other content changes and reorganization** Various other changes have been made to content and organization; the following are several:
  - earlier treatment of slope fields and l'Hôpital's rule (now in Chapter 4);
  - a new section on sigma notation and some of its properties (in Chapter 5);
  - earlier treatment of basic antidifferentiation ideas, including the method of substitution (in Chapter 5).

  For more information on changes to content and organization, please see the Web sites.

- **Exercises and answers** Many new exercises, of various types, have been added. Many users found the first edition short on "routine" exercises; many more have been added (but this edition still contains plenty of nonroutine problems). Other new exercises point to specific issues and examples raised in the narrative. A third new category is end-of-chapter exercises, which provide review and encourage synthesis. We have also included printed answers (not full-fledged solutions) to most of the odd-numbered exercises. (Students *and* teachers requested this change!)

- **End-of-chapter Interludes** Most chapters end with one or more Interludes—brief, project-oriented expositions designed for independent student work—addressing topics or questions that are "optional" or out of that chapter's main stream of development.

Philosophy

What should a beginning course in calculus be about? What topics and ideas should it contain? What are its most important goals and organizing principles? What level of mathematical rigor is appropriate? How should applications and theory be balanced?

Questions like these have been "live" for a very long time. For instance, debates over how and whether to teach related rates problems, and more generally over the role of applications, sprang up in the 1850s; for a delightful exposition, see "The Lengthening Shadow: The Story of Related Rates," by Bill Austin, Don Barry, and David Berman, *Mathematics Magazine*, February 2000. The relatively recent discussions on "calculus reform" continue an old tradition of questioning and revision, though in a new and different educational and technological environment. Although the phrase *calculus reform* now has little clear meaning, the continuing discussion it denotes has been useful both in directing attention to calculus courses and in identifying a few common threads of widespread (if not universal) agreement. These include calls for leaner, more concept-focused courses and for courses that better reflect modern technology in content and in pedagogy.

Focus on concepts; our strategy   Elementary calculus courses can serve various purposes: to introduce analytic mathematics, to support mathematical applications in the sciences, to provide a language for stating physical laws and deducing their consequences, to introduce mathematical ways of thinking. Whatever the goals of any particular course, the conclusion is the same: Concepts are fundamental. Whatever uses they make of the calculus, students need more than a compendium of manipulative techniques; they need conceptual understanding that is deep and flexible enough to accommodate diverse applications.

Our key strategy for improving conceptual understanding is combining, comparing, and moving among graphical, numerical, and algebraic representations of central concepts. By studying, representing, and manipulating ideas and objects in several forms, students gain better perspective on, and deeper understanding of, important concepts.

Audience and prerequisites

Our text is aimed at a broad mathematical audience, including (but not limited to) students majoring in mathematics, science, and engineering. Although we emphasize concepts more than detailed applications to specific disciplines, we regard a more conceptual calculus as also more applicable. To use calculus ideas and techniques effectively in any area, students must know what they are doing and why, not just how.

We expect that students have the "usual" precalculus preparation, including basic algebra and trigonometry and the rudiments of logarithmic and exponential functions. However, the appendices include reviews of these subjects, with routine exercises, written in a style that supports independent reading and self-study. (Appendices do not appear in the stand-alone Volume 2.)

Distinctive features

Several special features of our treatment deserve brief mention. For many more details and specific examples, see the *Instructor's Resource Manual*, described below.

- **Early differential equations**   Differential equations are mentioned early in Chapter 2, introduced informally (but at some length) in Chapter 4, and then revisited from time to time in later chapters. We believe that DEs are so basic and so important that students should see them early and often. Calculus I emphasizes the *idea* of a DE and its solutions. Several techniques for finding numerical and symbolic solutions are treated in Calculus II.

- **Graphs, then symbols**   Derivatives and integrals are introduced in a parallel manner, in which the graphical approach precedes the symbolic viewpoint. Thus, Chapter 1 stresses the *geometry* of the derivative; the *symbolic* approach enters in Chapter 2.

Similarly, Chapter 5 begins with a graphical definition of the integral and proceeds later to the Riemann sum definition.

- **The level of rigor**   Proving theorems in full generality is less valuable, we think, than helping students understand concretely what theorems say, why they're reasonable, and why they matter. Too often, fully rigorous proofs address questions that students are unprepared to ask.

   Still, we believe that introducing calculus students to the idea of proof—and to some especially important classical proofs—is essential. We prove major results but emphasize only those proofs that we believe contribute significantly to understanding calculus concepts. In examples and problems, too, we pay attention to developing analytic skills and synthesizing mathematical ideas.

- **Foreshadowing**   Important concepts and associated vocabulary often appear more than once—first informally and later in more detail. The geometric idea of concavity, for instance, is mentioned in Section 1.1 and then revisited several times. Instructors should know that it is *not* necessary to cover these ideas in full rigor at their first appearance.

The role of technology

Combining graphical, numerical, and symbolic views of important calculus ideas is forbiddingly laborious by hand. Technology makes this useful combination feasible.

   We see computers and graphing calculators as valuable and practical tools, but computing is not an end in itself in our course. We want students to focus on key calculus ideas, not on machines, and so we encourage thoughtful and balanced uses of computing. We believe, for instance, that students benefit substantially from frequent, careful study of well-drawn graphs; thus, we often encourage students to generate their own graphs (often by machine but sometimes by hand) to suggest guesses, gain insights, and check answers for plausibility.

Symbolic computing—boon or bane?   Symbolic computing has become widely available to students in the last few years; even inexpensive calculators such as the TI–89 now handle symbolic derivatives and even antiderivatives. What's left to teach when every student has such a machine? Although some may fear that sophisticated technology reduces calculus to mindless button-pushing, we find exactly the opposite effect more common in practice. With machines available to handle some routine calculations, calculus courses can and should become more, not less, focused on concepts and structures. Problems that are posed graphically, for instance, are essentially impervious to technology. More traditional, find-the-derivative-type problems, by contrast, are more likely to need new reasons for being.

   We make no assumption either way about students' access to symbolic computing. Given the wide variety of problems and viewpoints we present, we think our text makes sense in almost any technological environment.

Symbol manipulation and hand computation   We take the symbolic point of view in our title just as seriously as the graphical and the numerical, and we cover the "usual" methods of formal differentiation and antidifferentiation. Why do we do so?

   The fact that machines can "do" symbolic manipulations does not render by-hand operations obsolete. Nor are skills and ideas truly in opposition, as is sometimes asserted. On the contrary, a reasonable modicum of manipulative practice builds and supports conceptual understanding. Hand computation helps fix ideas, illustrates concepts concretely, builds symbolic intuition, and gives students a sense of mastery. It does not follow, however, that harder, more "baroque" computational problems are always better; we deemphasize convoluted symbolic calculations in order to free time and energy for work on concepts.

About the exercises

However clear its exposition, a textbook's exercises generate most of students' mathematical activity and occupy most of their time. The exercises we assign tell students concretely what we think they should know and do. Both routine drills and challenging theoretical exercises are standard in calculus texts, and our text contains some of each. More distinctive, perhaps, are the exercises that fall between these poles:

- Exercises that combine and compare algebraic, graphical, and numerical viewpoints and techniques.
- Exercises that test and stretch a student's understanding of concepts, definitions, and theorems.
- Exercises that develop students' problem-solving and mathematical reasoning skills.
- Exercises that use calculus as a *language* for interpreting and solving problems. Students "translate" problems into mathematical language, solve the resulting problems using the calculus tools, and reinterpret results in the original context.

''Basic'' and ''Further'' exercises   Each section has exercises of two types: "Basic" and "Further." Typical "Basic" exercises are relatively straightforward and focus on a single important idea. All students should aim to master most of these exercises. "Further" exercises are a little more ambitious; they may require the synthesis of several ideas, deeper or more sophisticated understanding of basic concepts, or better symbol manipulation skills. For more details and advice on assigning exercises, see the *Instructor's Resource Manual*, described below.

Exercises and technology   Because we view graphing and other technologies as tools for exploring and problem solving, not as ends in themselves, we seldom tag or label "technology exercises." We expect students themselves to decide in context which tools are most appropriate and how and when to employ them.

Annotated Table of Contents, Volume 1

Brief chapter-by-chapter information on Volume 1 follows. (See the Web sites for more detailed descriptions and for information about Volume 3.)

Chapter 1: Functions and Derivatives: The Graphical View   We move quickly to the subject of derivatives and rate functions, giving the graphical viewpoint first. Standard elementary functions appear as examples, but we don't—in this chapter—use nontrivial symbolic properties of these functions. The "Field Guide" section briefly reviews the standard elementary functions but concentrates mainly on their graphical properties.

Chapter 2: Functions and Derivatives: The Symbolic View   We review and expand on ideas of Chapter 1, but now in the symbolic setting. The ideas of antiderivative and differential equation appear for the first time—systematic treatments appear later. Sections 2.5, 2.6, and 2.7 describe some classical applications (linear motion, growth, and oscillation) related to finding derivatives and (in the simplest cases) antiderivatives of polynomial, exponential, and trigonometric functions.

Chapter 3: New Derivatives from Old   This chapter treats "combinatorial" ideas: producing new functions and new derivatives from old functions and *their* derivatives. We also cover implicit differentiation and derivatives of inverse functions and then apply these ideas to inverse trigonometric functions.

Chapter 4: Using the Derivative   This chapter is a selection of applications, extensions, and uses of the derivative and other ideas from earlier chapters. Some sections are independent of each other; not all are required for later work. Sections 4.8 and 4.9 (on continuity and differentiability) offer glimpses at some of the theoretical side of calculus. The mean value theorem, for instance, is discussed and proved fairly carefully in Section 4.10.

Chapter 5: The Integral   The first few sections introduce the definite integral *geometrically*, as signed area. Then the integral is linked to antiderivatives via the fundamental theorem. Substitution and use of tables, the most basic methods of finding antiderivatives, are covered here; much more discussion of antidifferentiation is given in Chapter 8. The general treatment of integrals mirrors that of derivatives: We start with geometric intuition and proceed to the limit-based analytic definition.

Annotated Table of Contents, Volume 2

Brief chapter-by-chapter information on Volume 2 follows. (See the Web sites for more detailed descriptions and for information about Volume 3.)

Chapter 5: The Integral   This chapter appears in both volumes for instructors' convenience.

Chapter 6: Numerical Integration   This chapter treats basic numerical views of the integral. Section 6.1 presents and compares different types of approximating sums. Some error-bound analysis appears in Section 6.2 (which some instructors may omit). Section 6.3, on Euler's method, applies similar numerical ideas to solve differential equations numerically. Interludes treat Simpson's rule and Gaussian quadrature.

Chapter 7: Using the Integral   This chapter treats applications, extensions, and uses of the integral—geometric, physical, mathematical, and economic. (As in Chapter 4, instructors can choose among topics.) Section 7.1 offers a variety of applications, including arc length, chosen partly to show the integral as a general tool for measuring things. Later sections and an Interlude treat volumes, work, separable DEs, present value, mass, and center of mass. Some integrals that arise might be done with the help of numerical methods, tables, or computers.

Chapter 8: Symbolic Antidifferentiation Techniques   The chapter contains a fairly standard menu of techniques. Even though antidifferentiation is readily handled by computer nowadays, we think that some practice of this sort is useful for building symbol manipulation skills of the sort many students will need in later courses; such practice also requires students to recognize and grapple with the structure of various classes of functions. Interludes treat functions *defined* by integration and methods for solving first-order linear DEs.

Chapter 9: Function Approximation   The chapter itself is entirely new to this edition, although most of its contents appeared (elsewhere) in the first edition. Moreover, the general theme of approximating one function with another runs through both volumes (e.g., in connection with differentiable functions being "almost" linear). Sections 9.1 and 9.2, on Taylor polynomials and Taylor's theorem, are intended to help disentangle these important ideas from questions of series convergence and divergence—topics that are sometimes conflated. Section 9.3 introduces Fourier approximation—an integral-based counterpart to Taylor approximation. An Interlude treats splines.

Chapter 10: Improper Integrals Improper integrals are both practically useful and closely analogous to infinite series—hence our relatively extended treatment. Many texts cover the subject in only one section, but we hope to use improper integrals to build intuition for the difficult ideas of infinite series, coming in Chapter 11. Section 10.3 applies improper integrals to probability, with emphasis on the normal probability distribution.

Chapter 11: Infinite Series Unlike some reform texts, we do treat convergence and divergence of numerical series but in a somewhat nontraditional way. We stress (i) the analogy with improper integrals; (ii) concrete (sometimes graphical) treatment of partial sums; and (iii) numerical estimation of limits. We think these strategies help make this difficult subject more concrete and accessible than it often is when the principal concern is the abstract question of convergence or divergence. The chapter ends with power series— but, with the new Chapter 9, students will already have had some exposure to this idea. A brief Interlude explores Fourier series.

Chapter V: Vectors and Polar Coordinates This chapter is a basic introduction to vectors in the plane, vector notation, vector-valued functions and derivatives, and to the polar coordinate view of the plane and of plane curves. (Students who continue to Calculus III will see this material in more detail later—this treatment is aimed mainly at students who will not continue in calculus.)

Chapter M: Multivariable Calculus: A First Look This chapter is a basic introduction to rudiments of functions of several variables. Some instructors may choose to cover it as an alternative to (or, if time permits) in addition to infinite series. Some instructors might choose material on partial derivatives but not on multiple integrals. Like Chapter V, this chapter is aimed mainly at students who will not continue in calculus.

Selections from Volume 1 Several sections from Volume 1 are reprinted in Volume 2 for instructors' convenience. They cover (1) inverse functions and their derivatives (including inverse trigonometric functions); (2) limits involving infinity and l'Hôpital's rule; (3) parametric curves.

Supplements to the text

The following aids for students and instructors are available from the publisher:

- **Student Solutions Manual** In three volumes, these manuals contain solutions, prepared by the authors, to all odd-numbered exercises. (Brief answers to selected exercises are available in the back of the text.)

- **Technology Projects and Lab Manual** Written by Glen Van Brummelen and Michael Caraco of Bennington College, this manual enables students and instructors to explore a wide range of technology products available to mathematicians and engineers. It offers the foundations for innovative projects and lab activities as well as tutorials and hints for using a variety of graphing calculators and math software packages, such as *Maple* and *Mathematica*. As students complete each lab, they learn basic commands and reinforce key calculus concepts.

- **Navigating Calculus** CD–ROM The companion CD–ROM, *Navigating Calculus*, authored by Jason Brown of Dalhousie University in Nova Scotia, and by Arnold Ostebee and Paul Zorn, is keyed closely to the book's table of contents and covers both single variable and multivariable material. *Navigating Calculus* contains a variety of useful activities, tools, and resources, including a powerful graphing calculator utility, a glossary with examples, and many interactive activities that deepen students' understanding of calculus fundamentals. This learning aid is accompanied by the *Navigating Calculus Workbook* written by Stephen Kokoska of Bloomsburg University of Pennsylvania. This workbook is designed to help both instructors and students fully utilize *Navigating Calculus* by offering guided instruction through the workings of the CD–ROM and providing examples and exercises that supplement those on the CD–ROM itself.

- **Web-based learning resources**   Additional learning tools can be found at the following Web site:

  www.harcourtcollege.com/math/oz2/

Instructors who adopt this text may receive the following items free of charge:

- **Annotated Instructor's Edition**   This special edition includes the full text of the student version together with marginal annotations that offer pedagogical hints and suggestions.

- **Instructor's Resource Manual**   This manual offers a variety of additional teaching resources, including (1) advice on using technology, facilitating group work, assigning homework, and testing; (2) sample lecture plans; (3) section-by-section notes; and (4) writing assignments.

- **Instructor's Solutions Manual**   In three volumes, the manual contains complete solutions, prepared by the authors, for all exercises.

- **Test Bank**   Prepared by Scott Inch of Bloomsburg University of Pennsylvania, this resource contains more than 500 questions covering multiple viewpoints and levels of difficulty. Answers to all questions are included.

- **Computerized Test Bank**   Available in Windows and Macintosh formats, this resource combines the items of the Test Bank with convenient features that allow instructors to prepare quizzes and examinations quickly and easily. Instructors can edit questions, add their own, administer tests over a computer network, and record student grades with the gradebook software.

- **Electronic Overhead Transparencies**   These may be downloaded in PDF format from the instructor's resource section of the Harcourt Web site. The set includes selected figures from both expository and exercise portions of the text.

For more information   For more information, please see the following Web sites:

  www.stolaf.edu/people/zorn/ozcalc/
  www.harcourtcollege.com/math/calc/oz2/

Advice from you

We appreciate hearing instructors' comments, suggestions, and advice on this edition. Our physical and e-mail addresses are below.

Arnold Ostebee and Paul Zorn
Department of Mathematics
St. Olaf College
1520 St. Olaf Avenue
Northfield, Minnesota   55057-1098

e-mail: **ostebee@stolaf.edu**   **zorn@stolaf.edu**

Acknowledgments

This text owes its existence to (literally) countless professors, students, publishing company professionals, friends, advisors, critics, "competitors, " family, and others. (These categories are not mutually exclusive!) It is a pleasure to acknowledge by name some—but, necessarily, only some—of the people who attended this text through its long gestation, birth, and publication.

Various versions of the manuscript were meticulously reviewed and suggestions made and errors caught by Mary Kay Abbey, Montgomery College – Takoma Park; Roy Alston, Stephen F. Austin State University; David Austin, Grand Valley State University; William Barnier, Sonoma State University; Neil Berger, University of Illinois – Chicago; Kelly Black, University of New Hampshire; Przemyslaw Bogacki, Old Dominion University; Jack Bookman, Duke University; Raouf N. Boules, Towson University; Holly Broesamle, Oakland Community College; Thomas Bullock, Brookdale Community College; Veena Chadha, University of Wisconsin – Eau Claire; Joseph Conrad, Solano Community College; Vittoria Cosentino, Metro Community College – Omaha; Ruth Dover, Illinois Math and Science Academy; Mary Ellen Davis, Georgia Perimeter College; David A. Edwards, University of Georgia; John Emert, Ball State University; Doug Ensley, Shippensburg University; Johanna Halsey, Dutchess Community College; Donald Hartig, California Polytechnic State University; Dorothy Hawkes, Solano Community College; Judy Holdener, Kenyon College; Linda Horner, Broward Community College – North; Henry Hosek, Purdue University Calumet; Alan Jian, Solano Community College; Matthias Kawski, Arizona State University; David Keller, Kirkwood Community College; James W. Lea, Middle Tennessee State University; Glenn Ledder, University of Nebraska; Mickey Levendusky, University of Arizona; Betty Liu, Monmouth University; Richard Maher, Loyola University of Chicago; Augustine Maison, Eastern Kentucky University; David Meredith, San Francisco State University; Edward S. Miller, Lewis–Clark State College; Charles Oelsner, Manlius Pebble Hill School; David Olson, Michigan Technological University; Sergei Ovchinnikov, San Francisco State University; Neville Robbins, San Francisco State University; Doug Shaw, University of Northern Iowa; Murray Siegel, Sam Houston State University; Marlene Sims, Kennesaw State University; Joanne Snow, St. Mary's College; Jerry Stonewater, Miami University of Ohio; K.D. Taylor, Utah Valley State College; Bruce Teague, Santa Fe Community College; and Kenneth Word, Central Texas College.

We are indebted, for many and different reasons, to members of Harcourt College Publishing's staff, including (alphabetically) Lisa Adamitis, Art Director; Carol Bleistine, Manager for Art & Design; Julie Conover, Executive Marketing Strategist; Alexa Epstein,

Manager of Technology; Bianca Huff and Pamela Meyers, Editorial Assistants; Alicia Jackson, Senior Production Manager; Amanda Loch, Technology Editor; Leslie Lahr, Developmental Editor; and Angus McDonald, Executive Editor.

We have learned much from professional colleagues, here at St. Olaf, elsewhere in the United States and around the world. Some are our teachers; some are departmental colleagues; some course-tested preliminary versions of these volumes at their own schools; some reviewed our grant proposals; some reviewed our project; some invited us to review theirs; some invited us to speak on panels or at workshops; some spoke on panels we organized; some helped us run our workshops; some are "calculus reform competitors"; some are advisers; some are critics. We list a small sample below, in alphabetical order. Regardless of category, we thank them all. Janet Andersen, Nazanin Azarnia, Russell Blythe, Mike Bolduan, Judith Cederberg, Caspar Curjel, Tom Dick, Ed Dubinsky, Wade Ellis, Bob Foote, Bonnie Gold, Michael Henle, Beth Hentges, Deborah Hughes Hallett, Allen Holmes, Paul Humke, Zaven Karian, John Kenelly, Dan Kennedy Charley Kerr, Harvey Keynes, Steve Kuhn, John Kurtzke, Reg Laursen, Sergio Loch, Joe May, Mike May, Bill McCallum, Richard Mercer, Eric Muller, Steve Monk, Lang Moore, Steven Olson, Lenore Parens, John Peterson, Tommy Ratliff, Wayne Roberts, Don Small, David Smith, Keith Stroyan, Douglas Swan, Tom Tucker, Jerry Uhl, Ted Vessey, and Roger Woods.

We thank St. Olaf College in general, and our departmental colleagues in particular, for their advice, support, good humor, and (sometimes) forbearance during the many years of this project's development and progress to a second edition. Countless students, here and at other institutions, also offered generous advice, praise, and criticism—all of it useful.

Our families, finally, deserve our deepest thanks. They have coped cheerfully with peculiar hours, extended absences, mental distraction, blizzards of paper, missed meals, and every other vagary that such a project entails. Without their love and sacrifice, we would never have begun—let alone completed—this project.

HOW TO USE THIS BOOK: NOTES FOR STUDENTS

All authors want their books to be *used*: read, studied, thought about, puzzled over, reread, underlined, disputed, understood, and, ultimately, enjoyed. So do we.

That might go without saying for some books—beach novels, user manuals, field guides, and so on—but it may need repeating for a calculus textbook. We know as teachers (and remember as students) that mathematics textbooks are too often read *backwards*: faced with Exercise 231(b) on page 1638, we've all shuffled backwards through the pages in search of something similar. (Very often, moreover, our searches were rewarded.)

A calculus textbook isn't a novel. It's a peculiar hybrid of encyclopedia, dictionary, atlas, anthology, daily newspaper, shop manual, *and* novel—not exactly light reading, but essential reading nevertheless. Ideally, a calculus book should be read in *all* directions: left to right, top to bottom, back to front, and even front to back. That's a tall order. Here are some suggestions for coping with it.

Read the narrative   Each section's narrative is designed to be read—perhaps more than once—from beginning to end. The examples, in particular, are supposed to illustrate ideas and make them concrete—not just to serve as templates for homework exercises.

Read the examples   Examples are, if anything, more important than theorems, remarks, and other "talk." *We* use examples both to show already-familiar calculus ideas "in action" and to set the stage for new ideas. *You* should use them to learn new ideas and review old ones.

Read the pictures   We're serious about the "graphical points of view" mentioned in our title. The pictures in this book are not "illustrations" or "decorations." Pictures are everywhere in this book—sometimes even in the middle of sentences. That's intentional: Graphs are an important part of the language of calculus. An ability to think "pictorially"—as well as symbolically and numerically—about mathematical ideas may be the most important benefit calculus can offer.

Read with a calculator and pencil   This book is full of requests ➔ to check a calculation, sketch a graph, or "convince yourself" that something makes sense. Take these requests

Sometimes they're put in margin notes, like this one.

seriously: Mastering mathematical ideas takes more than reading; it takes doing, drawing, and thinking.

Read the language   Mathematics is not a "natural language" like English or French, but it has its own vocabulary and rules of usage. Calculus, especially, relies on careful use of technical language. Words like rate, amount, stationary point, and root have precise, agreed-upon mathematical meanings. Understanding such words goes a long way toward understanding the mathematics they convey; misunderstanding the words leads inevitably to confusion. When in doubt, consult the index.

Read the Appendices   The human appendix generally lies unnoticed—unless trouble starts, when it's taken out and thrown away. Don't treat this book's appendices that way. Though perhaps slightly enlarged, they're full of healthy matter: reviews of precalculus topics, help with "story problems," proofs of various kinds, and more. Used as directed, the appendices will help appreciably in digesting the material.

Read the Instructors' Preface   Get a jump on your teacher.

In short: *Read the book*.

A last note   Why study calculus at all? There are plenty of good, practical, and "educational" reasons: because it's good for applications, because higher mathematics requires it, because it's good mental training, because other majors require it, because jobs require it. Here's another reason to study calculus: because calculus is among our species' deepest, richest, farthest-reaching, and most beautiful intellectual achievements. We hope this book will help you see it in that spirit.

A last request   Last, a request. We sincerely appreciate—and take very seriously—students' opinions, suggestions, and advice. (Changes to this edition of the text stem partly from students' views of the first edition.) We invite you to offer your advice, either through your teacher or by writing us directly. Our addresses appear below.

Arnold Ostebee and Paul Zorn
Department of Mathematics
St. Olaf College
1520 St. Olaf Avenue
Northfield, Minnesota 55057-1098
e-mail: **ostebee@stolaf.edu   zorn@stolaf.edu**

Contents

THE INTEGRAL

5.1 AREAS AND INTEGRALS

Two big problems The **tangent-line problem** and the **area problem** (our next main topic) are the two main geometric problems of calculus. These problems have been "big" for millennia: The ancient Greeks worked hard and ingeniously at both. Had Greek mathematicians had our algebraic advantages, they would have gone farther than they did.

We've seen that the idea of derivative, together with the rules for computing derivatives, solves the tangent-line problem. For many functions, even complicated ones, it's now easy to describe the tangent line at any point on the graph. Now we turn to the area problem and develop a new calculus tool—the integral—to solve it.

The **area problem** is about measuring the area of a region in the plane. For some special regions (like rectangles, triangles, and circles), well-known formulas do the job. Area problems are much more challenging (and interesting) for more general regions, especially for regions bounded by graphs of functions. The most important area problem of this type is to measure an area like the shaded one in Figure 1.

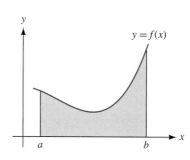

FIGURE 1
What's the shaded area?

Here is a careful statement of the problem:

Find the area of the region bounded above by the graph of f, below by the x-axis, on the left by the vertical line x = a, and on the right by the vertical line x = b.

That's a mouthful; we'll usually say just "the area under the graph of *f* from *a* to *b*." (When doubt seems possible we'll use the full form.)

Signed area In calculus it is useful to consider **signed area** as opposed to area in the everyday sense. Signed area has the following crucial feature:

Area below the x-axis counts as negative.

302

In Figure 1 all the shaded area is above the x-axis, and so no sign question arises. However, a circle of radius 1 centered at the origin, for instance, has ordinary area π but *signed* area 0. ➡

Draw your own circle to convince yourself.

Our first definition of the **integral** is in terms of signed area:

> **DEFINITION (The integral as signed area)**   Let f be a function defined for $a \leq x \leq b$. Either of the equivalent expressions
>
> $$\int_a^b f \quad \text{or} \quad \int_a^b f(x)\,dx$$
>
> denotes the signed area bounded by $x = a$, $x = b$, $y = f(x)$, and the x-axis.

Here are some comments on this crucial definition:

- **In words**   We read "the **integral** of f from a to b" for either $\int_a^b f$ or $\int_a^b f(x)\,dx$. The function f is called the **integrand**, and $[a, b]$ is the **interval of integration**. The numbers a and b are sometimes called **limits of integration**. (This phrase is a little unfortunate because "limit" has other important meanings in calculus.)

- **Which notation?**   Both notations $\int_a^b f$ and $\int_a^b f(x)\,dx$ denote the same area. The first form looks a little simpler; for now, we'll usually choose it. The form containing dx has advantages that will sometimes matter later on.

- **More formality later**   In Section 5.6 we'll give another, more "mathematical" definition of the integral as a certain limit. (We did something similar for the derivative, defining it first geometrically and later as a limit.)

- **Is any function OK?**   Does the integral $\int_a^b f$ make sense for *every* function f defined on an interval $[a, b]$? Not quite—some very ill-behaved, highly discontinuous functions fail to have integrals. But such functions are rare in calculus courses. Elementary functions, for instance, all have integrals on any closed interval $[a, b]$ on which they're defined.

- **How to calculate integrals?**   So far, we've only *defined* the symbolic expression $\int_a^b f$ to denote a certain signed area. The problem of *calculating* areas remains. We start that project in the next example.

EXAMPLE 1   Several shaded regions labeled as integrals appear in Figure 2. Use familiar area formulas to evaluate each integral.

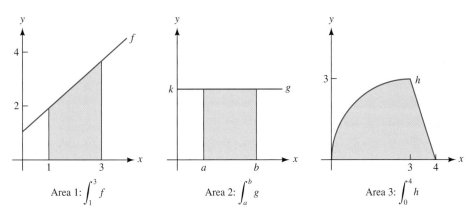

Area 1: $\int_1^3 f$ Area 2: $\int_a^b g$ Area 3: $\int_0^4 h$

FIGURE 2
What's the shaded area?

A trapezoid with base b and heights h_1 and h_2 has area $b(h_1 + h_2)/2$.

Solution   All three areas are easy to find. Area 1 is a trapezoid with base 2 and vertical sides 2 and 4, so its area ← is 6. Area 2 is a rectangle; its area is base × height = $(b - a) \cdot k$. Area 3 combines a quarter circle (of radius 3) and a triangle; a close look reveals total area $9\pi/4 + 3/2$, or about 8.569. Here are our results in integral notation:

$$\int_1^3 f = 6; \qquad \int_a^b g = k(b - a); \qquad \int_0^4 h = \frac{9\pi}{4} + \frac{3}{2}.$$

$\blacksquare$

Must area exist?   Signed area makes good sense for regions bounded by simple graphs, such as lines, parabolas, sinusoids, and so on. The situation is less clear for functions whose graphs are very irregular or ragged. But can the nasty curve in Figure 3 sensibly be said to bound area?

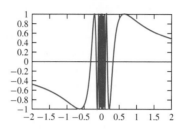

FIGURE 3
$y = \sin(\frac{1}{x})$

The answer turns out (for this curve, anyway) to be yes, but defining area (let alone computing it) takes special care in such unpromising circumstances. In Section 5.6, we'll redefine the integral using limits, partly in order to handle tough cases like this one.

Signed area: positive and negative contributions   Integrals measure *signed* area, so if an integrand f takes negative values in an interval $[a, b]$, then the integral $\int_a^b f$ may also be negative.

EXAMPLE 2   Let $f(x) = 1 - x^2$. Find or estimate values for the integrals $I_1 = \int_0^2 f$ and $I_2 = \int_{-2}^2 f$.

Solution   Figure 4 shows the areas in question; note that each grid square has area one.

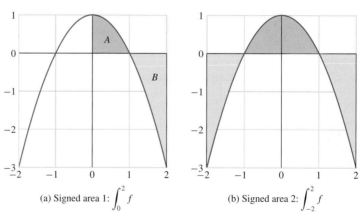

(a) Signed area 1: $\int_0^2 f$

(b) Signed area 2: $\int_{-2}^2 f$

FIGURE 4
Some areas defined by $y = 1 - x^2$

Computing I_1 and I_2 *exactly* is difficult for us now, but *estimating* answers is easy. An eyeball estimate suggests that A has signed area around 2/3, while B has signed area about $-3/2$. These guesses imply that

$$I_1 = \int_0^2 f \approx \frac{2}{3} - \frac{3}{2} = -\frac{5}{6}.$$

To estimate I_2 we can just double I_1:

$$I_2 = 2 \cdot I_1 \approx -\frac{5}{3}.$$

EXAMPLE 3   Let $g(x) = x^3$. Find or estimate $\int_{-1}^{1} g(x)\,dx$ and $\int_0^1 g(x)\,dx$.

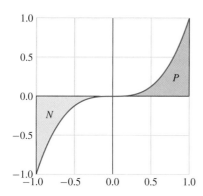

The shaded regions N and P have equal area.

FIGURE 5
Integrating an odd function: $g(x) = x^3$

The graph's symmetry means that the positive and negative areas P and N cancel, so

$$\int_{-1}^{1} g(x)\,dx = 0.$$

Symmetry considerations don't help with $\int_0^1 g$, the area of P, but a close look suggests that P's area is about that of one grid square, or 0.25, so

$$\int_0^1 g(x)\,dx \approx 0.25.$$

(We'll see soon that—and why—this estimate is actually *exact*.)

An odd property of integrals   The first part of Example 3 illustrates a general and sometimes useful fact:

> **FACT**   Let f be an *odd* function and $[-a, a]$ an interval that's symmetric about $x = 0$. Then $\int_{-a}^{a} f = 0.$

The Fact implies, for instance, that each of the following integrals is zero:

$$\int_{-5}^{5} \sin x \, dx; \quad \int_{-3}^{3} x^5 \, dx; \quad \int_{-1}^{1} \tan x \, dx; \quad \int_{-42}^{42} (x^3 + 3x + x^5) \, dx.$$

Properties of the integral

Thinking of the integral as signed area makes many of the integral's important properties simple and natural. For example, Figure 6 convincingly illustrates one simple but useful property of integrals:

If $a < c < b$, then $\int_a^b f = \int_a^c f + \int_c^b f$.

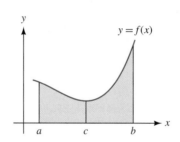

The full shaded area is the sum of two pieces.

FIGURE 6
Adding areas: $\int_a^b f = \int_a^c f + \int_c^b f$

The following theorem collects several basic properties of the integral, all in the spirit of Figure 6. (Fully rigorous proofs—even of these relatively "obvious" claims—depend on precise definitions of the sort we'll give later in this chapter.) Requiring that f and g be continuous guarantees that all the integrals in question exist.

THEOREM 1 (New integrals from old)   Let f and g be continuous functions on $[a, b]$; let k denote a real constant. Then

(i) **(Sum rule)** $\displaystyle\int_a^b \left(f(x) \pm g(x) \right) dx = \int_a^b f(x) \, dx \pm \int_a^b g(x) \, dx.$

(ii) **(Constant multiple rule)** $\displaystyle\int_a^b k f(x) \, dx = k \int_a^b f(x) \, dx.$

(iii) **(Smaller integrand, smaller integral)**   If $f(x) \le g(x)$ for all x in $[a, b]$, then $\displaystyle\int_a^b f(x) \, dx \le \int_a^b g(x) \, dx.$

(iv) **(Splitting the interval)**   If $a < c < b$, then
$$\int_a^b f(x) \, dx = \int_a^c f(x) \, dx + \int_c^b f(x) \, dx.$$

Here are some comments on parts of the theorem:

• **Like the derivative**   Properties (i) and (ii) resemble properties of the derivative—the integral "respects" sums and constant multiples. Figure 7 illustrates the constant multiple rule for integrals.

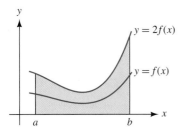

The total shaded area is twice the lower shaded area.
Look at narrow vertical strips to convince yourself.

FIGURE 7
The constant multiple rule: why $\int_a^b 2f = 2\int_a^b f$

• **Bounding integrals**   Property (iii) of the theorem is often used in the special case where either f or g is a *constant* function. Here is one useful version of the idea.

> **FACT (Bounding an integral)**   Suppose that, for some numbers m and M, the inequality $m \le f(x) \le M$ holds for all x in $[a, b]$. Then
>
> $$m \cdot (b - a) = \int_a^b m \, dx \le \int_a^b f(x) \, dx \le \int_a^b M \, dx = M \cdot (b - a).$$

Figure 8 illustrates the point; the numbers m and M are lower and upper bounds for f on $[a, b]$.

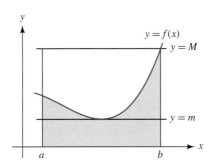

FIGURE 8
Bounding areas: why $m(b - a) \le \int_a^b f \le M(b - a)$

The picture shows that the value of the integral (the shaded area) lies between the areas of two rectangles based on the x-axis—the larger one bounded above by $y = M$ and the smaller by $y = m$.

The following examples show Theorem 1 in action.

EXAMPLE 4   The integral $I = \int_1^3 f(x) \, dx$, where $f(x) = 1/x$, measures the shaded area in Figure 9. Use the three linear functions g_1, g_2, and g_3 (also shown) to estimate I.

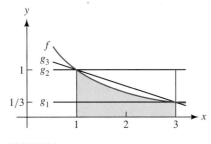

FIGURE 9
Three estimates for $\int_1^3 f$

$g_1(x) = 1/3$ *because*
$f(3) = 1/3$.

Solution   The shaded area isn't a rectangle or any other simple shape, so we can't find I exactly (for the moment). But we can estimate I using part (iii) of Theorem 1. As the picture shows, $f(x) \geq g_1(x) = 1/3$ for all x in $[1, 3]$. ◄ Therefore, by Theorem 1,

$$\int_1^3 f(x)\,dx \geq \int_1^3 g_1(x)\,dx = \frac{2}{3}.$$

(The right-hand integral is the area of a rectangle.)
    Similarly, $f(x) \leq g_2(x) = 1$ for all x in $[1, 3]$, and so

$$\int_1^3 f(x)\,dx \leq \int_1^3 g_2(x)\,dx = 2.$$

Thus, 2 is an upper bound for I, but a crude one, as the picture shows. Using g_3 in place of g_2 gives a *better* upper bound:

$$\int_1^3 f(x)\,dx \leq \int_1^3 g_3(x)\,dx = \frac{4}{3}.$$

(The right-hand integral corresponds to a *trapezoidal* area, and so an elementary area formula applies.) Combining these results gives lower and upper bounds for I:
$2/3 \leq I \leq 4/3$.   ∎

Draw them to convince yourself.

EXAMPLE 5   Find $\int_0^{\pi/2} \sin x\,dx$ and $\int_0^\pi (3\sin x + 2\cos x)\,dx$ by using the fact (which we'll prove later) that $\int_0^\pi \sin x\,dx = 2$.

Solution   Symmetry of the sine and cosine graphs ◄ shows that

$$\int_0^{\pi/2} \sin x\,dx = \frac{1}{2}\int_0^\pi \sin x\,dx = 1; \qquad \int_0^\pi \cos x\,dx = 0.$$

Now the sum and constant multiple rules for integrals tell us that

$$\int_0^\pi (3\sin x + 2\cos x)\,dx = 3\int_0^\pi \sin x\,dx + 2\int_0^\pi \cos x\,dx = 6 + 0 = 6.$$

A look at of Figure 10 makes the general size of the answer plausible.

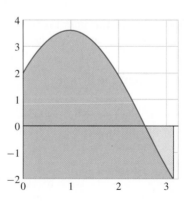

FIGURE 10
Estimating $\int_0^\pi (3\sin x + 2\cos x)\,dx$

Counting rectangles (each has area 1) shows that the signed area is indeed around 6.   ∎

EXAMPLE 6 For any positive constant b,

$$\int_0^b 1 \, dx = b \qquad \text{and} \qquad \int_0^b x \, dx = \frac{b^2}{2}.$$

Explain why. Then use Theorem 1 to find $\int_0^b (Cx + D) \, dx$ for any constants C and D.

Solution Graphs of $f(x) = 1$ and $g(x) = x$ over the interval $[0, b]$ show that the first two integrals have the claimed values. (Draw your own graphs to see why.) To finish, we use the sum and constant multiple rules:

$$\int_0^b (Cx + D) \, dx = \int_0^b Cx \, dx + \int_0^b D \, dx$$

$$= C \int_0^b x \, dx + \int_0^b D \, dx$$

$$= C \frac{b^2}{2} + Db. \qquad \blacksquare$$

Interpreting integrals

So far we've interpreted the integral $\int_a^b f$ in geometric terms, as signed area. But the integral has other useful interpretations.

Average values and integrals The rectangle in Figure 11 (below the line $y = \text{average}$) is chosen to have the same area as the shaded region.

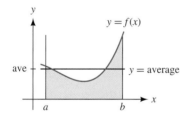

The shaded region has the same area as the rectangle under the line $y = \text{average}$.

FIGURE 11
Average value of a function

The height of that rectangle is called the **average value** of the function f over the interval $[a, b]$. The following definition puts this in analytic language:

> **DEFINITION** Let f be defined on an interval $[a, b]$. The **average value** of f over $[a, b]$ is defined by
>
> $$\frac{\int_a^b f(x) \, dx}{b - a} = \frac{\text{signed area}}{\text{length of interval}}.$$

The name is apt because, like other averages, the average value represents a sort of "typical value" of f for a range of inputs.

The ideas of integral and average value are closely linked. The preceding picture summarizes this two-way relationship: The average value is the *height* of a certain rectangle, while the integral is the *area* of the same rectangle.

Speed and distance   If a function $f(t)$ represents the *speed* of a moving object at time t, then the integral $\int_a^b f(t)\,dt$ represents the *distance* traveled by the object over the time interval $a \le t \le b$. Why is this true?

If the speed function $f(t)$ happens to be a constant, say k, then

$$\int_a^b f(t)\,dt = k \cdot (b-a) = \text{speed} \times \text{time} = \text{distance},$$

as claimed. Remarkably, integrating the speed function gives distance traveled even if $f(t)$ *isn't* constant. The next example isn't a proof, but it should help make the result believable.

EXAMPLE 7   The two graphs in Figure 12 show speed functions s_A and s_B for two cars A and B.

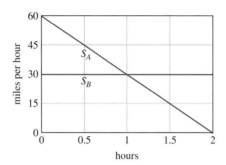

FIGURE 12
Speeds of cars A and B

What is the average value of each function over the interval $[0, 2]$? How far does each car travel over the two hours?

Solution   A look at areas shows readily that

$$\int_0^2 s_A(t)\,dt = 60 = \int_0^2 s_B(t)\,dt.$$

Thus, using the preceding definition, both speed functions have the *same* average value, 30 mph, over the interval $[0, 2]$. We expect, therefore, that the two cars cover the same total distance—60 miles—over the 2-hour period.   ■

Velocity and displacement   Speed and distance are *positive* quantities by definition. Velocity and displacement, on the other hand, can be either positive or negative, depending on the direction of motion. For an object moving along a line, for instance, we might take right as the positive direction and left as the negative direction. These conventions mean that integration of a velocity function $v(t)$ gives slightly different information from integration of a speed function. If $v(t)$ tells the *velocity* of an object moving along a left–right line at time t, then $\int_a^b f(t)\,dt$ represents the *displacement* that occurs over the time interval $a \le t \le b$. This displacement is *positive* if the net effect is *rightward* motion; otherwise, the displacement is negative.

Integrating from right to left: a technicality   The English description of $\int_a^b f(x)\,dx$— "the integral of f from $x = a$ to $x = b$"—implies a *direction*: x starts at a and ends at b. ◄ It's usually natural to think of x as moving from left to right. Up to now, we've done just that: In every integral $\int_a^b f$ we've treated so far, we've assumed or stated that $a \le b$. For example, we've discussed $\int_0^\pi \sin x\,dx$ but not $\int_\pi^0 \sin x\,dx$.

The variable name x isn't sacred; t is another popular choice.

The latter sort of "right-to-left" integrals *do* sometimes arise in calculations. The following convention handles all the possibilities with a minimum of fuss:

DEFINITION If $b < a$, then $\displaystyle\int_a^b f(x)\,dx = -\int_b^a f(x)\,dx$.

Why does the definition make sense? The velocity–displacement context gives a physical reason: If $f(t)$ represents velocity at time t, and $a < b$, then $\int_a^b f(t)\,dt$ gives the displacement from time a to time b. The integral $\int_b^a f(t)\,dt$ is thus the displacement over the "reversed" interval, and so is naturally opposite in sign to the first result.

The story so far, and a look ahead

In this section we defined the integral $\int_a^b f$ as a certain *area*. Then we observed some properties of the integral that follow directly from the area interpretation. What's missing so far is any general technique for *measuring* areas. In the simplest cases (such as rectangles and triangles) elementary area formulas may help, but these formulas apply *only* to the simplest functions. In the rest of this chapter we offer two different practical methods of calculating integrals. In Sections 5.2 and 5.3 we'll relate the new idea of the integral to the older (and better understood) idea of the derivative. This key connection between derivative and integral, called the **fundamental theorem of calculus**, is among the most important ideas of mathematics. In Section 5.6 we'll give another definition of the integral, this time as a limit of approximating sums.

BASIC EXERCISES

1. Let g be the function shown graphically below. When asked to estimate $\int_1^2 g(x)\,dx$, a group of calculus students submitted the following answers: -4, 4, 45, and 450. Only one of these responses is reasonable; the others are "obviously" incorrect. Which is the reasonable one? Why?

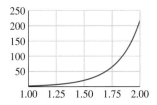

In Exercises 2–4, f is the function shown below.

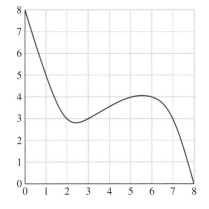

2. Which of the following is the best estimate of $\int_1^6 f(x)\,dx$: -24, 9, 20, 38? Justify your answer.

3. Find positive integers A and B such that $A \le \int_3^7 f(x)\,dx \le B$. Explain how you know that the values of A and B you chose have the desired properties.

4. $\displaystyle\int_6^8 f(x)\,dx \approx 4$. Does this approximation overestimate or underestimate the exact value of the integral? Justify your answer.

In Exercises 5–14, evaluate the integral assuming that f is the function shown below. [NOTE: The graph of f consists of two straight lines and two one-quarter circles.]

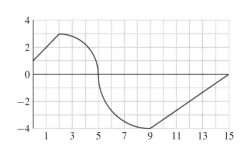

5. $\int_0^2 f(x)\,dx$

6. $\int_2^5 f(x)\,dx$

7. $\int_0^5 f(x)\,dx$

8. $\int_5^9 f(x)\,dx$

9. $\int_4^4 f(x)\,dx$

10. $\int_0^{15} f(x)\,dx$

11. $\int_0^{15} |f(x)|\,dx$

12. $\int_{15}^9 f(x)\,dx$

13. $\int_{12}^{15} f(x)\,dx$

14. $\int_9^{12} f(x)\,dx$

In Exercises 15–22, evaluate the definite integral using a graph of the integrand over the interval of integration.

15. $\int_{-4}^{-2} 3\,dt$

16. $\int_{-3}^4 \sqrt{2}\,du$

17. $\int_2^5 u\,du$

18. $\int_{-2}^1 w\,dw$

19. $\int_{-3}^3 |x|\,dx$

20. $\int_{-3}^3 |1+x|\,dx$

21. $\int_{-2}^2 \sqrt{4-s^2}\,ds$

22. $\int_{-3}^0 \sqrt{9-w^2}\,dw$

In Exercises 23–28, use the fact that $\int_0^\pi u^2\,du = \pi^3/3$ to evaluate the definite integral.

23. $\int_\pi^0 r^2\,dr$

24. $\int_0^\pi (3+4x^2)\,dx$

25. $\int_0^\pi (\pi^2 - 3t^2)\,dt$

26. $\int_{-\pi}^\pi w^2\,dw$

27. $\int_\pi^{2\pi} (v-\pi)^2\,dv$

28. $\int_0^\pi (\pi x + x^2)\,dx$

29. Find real numbers A and B such that $A \le \int_1^3 (4w - w^2)\,dw \le B$.

30. Show that $2\pi \le \int_0^{2\pi} \sqrt{1+3\cos^2 x}\,dx \le 4\pi$.

31. Find real numbers A and B such that $A \le \int_1^3 \frac{dt}{1+t^2} \le B$.

32. Find real numbers A and B such that $A \le \int_0^{\pi/4} \frac{d\theta}{1+\tan\theta} \le B$.

In Exercises 33–36, evaluate the integral assuming that $\int_{-2}^5 f(x)\,dx = 18$, that $\int_{-2}^5 g(x)\,dx = 5$, and that $\int_{-2}^5 h(x)\,dx = -11$.

33. $\int_{-2}^5 \big(3f(x) + 4g(x)\big)\,dx$

34. $\int_{-2}^5 \big(3h(x) - 4f(x)\big)\,dx$

35. $\int_5^{-2} f(x)\,dx$

36. $\int_{-2}^5 \big(h(x) + 1\big)\,dx$

FURTHER EXERCISES

37. Four students disagree on the value of the integral $\int_0^{\pi/2} \cos^8 x\,dx$. Jack argues for $\pi \approx 3.14$, Joan for $35\pi/256 \approx 0.43$, Ed for $2\pi/9 - 1 \approx -0.30$, and Lesley for $\pi/4 \approx 0.79$. Use the graph below to determine who is right. (One *is* right!) Explain how you know that the other values are incorrect.

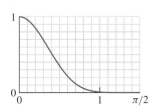

38. Suppose that a car travels on an east–west road with eastward velocity $v(t) = 60 - 20t$ mph at time t hours.

(a) Evaluate $\int_0^4 v(t)\,dt$. Interpret the answer in car talk.

(b) Find the average value of $v(t)$ on the interval $[0, 4]$. Show your result graphically.

(c) Let $s(t)$ be the car's speed at time t. Evaluate $\int_0^4 s(t)\,dt$ and interpret the answer in car talk.

(d) What is the car's average speed between $t = 0$ and $t = 4$?

39. Explain why $\int_1^2 x^3\,dx = \int_3^4 (x-2)^3\,dx$.

40. Explain why $\int_1^2 x^3\,dx = \int_{-3}^{-2} (x+4)^3\,dx$.

In Exercises 41–48, evaluate the integral assuming that
(i) $\int_0^2 f(x)\,dx = 2$; (ii) $\int_1^2 f(x)\,dx = -1$; (iii) $\int_2^4 f(x)\,dx = 7$.

41. $\int_1^4 f(x)\,dx$

42. $\int_0^4 3f(x)\,dx$

43. $\int_0^1 f(x)\,dx$

44. $\int_0^1 \left(f(x) + 1 \right) dx$

45. $\int_2^4 \left(f(x) - 2 \right) dx$

46. $\int_0^4 \left(3 - 2f(x) \right) dx$

47. Explain why f must be negative somewhere in the interval $[1, 2]$. [HINT: Sketch the graph of a function that has properties (i)–(iii).]

48. Explain why $f(x) \geq 3$ somewhere in the interval $[2, 4]$.

Evaluate the integrals in Exercises 49–52 using graphical arguments.

49. $\int_0^{\pi} \cos x\,dx$

50. $\int_{\pi/2}^{3\pi/2} \sin x\,dx$

51. $\int_{-1}^{1} (4x^3 - 2x)\,dx$

52. $\int_{-2}^{2} (7x^5 + 3)\,dx$

Evaluate the integrals in Exercises 53–60 using graphical arguments and the fact that $\int_0^{\pi} \sin x\,dx = 2$.

53. $\int_0^{2\pi} \sin x\,dx$

54. $\int_0^{2\pi} |\sin x|\,dx$

55. $\int_0^{\pi} (1 + \sin x)\,dx$

56. $\int_{-\pi/2}^{\pi/2} \cos x\,dx$

57. $\int_0^{\pi/2} (x + \cos x)\,dx$

58. $\int_0^{100\pi} \sin x\,dx$

59. $\int_0^{100\pi} |\sin x|\,dx$

60. $\int_0^{100\pi} \cos x\,dx$

61. Let $f(x) = \frac{1}{2}x + \cos x$. Show that $1.3 \leq \int_0^3 f(x)\,dx \leq 3.5$.

62. Show that $\pi/6 \leq \int_{\pi/6}^{\pi/2} \sin x\,dx \leq \pi/3$.

63. Show that $-\pi/3 \leq \int_{2\pi/3}^{\pi} \cos x\,dx \leq -\pi/6$.

64. Show that $4.5 \leq \int_1^3 e^x \sin x\,dx \leq 15$.

65. Sketch the graph of a function f with the property that

$$\left| \int_1^5 f(x)\,dx \right| < \int_1^5 |f(x)|\,dx.$$

Explain why the function f you sketched has this property.

66. Suppose that f is a continuous function. If the average value of f over the interval $[0, 1]$ is 2 and the average value of f over the interval $[1, 3]$ is 4, what is the average value of f over the interval $[0, 3]$?

67. Suppose that f is a continuous function. If the average value of f over the interval $[-3, 1]$ is 2 and the average value of f over the interval $[-3, 7]$ is 5, what is the average value of f over the interval $[1, 7]$?

68. Suppose that f is continuous on $[a, b]$. Show that
$$\left| \int_a^b f(x)\,dx \right| \leq \int_a^b |f(x)|\,dx.$$
[HINT: $-|f(x)| \leq f(x) \leq |f(x)|$.]

69. Let $f(x) = x^2$ and $g(x) = 2x$.

(a) Sketch the region in the xy-plane bounded by the curves $y = f(x)$ and $y = g(x)$.

(b) Explain why the area of the region in part (a) is $\int_0^2 g(x)\,dx - \int_0^2 f(x)\,dx$.

70. Let $f(x) = \sqrt{x}$ and $g(x) = x$.

(a) Sketch the region in the xy-plane bounded by the curves $y = f(x)$ and $y = g(x)$.

(b) Express the area of the region in part (a) using integrals.

71. Let $f(x) = x + 6$ and $g(x) = x^3$.

(a) Sketch the region in the xy-plane bounded by the curves $y = f(x)$, $y = g(x)$, and the line $x = 0$.

(b) Express the area of the region in part (a) using integrals.

72. Let $f(x) = \ln x$ and $g(x) = 1 - x$.

(a) Sketch the region in the xy-plane bounded by the curves $y = f(x)$, $y = g(x)$, and the line $x = e$.

(b) Express the area of the region in part (a) using integrals.

73. Archimedes (ca. 250 B.C.E.) proved that the area under a parabolic arch is $2bh/3$, where b is the width of the base of the arch and h is the height. [NOTE: Archimedes actually proved a more general theorem about the area of any region cut off from a parabola by a line.]

(a) Use Archimedes's result to show that $\int_{-a}^{a} x^2 \, dx = 2a^3/3$. [HINT: Draw the curve $y = x^2$ and the line $y = a^2$.]

(b) Use part (a) and the fact that x^2 is an even function to show that $\int_{0}^{a} x^2 \, dx = a^3/3$.

[HINT: $\int_{-a}^{a} x^2 \, dx = \int_{-a}^{0} x^2 \, dx + \int_{0}^{a} x^2 \, dx$.]

(c) Use part (b) to show that $\int_{a}^{b} x^2 \, dx = (b^3 - a^3)/3$.

[HINT: $\int_{a}^{b} x^2 \, dx = \int_{0}^{b} x^2 \, dx - \int_{0}^{a} x^2 \, dx$.]

74. Let $f(x) = 2x + 3$.

(a) Sketch a graph of f in the viewing window $[0, 3] \times [0, 9]$.

(b) Evaluate $\int_{1}^{3} f(x) \, dx$. Shade the region of the part (a) graph represented by this integral.

(c) Show that $f^{-1}(y) = (y - 3)/2$.

(d) Evaluate $\int_{5}^{9} f^{-1}(y) \, dy$ and shade the region of the part (a) graph represented by this integral.

(e) Verify that

$$\int_{1}^{3} f(x) \, dx = 3f(3) - f(1) - \int_{f(1)}^{f(3)} f^{-1}(y) \, dy.$$

75. Suppose that f is a continuous function and invertible on $[a, b]$. Then it is true that

$$\int_{a}^{b} f(x) \, dx = bf(b) - af(a) - \int_{f(a)}^{f(b)} f^{-1}(y) \, dy.$$

Give a graphical proof of this result if $0 < a < b$, and $0 < f(a) < f(b)$.
[HINT: Draw a sketch of f in the viewing window $[0, b] \times [0, f(b)]$.]

76. Use the result stated in Exercise 75 to show that $\int_{1}^{e} \ln x \, dx = e - \int_{0}^{1} e^x \, dx$.

77. Use the result stated in Exercise 75 to show that $\int_{a}^{b} \sqrt{x} \, dx = b^{3/2} - a^{3/2} - \int_{\sqrt{a}}^{\sqrt{b}} x^2 \, dx$.

78. Use the results stated in Exercises 73 and 75 to show that $\int_{0}^{a} \sqrt{x} \, dx = 2a^{3/2}/3$.

5.2 THE AREA FUNCTION

In Section 5.1 we defined the integral $\int_{a}^{b} f$ to be the signed area of the region from $x = a$ to $x = b$, bounded above or below by the graph of f. For given f, a, and b, this area is a certain fixed *number* (positive, negative, or zero). In this section we define and study a *function* based on the signed area bounded by the graph of f. This **area function**, which we'll denote by A_f, is "built" from f using the integral.

We've played the new-function-from-old game before. In the most important case, we built the derivative function f' from an "original" function f and then studied what each tells about the other. Here we repeat the process, this time with f and A_f. The relationship between f and A_f will prove remarkably similar to that between f' and f.

Defining the area function

First comes the formal definition:

> **DEFINITION** Let f be a function and a any point of its domain. For any input x, the **area function** A_f is defined by the rule
>
> $$A_f(x) = \int_{a}^{x} f(t) \, dt.$$
>
> In words, $A_f(x)$ is the signed area defined by f from a to x.

A "generic" picture (Figure 1) illustrates the idea.

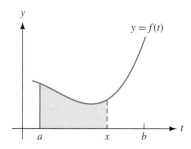

FIGURE 1
The area function: $A_f(x)$ is the shaded area

$A_f(x)$ represents *signed* area—here, all the shaded area counts as positive.

The area function idea is best approached through pictorial examples. First, some brief remarks:

- **Variable names: why so many?** Why did we use another variable, t, in defining the expression $A_f(x) = \int_a^x f(t)\,dt$? We did it because the letter x was already taken—we used x to denote the right-hand boundary of the defining region. We introduced t to avoid using x in two different ways. Note that x, not t, is the input variable to the function A_f.

- **The domain of A_f** For which inputs x does the rule $A_f(x) = \int_a^x f$ make sense? Barring bad behavior (such as discontinuities of f), the new function A_f has the same domain as f itself. In particular, $A_f(x)$ makes sense even if x is to the *left* of a; in this case we use the sign convention $\int_a^x f = -\int_x^a f$ (which we discussed in Section 5.1).

- **The choice of a** The role of a is to fix one edge (the left edge in Figure 1) of the region whose area gives $A_f(x)$. The other edge varies freely. As the definition says, *any a in the domain of f is a legal choice.* We'll soon see that different choices of a give different (but not *much* different) functions A_f.

EXAMPLE 1 Let $f(x) = 3$ and $a = 0$. Describe the area function $A_f(x) = \int_0^x f(t)\,dt$ for positive inputs x. Does A_f have a simple algebraic formula?

Solution The formula $f(x) = 3$ implies that $A_f(x) = 3x$. Figure 2(a) shows why. ➡ Figure 2(b) shows graphs of f and A_f together.

A rectangle with height 3 and base x has area $3x$.

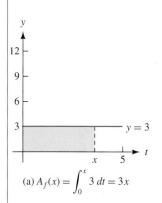

(a) $A_f(x) = \displaystyle\int_0^x 3\,dt = 3x$

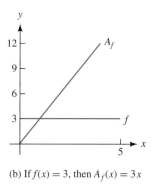

(b) If $f(x) = 3$, then $A_f(x) = 3x$

Note the horizontal axes: t is the input variable in (a) while x is the variable in (b).

FIGURE 2

Notice, finally, that $A_f(x) = 3x$ is an *antiderivative* of $f(x) = 3$. This is no accident—we'll see the same pattern in later examples. ∎

EXAMPLE 2 Let $f(x) = 3$ and $a = 0$, as in Example 1. Discuss the area function $A_f(x) = \int_0^x f(t)\,dt$ for *negative* values of x. Does the formula $A_f(x) = 3x$ from Example 1 hold here as well?

Solution Yes it does—the formula $A_f(x) = 3x$ holds for *all* inputs x. The key step is a careful look at signs; see Figure 3:

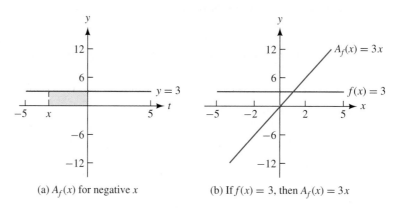

(a) $A_f(x)$ for negative x (b) If $f(x) = 3$, then $A_f(x) = 3x$

FIGURE 3

The key to the formula is to recall how we handle "right-to-left" integrals. Since $x < 0$,

$$\int_0^x f = -\int_x^0 f = -\,(\text{shaded area}).$$

Because $x < 0$ and the rectangle has height 3, the product $3x$ gives the *opposite* of the shaded area. Combining this result and that of Example 1 shows that if $f(x) = 3$ and $a = 0$, then $A_f(x) = \int_0^x f = 3x$ for *all* inputs x, positive or negative. ∎

A happy moral The geometric meaning of $\int_a^x f$ is clear if $x \geq a$. If $x < a$, the situation looks trickier at first glance, but the moral of Examples 1 and 2 is that, when all the dust settles, the best outcome is revealed:

The same formula for $A_f(x)$ works for all x.

The same principle holds for any well-behaved function f, so we'll seldom worry about (or even mention) the issue of whether $x < a$ or $x \geq a$.

EXAMPLE 3 Consider the linear function $f(x) = x$ and the area function $A_f(x) = \int_0^x t\,dt$. Find the formula for A_f in this case.

Solution As Figure 4(a) shows, the region that defines $A_f(x)$ for a positive input x is a triangle with base x and height x.

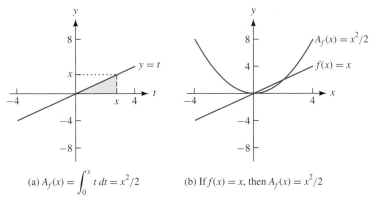

(a) $A_f(x) = \displaystyle\int_0^x t\, dt = x^2/2$ (b) If $f(x) = x$, then $A_f(x) = x^2/2$

FIGURE 4

The formula we want is therefore

$$A_f(x) = \frac{\text{base} \times \text{height}}{2} = \frac{x^2}{2}.$$

(The same formula holds for negative x, for reasons just mentioned.) Graphs of $f(x)$ and $A_f(x)$ appear together in Figure 4(b). Once again, A_f turns out to be an antiderivative of f. ■

EXAMPLE 4   (Another f, same a)   Consider the new linear function $f(x) = 2 - x$. Find the new area function $A_f(x) = \int_0^x f$. Is A_f *again* an antiderivative of f?

Solution   The usual signed area picture (Figure 5) applies; note that area below the x-axis counts as negative.

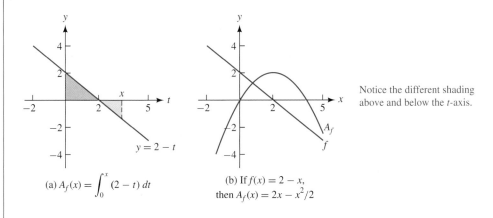

(a) $A_f(x) = \displaystyle\int_0^x (2-t)\, dt$ (b) If $f(x) = 2 - x$, then $A_f(x) = 2x - x^2/2$

Notice the different shading above and below the t-axis.

FIGURE 5

With a little care, one can use triangles (or trapezoids, if $x < 2$) to find the signed area shown shaded. In Figure 5(a) the upper triangle has area 2, and the lower triangle has area $(x-2)^2/2$, so the total signed area is

$$A_f(x) = 2 - \frac{(x-2)^2}{2} = 2x - \frac{x^2}{2}.$$

Figure 5(a) shows a value of x greater than 2. The *picture* is different if $0 < x < 2$, but the resulting formula is the same. Figure 5(b) shows graphs of both f and A_f.

Another (perhaps easier) approach to the problem is to use additive properties of the integral (from Theorem 1, page 306) to combine results from earlier examples:

$$A_f(x) = \int_0^x (2 - t)\, dt = \int_0^x 2\, dt - \int_0^x t\, dt = 2x - \frac{x^2}{2}.$$

Either way, the conclusion is the same: $A_f(x) = 2x - x^2/2$ is an antiderivative of $f(x) = 2 - x$. ∎

EXAMPLE 5   (Same f, another a)   Consider the same linear function, $f(x) = 2 - x$, as in Example 4, but now with $a = 1$, so that $A_f(x) = \int_1^x f$. Find a formula for the new area function.

Solution   The only difference from Example 4 concerns the lower endpoint of integration, so the new function A_f differs from the old one only by the area of the trapezoidal region under the f-graph from $x = 0$ to $x = 1$. This area is $3/2$; see Figure 6. In symbols,

$$A_f(x) = \int_1^x f(t)\, dt = \int_0^x f(t)\, dt - \int_0^1 f(t)\, dt$$
$$= 2x - \frac{x^2}{2} - \frac{3}{2}.$$

The graphs in Figure 6 support this result.

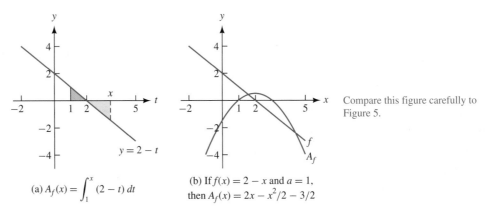

(a) $A_f(x) = \int_1^x (2 - t)\, dt$

(b) If $f(x) = 2 - x$ and $a = 1$, then $A_f(x) = 2x - x^2/2 - 3/2$

Compare this figure carefully to Figure 5.

FIGURE 6

As in Example 4, the area function $A_f(x) = 2x - x^2/2 - 3/2$ is an antiderivative of the original function $f(x) = 2 - x$. This time, however, A_f is the antiderivative for which $A_f(1) = 0$. ◄ ∎

Two antiderivatives of the same function can differ only by an additive constant, as they do here.

In Examples 1–5 we used an elementary area formula to find an explicit algebraic expression for the area function. In the next example, no such elementary area formula is available.

EXAMPLE 6 Let $f(x) = 1/x$ and $A_f(x) = \int_1^x f(t)\,dt$. Discuss the area function A_f. Does it look familiar?

Solution Figure 7(a) shows $f(x)$; the shaded area represents $A_f(3)$.

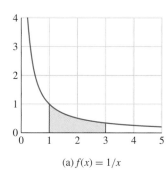

(a) $f(x) = 1/x$

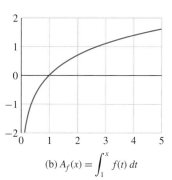

(b) $A_f(x) = \int_1^x f(t)\,dt$

FIGURE 7

There's no simple area formula, but we can *estimate* values of A_f using the grid. The shaded area, for instance, seems to be a little more than one unit, so $A_f(3)$ is a little more than 1. By making and plotting many such estimates we could—laboriously—produce a graph of A_f; one appears in Figure 7(b). ➡

Note that $A_f(x) < 0$ for $x < 1$.

The graph of A_f might look familiar; in fact, it turns out that $A_f(x) = \ln x$. We'll explain this striking result in the next section, but we can see right now that it's consistent with results of earlier examples, because $A_f(x) = \ln x$ is an antiderivative of $f(x) = 1/x$. ∎

Properties of A_f: patterns in the examples The examples illustrate what's most important about an area function A_f—its relationship to the "original" function f from which it's built. In the following list we assemble several properties of this relationship that can be seen in one or more of the preceding examples. For simplicity we assume throughout that f is a continuous function defined for all real inputs; doing so ensures that A_f is also continuous and defined for all inputs.

- $A_f(a) = 0$. (See *any* example.)

- Where f is positive, A_f is increasing. Where f is negative, A_f is decreasing. (See Examples 3–5.)

- Where f is zero, A_f has a stationary point. (See Examples 3–5.)

- Where f is increasing, A_f is concave up. Where f is decreasing, A_f is concave down. (See Examples 3–6.)

These principles hold in general (not only for our example functions). We won't prove them here because all of them follow from the fundamental theorem of calculus—the main result of the next section.

BASIC EXERCISES

1. Let $f(x) = x$ and $A_f(x) = \int_0^x f(t)\,dt$. It was shown in the text that $A_f(x) = x^2/2$ if $x > 0$. Show that this formula is correct even if $x < 0$.

2. Let $f(x) = x$ and $A_f(x) = \int_4^x f(t)\,dt$. Show that $A_f(x) = x^2/2 - 8$ for *all* values of x.

In Exercises 3–8, $A_f(x) = \int_0^x f(t)\, dt$, where f is the function graphed below.

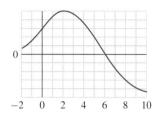

3. Which is larger: $A_f(1)$ or $A_f(5)$? Justify your answer.

4. Which is larger: $A_f(7)$ or $A_f(10)$? Justify your answer.

5. Explain why $A_f(-1) < 0$.

6. Is $A_f(8) > 0$? Justify your answer.

7. Which is larger: $A_f(-2)$ or $A_f(-1)$? Justify your answer.

8. Let $F(x) = \int_{-2}^x f(t)\, dt$. Explain why $A_f(x) = F(x) + C$, where C is a negative constant.

In Exercises 9–14, let $F(x) = \int_0^x f(t)\, dt$ and $G(x) = \int_5^x f(t)\, dt$. Furthermore, suppose that $f(t) > 1$ for all $t \geq 0$.

9. Explain why $F(7) > F(1)$.

10. Is $G(9) > G(5)$? Justify your answer.

11. Explain why $F(x) > F(0)$ for all $x > 0$.

12. **(a)** Explain why $G(x) = F(x) + C$, where C is a constant.
 (b) Is the constant C in part (a) positive? Justify your answer.

13. Explain why $F(6) \geq 6$.

14. Is it possible that $G(10) = 3$? Justify your answer.

In Exercises 15–20, let $F(x) = \int_0^x f(t)\, dt$ and $G(x) = \int_4^x f(t)\, dt$. Furthermore, suppose that $f(t) < -2$ for all $t \geq 0$.

15. Explain why $F(6) < F(2)$.

16. Is $G(3) > G(5)$? Justify your answer.

17. Explain why $F(x) < F(1)$ for all $x > 1$.

18. **(a)** Explain why $F(x) = G(x) + C$, where C is a constant.
 (b) Is the constant C in part (a) positive? Justify your answer.

19. Is it possible that $F(4) = -7$? Justify your answer.

20. Is it possible that $G(10) = -14$? Justify your answer.

In Exercises 21–28, let $f(x) = 3x + 2$, $F(x) = \int_0^x f(t)\, dt$, $G(x) = \int_1^x f(t)\, dt$, and $H(x) = \int_{-2}^x f(t)\, dt$.

21. Find formulas for F, G, and H.

22. Find the number C such that $G(x) = F(x) + C$.

23. Find a formula for $F(x) - H(x)$.

24. Where is F increasing? What is true about f on these intervals?

25. Where is F decreasing? What is true about f on these intervals?

26. Is F concave up anywere? If so, what is true about f on these intervals?

27. Is F concave down anywhere? If so, what is true about f on these intervals?

28. Are F, G, and H all antiderivatives of f? Justify your answer.

29–36. Repeat Exercises 21–28 using $f(x) = 2 - 3x$.

37. Suppose that F is an antiderivative of a differentiable function f.
 (a) If f is negative on $[a, b]$, what is true about F?
 (b) If f' is positive on $[a, b]$, what is true about F?

38. Suppose that $G' = g$ and that g is a differentiable function.
 (a) If G is increasing on $[a, b]$, what is true about g?
 (b) If G is concave down on $[a, b]$, what is true about g'?
 (c) Suppose that H is another function with the property $H' = g$. What relationship exists between G and H?

FURTHER EXERCISES

In Exercises 39–44, let $f(x) = 2 - |x|$ and $A_f(x) = \int_0^x f(t)\, dt$.

39. **(a)** Find a formula for A_f.
  [HINT: Write A_f as a piecewise-defined function.]
 (b) Plot f and A_f on the same axes.

40. Where is A_f increasing? What is true about f on these intervals?

41. Where is A_f decreasing? What is true about f on these intervals?

42. Where is A_f concave up? What is true about f on these intervals?

43. Where is A_f concave down? What is true about f on these intervals?

44. Where does A_f have inflection points? What is true about f at these points?

45–50. Repeat Exercises 39–44 assuming $A_f(x) = \int_3^x f(t)\, dt$.

In Exercises 51–56, $G(x) = \int_{-3}^{x} f(t)\,dt$ *and* $H(x) = \int_{2}^{x} f(t)\,dt,$
where f is the function graphed below.
[NOTE: The graph of f is made up of straight lines and a semicircle.]

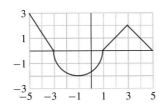

51. How are the values of $G(x)$ and $H(x)$ related? Give a graphical explanation of this relationship.

52. On which subintervals of $[-5, 5]$, if any, is H increasing?

53. Explain why G has a local minimum at $x = 1$.

54. Does H have a local maximum at $x = -3$? Justify your answer.

55. Where in the interval $[-5, -2]$ does G achieve its minimum value? Its maximum value? What are these values?

56. Where in the interval $[-5, 5]$ does H achieve its minimum value? Its maximum value? What are these values?

In Exercises 57–60, suppose that f is a continuous function and that $\int_{0}^{x} f(t)\,dt = \sin(x^2).$

57. Evaluate $\int_{0}^{\sqrt{\pi/2}} f(t)\,dt.$

58. Show that $\int_{\sqrt{\pi/2}}^{x} f(t)\,dt = \sin(x^2) - 1.$

59. Find an expression for $\int_{-\sqrt{3\pi/2}}^{x} f(t)\,dt.$

60. Find an expression for $\int_{0}^{2x} f(t)\,dt.$

In Exercises 61–64, rewrite the expression in terms of F, where $F(t) = \int_{-1}^{t} \sqrt{1 + x^4}\,dx.$

61. $\int_{-1}^{4} \sqrt{1 + u^4}\,du$

62. $\int_{0}^{-1} \sqrt{1 + z^4}\,dz$

63. $\int_{-2}^{3} \sqrt{1 + t^4}\,dt$

64. $\int_{0}^{1} \sqrt{1 + u^4}\,du$

65. Let $F(x) = \int_{a}^{x} f(t)\,dt$ and $G(x) = \int_{a}^{x} g(t)\,dt$, where f and g are continuous functions. Suppose that $f(x) \le g(x)$ for all $x \ge a$. Show that $F(x) \le G(x)$ for all $x \ge a$.

66. Let $A_f(x) = \int_{c}^{x} f(t)\,dt$, where c is a constant and f is a continuous function.

   (a) Prove that if f is positive on $[a, b]$, then A_f is increasing on $[a, b]$. [HINT: Let y and z be numbers such that $a \le y < z \le b$. Compare $A_f(y)$ and $A_f(z)$.]

   (b) Use part (a) to show that if f is negative on $[a, b]$, then A_f is decreasing on $[a, b]$. [HINT: The function g defined by $g(x) = -f(x)$ is positive wherever f is negative.]

   (c) Use parts (a) and (b) to show that A_f has a local maximum or a local minimum wherever f changes sign.

67. (a) Use the figure below to show that $\int_{0}^{x} \sqrt{1 - t^2}\,dt = \frac{1}{2}x\sqrt{1 - x^2} + \frac{1}{2}\arcsin x$ if $0 \le x \le 1$. [HINT: The area of a circular sector of radius r and angle θ is $r^2\theta/2$.]

   (b) Show that $\int_{0}^{x} \sqrt{1 - t^2}\,dt = \frac{1}{2}x\sqrt{1 - x^2} + \frac{1}{2}\arcsin x$ if $-1 \le x \le 0$.

   (c) Let $f(x) = \sqrt{1 - x^2}$ and let $A_f(x) = \int_{0}^{x} f(t)\,dt$. Show that A_f is an antiderivative of f.

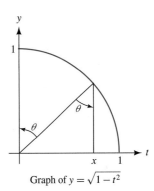

Graph of $y = \sqrt{1 - t^2}$

68. Let $f(x) = \sqrt{1 - x^2}$ and let $A_f(x) = \int_{-1/2}^{x} f(t)\,dt$.

   (a) Use the result stated in part (a) of Exercise 67 to find a formula for A_f.

   (b) Is A_f an antiderivative of f? Justify your answer.

69. Let $a > 0$ be a constant. Use a sketch similar to that given in Exercise 67 to find an expression for $\int_{0}^{x} \sqrt{a^2 - t^2}\,dt$ if $-a \le x \le a$. [HINT: Check that your answer produces the correct value if $x = a$.]

70. Let $f(x) = x^2$ and $A_f(x) = \int_{0}^{x} f(t)\,dt$.

   (a) Use the result of Exercise 73 in Section 5.1 to find a formula for $A_f(x)$.

   (b) Is A_f an antiderivative of f? Justify your answer.

71. Let $f(x) = x^2$ and $A_f(x) = \int_{-3}^{x} f(t)\,dt$.

   (a) Use the result of Exercise 73 in Section 5.1 to find a formula for $A_f(x)$.

   (b) Is A_f an antiderivative of f? Justify your answer.

5.3 THE FUNDAMENTAL THEOREM OF CALCULUS

What the theorem says

In the preceding section we constructed the area function A_f from an "original" function f and a base point a. All of the evidence amassed so far, both from examples and from the list on page 319 of relations between f and A_f, points to one conclusion:

THEOREM 2 (The fundamental theorem of calculus, informal version)
For any well-behaved function f and any base point a, A_f is an antiderivative of f.

The theorem's grand-sounding title is fully justified. The formal statement of this fact is the single most important theorem of elementary calculus. Here it is, complete with mathematical fine print:

THEOREM 3 (The fundamental theorem of calculus, formal version)
Let f be a continuous function defined on an open interval I containing a. The function A_f with rule

$$A_f(x) = \int_a^x f(t)\, dt$$

is defined for every x in I, and

$$\frac{d}{dx}\left(A_f(x)\right) = f(x).$$

We'll say much more about the FTC (our shorthand) and provide a proof at the end of this section. First, some remarks.

- **Why continuous?**   Requiring f to be continuous ensures that the integral $\int_a^x f$ exists. If f is discontinuous (if it blows up or has gaps, for instance), then A_f may have problems, too.

- **Why is I open?**   Recall that *open* intervals, such as $(-3, 10)$ and (∞, ∞), are intervals that don't contain endpoints. Working on an open interval I is a technical convenience; doing so ensures that there's "room" on both sides of a. ◂

This hypothesis rarely causes trouble in elementary calculus. We include it mainly for completeness.

- **In other symbols**   The conclusion of the FTC is sometimes written as follows:

$$\frac{d}{dx}\left(\int_a^x f(t)\, dt\right) = f(x).$$

This form avoids the symbol A_f, but the density of symbols on the left side may be mystifying. ◂

Which form seems clearer to you? Do you see that both forms say the same thing?

- **In graphical words**   The FTC means, graphically, that the *rate* of change of the area function is the *height* of the original function.

EXAMPLE 1   Let $f(x) = 2x \sin(x^2)$ and let $A_f(x) = \int_0^x f(t)\, dt$. Find a symbolic formula for A_f; interpret results graphically.

Solution   In the simple examples of the preceding section we could use elementary area formulas to find explicit formulas for A_f. Here, simple geometric formulas aren't enough. How, for instance, might we find the exact value of $A_f(2)$, the signed area shown shaded in Figure 1?

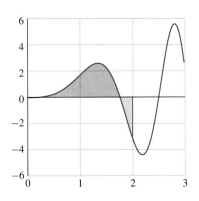

$A_f(2)$ is the net shaded area.

FIGURE 1
Finding $A_f(2)$ for $f(x) = 2x\sin(x^2)$

The FTC solves our problem. It says that A_f is *some* antiderivative of f; we need only find the right one.

Let's observe first that for any constant C, the function $F(x) = -\cos(x^2) + C$ is an antiderivative of $f(x) = 2x\sin(x^2)$. (This might be hard to guess, but it's easy to check.) ➧ We know too that *every* antiderivative of f—including A_f, the one we want—has this form. Thus, we must have

Do check; differentiate A_f.

$$A_f(x) = -\cos(x^2) + C$$

for some constant C. To choose the *right* value of C we use the fact that $A_f(0) = 0$. ➧ Thus,

Why? Because $\int_0^0 f = 0$ for any function f.

$$0 = A_f(0) = -\cos 0 + C \implies C = 1,$$

and we've found our formula: $A_f(x) = -\cos(x^2) + 1$. (In particular, $A_f(2) = -\cos(4) + 1 \approx 1.65364$.) The graphs of f and A_f in Figure 2 are consistent with our conclusion—the A_f-graph does appear (qualitatively, at least) to describe the behavior of signed area under the f-graph, based at 0.

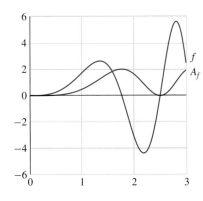

A_f is an antiderivative for f. For instance, A_f is stationary where f has x-intercepts.

FIGURE 2
If $f(x) = 2x\sin(x^2)$, then $A_f(x) = -\cos(x^2) + 1$ ■

Finding *all* antiderivatives of a function

In Example 1 we claimed that any two antiderivatives of the function $f(x) = 2x\sin(x^2)$ differ only by an additive constant. We've discussed this principle before (in Sections 2.4 and 4.9), but it's important enough here to deserve repeating:

> **FACT**   Suppose that $F'(x) = G'(x)$ for all x in an interval I. Then, for some constant C, $F(x) = G(x) + C$ for all x in I.

The Fact means, in practice, that finding *one* antiderivative for a function f on an interval I is tantamount to finding *all* antiderivatives: If $F(x)$ is *any* antiderivative, then every other antiderivative is of the form $F(x) + C$. (We gave more details on this Fact in Section 4.11.)

Continuous functions have antiderivatives.   The fundamental theorem guarantees (among other things) that every continuous function f on I has an antiderivative on I—namely, the area function A_f.

For the usual, nicely behaved functions of elementary calculus, this is no great surprise. For naughtier functions, such as the one in Figure 3, the situation is murkier.

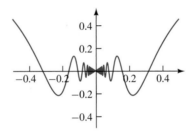

FIGURE 3

Does $f(x) = x\sin(\frac{1}{x})$ have an antiderivative?

The fundamental theorem says that every continuous function f, naughty or nice *has* an antiderivative. (*Finding* an antiderivative in symbolic form may be difficult or impossible; the FTC offers no guarantees there.)

Why the FTC is fundamental

The FTC deserves its high-sounding name for both theoretical and practical reasons. It's fundamental *theoretically* because it connects the two main concepts of calculus: the derivative and the integral. Each is a sort of inverse of the other.

The FTC's *practical* consequences are at least as important; implicitly or explicitly, we'll use them again and again. Most important, the FTC leads to a practical method of calculating specific integrals. ◄

So far, we've relied entirely on elementary area formulas.

Another version of the FTC   Before proving the FTC, let's use it to prove another theorem closely related to the FTC, but handier in computations.

> **THEOREM 4 (Fundamental theorem, second version)**   Let f be continuous on $[a, b]$, and let F be *any* antiderivative of f. Then
>
> $$\int_a^b f(x)\,dx = F(b) - F(a).$$

Proof.   This result follows readily from the original FTC. If F is any antiderivative of f, then $F(x)$ can differ from $A_f(x)$ (which is another antiderivative of f, according to the FTC) only by some constant C. In other words, $F(x) = A_f(x) + C$. It follows that

$$F(b) - F(a) = \big(A_f(b) + C\big) - \big(A_f(a) + C\big)$$
$$= A_f(b) - A_f(a)$$
$$= \int_a^b f - \int_a^a f = \int_a^b f,$$

just as claimed.   □

Calculating areas with antiderivatives

With Theorem 4 available it's a routine matter to calculate many signed areas.

EXAMPLE 2   In Section 5.1 we estimated—but didn't prove—that the shaded regions N and P in Figure 4 have signed areas -0.25 and 0.25, respectively. Prove it now.

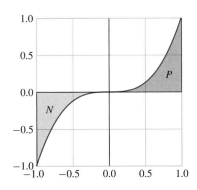

FIGURE 4
Two areas bounded by $y = x^3$

Solution   Let $f(x) = x^3$. The signed areas N and P are given by integrals:

$$N = \int_{-1}^0 f(x)\,dx; \qquad P = \int_0^1 f(x)\,dx.$$

Now $F(x) = x^4/4$ is an antiderivative of $f(x)$, and so Theorem 4 applies:

$$\int_{-1}^0 f(x)\,dx = F(0) - F(-1) = -\frac{1}{4}; \qquad \int_0^1 f(x)\,dx = F(1) - F(0) = \frac{1}{4},$$

as claimed earlier.   ■

Bracket notation   The **bracket notation** offers a convenient shorthand for calculating integrals.

$$\text{For any function } F, \ F(x)\Big]_a^b \text{ means } F(b) - F(a).$$

Watch for the bracket notation in the next example.

EXAMPLE 3   Evaluate $\int_0^1 x^2\,dx$ and $\int_0^1 x^{10}\,dx$. Interpret each integral as an area.

Solution   By Theorem 4 it's enough to find *any* antiderivative, plug in endpoints, and subtract. We do just that:

$$\int_0^1 x^2\,dx = \frac{x^3}{3}\bigg]_0^1 = \frac{1}{3}; \qquad \int_0^1 x^{10}\,dx = \frac{x^{11}}{11}\bigg]_0^1 = \frac{1}{11}.$$

Each integral measures an area. The lightly and darkly shaded areas in Figure 5 represent the integrals just calculated.

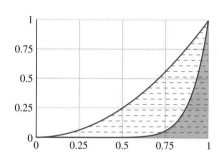

FIGURE 5
Areas under $y = x^{10}$ and $y = x^2$

■

EXAMPLE 4   Calculate the shaded area in Figure 6.

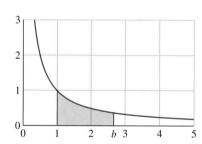

In the case shown, $b = e$ and the shaded area is 1.

FIGURE 6
Area under $y = \frac{1}{x}$ from $x = 1$ to $x = b$

Which value of b makes the shaded area 1?

Solution   The shaded area is the value of the integral $\int_1^b 1/x\,dx$. Recall that the function $\ln x$ is an antiderivative for $1/x$. ◄ Now Theorem 4 says that

$$\int_1^b \frac{1}{x}\,dx = \ln x\Big]_1^b = \ln b - \ln 1 = \ln b.$$

If $b = e$ (as in the picture), the area is $\ln e = 1$.

We discovered this in Section 2.6.

■

From rates to amounts with the FTC

The second version of the fundamental theorem can be restated slightly (without changing its meaning) to read as follows:

> **FACT**   Let f be a function defined on $[a, b]$, with continuous derivative f'. Then
> $$\int_a^b f'(t) \, dt = f(b) - f(a).$$

In words, the Fact says this:

> *Integrating f' (the* rate *function) over $[a, b]$ gives the change in f (the* amount *function) over the same interval.*

This Fact has important practical uses; with it, we can deduce *amount* information from *rate* information.

EXAMPLE 5   Demand for electricity varies predictably over the course of a day. Drawing on experience, engineers in a small town (a bedroom suburb, perhaps) use the formula

$$r(t) = 4 + \sin(0.263t + 4.7) + \cos(0.526t + 9.4)$$

to model the town's typical rate of consumption, in megawatts, t hours after midnight. Figure 7 shows the function r.

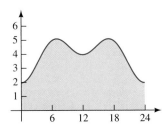

FIGURE 7
**Rate of power use (megawatts)
over 24 hours**

Is the model reasonable? What does the shaded area represent?

Solution   The graph looks generally believable. Residential demand peaks in the morning, declines during working hours, and rises again in the evening. The formula for r is complicated, but the basic ingredients—periodic functions and constants—look right.

   The shaded area is the integral $\int_0^{24} r(t) \, dt$. To understand it in context, let $a(t)$ be the total *amount* of electricity consumed up to time t, starting at midnight. ➡ Then $a'(t) = r(t)$, and by the preceding Fact,

By definition, therefore, $a = 0$ when $t = 0$.

$$\int_0^{24} r(t) \, dt = a(24) - a(0) = a(24) = \text{total daily consumption}.$$

To evaluate this integral numerically we need an antiderivative for $r(t)$. The messy formula for $r(t)$ complicates the search, but we might eventually guess (and it's easy to *check* using derivative rules) that

$$4t - \frac{\cos(0.263t + 4.7)}{0.263} + \frac{\sin(0.526t + 9.4)}{0.526}$$

is an antiderivative for $r(t)$. Thanks to Theorem 4, the rest is straight (though slightly messy) calculation:

$$\int_0^{24} r(t)\, dt = 4t - \frac{\cos(0.263t + 4.7)}{0.263} + \frac{\sin(0.526t + 9.4)}{0.526} \Bigg]_0^{24}$$

$$\approx 95.781 \text{ megawatt hours.} \qquad \blacksquare$$

When *not* to use the FTC

The FTC, useful as it is, doesn't solve *all* our integral problems. Evaluating integrals by antidifferentiation requires first finding a usable antiderivative. For a surprising number of integrands, even apparently simple ones, finding an antiderivative formula is hard or even impossible. None of the following functions, for instance, has an elementary antiderivative:

$$\sin(x^2); \qquad \frac{x}{\ln x}; \qquad 3 + \sin x + 0.3 \arcsin(\sin(7x)).$$

Nevertheless, all have perfectly sensible integrals over finite intervals. For instance, the shaded area in Figure 8 shows what $\int_0^{10} f$ means for the last of the preceding functions.

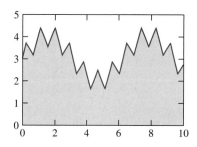

Can you estimate the shaded area? The bounding box has total area 50.

FIGURE 8
An FTC-resistant integral: $\int_0^{10} \left(3 + \sin x + 0.3 \arcsin(\sin(7x))\right)\ dx$

With or without an antiderivative formula, the integral in Figure 8 certainly exists. Soon we'll study techniques that *don't* involve antiderivatives for accurately *estimating* integrals like this one.

Proving the FTC

The FTC is certainly useful, and by now, we hope, it's plausible. But why is it *true*?

The idea of the proof Recall the setup: f is a continuous function, defined on an interval I containing $x = a$, and

$$A_f(x) = \int_a^x f(t)\, dt.$$

The theorem asserts that, for x in I,

$$\frac{d}{dx} A_f(x) = f(x).$$

By the definition of derivative

$$\frac{d}{dx} A_f(x) = \lim_{h \to 0} \frac{A_f(x+h) - A_f(x)}{h};$$

We'll work directly with the difference quotient to prove that the last limit above is $f(x)$. Figure 9 will prove useful.

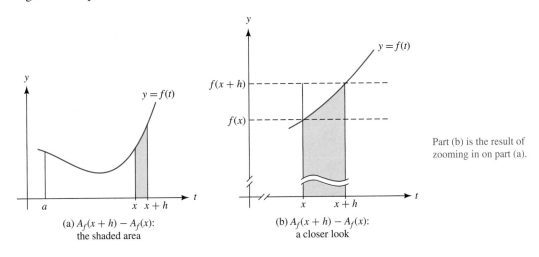

(a) $A_f(x+h) - A_f(x)$:
the shaded area

(b) $A_f(x+h) - A_f(x)$:
a closer look

Part (b) is the result of zooming in on part (a).

FIGURE 9
Proving the FTC

Notice first that $A_f(x+h) - A_f(x) = \int_a^{x+h} f - \int_a^x f = \int_x^{x+h} f$; Figure 9(a) illustrates why: the shaded region ($\int_x^{x+h} f$) is the result of subtracting the smaller region ($\int_a^x f$) from the larger ($\int_a^{x+h} f$).

Figure 9(b) shows that, for small positive values of h,

$$\int_x^{x+h} f \approx f(x) \cdot h.$$

(The last quantity is the area of the shorter rectangle.) Thus, for h near 0 we have

$$\frac{A_f(x+h) - A_f(x)}{h} \approx \frac{f(x) \cdot h}{h} = f(x),$$

and as $h \to 0$, the difference quotient tends to $f(x)$, as desired. $\qquad \square$

More on the proof: fine print   A rigorous proof can be constructed from the preceding informal argument. The missing ingredient is a precise approach to the approximation $\int_x^{x+h} f \approx f(x) \cdot h$. It helps if f happens to be increasing near x—as it is in Figure 9. In that case, we have

$$f(x) \cdot h \le A_f(x+h) - A_f(x) \le f(x+h) \cdot h.$$

(Note that $A_f(x+h) - A_f(x)$ is the shaded area.) Thus,

$$f(x) \le \frac{A_f(x+h) - A_f(x)}{h} \le f(x+h).$$

As $h \to 0$, the quantities on both left and right tend to $f(x)$ and so, therefore, must the difference quotient. (A similar approach works if f is not increasing, although some technical details differ.)

BASIC EXERCISES

In Exercises 1–8, $F(x) = \int_0^x f(t)\,dt$, where f is the function whose graph is shown below.

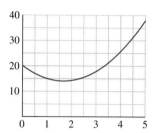

1. Is F increasing at $x = 3$? Justify your answer.

2. Is F decreasing at $x = 1$? Justify your answer.

3. Is F concave up at $x = 3$? Justify your answer.

4. Is F concave up at $x = 1$? Justify your answer.

5. Is $F'(4) > 20$? Justify your answer.

6. Is $F'(2) < 20$? Justify your answer.

7. Is $F(2) > 20$? Justify your answer.

8. Suppose that g is a function such that $g(0) = 2$ and $g'(x) = f(x)$ for all x. Explain how the graphs of F and g are related.

In Exercises 9–18, $F(x) = \int_0^x f(t)\,dt$, where f is the function graphed below. (The graph of f is made up of straight lines and a semicircle.)

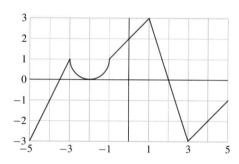

9. Identify all the stationary points of F in the interval $[-5, 5]$.

10. Identify all the critical points of F in the interval $[-5, 5]$.

11. Where in the interval $[-5, 5]$ is F decreasing? Justify your answer.

12. Where in the interval $[-5, 5]$ does F have local extrema? Justify your answer.

13. Evaluate $F(0)$, $F'(0)$ and $F''(0)$.

14. Evaluate $F(2)$, $F'(2)$ and $F''(2)$.

15. Find an equation of the line tangent to the graph $y = F(x)$ at $x = -1$.

16. Find an equation of the line tangent to the graph $y = F(x)$ at $x = 4$.

17. Identify all the inflection points of F in the interval $[-5, 5]$.

18. Let $G(x) = \int_0^x F(t)\,dt$. On which subintervals of $[-5, 5]$, if any, is G concave upward?

In Exercises 19–30, evaluate the definite integral using the fundamental theorem of calculus.

19. $\displaystyle\int_1^3 x\,dx$

20. $\displaystyle\int_0^4 \sqrt{x}\,dx$

21. $\displaystyle\int_{-1}^1 x^2\,dx$

22. $\displaystyle\int_{-2}^2 x^3\,dx$

23. $\displaystyle\int_{-1}^2 e^x\,dx$

24. $\displaystyle\int_0^\pi \cos x\,dx$

25. $\displaystyle\int_{\pi/2}^\pi \sin x\,dx$

26. $\displaystyle\int_1^e \frac{dx}{x}$

27. $\displaystyle\int_1^4 (5x + x^{3/2})\,dx$

28. $\displaystyle\int_{-2}^5 \frac{dx}{x+3}$

29. $\displaystyle\int_{-\pi/2}^\pi \sin(2x)\,dx$

30. $\displaystyle\int_{\pi/2}^{2\pi} \cos(3x)\,dx$

31. **(a)** Find an equation of the line tangent to the graph of $F(x) = \int_1^x \sqrt[3]{t^2 + 7}\,dt$ at $x = 1$.
   (b) Suppose that the tangent line in part (a) is used to compute an estimate of $F(1.1)$. Is this estimate too large? Justify your answer.

32. Suppose that $G(x) = \int_a^x \sin(t)\,dt$, a is a constant, and $G(\sqrt{\pi}) = 1$.
   (a) Find an equation of the line tangent to the graph $y = G(x)$ at $x = \sqrt{\pi}$.
   (b) Suppose that the tangent line in part (a) is used to compute an estimate of $G(1.8)$. Is this estimate too large? Justify your answer.

33. **(a)** Show that $\dfrac{d}{dx} \sin(x^2) = 2x \cos(x^2)$.
   (b) Use part (a) and the fundamental theorem of calculus to evaluate $\int_0^3 x \cos(x^2)\,dx$.

34. **(a)** Let $f(x) = e^{-x^3}$. Find a formula for $f'(x)$.
   (b) Use part (a) and the fundamental theorem of calculus to evaluate $\int_0^1 x^2 e^{-x^3}\,dx$.

FURTHER EXERCISES

In Exercises 35–42, $F(x) = \int_0^x f(t)\, dt$, *where* $f(t)$ *is the function shown below.*

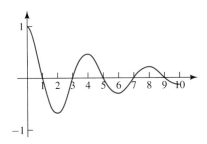

35. Does $F(x)$ have any local maxima in the interval $[0, 10]$? If so, where are they located?

36. Does $F(x)$ have any local minima in the interval $[0, 10]$? If so, where are they located?

37. At what value of x does $F(x)$ attain its minimum value on the interval $[1, 10]$?

38. On which subinterval(s) of $[0, 10]$, if any, is the graph of $F(x)$ concave up? Justify your answer.

39. Is $F'(2) < -0.4$? Justify your answer.

40. Is $F'(4) > 0.7$? Justify your answer.

41. Is $F''(7) > 1$? Justify your answer.

42. Is $F''(5) < -1$? Justify your answer.

43. Suppose that $\int_0^x f(t)\, dt = 3x^2 + e^x - \cos x$. Evaluate $f(2)$.

44. Suppose that $\int_{-1}^w g(u)\, du = \cos(\pi e^w)$. Evaluate $g(0)$.

In Exercises 45 and 46, f *is a function such that* $f'(x) = \sqrt[3]{1 + x^2}$ *and* $f(1) = 0$.

45. Explain why $f(x) = \int_1^x \sqrt[3]{1 + t^2}\, dt$.

46. Express $\int_0^3 \sqrt[3]{1 + x^2}\, dx$ in terms of f.

47. Suppose that $g(x) = b + \int_a^x f(t)\, dx$, where a and b are constants. Show that g is a solution of the IVP $g' = f$, $g(a) = b$.

48. Show that if $f(a) \neq 0$, then $\dfrac{d}{dx}\left[\int_a^x f(t)\, dt\right] \neq \int_a^x \left[\dfrac{d}{dt} f(t)\right] dt$.

49. Let $F(x) = \int_0^x \sqrt[3]{1 + t^2}\, dt$. Find a formula for $\dfrac{d}{dx}\left[F(x^2)\right]$.

50. Let $G(x) = \int_2^{x^3} e^{-t^2}\, dt$. Find a formula for $G'(x)$.

51. Suppose that f' is a continuous function, that $f(1) = 13$, and that $f(10) = 7$. Find the average value of f' over the interval $[1, 10]$.

52. Suppose that g' is a continuous function, that $g(-2) = 3$, and that the average value of g' over the interval $[-2, 5]$ is 1. Find $g(5)$.

53. Find the area of the region enclosed by one arch of the sine function.

54. Find the area of the region enclosed by the graphs $y = 2x$ and $y = x^2$ between $x = 0$ and $x = 2$.

In Exercises 55–58, $g(u) = \int_1^u f(x)\, dx$ *where* f *is the function graphed below.*

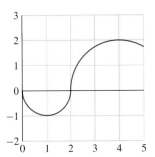

55. Evaluate $g'(4)$.

56. Does g have a tangent line at $x = 2$? Justify your answer.

57. Where in the interval $[0, 5]$ is g concave up? Justify your answer.

58. Is the average value of g' over the interval $[0, 3]$ greater than 1? Justify your answer.

In Exercises 59–63, $g(x) = \int_1^x f(t)\, dt$ *and* $h(x) = \int_3^x f(t)\, dt$, *where* f *is a function with the following properties:*

(i) $f'(x) \geq 0$ *if* $0 \leq x \leq 3$ *and* $6 \leq x \leq 7$; $f'(x) \leq 0$ *if* $3 \leq x \leq 6$.

(ii) $f''(x) \leq 0$ *if* $0 \leq x \leq 5$; $f''(x) \geq 0$ *if* $5 \leq x \leq 7$.

(iii) f *has the values shown below*

x	0	1	2	3	4	5	6	7
$f(x)$	0	2	4	5	4	0	-2	0

59. Evaluate $g'(4)$.

60. Where is the graph of g concave down? Justify your answer.

61. Find the value of x where $g(x)$ is maximum on the interval $[0, 7]$.

62. Find the value of x where $h(x)$ is minimum on the interval $[0, 7]$.

63. What is the average value of f' over the interval $[2, 5]$?

64. Find a polynomial function f with the following properties:
  (i) $f'(x) = ax^2 + bx$
  (ii) $f'(1) = 6$
  (iii) $f''(1) = 18$
  (iv) $\int_1^2 f(x)\,dx = 18$

65. A company is planning to phase out a product because demand for it is declining. Demand for the product is currently 800 units/month and is dropping by 10 units/month each month. To maintain customer and employee relations, the company has announced that it will continue to produce the product for one more year. At present, the company is producing the product at a rate of 900 units/month, and 1680 units of the product are in inventory.

  (a) Give expressions for the demand and production rates as functions of time t (take $t = 0$ as the present).

  (b) Let $D(t)$ be the demand rate for the product at time t. Explain why $\int_0^t D(s)\,ds$ is the total demand for the product between now and time t.

  (c) Give an expression for the inventory at time t.
  [HINT: The amount of the product in inventory is the difference between supply and demand at the end of t months.]

  (d) The company would like to reduce production at a constant rate of R units/month, with R chosen so the product inventory is zero at the end of 12 months. Find a suitable R.

66. The function $\ln x$ is sometimes *defined* for $x > 0$ by integration:

$$\ln x = \int_1^x \frac{dt}{t}.$$

Assume $a > 0$ and $x > 0$. Derive the identity $\ln(ax) = \ln a + \ln x$ by carrying out and justifying each of the following steps:

  (a) Show that $\left(\int_1^{ax} \frac{dt}{t} \right)' = \frac{1}{x}$.
  [HINT: Use the chain rule.]

  (b) Explain why the result in part (a) implies that $\ln(ax) = \ln x + C$.

  (c) By choosing an appropriate x, show that the value of the constant C in part (b) is $\ln a$.

5.4 FINDING ANTIDERIVATIVES; THE METHOD OF SUBSTITUTION

The fundamental theorem of calculus (FTC) makes it easy to evaluate many integrals: If f is a continuous function and F is any antiderivative of f, then

$$\int_a^b f(x)\,dx = F(x)\Big]_a^b = F(b) - F(a).$$

In effect, finding an explicit antiderivative function F transforms an integral problem into something much easier: plugging a and b into F.

To get anywhere with the FTC, of course, we need first to *find* a symbolic antiderivative for the given integrand f. Doing so is sometimes easy, sometimes hard, depending on f. (Finding symbolic *derivatives* is relatively easier, given the product, quotient, chain, and other "combinatorial" rules.)

In this section, we first briefly review some familiar antiderivatives and their properties; then we present one useful method, called **substitution**, for finding antiderivatives systematically. (In Chapter 8 we return in much more detail to the problem of symbolic antidifferentiation.)

Finding antiderivatives: general ideas

Notation and terminology We'll use several standard notations and technical terms in searching for antiderivatives.

- **Indefinite integrals** The expression

$$\int f(x)\,dx$$

is called an **indefinite integral**; f is the **integrand** and x is the **variable of integration.** For example, the equation

$$\int \cos x \, dx = \sin x + C$$

means that every function of the form $\sin x + C$ (where C, the **constant of integration**, may take any value) is an antiderivative of $\cos x$, and that *all* antiderivatives have this form. Notice one subtlety: Interpreted scrupulously, the indefinite integral $\int f(x)\,dx$ denotes not just one function but many—the "family" of all possible antiderivatives of f.

- **Definite integrals**   A **definite integral** is an expression of the form

$$\int_a^b f(x)\,dx,$$

with "definite" endpoints a and b (also known as **limits of integration**). Though they look much alike, definite and indefinite integrals are very different objects: the first is a real *number* and the second is family of *functions*.

Basic antiderivatives

In symbolic differentiation we start with a few known derivatives and combine them according to various rules to produce new derivatives. Antidifferentiation also requires a base of known formulas, to which more complicated problems can be reduced. The following table collects a few very basic antiderivative formulas: ➤

Many more antiderivative formulas appear in the back of this book.

Basic antiderivative formulas
$\int x^k \, dx = \dfrac{x^{k+1}}{k+1} + C \quad (\text{if } k \ne -1)$
$\int \dfrac{1}{x} \, dx = \ln|x| + C$
$\int e^x \, dx = e^x + C$
$\int a^x \, dx = \dfrac{a^x}{\ln a} + C \quad (\text{if } a \ne 1)$
$\int \sin x \, dx = -\cos x + C$
$\int \cos x \, dx = \sin x + C$
$\int \sec^2 x \, dx = \tan x + C$
$\int \dfrac{dx}{\sqrt{1-x^2}} = \arcsin x + C$
$\int \dfrac{dx}{1+x^2} = \arctan x + C$

Each formula is easily checked by differentiation—with one possible exception.

EXAMPLE 1   The formula $\int \dfrac{1}{x}\,dx = \ln|x| + C$ has an unexpected ingredient: the absolute value. Because $\ln x$ itself is an antiderivative for $1/x$ (as we showed in Section 2.6), why do we bother with the absolute value?

Solution   The answer has to do with domains. The function $f(x) = 1/x$ is defined for all $x \neq 0$; ideally, an antiderivative should have the same domain. Unfortunately, the ordinary logarithm function $\ln x$ accepts only *positive* inputs x. The related function $\ln|x|$ solves this problem neatly—it has both the correct domain (positive *and* negative numbers) and the correct derivative. The graphs in Figure 1 give convincing evidence.

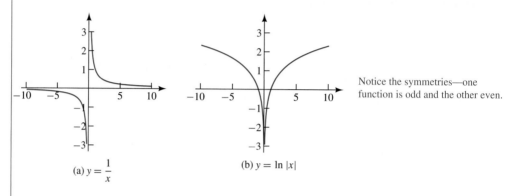

Notice the symmetries—one function is odd and the other even.

(a) $y = \dfrac{1}{x}$

(b) $y = \ln|x|$

FIGURE 1
Why $\ln|x|$ is an antiderivative for $\dfrac{1}{x}$

That $\ln|x|$ is indeed an antiderivative can be seen either in the graphs or by symbolic differentiation. ∎

Combining antiderivatives   Every derivative rule, read backwards, says something about antiderivatives. The rules are especially simple for sums and constant multiples. We stated the following Fact in Section 2.4, but using different notation:

FACT   Let f and g be continuous functions and k any constant. Then

$$\int \big(f(x) + g(x)\big)\,dx = \int f(x)\,dx + \int g(x)\,dx;$$

$$\int k\,f(x)\,dx = k \int f(x)\,dx.$$

EXAMPLE 2   Find $\int (3\cos x - 2e^x)\,dx$.

Solution   By the Fact,

$$\int (3\cos x - 2e^x)\,dx = \int 3\cos x\,dx - \int 2e^x\,dx = 3\int \cos x\,dx - 2\int e^x\,dx$$
$$= 3\sin x - 2e^x + C,$$

an answer we can check by differentiation. ∎

Creative guessing The next two examples illustrate educated guessing—a fully respectable but sometimes underappreciated technique in mathematics. Differentiation is easy; thus, most guesses are easy to check and incorrect guesses may be easy to fix.

EXAMPLE 3 Find $\int \cos(2x)\, dx$ by guessing.

Solution Since $\int \cos x\, dx = \sin x + C$, we might—just guessing—try $\sin(2x)$ as an antiderivative. Differentiation shows that we missed (but only by a factor of 2).

$$\big(\sin(2x)\big)' = \cos(2x) \cdot 2 \neq \cos(2x).$$

Only a tiny alteration is needed:

$$\left(\frac{\sin(2x)}{2}\right)' = \cos(2x) \implies \int \cos(2x)\, dx = \frac{\sin(2x)}{2} + C.$$

■

EXAMPLE 4 Find $\int \dfrac{6}{1+4x^2}\, dx$ by creative guessing.

Solution The integrand's form and a look at the table of basic antiderivatives suggests an arctangent, perhaps with some constants attached, as in $F(x) = A \arctan(Bx)$. Differentiation gives ➡

Check this; use the chain rule carefully.

$$F'(x) = \big(A \arctan(Bx)\big)' = \frac{A}{1+(Bx)^2} \cdot B = \frac{AB}{1+B^2x^2}.$$

Our desired antiderivative therefore has $AB = 6$ and $B^2 = 4$. Setting $A = 3$ and $B = 2$ does nicely; therefore, $F(x) = 3 \arctan(2x)$ is a suitable antiderivative.

This method of guessing—starting with **undetermined coefficients** A and B and then finding appropriate values for them—is common in calculus; we'll see it again.

■

Looks can deceive Checking guesses is easy and effective, but caution is needed—some functions (the trigonometric type, especially) can be written in more than one way.

EXAMPLE 5 Is $\int \sin(2x)\, dx = \sin^2(x) + C$?

Solution Let's check. By the chain rule,

$$\left(\sin^2(x) + C\right)' = 2 \sin x \cos x.$$

We hoped to get $\sin(2x)$, so we're tempted to answer no. In fact, the right answer is yes. One of the many trigonometric identities says that

$$2 \sin x \cos x = \sin(2x).$$

We stand corrected.

■

The method of substitution

The simplest and most important antidifferentiation technique is called, variously, **direct substitution**, **u-substitution**. or just plain **substitution**. ➡

For obscure reasons, the letter u is almost invariably used for substitutions. We'll follow convention.

We'll show through many examples both *how* substitution works and *that* it works; the examples will also suggest *why* it works, and at a strong link to the chain rule.

EXAMPLE 6   Find $\int 2x \, \cos(x^2) \, dx$.

Solution   Let $u = u(x) = x^2$; then

$$\frac{du}{dx} = 2x, \quad \text{so} \quad du = 2x \, dx.$$

(Whoa—can du and dx legally be separated this way? The short answer is yes; we'll return to the question later.) Substituting both u and du into the original integral gives

$$\int \cos(x^2) \, 2x \, dx = \int \cos u \, du.$$

The second integral is simpler than the first, and so we've made progress. To continue, we antidifferentiate in u and then substitute back:

$$\int \cos u \, du = \sin u + C = \sin(x^2) + C.$$

Is the answer right? An easy check, using the chain rule, shows that it is:

$$(\sin(x^2) + C)' = \cos(x^2) \cdot 2x.$$

That was easy, quick, and possibly dirty—separating du and dx might seem suspicious, for one thing. But the method—dirty or not—produced an answer we could easily check. ∎

EXAMPLE 7   Find $\int \dfrac{x}{1+x^2} \, dx$.

Solution   If $u(x) = 1 + x^2$, then $du/dx = 2x$ and $du = 2x \, dx$. Making the substitutions gives

$$\int \frac{x}{1+x^2} \, dx = \frac{1}{2} \int \frac{2x}{1+x^2} \, dx = \frac{1}{2} \int \frac{du}{u}.$$

Notice the first step: multiplying by 2 inside the integral sign "produced" the desired du. (We paid for this convenience by dividing by 2 *outside* the integral.)

In its new form the antiderivative problem is easier:

$$\frac{1}{2} \int \frac{du}{u} = \frac{1}{2} \ln |u| + C = \frac{1}{2} \ln(1+x^2) + C,$$

where we substituted $u = 1 + x^2$ in the last step. Checking the answer is another chain-rule exercise (left to you). ∎

Trading one integral for another   A successful u-substitution transforms one indefinite integral into another, the second simpler than the first. The process has three steps:

- **Substitute**   Choose a function $u = u(x)$, and write $du = u'(x) \, dx$. Then substitute u and du into the original integral $\int f(x) \, dx$ to produce a new integral of the form $\int g(u) \, du$.
- **Antidifferentiate in u**   Solve $\int g(u) \, du$ as an antidifferentiation problem in u; that is, find $G(u)$ such that $G'(u) = g(u)$.
- **Resubstitute**   Substitute *back* to eliminate u. The result, $F(x) = G(u(x))$, is *one* antiderivative of f; other antiderivatives have the form $F(x) + C$.

Differentials   The symbols du and dx are called the **differentials** of u and x, respectively. If $u = u(x)$, then du and dx are related in the expected way: $du = u'(x) \, dx$.

A rigorous (and rather subtle) theory of differentials exists in higher mathematics; we won't need it in this book. For us, the differential in an integral expression (dx in $\int f(x) \, dx$, for instance) is mainly an aid to memory: it reminds us of the variable of integration. In substitution problems, trading dx's for du's (or vice versa) is a useful bookkeeping device; it helps keep us honest when we change variables.

A substitution sampler

Substitution is easy enough in theory. In practice, finding the "right" substitution is both art and science. In both realms, practice (false starts and all) makes perfect. The following examples illustrate some of the possibilities and some useful tricks of the trade.

Choosing u wisely   Choosing u properly is the tricky part of substitution. Things go well when (as in the preceding examples) both u and du appear conveniently in the original integrand and when the new integral is simpler than the old.

EXAMPLE 8   Find $\displaystyle\int \sin^3 x \cos x \, dx$.

Solution   If $u = \sin x$, then $du = \cos x \, dx$. With u and du staring us in the face, prospects look good:

$$\int \sin^3 x \, \cos x \, dx = \int u^3 \, du = \frac{u^4}{4} + C = \frac{\sin^4 x}{4} + C.$$ ■

EXAMPLE 9   Find $\displaystyle\int \frac{x}{1+x^4} \, dx$.

Solution   No obvious substitution suggests itself. We might try $u = 1 + x^4$, but then $du = 4x^3 \, dx$, and nothing of the sort appears in the given integrand. A better choice, it turns out, is $u = x^2$. Then $du = 2x \, dx$, and

$$\int \frac{x}{1+x^4} \, dx = \frac{1}{2} \int \frac{2x}{1+x^4} \, dx = \frac{1}{2} \int \frac{du}{1+u^2}.$$

We've made progress—the last antiderivative is an old standard:

$$\frac{1}{2} \int \frac{du}{1+u^2} = \frac{1}{2} \arctan u + C = \frac{1}{2} \arctan(x^2) + C.$$

Checking the answer involves (as usual) the chain rule.   ■

Traveling constants   Multiplicative constants move freely (and legally) in and out of antiderivatives. We exploit this freedom in the next example, as we did in Examples 7 and 9.

EXAMPLE 10   Find $\displaystyle\int \frac{3x}{5x^2+7} \, dx$.

Solution   The numerator is almost (except for multiplicative constants) the derivative of the denominator; this brings $\int du/u$ to mind. If we set $u = 5x^2 + 7$, then $du = 10x \, dx$, and ➡   *Watch the constants carefully.*

$$\int \frac{3x}{5x^2+7} \, dx = \frac{3}{10} \int \frac{10x}{5x^2+7} \, dx = \frac{3}{10} \int \frac{du}{u}$$

$$= \frac{3}{10} \ln |u| + C = \frac{3}{10} \ln (5x^2+7) + C.$$

(Since $5x^2 + 7 > 0$ for all x, it's fair to ignore the absolute value.)   ■

Inverse substitutions: writing x and dx in terms of u and du   In the examples so far, we've written u and du in terms of x and dx. Sometimes it's simpler to write x and dx in terms of u and du. The next two examples illustrate this strategy.

EXAMPLE 11   Find $\displaystyle\int \frac{dx}{1+\sqrt{x}}$.

Solution   Let $u = \sqrt{x}$. (This is not the only possibility: $u = 1 + \sqrt{x}$ also works.) Then $x = u^2$ and $dx = 2u\,du$; substituting for x and dx gives

$$\int \frac{dx}{1+\sqrt{x}} = \int \frac{2u}{1+u}\,du.$$

The second integral is simpler than the first, but *another* substitution makes things simpler still. If $v = 1 + u$, then $dv = du$ and $u = v - 1$, so

$$\int \frac{2u}{1+u}\,du = \int \frac{2(v-1)}{v}\,dv = 2\int \left(1 - \frac{1}{v}\right)\,dv = 2(v - \ln|v|) + C.$$

Last, we substitute back, first for u and then for x:

$$2(v - \ln|v|) + C = 2(1 + u - \ln|1+u|) + C = 2 + 2\sqrt{x} - 2\ln|1+\sqrt{x}| + C.$$ ■

EXAMPLE 12   Find $\displaystyle\int \frac{x}{\sqrt{2x+3}}\,dx$.

Solution   If $u = \sqrt{2x+3}$, then $du = dx/\sqrt{2x+3}$, and substitution gives

$$\int \frac{x}{\sqrt{2x+3}}\,dx = \int x\,\frac{dx}{\sqrt{2x+3}} = \int x\,du.$$

So far so good, but what about that remaining x? To eliminate it, we write x in terms of u:

$$u = \sqrt{2x+3} \implies u^2 = 2x+3 \implies \frac{u^2-3}{2} = x.$$

Now we're on our way:

$$\int x\,du = \int \frac{u^2-3}{2}\,du = \frac{u^3}{6} - \frac{3u}{2} + C = \frac{(2x+3)^{3/2}}{6} - \frac{3\sqrt{2x+3}}{2} + C.$$

For symbolic variety, let's redo the problem a little differently, again starting with $u = \sqrt{2x+3}$. Then $u^2 = 2x+3$, and differentiating both sides gives $2u\,du = 2\,dx$; thus, $u\,du = dx$. Now we substitute for x, dx, and $\sqrt{2x+3}$—all at once:

$$\int \frac{x}{\sqrt{2x+3}}\,dx = \int \frac{u^2-3}{2u}\,u\,du = \int \frac{u^2-3}{2}\,du,$$

and the problem proceeds as before. ■

Substitution in definite integrals

The next example shows two ways of evaluating *definite* integrals by substitution.

EXAMPLE 13   Find $\displaystyle\int_0^{\sqrt{\pi/2}} 2x\,\cos(x^2)\,dx$.

Solution   We'll solve the problem in two different ways.

Method 1: Antidifferentiate in *x* and plug in endpoints   If $F(x)$ is any antiderivative of $2x\,\cos(x^2)$, then $F(\sqrt{\pi/2}) - F(0)$ is our answer. In fact, we already found a suitable F in Example 6, where we substituted $u = x^2$ to find $F(x) = \sin(x^2)$. ➤ Thus,

We need just one antiderivative so we set $C = 0$.

$$\int_0^{\sqrt{\pi/2}} 2x\,\cos(x^2)\,dx = \sin(x^2)\Big]_{x=0}^{x=\sqrt{\pi/2}} = \sin(\pi/2) - \sin 0 = 1.$$

(We wrote the symbol x in the bracket notation to emphasize which variable is in use.)

Method 2: Substitute to create a new definite integral   Again, let $u = x^2$ and $du = 2x\,dx$. At the endpoints $x = 0$ and $x = \sqrt{\pi/2}$, we have $u = 0$ and $u = \pi/2$, respectively. Now we substitute for u, du, *and for the endpoints*, to get a new *definite* integral:

$$\int_0^{\sqrt{\pi/2}} 2x\,\cos(x^2)\,dx = \int_0^{\pi/2} \cos u\,du.$$

The last integral is easy to calculate:

$$\int_0^{\pi/2} \cos u\,du = \sin u\Big]_{u=0}^{u=\pi/2} = \sin(\pi/2) - \sin 0 = 1,$$

the same result as before. ➤   ■

The u symbol in the bracket notation reminds us which variable is in play.

Another look   Notice carefully the different uses of substitution in the two solutions to Example 13. In the first solution we used substitution only to find an antiderivative of the given integrand with respect to the original variable x. In the second solution we used substitution to transform the original definite integral into a *new definite integral*, with new limits of integration and a new integrand.

Why substitution works

Substitution certainly seems to work. But *why* does it work? To see, we'll describe the process in general terms.

For any antiderivative problem $\int f(x)\,dx$, substitution means finding a function $u = u(x)$ such that $f(x)$ can be rewritten in the form

$$f(x) = g\big(u(x)\big)\,u'(x).$$

To say that substitution "works" means that

$$\int f(x)\,dx = \int g(u)\,du.$$

This means, in turn, that if G is a function for which $G'(u) = g(u)$, then $G\big(u(x)\big)$ is an antiderivative of f. The chain rule shows that this is indeed so:

$$\frac{d}{dx}\left(G\big(u(x)\big) \right) = G'\big(u(x)\big) \cdot u'(x) = g\big(u(x)\big) \cdot u'(x) = f(x),$$

so $G\big(u(x)\big)$ is indeed the antiderivative we seek.

Substitution works in *definite* integrals for essentially the same reason. The following theorem gives details.

> **THEOREM 5 (Change of variables in definite integrals)**   Let f, u, and g be continuous functions such that for all x in $[a, b]$,
>
> $$f(x) = g\big(u(x)\big) \cdot u'(x).$$
>
> Then
>
> $$\int_a^b f(x)\, dx = \int_{u(a)}^{u(b)} g(u)\, du.$$

Proof   Let G be an antiderivative of g. Then, by the FTC,

$$\int_{u(a)}^{u(b)} g(u)\, du = G\big(u(b)\big) - G\big(u(a)\big).$$

We showed just above that $G\big(u(x)\big)$ is an antiderivative for $f(x)$. Hence, by the FTC,

$$\int_a^b f(x)\, dx = G\big(u(b)\big) - G\big(u(a)\big),$$

so the two integrals are equal. $\square$

BASIC EXERCISES

In Exercises 1–10, evaluate the antiderivative using the indicated substitution. (Check your answers by differentiation.)

1. $\displaystyle\int (4x+3)^{-3}\, dx; \quad u = 4x+3$

2. $\displaystyle\int x\sqrt{1+x^2}\, dx; \quad u = 1+x^2$

3. $\displaystyle\int e^{\sin x} \cos x\, dx; \quad u = \sin x$

4. $\displaystyle\int \frac{(\ln x)^3}{x}\, dx; \quad u = \ln x$

5. $\displaystyle\int \frac{\arctan x}{1+x^2}\, dx; \quad u = \arctan x$

6. $\displaystyle\int \frac{\sin\left(\sqrt{x}\right)}{\sqrt{x}}\, dx; \quad u = \sqrt{x}$

7. $\displaystyle\int \frac{e^{1/x}}{x^2}\, dx; \quad u = 1/x$

8. $\displaystyle\int x^3\sqrt{9-x^2}\, dx; \quad u = 9-x^2$

9. $\displaystyle\int \frac{e^x}{1+e^{2x}}\, dx; \quad u = e^x$

10. $\displaystyle\int \frac{x}{1+x}\, dx; \quad u = 1+x$

In Exercises 11–14, first find real numbers a and b so that equality holds; then evaluate the definite integral.

11. $\displaystyle\int_{-2}^{1} \frac{x}{1+x^4}\, dx = \frac{1}{2}\int_a^b \frac{1}{1+u^2}\, du$

12. $\displaystyle\int_{-\sqrt{\pi/2}}^{\sqrt{\pi}} x\cos(3x^2)\, dx = \frac{1}{6}\int_a^b \cos u\, du$

13. $\displaystyle\int_0^3 \frac{x}{(2x^2+1)^3}\, dx = \frac{1}{4}\int_a^b u^{-3}\, du$

14. $\displaystyle\int_1^2 x^2 e^{x^3/4}\, dx = \frac{4}{3}\int_a^b e^u\, du$

FURTHER EXERCISES

15. Example 9 says that $\displaystyle\int \frac{x}{1+x^4}\, dx = \frac{1}{2}\arctan(x^2) + C$. Verify this by differentiation.

16. Example 12 says that $\displaystyle\int \frac{x}{\sqrt{2x+3}}\, dx = \frac{(2x+3)^{3/2}}{6} - \frac{3\sqrt{2x+3}}{2} + C$. Verify this by differentiation.

17. Use the substitution $u = 1 + \sqrt{x}$ to find the antiderivative $\displaystyle\int \frac{dx}{1+\sqrt{x}}$.

18. Let $a > 0$ be a constant. Derive the antiderivative formula $\displaystyle\int \frac{dx}{\sqrt{a^2-x^2}} = \arcsin(x/a) + C$ from the antiderivative formula $\displaystyle\int \frac{dx}{\sqrt{1-x^2}} = \arcsin x + C$.

19. Let a and $b > 0$ be constants. Use the substitution $u = a + b/x$ to find the antiderivative $\int \dfrac{dx}{ax^2 + bx}$.

20. Show that $\int \dfrac{dx}{1 + e^x} = x - \ln(1 + e^x) + C$.

[HINT: $\dfrac{1}{u(1+u)} = \dfrac{1}{u} - \dfrac{1}{1+u}$.]

21. Suppose that $\int_0^{12} g(x)\,dx = \pi$. Evaluate $\int_0^4 g(3x)\,dx$.

22. Let $I = \int \sec^2 x \tan x\,dx$.

(a) Use the substitution $u = \sec x$ to show that $I = \frac{1}{2}\sec^2 x + C$.

(b) Use the substitution $u = \tan x$ to show that $I = \frac{1}{2}\tan^2 x + C$.

(c) The expressions for I in parts (a) and (b) look different, but both are correct. Explain this apparent paradox.

In Exercises 23–68, evaluate the antiderivative. (Check your answers by differentiation.)

23. $\int \cos(2x + 3)\,dx$

24. $\int \sin(2 - 3x)\,dx$

25. $\int (3x - 2)^4\,dx$

26. $\int x \cos(1 - x^2)\,dx$

27. $\int \dfrac{2x^3}{1 + x^4}\,dx$

28. $\int x(3x + 2)^4\,dx$

29. $\int \dfrac{dx}{1 - 2x}$

30. $\int \sqrt{3x - 2}\,dx$

31. $\int x\sqrt{3 - 2x}\,dx$

32. $\int \dfrac{\ln x}{x}\,dx$

33. $\int \dfrac{\sqrt{1 + 1/x}}{x^2}\,dx$

34. $\int x e^{x^2}\,dx$

35. $\int x^3(x^4 - 1)^2\,dx$

36. $\int \dfrac{x^3}{1 + x^2}\,dx$

37. $\int \dfrac{x^2}{1 + x^6}\,dx$

38. $\int \dfrac{x}{\sqrt{1 - x^4}}\,dx$

39. $\int x^2\sqrt{4x^3 + 5}\,dx$

40. $\int \dfrac{x}{\sqrt{1 + x^2}}\,dx$

41. $\int \dfrac{x + 4}{x^2 + 1}\,dx$

42. $\int x(1 - x^2)^{15}\,dx$

43. $\int \dfrac{2x + 3}{(x^2 + 3x + 5)^4}\,dx$

44. $\int (x + 2)(x^2 + 4x + 5)^6\,dx$

45. $\int \dfrac{x + 1}{\sqrt[3]{3x^2 + 6x + 5}}\,dx$

46. $\int \dfrac{e^{2x}}{(1 + e^{2x})^3}\,dx$

47. $\int \dfrac{e^x}{(2e^x + 3)^2}\,dx$

48. $\int \dfrac{e^{\sqrt{x}}}{\sqrt{x}}\,dx$

49. $\int \dfrac{2x + 3}{(x + 1)^2}\,dx$

50. $\int \dfrac{x^2}{x - 3}\,dx$

51. $\int \tan x\,dx$

52. $\int \sec x \tan x\,dx$

53. $\int \dfrac{\arcsin x}{\sqrt{1 - x^2}}\,dx$

54. $\int \sec x \tan x\sqrt{1 + \sec x}\,dx$

55. $\int \dfrac{5x}{3x^2 + 4}\,dx$

56. $\int \dfrac{\cos x}{1 + \sin^2 x}\,dx$

57. $\int \dfrac{\sec^2 x}{\sqrt{1 - \tan^2 x}}\,dx$

58. $\int \tan^2 x \csc x\,dx$

59. $\int x \tan(x^2)\,dx$

60. $\int x^2 \sec^2(x^3)\,dx$

61. $\int \dfrac{\cos x}{\sin^4 x}\,dx$

62. $\int x^4 \sqrt[3]{x^5 + 6}\,dx$

63. $\int \dfrac{(1 + \sqrt{x})^3}{\sqrt{x}}\,dx$

64. $\int \dfrac{dx}{\sqrt{x}\,\left(\sqrt{x}+2\right)^3}$

65. $\int x^3\sqrt{x^2+2}\,dx$

66. $\int \sqrt{1+\sqrt{x}}\,dx$

67. $\int \dfrac{e^x}{e^{2x}+2e^x+1}\,dx$

68. $\int \dfrac{e^{\tan x}}{1-\sin^2 x}\,dx$

In Exercises 69–74, evaluate the definite integral.

69. $\int_0^2 \dfrac{x}{(1+x^2)^3}\,dx$

70. $\int_{-19}^8 \sqrt[3]{8-x}\,dx$

71. $\int_1^e \dfrac{\sin(\ln x)}{x}\,dx$

72. $\int_e^{4e} \dfrac{dx}{x\sqrt{\ln x}}$

73. $\int_0^\pi \sin^3 x \cos x\,dx$

74. $\int_{-\pi/2}^\pi e^{\cos x}\sin x\,dx$

75. Suppose that the function f is continuous on the interval $[-1, 1]$. Use the substitution $u = \pi - x$ to show that

$$\int_0^\pi x f(\sin x)\,dx = \frac{\pi}{2}\int_0^\pi f(\sin x)\,dx.$$

The substitution $u^n = ax+b$ can sometimes be used to find antiderivatives of expressions involving the form $\sqrt[n]{ax+b}$. Use this substitution to evaluate the antiderivatives in Exercises 76 and 77.

76. $\int x\sqrt{2x+1}\,dx$

77. $\int \dfrac{dx}{\sqrt{x}+\sqrt[3]{x}}$

5.5 INTEGRAL AIDS: TABLES AND COMPUTERS

In Section 5.4 we found various symbolic antiderivatives by the method of substitution. The method worked well in the examples given, but honesty compels a tiny authorial admission: we *chose* examples for which substitution works well. For many functions—even some that look quite simple—the search for symbolic antiderivatives can be difficult, tricky, and laborious. In practice, calculus users freely resort to integral tables, computer software, roommates, or any other aids that come to hand.

Like all power tools, tables and mathematical software require careful handling. In this section, we illustrate some of their possibilities and pitfalls.

Integral tables

"Antiderivative table" would be a better name.

Standard scientific reference works contain extensive integral tables. ◄ The *CRC Handbook of Chemistry and Physics,* 48th edition, for instance, lists nearly 600 different antiderivative formulas on more than 40 pages. (The CRC tables are relatively modest—other books contain hundreds of pages of integral formulas.) One of them reads as follows: ◄

We added the absolute value signs.

FACT (Formula 403 from the *CRC Handbook*)

$$\int \frac{dx}{a+be^{px}} = \frac{x}{a} - \frac{1}{ap}\ln\left|a+be^{px}\right|$$

Notice the lack of a constant of integration on the right side; supplying one every time would be typographically tiresome.

Parameters and pattern matching Integral tables use **parameters**—letters other than the variable of integration—to stand for numerical constants that can vary from problem to problem. With this device, a single integral formula can solve an entire family of similar problems. The next few examples concern the parameters in the preceding formula.

EXAMPLE 1   Verify Formula 403 by differentiation.

Solution   We differentiate the right side with respect to x; all other letters are constants. Try to follow the steps:

$$\left(\frac{x}{a} - \frac{1}{ap} \ln |a + be^{px}| \right)' = \frac{1}{a} - \frac{1}{ap} \frac{bpe^{px}}{a + be^{px}}$$

$$= \frac{1}{a} - \frac{1}{a} \frac{be^{px} + a - a}{a + be^{px}}$$

$$= \frac{1}{a} - \frac{1}{a} \left(1 - \frac{a}{a + be^{px}} \right)$$

$$= \frac{1}{a + be^{px}}.$$

That's the result we wanted; the *CRC Handbook* got it right.   ∎

Getting it right.   Not every equation in published integral tables is correct. Integral tables change over time as new antiderivative formulas come into favor and old ones fall into disuse. With each new version of an integral table, new typographical errors may creep in; some errors have lain undiscovered for years.

Standard integral tables are sometimes used as benchmarks to check mathematical software programs for accuracy. This process sometimes uncovers errors—not in the software but in the tables. Some integral tables have been found to have error rates in the double-digit percentage range.

EXAMPLE 2   Use Formula 403 to find $\displaystyle\int \frac{5}{3 - 2e^{-x}}\, dx$.

Solution   Writing

$$\int \frac{5}{3 - 2e^{-x}}\, dx = 5 \int \frac{dx}{3 - 2e^{-x}}$$

makes the parameter values in the formula directly apparent: $a = 3$, $b = -2$, $p = -1$. ➤   *Don't forget the 5 out front.*
The result is

$$\int \frac{5}{3 - 2e^{-x}}\, dx = 5 \left(\frac{x}{3} + \frac{1}{3} \ln |3 - 2e^{-x}| \right) + C.$$ ∎

Making the shoe fit   Antiderivatives that don't appear to fit an integral template can sometimes be *made* to do so. A *u*-substitution, some symbolic algebra, or a combination of both often effects this Cinderella-like transformation. (Sometimes, the shoe just *won't* fit.)

EXAMPLE 3   Find $I = \displaystyle\int \frac{\cos x}{5 + 2e^{3\sin x}}\, dx$.

Solution   As it stands the integral resembles Formula 403 only vaguely. But substituting $u = \sin x$ and $du = \cos x\, dx$ reveals the familiar pattern:

$$\int \frac{\cos x}{5 + 2e^{3\sin x}}\, dx = \int \frac{du}{5 + 2e^{3u}}.$$

Setting $a = 5$, $b = 2$, and $p = 3$ completes the problem:

$$I = \frac{u}{5} - \frac{\ln\left|5 + 2e^{3u}\right|}{15} + C = \frac{\sin x}{5} - \frac{\ln\left|5 + 2e^{3\sin x}\right|}{15} + C.$$ ■

EXAMPLE 4 Find $I = \displaystyle\int \frac{e^x}{3e^{-x} + 2e^x}\, dx.$

Solution A touch of exponential algebra (multiplying top and bottom by e^{-x}) puts the integrand into the necessary form:

$$\frac{e^x}{3e^{-x} + 2e^x} = \frac{e^x e^{-x}}{e^{-x}\left(3e^{-x} + 2e^x\right)} = \frac{1}{2 + 3e^{-2x}}.$$

Now we can use Formula 403, with $a = 2$, $b = 3$, $p = -2$:

$$I = \int \frac{dx}{2 + 3e^{-2x}} = \frac{x}{2} + \frac{1}{4}\ln\left|2 + 3e^{-2x}\right| + C.$$ ■

Handling quadratics: completing the square Quadratic expressions (of the form $ax^2 + bx + c$) appear often in antiderivative problems. Completing the square sometimes helps shoehorn such integrands into an integral table's template. ◄ We illustrate with two examples.

See Appendix D for reminders on completing the square.

EXAMPLE 5 Find $I = \displaystyle\int \frac{dx}{x^2 + 4x + 5}.$

Solution Rummaging through an integral table (such as the one at the end of this book) turns up two likely candidates:

$$\int \frac{dx}{x^2 + a^2} = \frac{1}{a}\arctan\left(\frac{x}{a}\right) + C; \qquad \int \frac{dx}{x^2 - a^2} = \frac{1}{2a}\ln\left|\frac{x - a}{x + a}\right| + C.$$

Which formula, if either, applies? To decide we'll complete the square in the denominator:

$$x^2 + 4x + 5 = (x^2 + 4x + 4) + (5 - 4) = (x + 2)^2 + 1.$$

This, in turn, suggests the substitution $u = x + 2$, $du = dx$. Here's the result:

$$\int \frac{dx}{x^2 + 4x + 5} = \int \frac{du}{u^2 + 1} = \arctan u + C = \arctan(x + 2) + C.$$ ■

EXAMPLE 6 Find $I = \displaystyle\int \frac{dx}{x^2 + 4x + 3}.$

Solution The same two possibilities suggest themselves as in Example 5. Again we complete the square:

$$x^2 + 4x + 3 = (x^2 + 4x + 4) + (3 - 4) = (x + 2)^2 - 1.$$

The rest follows easily. Substitution of $u = x + 2$ and $du = dx$ gives

$$\int \frac{dx}{x^2 + 4x + 3} = \int \frac{du}{u^2 - 1} = \frac{1}{2}\ln\left|\frac{u - 1}{u + 1}\right| + C = \frac{1}{2}\ln\left|\frac{x + 1}{x + 3}\right| + C.$$ ■

A graphical reality check

Results obtained by one method should sometimes be tested by another. Symbolic antiderivatives especially deserve such attention. Functions, after all, are more than purely formal expressions; they're graphical and numerical objects as well. If symbolic operations really make sense, they should stand up to graphical scrutiny. In other words, symbolically produced antiderivatives should "look right." Now and then, we'll check to make sure that they do.

Examples 5 and 6 invite such a look. Their symbolic results,

$$\int \frac{dx}{x^2+4x+5} = \arctan(x+2); \qquad \int \frac{dx}{x^2+4x+3} = \frac{1}{2}\ln\left|\frac{x+1}{x+3}\right|,$$

are a bit mysterious. Although the integrands are almost identical, the antiderivatives look surprisingly different. Could there be some mistake?

Let's approach this graphically. Figure 1 shows graphs of the two integrands.

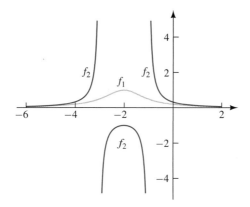

FIGURE 1

$$f_1(x) = \frac{1}{x^2+4x+5} \text{ and } f_2(x) = \frac{1}{x^2+4x+3}$$

Now the mystery begins to dissolve. Despite their typographical similarity, the two integrands are quite different: f_1 behaves tamely, while f_2 has vertical asymptotes at $x = -3$ and $x = -1$. As $x \to \pm\infty$, on the other hand, both f_1 and f_2 tend to zero.

Plots of the antiderivatives F_1 and F_2 (Figure 2) show that they, too, behave as they should.

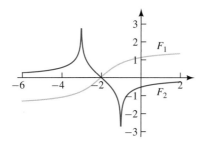

FIGURE 2

$$F_1(x) = \arctan(x+2) \text{ and } F_2(x) = 0.5\ln\left|\frac{x+1}{x+3}\right|$$

Notice these features of the graphs:

- **Asymptotes** Both f_1 and its antiderivative F_1 are tamely behaved. By contrast, f_2 and its antiderivative F_2 have vertical asymptotes at $x = -3$ and $x = -1$.

- **Long-run behavior** Both antiderivatives F_1 and F_2 appear to tend toward the horizontal as $x \to \pm\infty$. This is to be expected because the respective derivatives f_1 and f_2 both tend to zero in the long run.

Reduction formulas: one antiderivative in terms of another Some formulas in integral tables have antiderivatives on *both* sides of an equation. Here's another entry from the *CRC Handbook*:

FACT (Formula 311) If $n \neq 1$, then

$$\int \tan^n x \, dx = \frac{\tan^{n-1} x}{n-1} - \int \tan^{n-2} x \, dx.$$

Applying such a **reduction formula** (perhaps repeatedly) transforms one integral problem into another, the second simpler than the first.

EXAMPLE 7 Find $\int \tan^4 x \, dx$.

Solution Formula 311, with $n = 4$, says that

$$\int \tan^4 x \, dx = \frac{\tan^3 x}{3} - \int \tan^2 x \, dx.$$

The last integral may itself appear in an integral table. If so, we're finished. If not, we can apply Formula 311 *again* (this time with $n = 2$).

$$\int \tan^2 x \, dx = \frac{\tan x}{1} - \int \tan^0 x \, dx = \tan x - x.$$

Combining the results we obtain

$$\int \tan^4 x \, dx = \frac{\tan^3 x}{3} - \tan x + x + C.$$ ∎

Antidifferentiation by machine

Old mathematical handbooks contain vast tables of trigonometric, exponential, and logarithmic function values. Scientific calculators make such tables obsolete. To a lesser (but significant) degree, sophisticated mathematical software effectively replaces integral tables. *Maple*, *Mathematica*, the TI–89, and other computing tools now "know" not just integral formulas but—much more challenging—how and when to apply them.

Symbolic camouflage Equivalent mathematical expressions can appear in radically different symbolic forms. Trigonometric expressions are especially tricky; the many "trigonometric identities" offer almost unlimited possibilities for disguise.

Computer symbolic operations often produce variant or unexpected forms of symbolic results. (What's convenient for a computer may be baffling to you and me—and vice versa.)

EXAMPLE 8 According to one integral table,

$$\int \sin^2 x \, dx = \frac{x}{2} - \frac{\sin(2x)}{4}.$$

Mathematica agrees, but here's what *Maple* says:

```
> int( sin(x)^2, x );

          - 1/2 cos(x) sin(x) + 1/2 x
```

Who's right?

Solution For once, *everybody* is right (we'll forgive everyone the omitted constant of integration). This time, the reason is simple (sometimes it's harder). Stirring the trigonometric identity $\sin(2x) = 2\sin x \cos x$ into the first formula readily produces *Maple*'s answer. (See for yourself.) ∎

When nothing works

Sometimes nothing seems to work. Symbolic methods, integral tables, algebraic tricks, and software may *all* fail to find an antiderivative in elementary form. Two reasons are possible: Either we haven't (or a computer hasn't) searched cleverly enough, or there *is* no elementary antiderivative. Some elementary functions, even simple ones, don't *have* antiderivatives with elementary formulas. None of the following functions does, for instance:

$$e^{x^2}, \qquad \sin(x^2), \qquad \frac{\sin x}{x}, \qquad \frac{x}{\ln x}.$$

Like all continuous functions, these do have antiderivatives, but not elementary ones.

The bright side Even when symbolic methods fail, all is not lost. Numerical methods, such as those we'll study in the next chapter, often succeed. For example, numerical methods will nicely handle *definite* integrals of all the preceding "problematic" functions.

BASIC EXERCISES

Evaluate the antiderivatives in Exercises 1–10 using basic antiderivative formulas and the table of integrals in the back of the book.

1. $\int \dfrac{dx}{3 + 2e^{5x}}$

2. $\int \dfrac{dx}{x(2x+3)}$

3. $\int \dfrac{dx}{x^2(3-x)}$

4. $\int x^2 \ln x \, dx$

5. $\int \dfrac{dx}{x\sqrt{2x+1}}$

6. $\int x \sin(2x) \, dx$

7. $\int e^{2x} \cos(3x) \, dx$

8. $\int \dfrac{dx}{4 - x^2}$

9. $\int \dfrac{dx}{x\sqrt{3x-2}}$

10. $\int \dfrac{dx}{4x^2 - 1}$

FURTHER EXERCISES

Evaluate the antiderivatives in Exercises 11–40 using a table of integrals or a software package. Making simple u-substitutions or completing the square first may help.

11. $\int \dfrac{x^2}{1+x^2} \, dx$ [HINT: $\dfrac{x^2}{1+x^2} = 1 - \dfrac{1}{1+x^2}$.]

12. $\int \dfrac{4x^2 + 3x + 2}{x+1} \, dx$ [HINT: $\dfrac{4x^2 + 3x + 2}{x+1} = 4x - 1 + \dfrac{3}{x+1}$.]

13. $\int \dfrac{x-1}{x+1} \, dx$ [HINT: $x - 1 = (x+1) - 2$.]

14. $\displaystyle\int \frac{dx}{\sqrt{x-1}+\sqrt{x+1}}$

[HINT: $\left(\sqrt{x-1}+\sqrt{x+1}\right)\left(\sqrt{x-1}-\sqrt{x+1}\right)=-2$.]

15. $\displaystyle\int \frac{dx}{1+\sin x}$

[HINT: $(1+\sin x)(1-\sin x)=1-\sin^2 x=\cos^2 x$.]

16. $\displaystyle\int \tan^4(2x)\,dx$

17. $\displaystyle\int \frac{dx}{1+9x^2}$

18. $\displaystyle\int \frac{dx}{9+x^2}$

19. $\displaystyle\int \tan^3(5x)\,dx$

20. $\displaystyle\int x^2 e^{3x}\,dx$

21. $\displaystyle\int \frac{2x+3}{4x+5}\,dx$

22. $\displaystyle\int x^2\sqrt{1-3x}\,dx$

23. $\displaystyle\int \frac{4x+5}{(2x+3)^2}\,dx$

24. $\displaystyle\int \frac{dx}{(4x^2-9)^2}$

25. $\displaystyle\int \frac{x+2}{2+x^2}\,dx$

26. $\displaystyle\int \frac{3}{\sqrt{6x+x^2}}\,dx$

27. $\displaystyle\int \frac{5}{4x^2+20x+16}\,dx$

28. $\displaystyle\int \frac{dx}{\sqrt{x^2+2x+26}}$

29. $\displaystyle\int x^3\cos(x^2)\,dx$

30. $\displaystyle\int \frac{x}{\sqrt{x^2+4x+3}}\,dx$

31. $\displaystyle\int \frac{dx}{(x^2+3x+2)^2}$

32. $\displaystyle\int x^2(\cos x+3\sin x)\,dx$

33. $\displaystyle\int \frac{e^x}{e^{2x}-2e^x+5}\,dx$

34. $\displaystyle\int \cos x\sin x\sin^2(2\cos^2 x+1)\,dx$

35. $\displaystyle\int \sqrt{x^2+4x+1}\,dx$

36. $\displaystyle\int \frac{\cos x}{3\sin^2 x-11\sin x-4}\,dx$

37. $\displaystyle\int \frac{\cos x\sin x}{(\cos x-4)(3\cos x+1)}\,dx$

38. $\displaystyle\int \frac{e^{2x}}{\sqrt{e^{2x}-e^x+1}}\,dx$

39. $\displaystyle\int x\sin(3x+4)\,dx$

40. $\displaystyle\int x^3 e^{-x^2}\,dx$

5.6 APPROXIMATING SUMS: THE INTEGRAL AS A LIMIT

The integral $\int_a^b f$, as defined so far, is the signed area of the region bounded by the graph of f from $x=a$ to $x=b$. In Section 5.3 we saw, thanks to the FTC, that we can readily calculate any integral for which we can find an antiderivative. Sections 5.4 and 5.5 introduced some basic techniques for finding antiderivatives; we'll see more techniques in Chapter 8.

Integration and antidifferentiation are close cousins, but they are *not* the same thing. Many integrals (even simple-looking ones such as $\int_0^1 \sin(x^2)\,dx$), cannot be found using antidifferentiation because no convenient antiderivative formula exists. For such integrals we need another strategy.

For the rest of this chapter we take a different approach to the integral, defining it formally as a certain limit of "approximating sums." The limit-based definition offers new and useful perspectives on the integral and what it means. It also lends itself well—better than the FTC—to the problem of estimating integrals numerically.

This section closely mirrors Section 2.1, where we defined the derivative as a limit. There, as here, the general problem was to translate an intuitively reasonable geometric idea (the slope of a tangent line for derivatives; the area under a graph for integrals) into precise analytic language. The idea of limit is crucial for both the derivative and the integral—it links approximations to exact values.

Approximating sums

The definition of integral as a limit involves some formidable-looking terminology and notation. The underlying ideas, on the other hand, are natural and straightforward, especially when understood graphically. We'll approach the definition via a leisurely example, introducing terms and ideas as we go.

EXAMPLE 1 Figure 1 shows part of the graph of $f(x) = x^3 - 3x^2 + 8$; the shaded area represents the integral $\int_{0.5}^{3.5} f(x)\,dx$. Use the FTC to evaluate the integral *exactly*. Then use approximating sums (we explain them below) to *estimate* the same integral.

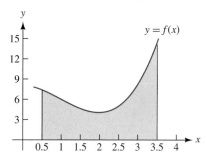

FIGURE 1
What's the area?

Solution An exact answer is easy to find by antidifferentiation: ➤   *Check the (easy) steps.*

$$\int_{0.5}^{3.5} \left(x^3 - 3x^2 + 8\right) dx = \left(\frac{x^4}{4} - x^3 + 8x\right)\Bigg]_{0.5}^{3.5} = \frac{75}{4} = 18.75.$$

But suppose we hadn't known the FTC or couldn't find an antiderivative? How might we *estimate* the area in question? Figure 2 suggests four reasonable strategies.

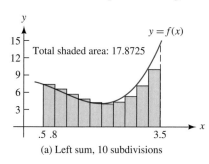

(a) Left sum, 10 subdivisions

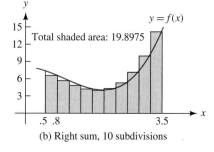

(b) Right sum, 10 subdivisions

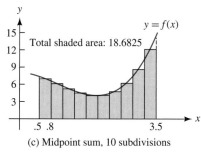

(c) Midpoint sum, 10 subdivisions

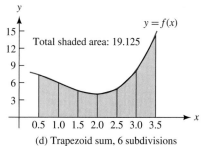

(d) Trapezoid sum, 6 subdivisions

FIGURE 2
Four summing strategies

In each case we approximated the desired area with a sum of simpler areas, either rectangles or trapezoids; their vertical heights are determined by values of the integrand function, as shown. (We used only six trapezoids to show more clearly the difference between the integral and the trapezoid estimate.) Adding up the areas of these simpler figures gives, in each case, a natural estimate to the desired area. ◄   ■

Which estimate looks *best to you? Which one* is *best?*

Calculating the approximations

How, exactly, did we calculate the preceding area estimates? A close look at each picture gives the answer.

Consider first the **left approximating sum with ten subdivisions** (Figure 2(a)), which we'll denote by L_{10}. There are ten shaded rectangles, each with base 0.3 (that is, $(3.5 - 0.5)/10$). The height of each rectangle is the value of the function f at the *left* edge of the base subdivision (hence the name "left sum"). The second rectangle, for example, has the *x*-interval $[0.8, 1.1]$ for its base; its height is $f(0.8)$ (the value of f at the *left* endpoint). Because $f(x) = x^3 - 3x^2 + 8$, $f(0.8) = 6.592$. Therefore, for the second rectangle,

$$\text{Area} = 0.3 \cdot 6.592 = 1.9776.$$

The *total* area of all ten left rectangles is the sum

$$
\begin{aligned}
L_{10} &= f(0.5) \cdot 0.3 + f(0.8) \cdot 0.3 + f(1.1) \cdot 0.3 + \cdots + f(3.2) \cdot 0.3 \\
&= 7.375 \cdot 0.3 + 6.592 \cdot 0.3 + 5.701 \cdot 0.3 + \cdots + 7.159 \cdot 0.3 + 10.048 \cdot 0.3 = 17.8725.
\end{aligned}
$$

Think this sentence through carefully. Do you see how the picture says the same thing?

The **right approximating sum with ten subdivisions** (in Figure 2(b)), denoted by R_{10}, differs only slightly from L_{10}—this time, the height of each rectangle is the value of f at the *right* endpoint of the base interval. ◄ Hence, the total area of all ten "right" rectangles is the sum

$$
\begin{aligned}
R_{10} &= f(0.8) \cdot 0.3 + f(1.1) \cdot 0.3 + f(1.4) \cdot 0.3 + \cdots + f(3.5) \cdot 0.3 \\
&= 6.592 \cdot 0.3 + 5.701 \cdot 0.3 + \cdots + 10.048 \cdot 0.3 + 14.125 \cdot 0.3 = 19.8975.
\end{aligned}
$$

For M_{10}, the **midpoint approximating sum with ten subdivisions** (see Figure 2(c)), heights are evaluated at the *midpoint* of each subinterval. Therefore,

$$
\begin{aligned}
M_{10} &= f(0.65) \cdot 0.3 + f(0.95) \cdot 0.3 + f(1.25) \cdot 0.3 + \cdots + f(3.35) \cdot 0.3 \\
&= 7.007 \cdot 0.3 + 6.150 \cdot 0.3 + 5.266 \cdot 0.3 + \cdots + 11.928 \cdot 0.3 = 18.6825.
\end{aligned}
$$

Figure 2(d) shows a **trapezoid approximating sum**, denoted by T_n. Calculating T_n is no harder than calculating L_n and R_n, thanks to a key idea: The area of each trapezoid in an approximating sum is the *average* of the left- and right-rule rectangles on the same subinterval. ◄ The same averaging applies to T_n itself: For any n, $T_n = (L_n + R_n)/2$. In the case at hand,

Make a sketch to see why.

$$T_6 = \frac{L_6 + R_6}{2} = \frac{17.4375 + 20.8125}{2} = 19.125.$$

Getting better all the time

As the two left-rule pictures in Figure 3 (which still pertain to $\int_{0.5}^{3.5} f(x)\,dx$) suggest, we'd expect approximating sums (of any type) to approximate an integral more closely as the number of subdivisions increases.

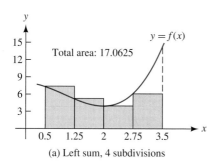
(a) Left sum, 4 subdivisions

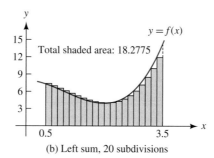
(b) Left sum, 20 subdivisions

FIGURE 3
Two left sums

Neither approximation looks perfect, but the error committed by L_{20} appears much less than that committed by L_4. Numerical values of various sums (all computed by machine, of course) give the same message:

\multicolumn{5}{c}{**Approximating sums for the integral $\int_{0.5}^{3.5} f(x)\,dx$**}
n	L_n	R_n	M_n	T_n
2	17.0625	27.1875	17.0625	22.1250
4	17.0625	22.1250	18.3281	19.5938
8	17.6953	20.2266	18.6445	18.9609
16	18.1699	19.4355	18.7236	18.8027
32	18.4468	19.0796	18.7434	18.7632
64	18.5951	18.9115	18.7484	18.7533
128	18.6717	18.8299	18.7496	18.7508
256	18.7107	18.7898	18.7499	18.7502

As n increases (down the columns), all four types of sums approach 18.75, the answer we found earlier by antidifferentiation.

Riemann sums and the limit definition   Left, right, and midpoint approximating sums are all examples of **Riemann sums**. ➡ Riemann's idea of an approximating sum is quite general. He allows rectangles with unequal bases, and their heights can be determined anywhere— even randomly—within their respective subintervals. Roughly speaking, any "rectangular" approximating sum is a Riemann sum. Figure 4 shows an "irregular" Riemann sum for our by-now-familiar integral.

Named for the German mathematician G. F. B. Riemann (1826–66), who gave the first modern definition of the integral.

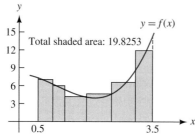
A Riemann sum, 6 (unequal) subdivisions

FIGURE 4
An irregular Riemann sum

We won't normally use such irregular sums, but the formal definition of Riemann sum (which allows for such sums) follows for the record. We'll need the following ingredients: a function f, an interval $[a, b]$, a **partition** of $[a, b]$, and a **sampling point** in each subinterval. Here is the recipe:

DEFINITION Let the interval $[a, b]$ be partitioned into n subintervals by any $n + 1$ points

$$a = x_0 < x_1 < x_2 < \cdots < x_{n-1} < x_n = b,$$

and let $\Delta x_i = x_i - x_{i-1}$ denote the width of the ith subinterval. Within each subinterval $[x_{i-1}, x_i]$, choose any **sampling point** c_i. The sum

$$S_n = f(c_1)\Delta x_1 + f(c_2)\Delta x_2 + \cdots + f(c_n)\Delta x_n$$

is a **Riemann sum with n subdivisions** for f on $[a, b]$.

Left, right, and midpoint sums (all shown in Figure 2) all fit this definition nicely. Each is built from a **regular partition** of $[a, b]$ (one with subintervals of *equal* length) and some simple, consistent scheme for choosing the sampling points c_i. *Trapezoid* approximating sums don't quite fit the definition, because the summands correspond to trapezoids, not rectangles. (Trapezoid sums still do a good job of approximating integrals, as we'll see.)

EXAMPLE 2 For the integral $I = \displaystyle\int_1^2 \frac{1}{x}\, dx$, find the Riemann sums L_4, R_4, and M_4. How well do they approximate the true value of I?

Solution To partition $[1, 2]$ into four subdivisions, each of length $\Delta x = 1/4$, we use

$$a = 1 < \frac{5}{4} < \frac{3}{2} < \frac{7}{4} < 2 = b.$$

Then

$$L_4 = \frac{f(1)}{4} + \frac{f(5/4)}{4} + \frac{f(3/2)}{4} + \frac{f(7/4)}{4} = \left(1 + \frac{4}{5} + \frac{2}{3} + \frac{4}{7}\right)\frac{1}{4} \approx 0.7595;$$

$$R_4 = \frac{f(5/4)}{4} + \frac{f(3/2)}{4} + \frac{f(7/4)}{4} + \frac{f(2)}{4} = \left(\frac{4}{5} + \frac{2}{3} + \frac{4}{7} + \frac{1}{2}\right)\frac{1}{4} \approx 0.6345.$$

For L_4 and R_4 the sampling points in each subinterval were the left and right endpoints. For M_4 the sampling points are *midpoints* of the subintervals:

$$M_4 = \frac{f(9/8)}{4} + \frac{f(11/8)}{4} + \frac{f(13/8)}{4} + \frac{f(15/8)}{4} = \left(\frac{8}{9} + \frac{8}{11} + \frac{8}{13} + \frac{8}{15}\right)\frac{1}{4} \approx 0.6912.$$

Check for yourself by antidifferentiation.

For comparison, the exact value of I ◄ is $\ln 2 \approx 0.6931$. ∎

Sigma notation For economy and clarity, Riemann sums are usually written using **sigma** (Σ) **notation**. For the Riemann sum defined above, for instance, we could write

$$S_n = f(c_1)\Delta x_1 + f(c_2)\Delta x_2 + \cdots + f(c_n)\Delta x_n = \sum_{i=1}^{n} f(c_i)\,\Delta x_i.$$

The sum is taken as i, the **index variable**, ranges from 1 to n. (We'll say more about sigma notation in the next section; for the moment it's just a convenient shorthand.)

EXAMPLE 3   Find R_{10}, the right Riemann sum with ten subdivisions for the integral $I = \int_0^{\pi} \sin x \, dx$. (The exact value is 2.) How do R_{10} and R_{100} appear in sigma notation?

Solution   A ten-member partition cuts the interval $[0, \pi]$ into ten equal subintervals. Since the interval has length π, each subinterval has width $\Delta x = \pi/10$, and the partition points are $0, \pi/10, 2\pi/10, \ldots, 9\pi/10, \pi$. Therefore, R_{10} has the form

$$R_{10} = \sin\left(\frac{\pi}{10}\right)\frac{\pi}{10} + \sin\left(\frac{2\pi}{10}\right)\frac{\pi}{10} + \cdots + \sin\left(\frac{9\pi}{10}\right)\frac{\pi}{10} + \sin(\pi)\frac{\pi}{10}.$$

To write the ten-member sum in sigma form, we set $c_i = i\pi/10$; observe that $\Delta x_i = \pi/10$ for all i (all subdivisions have the same width), and let i run from one to ten. This gives

$$R_{10} = \sum_{i=1}^{10} f(c_i)\,\Delta x_i = \sum_{i=1}^{10} \sin(i\pi/10)\frac{\pi}{10}.$$

For the right sum R_{100}, we have $c_1 = \pi/100$, $c_2 = 2\pi/100, \ldots, c_i = i\pi/100$, and $\Delta x = \pi/100$. In sigma form, this is

$$R_{100} = \sum_{i=1}^{100} f(c_i)\,\Delta x_i = \sum_{i=1}^{100} \sin(i\pi/100)\frac{\pi}{100}.$$

The sums R_{10} and R_{100} would be tedious to evaluate by hand, but a computer makes short work of both: $R_{10} \approx 1.9835$ and $R_{100} \approx 1.9998$. As we'd expect, R_{10} is close to the exact value, 2, and R_{100} is closer still.   ■

The integral defined   Graphical intuition and numerical evidence ➤ suggest that, for the integral $I = \int_{0.5}^{3.5} f(x)\,dx$, all of the approximating sums—L_n, R_n, M_n, and T_n—approach the true value of I as n increases. The same holds for other integrals; the following limit-based definition makes these ideas precise:

See the table on page 351.

> **DEFINITION**   Let a function $f(x)$ be defined on the interval $[a, b]$. The **integral** of f over $[a, b]$, denoted $\int_a^b f(x)\,dx$, is the number, if one exists, to which all Riemann sums S_n tend as n tends to infinity and as the widths of all subdivisions tend to zero. In symbols:
>
> $$\int_a^b f(x)\,dx = \lim_{n\to\infty} S_n = \lim_{n\to\infty} \sum_{i=1}^{n} f(c_i)\,\Delta x_i.$$

This important definition deserves some comment:

- **Must the limit exist?**   In a word, no. A function for which the limit *does* exists is called **integrable** on $[a, b]$. In practice, being integrable on $[a, b]$ is "easy": almost every calculus-style function that's defined and bounded on $[a, b]$ is integrable. It isn't even really necessary that f be continuous on $[a, b]$—a few jumps or breaks in the f-graph are not fatal to the integral. The surprise, if any, is the difficulty of finding a *nonintegrable* function. ➤

- **Well-behaved functions**   The limit in the definition, taken at face value, is a slippery customer. The ramifications of permitting arbitrary partitions and sampling points, for example, can be tricky. Luckily, the matter is much simpler for the well-behaved functions usually met in calculus courses. For such functions, almost any respectable sort of approximating sum does what we'd expect—approaches the true value of the integral as n tends to infinity.

Here's one: $f(x) = 1$ if x is rational; $f(x) = 0$ if x is irrational.

- **What's dx?**   The limit definition helps explain the somewhat mysterious dx in the notation $\int_a^b f(x)\,dx$. Consider, for example, a right approximating sum with n equal subdivisions, each of length Δx. Then, by definition,

$$\int_a^b f(x)\,dx = \lim_{n\to\infty} \sum_{i=1}^{n} f(x_i)\,\Delta x.$$

We've seen this before:
$dy/dx = \lim_{\Delta x \to 0} \Delta y/\Delta x.$

Now the left side looks a lot like the right; the dx on the left (sometimes called a **differential**) corresponds naturally to the Δx (a finite difference) on the right. ← (The differential dx is sometimes described as an "infinitesimal" analogue of the finite difference Δx.)

EXAMPLE 4   The following table shows several approximating sums (rounded to 4 decimals) for the integral $I = \int_0^1 \sin(x^2)\,dx$. Can we guess a value for I? Can we tell I *exactly*?

n	L_n	R_n	T_n	M_n
8	0.2591	0.3643	0.3117	0.3096
16	0.2843	0.3369	0.3106	0.3101
32	0.2972	0.3235	0.3104	0.3102
64	0.3037	0.3169	0.3103	0.3103

Solution   The numbers don't identify a single obvious value for I, but something in the vicinity of 0.31 looks plausible. As we said above, $\sin(x^2)$ has no elementary antiderivative formula, and so the FTC can't help us find an exact value for I. ∎

Using the definition   The definition of $\int_a^b f(x)\,dx$ as a limit of approximating sums helps explain what an integral is and why certain sums give good approximations to integrals. But we'll seldom use the definition itself to *calculate* integrals exactly. ← Instead, we'll rely on antiderivatives when we can find them and settle for approximations when we can't.

But we will use the definition directly in examples in the next section.

BASIC EXERCISES

In Exercises 1–6, g is the function graphed below.

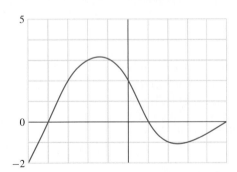

1. (a) Estimate the value of $\int_{-5}^{5} g(x)\,dx$ by evaluating a left sum with five equal subintervals.

   (b) Draw a sketch that illustrates the sum you computed in part (a).

2. (a) Estimate the value of $\int_{-5}^{5} g(x)\,dx$ by evaluating a right sum with five equal subintervals.

   (b) Draw a sketch that illustrates the sum you computed in part (a).

3. (a) Estimate the value of $\int_{-5}^{5} g(x)\,dx$ by evaluating a midpoint sum with five equal subintervals.

   (b) Draw a sketch that illustrates the sum you computed in part (a).

4. (a) Estimate the value of $\int_{-5}^{5} g(x)\,dx$ by evaluating a trapezoid sum with five equal subintervals.

   (b) Draw a sketch that illustrates the sum you computed in part (a).

5. (a) Estimate the value of $\int_{-1}^{4} g(x)\,dx$ by evaluating a left sum with five equal subintervals.

(b) Draw a sketch that illustrates the sum you computed in part (a).

6. (a) Estimate the value of $\int_{-3}^{2} g(x)\,dx$ by evaluating a right sum with five equal subintervals.

(b) Draw a sketch that illustrates the sum you computed in part (a).

In Exercises 7–14, f is the function graphed below.

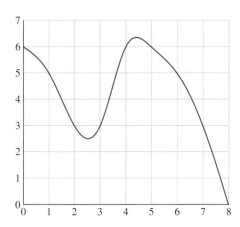

7. Estimate the value of $\int_{1}^{7} f(x)\,dx$ using a right sum with three equal subintervals.

8. Estimate the value of $\int_{1}^{7} f(x)\,dx$ using a left sum with three equal subintervals.

9. Estimate the value of $\int_{1}^{7} f(x)\,dx$ using a midpoint sum with three equal subintervals.

10. Estimate the value of $\int_{1}^{7} f(x)\,dx$ using a trapezoid sum with three equal subintervals.

11. Estimate the value of $\int_{0}^{8} f(x)\,dx$ using a right sum with four equal subintervals.

12. Estimate the value of $\int_{0}^{8} f(x)\,dx$ using a left sum with four equal subintervals.

13. Estimate the value of $\int_{0}^{8} f(x)\,dx$ using a midpoint sum with four equal subintervals.

14. Estimate the value of $\int_{0}^{8} f(x)\,dx$ using a trapezoid sum with four equal subintervals.

In Exercises 15–18, draw a sketch that illustrates the approximations L_4, R_4, M_4, and T_4. Then compute each approximation and compare it with the exact value of the integral computed using the FTC.

15. $\int_{-1}^{3} (x+1)\,dx$

16. $\int_{-1}^{3} x^2\,dx$

17. $\int_{-2}^{2} x^3\,dx$

18. $\int_{-\pi/2}^{\pi/2} \sin x\,dx$

FURTHER EXERCISES

In Exercises 19–24, construct a table like the one shown on page 351 for the integral. Do the values in the table appear to converge to the exact value of the integral (evaluated using the FTC)?

19. $\int_{-1}^{2} x\,dx$

20. $\int_{-1}^{2} x^2\,dx$

21. $\int_{0}^{1} x^3\,dx$

22. $\int_{1}^{4} \sqrt{x}\,dx$

23. $\int_{0}^{\pi} \sin x\,dx$

24. $\int_{0}^{1} e^x\,dx$

In Exercises 25–28, f is a continuous function with the values shown in the following table.

x	1	2	3	4	5	6
$f(x)$	0.14	0.21	0.28	0.36	0.44	0.54

25. Estimate $\int_{2}^{5} f(x)\,dx$ using a left sum with three equal subintervals.

26. Estimate $\int_{2}^{5} f(x)\,dx$ using a trapezoid sum with three equal subintervals.

27. Estimate $\int_{1}^{5} f(x)\,dx$ using a right sum with two equal subintervals.

28. Estimate $\int_{1}^{5} f(x)\,dx$ using a midpoint sum with two equal subintervals.

29. Find a definite integral for which $5(f(5)+f(10)+f(15))$ is
   (a) A left Riemann sum with three equal subintervals.
   (b) A right Riemann sum with three equal subintervals.
   (c) A midpoint Riemann sum with three equal subintervals.

30. Find a definite integral for which $S = 2(f(2)+f(4)+f(6)+f(8))$ is
   (a) A left Riemann sum with four equal subintervals.
   (b) A right Riemann sum with four equal subintervals.
   (c) A midpoint Riemann sum with four equal subintervals.

31. Explain why $f(2) \cdot 2 + f(4) \cdot 3$ is a Riemann sum approximation to $\int_2^7 f(x)\,dx$. [HINT: Draw a picture.]

32. Is $f(1) \cdot 3 + f(4) \cdot 2$ is a Riemann sum approximation to $\int_1^6 f(x)\,dx$? Justify your answer.

33. Is $f(3) \cdot 1 + f(4) \cdot 4$ is a Riemann sum approximation to $\int_0^5 f(x)\,dx$? Justify your answer.

34. Is $f(3) \cdot 4 + f(5) \cdot 4$ is a Riemann sum approximation to $\int_0^7 f(x)\,dx$? Justify your answer.

5.7 WORKING WITH SUMS

In Section 5.6 we defined the integral as a limit of approximating sums:

$$\int_a^b f(x)\,dx = \lim_{n \to \infty} \sum_{i=1}^{n} f(c_i)\,\Delta x_i.$$

As we said there, we'll seldom *calculate* integrals using the definition as a limit; whenever possible, we use antiderivatives instead. (Derivatives, too, are only rarely calculated directly from the definition.) But, for integrals as for derivatives, *understanding* the definition is essential to understanding what integrals are, what they measure, and what they say about functions.

In this section we focus on approximating sums themselves—their definitions, meanings, and uses. We'll also show how the limit-of-sums definition *can* be used (with effort, and for carefully chosen functions) to calculate some basic integrals.

Sums and sigma notation

The following equations illustrate some uses of the sigma notation:

$$\sum_{j=1}^{10} j^2 = 1^2 + 2^2 + \cdots + 10^2 = 385;$$

$$\sum_{k=1}^{10} (11-k)^2 = 10^2 + 9^2 + \cdots + 1^2 = 385;$$

$$\sum_{l=1}^{100} \frac{\sin(3+l/100)}{100} = \frac{\sin(3.01)+\sin(3.02)+\cdots+\sin(4)}{100} \approx -0.3408.$$

What do the examples illustrate?

- **Different forms, same answer**   Slightly different sigma notations may produce the same result. For instance, we used j, k, and l as index variable names above, but other letters could have been used. Notice also that the first and second sums, which look quite different in sigma form, produce exactly the *same* summands.

- **Finding values**   Many-term sums can be tedious to calculate by hand, but they're easily found by computer. In *Maple*, for instance, a command of the form `sum(j^2,j=1..10)` finds the first sum above; so would `sum(i^2,i=1..10)`.

- **Algebraic properties**   The arithmetic properties of real numbers apply to many-term sums as well. For instance, the equations

$$(a_1 + b_1) + \cdots + (a_n + b_n) = (a_1 + \cdots + a_n) + (b_1 + \cdots + b_n)$$

and

$$a_1 \cdot c + a_2 \cdot c + \cdots + a_n \cdot c = c\,(a_1 + a_2 + \cdots + a_n)$$

(c is a constant) follow from the commutative and distributive properties of real numbers. Here are "sigma versions" of these facts:

Commutativity: $$\sum_{i=1}^{n}(a_i + b_i) = \sum_{i=1}^{n} a_i + \sum_{i=1}^{n} b_i$$

Distributivity: $$\sum_{i=1}^{n} c\, a_i = c \sum_{i=1}^{n} a_i$$

- **Sums vs. integrals**   Sums and integrals have similar properties. The commutative and distributive rules for sums, for instance, have close analogues for integrals:

$$\int_a^b (f + g) = \int_a^b f + \int_a^b g; \qquad \int_a^b c\, f = c \int_a^b f.$$

The similarity is no accident—integrals are *limits* of sums and so inherit many of their traits from sums. (Formal proofs of integral properties start with analogous properties of sums.)

EXAMPLE 1   Write sigma forms of the sums R_{10}, L_{10}, and M_{10} for $\int_2^5 \ln x\, dx$.

Solution   The interval $[2, 5]$ has length 3, and so a 10-subinterval partition has endpoints $x_0 = 2.0$, $x_1 = 2.3$, $x_2 = 2.6$, ... $x_9 = 4.7$, $x_{10} = 5.0$. In general, $x_i = 2 + 0.3i$. This gives

$$R_{10} = \big(\ln(2.3) + \ln(2.6) + \cdots + \ln(5.0)\big) \times 0.3 = \sum_{i=1}^{10} \ln(2 + 0.3i) \times 0.3;$$

$$L_{10} = \big(\ln(2.0) + \ln(2.3) + \cdots + \ln(4.7)\big) \times 0.3 = \sum_{i=0}^{9} \ln(2 + 0.3i) \times 0.3.$$

For the *midpoint* sum, the sampling points $(2.15, 2.45, \ldots, 4.85)$ have the form $2.15 + 0.3i$; therefore,

$$M_{10} = \big(\ln(2.15) + \ln(2.45) + \cdots + \ln(4.85)\big) \times 0.3 = \sum_{i=0}^{9} \ln(2.15 + 0.3i) \times 0.3.$$

The sums are easily found by machine: $R_{10} \approx 3.7961$; $L_{10} \approx 3.5212$; $M_{10} \approx 3.6620$.   ∎

Closed forms   The following equations illustrate what are called **closed forms:** ➡   *c is a constant*

$$\sum_{k=1}^{n} c = c + c + \cdots + c = cn;$$

$$\sum_{k=1}^{n} k = 1 + 2 + \cdots + n = \frac{n(n+1)}{2};$$

$$\sum_{k=1}^{n} k^2 = 1^2 + 2^2 + \cdots + n^2 = \frac{n(n+1)(2n+1)}{6}.$$

The "closed forms" are the expressions on the right—easily computable expressions that calculate the less obvious sums to the left.

Formulas like these let us trade repetitive calculations for simple arithmetic. The third line says, for instance, that

$$1^2 + 2^2 + 3^2 + \cdots + 1000^2 = \frac{1000 \cdot 1001 \cdot 2001}{6} = 333{,}833{,}500;$$

calculating the left side directly would be no fun! (In a later example we'll use the same formula to calculate an integral.)

The first identity above is obvious at a glance. The third form is subtler; a formal proof appears in Appendix H, and we take another look in Example 3. The second identity is simpler, and interesting in its own right. Figure 1 (a "proof without words") suggests why it holds.

The figure shows *two* copies (one gray and one white) of the sum $1 + 2 + \cdots + n$; they "fit together" to form an $(n + 1) \times n$ rectangle, and the desired equation follows. (The figure shows $n = 7$, but the same idea works for every n.) The following calculation uses the same argument, but dresses it up in sigma notation. (If the calculation looks a bit magical, refer again to the picture.)

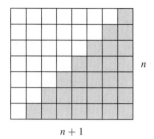

n

$n + 1$

FIGURE 1
Why

$$1 + 2 + \cdots + n = \frac{n(n+1)}{2}$$

EXAMPLE 2   Show that $1 + 2 + 3 + \cdots + n = \dfrac{n(n+1)}{2}$ for all positive integers n.

Solution   If $S = 1 + 2 + 3 + \cdots + n$, then

$$2S = (1 + 2 + \cdots + n) + (n + (n-1) + \cdots + 1) \qquad \text{(writing } S \text{ twice)}$$

$$= \sum_{k=1}^{n} k + \sum_{k=1}^{n} (n + 1 - k) \qquad \text{(rewriting in sigma form)}$$

$$= \sum_{k=1}^{n} (k + n + 1 - k) \qquad \text{(combining the sums)}$$

$$= \sum_{k=1}^{n} (n + 1) = n(n + 1),$$

as we aimed to show.   ∎

We haven't forgotten about calculus—we'll use the identity in a moment to calculate an integral.

EXAMPLE 3   We claimed above that, for all n,

$$1^2 + 2^2 + 3^2 + \cdots + n^2 = \frac{n(n+1)(2n+1)}{6}.$$

Why does the formula hold? How might one guess it? ◂

Solution   Let $s(n)$ denote the expression on the right side above. Clearly, $s(1) = 1$, and so the formula "works" for $n = 1$.

Now consider $s(n) - s(n-1)$. If the formula $s(n)$ is correct, we should have

$$s(n) - s(n-1) = (1^2 + 2^2 + \cdots + (n-1)^2 + n^2) - (1^2 + 2^2 + \cdots + (n-1)^2) = n^2.$$

By definition,

$$s(n) - s(n-1) = \frac{n(n+1)(2n+1)}{6} - \frac{(n-1)n(2n-1)}{6},$$

and some basic but careful algebra (multiply out both numerators; then combine results) shows that the right side does indeed reduce to n^2. This result, together with the fact that $s(1) = 1$, shows that the formula $s(n)$ "works" for all positive integers n.

Where did the formula $s(n)$ in Example 3 come from? Could we have guessed it? Not easily, perhaps, but we suggest some strategies in this section's exercises.   ∎

Integrals as limits: a direct calculation

By definition, $\displaystyle\int_a^b f(x)\,dx = \lim_{n\to\infty}\sum_{i=1}^{n} f(c_i)\,\Delta x_i$. The definition is inconvenient to use directly for most integrals. But a few simple cases are exceptional—and worthwhile in showing what the definition really means.

EXAMPLE 4 Use the limit definition to find $\displaystyle\int_0^1 x^2\,dx$.

Solution We'll work with right sums R_n. A regular partition of $[0, 1]$ into n pieces has endpoints $0, 1/n, 2/n, \ldots, 1$. Thus, $x_i = i/n$ and $\Delta x_i = 1/n$ for all i, so

$$R_n = \sum_{i=1}^{n} f(x_i)\Delta x = \sum_{i=1}^{n} f\left(\frac{i}{n}\right)\frac{1}{n} = \sum_{i=1}^{n}\left(\frac{i}{n}\right)^2\frac{1}{n}$$

$$= \sum_{i=1}^{n} i^2\frac{1}{n^3} = \frac{1}{n^3}\sum_{i=1}^{n} i^2.$$

(The last step works because n is constant with respect to i.) Next we use our closed form for the sum of squares:

$$R_n = \frac{1}{n^3}\sum_{i=1}^{n} i^2 = \frac{1}{n^3}\frac{n(n+1)(2n+1)}{6} = \frac{n(n+1)(2n+1)}{6n^3} = \frac{2n^3 + 3n^2 + n}{6n^3}.$$

Last, we find the limit: ➡

$$\int_0^1 x^2\,dx = \lim_{n\to\infty}\frac{2n^3 + 3n^2 + n}{6n^3} = \frac{1}{3}.$$

We found similar limits in Section 4.2.

We could have found the same answer, of course, by antidifferentiation. The point is that the two methods agree. ∎

Three views of approximating sums

Approximating sums are *discrete* approximations to integrals of *continuously* varying functions. This relationship can be interpreted in several useful ways. We discuss three.

1. Polygonal approximations Geometrically, approximating sums represent simpler, straight-edged versions of regions with curved edges. ➡ The shaded region in Figure 2, for example, could be called a **polygonal approximation** to the region under the f-graph from a to b.

Most of the pictures in this section and the last reflect this point of view.

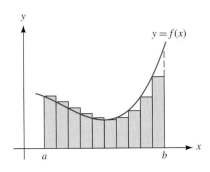

FIGURE 2

A polygonal approximation to a curved area

The total area of the polygonal region is the value of a left sum; it's apparently a reasonable approximation to $\int_a^b f$, the true area under the graph.

2. Average values   The average of n *numbers* is their sum divided by n. How is this notion related to the average value of a *function* f on an interval $[a, b]$? ◄

We defined average value in Section 5.1.

The relation is quite close—thanks to approximating sums. If, say, $c_1, \ldots, c_n$ are the midpoints of n equal subdivisions of $[a, b]$, then

$$\frac{f(c_1) + f(c_2) + f(c_3) + \cdots + f(c_n)}{n}$$

is the ordinary average of n equally-spaced sample values of f. The midpoint sum M_n is *almost* identical:

$$M_n = \left(f(c_1) + f(c_2) + f(c_3) + \cdots + f(c_n)\right) \cdot \frac{b - a}{n};$$

it differs from the preceding average only by the factor $(b - a)$. Now as $n \to \infty$, we have $M_n \to \int_a^b f(x)\,dx$, and so

$$\left(f(c_1) + f(c_2) + \cdots + f(c_n)\right)\frac{1}{n} \longrightarrow \frac{\int_a^b f(x)\,dx}{b - a},$$

which is just the average value of f on $[a, b]$, as defined earlier. Here is our conclusion:

The average value f over $[a, b]$ is a limit of averages taken over n-member subsets of $[a, b]$.

3. Weighted sums   An expression of the form

$$f(c_1)w_1 + f(c_2)w_2 + \cdots + f(c_n)w_n,$$

where the w_i are any positive numbers (the **weights**), is called a **weighted sum** of the function values $f(c_i)$. If $w_i = 1/n$ for all i, then the weighted sum is the *ordinary* average of the $f(c_i)$. Every Riemann sum $f(c_1)\Delta x_1 + \cdots f(c_n)\Delta x_n$ is a weighted sum in this sense, with weight $w_i = \Delta x_i$ and the c_i chosen depending on the type of sum at hand.

> **EXAMPLE 5**   Write L_n, R_n, and T_n as weighted sums. What are the weights?
>
> **Solution**   Let x_0 through x_n define a regular partition of $[a, b]$. Then $\Delta x = (b - a)/n$, and so
>
> $$R_n = f(x_1)\Delta x + f(x_2)\Delta x + \cdots + f(x_n)\Delta x;$$
> $$L_n = f(x_0)\Delta x + f(x_1)\Delta x + \cdots + f(x_{n-1})\Delta x.$$
>
> Thus, both R_n and L_n are weighted sums (of different sets of values); all the weights are equal. Trapezoid sums T_n are also weighted sums, but with a little twist. Since T_n is the average $(L_n + R_n)/2$, we have
>
> $$T_n = \left(\frac{f(x_0) + f(x_1)}{2} + \frac{f(x_1) + f(x_2)}{2} + \cdots + \frac{f(x_{n-1}) + f(x_n)}{2}\right)\Delta x$$
> $$= \left(\frac{f(x_0)}{2} + f(x_1) + \cdots + f(x_{n-1}) + \frac{f(x_n)}{2}\right)\Delta x.$$
>
> Thus, T_n is an *unequally* weighted sum of values of f at $n + 1$ (rather than n) of the partition points x_i; the first and last values have *half* the weight of the others.   ∎

Using approximating sums

Approximating sums are good for more than formally defining the integral. Indeed, compared with symbolic integrals approximating sums are often simpler conceptually, handier in applications, and easier to calculate—especially for machines. For the many integrals that involve complicated formulas or no formulas at all, approximating sums may be the only practical resort.

EXAMPLE 6 If $v(t)$ is the speed (in miles per hour) of a car at time t (in hours), then $\int_a^b v(t)\,dt$ = miles traveled from $t = a$ to $t = b$. To measure distance traveled from $t = 0$ to $t = 1$, we need the integral $\int_0^1 v(t)\,dt$.

Real cars don't adhere to explicit speed formulas. What we *can* observe, practically speaking, are numerical speed data such as those following. (Notice the irregular time intervals; real data sets *often* have gaps.)

Speed readings over one hour										
Time (min)	0	5	10	15	20	30	35	40	45	55	60
Speed (mph)	35	38	36	57	71	55	51	23	10	27	35

How far did the car travel in 1 hour?

Solution Without more information (e.g., about speeds *between* the times measured), we can't know exactly. But we can use approximating sums to estimate $\int_0^1 v(t)\,dt$. The tabulated speed data are mathematically compatible with many possible speed functions. One possibility is drawn by "connecting the dots" (see Figure 3(a)): ➤

We converted time from minutes to hours.

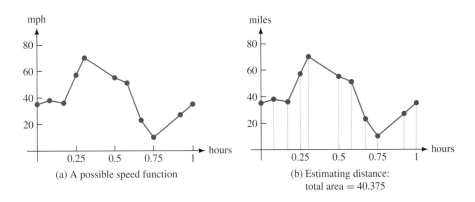

(a) A possible speed function

(b) Estimating distance:
total area = 40.375

FIGURE 3
Estimating distance from speed data

The total bounded area (Figure 3(b)) can be computed using ten trapezoids. (We spare you the arithmetic.) We get 40.375—a reasonable estimate for distance traveled over the hour.

The piecewise-linear speed function just plotted isn't the only possibility. Many curves can be drawn through the known data points, but some are physically more likely than others. The important point is that the trapezoid estimate depends only on the 11 observed data points, not on the particular curve drawn through them. Figure 4 illustrates this idea.

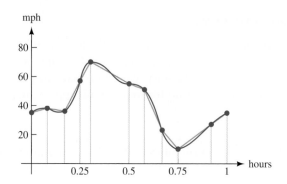

FIGURE 4
New speed function; same trapezoidal area

It shows a new speed function (better or worse than the old?) but the same trapezoid estimate (40.375 miles) for distance. ∎

BASIC EXERCISES

In Exercises 1–4, express the sum using Σ-notation.

1. $1 + 2 + 3 + \cdots + 49 + 50$

2. $\sqrt[3]{2} + \sqrt[3]{4} + \sqrt[3]{6} + \cdots + \sqrt[3]{98} + \sqrt[3]{100}$

3. $1^2 + 3^2 + 5^2 + \cdots + 97^2 + 99^2$

4. $2^1 + 2^2 + 2^3 + \cdots + 2^{14} + 2^{15}$

In Exercises 5–12, evaluate the sum.

5. $\displaystyle\sum_{k=1}^{5} k^{-1}$

6. $\displaystyle\sum_{k=1}^{4} \sqrt{k}$

7. $\displaystyle\sum_{k=1}^{6} \sin(k\pi/4)$

8. $\displaystyle\sum_{k=0}^{5} \cos(k\pi/2)$

9. $\displaystyle\sum_{k=1}^{17} k$

10. $\displaystyle\sum_{k=1}^{21} k^2$

11. $\displaystyle\sum_{k=0}^{9} (k+1)^2$

12. $\displaystyle\sum_{k=1}^{19} (k^2 + 2k)$

13. **(a)** Use Σ-notation to express the right sum approximation to $\displaystyle\int_{0}^{10} x^2\, dx$ with $n = 10$ equal subintervals.

   (b) Repeat part (a) using $n = 5$.

   (c) Repeat part (a) using $n = 20$.

   (d) Repeat part (a) using $n = 50$.

14. Repeat Exercise 13 for the left sum approximation.

15. **(a)** Use Σ-notation to express the left sum approximation to $\displaystyle\int_{1}^{11} x^2\, dx$ with $n = 10$ equal subintervals.

   (b) Repeat part (a) using $n = 5$.

   (c) Repeat part (a) using $n = 20$.

   (d) Repeat part (a) using $n = 50$.

16. Repeat Exercise 15 for the right sum approximation.

17. **(a)** Use Σ-notation to express the left sum approximation to $\displaystyle\int_{0}^{1} \sqrt[3]{x}\, dx$ with $n = 10$ equal subintervals.

   (b) Repeat part (a) using $n = 5$.

   (c) Repeat part (a) using $n = 20$.

   (d) Repeat part (a) using $n = 50$.

18. Repeat Exercise 17 for the right sum approximation.

19. **(a)** Use Σ-notation to express the right sum approximation to $\displaystyle\int_{1}^{2} \sqrt[3]{x}\, dx$ with $n = 10$ equal subintervals.

   (b) Repeat part (a) using $n = 5$.

   (c) Repeat part (a) using $n = 20$.

   (d) Repeat part (a) using $n = 50$.

20. Repeat Exercise 19 for the left sum approximation.

21. **(a)** Use Σ-notation to express the right sum approximation to $\displaystyle\int_{3}^{7} \sin(x^2)\, dx$ with $n = 10$ equal subintervals.

   (b) Repeat part (a) using $n = 5$.

   (c) Repeat part (a) using $n = 20$.

   (d) Repeat part (a) using $n = 50$.

22. Repeat Exercise 21 for the left sum approximation.

23. (a) Use Σ-notation to express the midpoint sum approximation to $\int_0^5 \sqrt{x}\,dx$ with $n = 10$ equal subintervals.

(b) Repeat part (a) using $n = 5$.

(c) Repeat part (a) using $n = 20$.

(d) Repeat part (a) using $n = 50$.

24. (a) Use Σ-notation to express the midpoint sum approximation to $\int_2^{12} \sqrt{x}\,dx$ with $n = 10$ equal subintervals.

(b) Repeat part (a) using $n = 5$.

(c) Repeat part (a) using $n = 20$.

(d) Repeat part (a) using $n = 50$.

FURTHER EXERCISES

25. Evaluate $\displaystyle\lim_{n\to\infty} \frac{1+2+3+\cdots+n}{n^2}$.

26. Evaluate $\displaystyle\lim_{n\to\infty} \frac{1^2+2^2+3^2+\cdots+n^2}{n^3}$.

27. Explain why $\displaystyle\lim_{n\to\infty} \frac{1}{n}\sum_{k=1}^{n}(4+k/n)^3 = \int_4^5 x^3\,dx$.

28. Explain why $\displaystyle\lim_{n\to\infty} \frac{1}{n}\sum_{k=1}^{n}(4+k/n)^3 = \int_0^1 (4+x)^3\,dx$.

29. Explain why $\displaystyle\lim_{n\to\infty} \frac{3}{n}\sum_{k=1}^{n}\sqrt{1+3k/n} = \int_1^4 \sqrt{x}\,dx$.

30. Explain why $\displaystyle\lim_{n\to\infty} \frac{3}{n}\sum_{k=0}^{n-1}\sqrt{1+3k/n} = \int_1^4 \sqrt{x}\,dx$.

31. Explain why $\displaystyle\lim_{n\to\infty} \frac{3}{n}\sum_{k=1}^{n}\sqrt{1+3(k-1)/n} = \int_1^4 \sqrt{x}\,dx$.

32. Explain why $\displaystyle\lim_{n\to\infty} \frac{3}{n}\sum_{k=0}^{n-1}\sqrt{1+3(2k+1)/2n} = \int_1^4 \sqrt{x}\,dx$.

33. Explain why $\displaystyle\lim_{n\to\infty} \frac{3}{n}\sum_{k=1}^{n}\sqrt{1+3(2k-1)/2n} = \int_1^4 \sqrt{x}\,dx$.

34. Evaluate $\int_0^3 x\,dx$ by computing the limit of right sums R_n.

35. Evaluate $\int_2^5 x\,dx$ by computing the limit of left sums L_n.

36. Evaluate $\int_1^4 x\,dx$ by computing the limit of midpoint sums M_n.

37. In Example 4 the integral $\int_0^1 x^2\,dx$ was evaluated by taking a limit of right sums. Find this integral by evaluating a limit of left sums L_n.

38. In Example 4 the integral $\int_0^1 x^2\,dx$ was evaluated by taking a limit of right sums. Find this integral by evaluating a limit of midpoint sums M_n.

39. Find a definite integral for which $\dfrac{2}{100}\sum_{k=1}^{100}\sin\left(\dfrac{2k}{100}\right)$ is a right sum approximation.

40. Find a definite integral for which $\dfrac{2}{10}\sum_{k=1}^{40}\cos\left(\dfrac{2k-1}{10}\right)$ is a midpoint sum approximation.

41. Evaluate $\displaystyle\lim_{n\to\infty} \frac{1}{n}\sum_{j=1}^{n}\left(\frac{j}{n}\right)^3$ by expressing it as a definite integral and then evaluating this integral using the fundamental theorem of calculus.

42. Evaluate $\displaystyle\lim_{n\to\infty} \frac{2}{n}\sum_{j=1}^{n}\left(\frac{2j}{n}\right)^3$ by expressing it as a definite integral and then evaluating this integral using the fundamental theorem of calculus.

43. Explain why $\displaystyle\lim_{n\to\infty} \frac{1}{n}\sum_{k=1}^{n}(2+5k/n)^3$ is equal to the average value of x^3 over the interval $[2, 7]$.

44. Explain why $\displaystyle\lim_{n\to\infty} \frac{1}{n}\sum_{k=1}^{n}\sqrt{2(2k-1)/n}$ is equal to the average value of $\sqrt{x}$ over the interval $[0, 4]$.

45. The following table gives speedometer readings at various times over a 1-hour interval.

Speed readings over one hour						
Time (min)	0	10	20	30	40	50	60
Speed (mph)	42	38	36	57	0	55	51

(a) Draw a plausible speed graph for the 1-hour period.

(b) Estimate the total distance traveled using

  (i) A trapezoid approximating sum, six subdivisions.

  (ii) A left approximating sum, six subdivisions.

  (iii) A midpoint approximating sum, three subdivisions. Which answer is most convincing? Why?

(c) Draw a plausible distance graph for the 1-hour period.

46. At time t (measured in hours), $0 \le t \le 24$, a firm uses electricity at the rate of $E(t)$ kilowatts. The power company's rate schedule indicates that the cost per kilowatt-hour at time t is $c(t)$ dollars. Assume that both E and c are continuous functions.

(a) Set up an N-term left sum that approximates the cost of the electricity consumed in a 24-hour period.

(b) What definite integral is approximated by the sum you found in part (a)?

47. This exercise outlines a proof that $\displaystyle\sum_{k=1}^{n} k = \frac{n(n+1)}{2}$.

(a) Show that $\displaystyle\sum_{k=1}^{n}\left((k+1)^2 - k^2\right) = (n+1)^2 - 1$.

(b) Show that $(k+1)^2 - k^2 = 2k + 1$.

(c) Combining the results in parts (a) and (b) leads to the identity $\sum_{k=1}^{n}(2k+1) = (n+1)^2 - 1$. Use this identify to find a closed form expression for $\sum_{k=1}^{n} k$.

48. This exercise outlines a proof that $\sum_{k=1}^{n} k^2 = \dfrac{n(n+1)(2n+1)}{6} = \frac{1}{3}n^3 + \frac{1}{2}n^2 + \frac{1}{6}n$.

(a) Show that $\sum_{k=1}^{n}\left((k+1)^3 - k^3\right) = (n+1)^3 - 1$.

(b) Show that $(k+1)^3 - k^3 = 3k^2 + 3k + 1$.

(c) Use parts (a) and (b) together with the identity proved in the previous exercise to find the desired closed form expression for $\sum_{k=1}^{n} k^2$.

49. Prove that $\sum_{k=1}^{n} k^3 = \frac{1}{4}n^2(n+1)^2$.

[HINT: See the previous exercise.]

SUMMARY

This chapter introduced the **integral**. The derivative and the integral are the two most important concepts of calculus. Remarkably, they're closely related; the **fundamental theorem of calculus** describes how and why.

The integral as signed area   Section 5.1 introduced the integral graphically, as measuring **signed area**. ("Signed" means that area below the x-axis counts negatively.) The graphical viewpoint, though insufficient for *exact* calculations, explains clearly what the integral is and why some of its elementary properties hold.

The area function   Given a function f and a starting point $x = a$, we defined a related **area function** A_f:

$$A_f(x) = \int_a^x f = \text{signed area bounded by } f \text{ from } a \text{ to } x.$$

Simple examples suggest that the area function is an **antiderivative** for f. This idea, suitably embellished, is the simplest form of the **fundamental theorem of calculus**.

The fundamental theorem   Put more formally, the fundamental theorem says that if A_f is defined as we just stated, then

$$\frac{d}{dx}\left(A_f(x)\right) = f(x).$$

This result connects derivatives and integrals, the two main calculus ideas. For that reason, it deserves to be called *fundamental*.

The same idea, restated in another form, is called the **second version of the fundamental theorem**. If F is *any* antiderivative of f, then

$$\int_a^b f(x)\,dx = F(b) - F(a).$$

In other words,

To find $\int_a^b f$, find any antiderivative, plug in endpoints, and subtract.

Antidifferentiation by substitution   The connection between integration and antidifferentiation means that it's worth trying to find symbolic antiderivatives, even of complicated functions. The simplest and most important antidifferentiation technique is called **substitution**. Antidifferentiation by any method is nothing more than differentiation in reverse. Substitution, in particular, is the chain rule running backward. We explained this connection in Section 5.4 and used it repeatedly to find antiderivatives.

Using integral tables and computers Finding symbolic antiderivatives from scratch can be difficult. In practice, calculus users freely resort to integral tables, computer software, and other tools. Scientific reference works contain extensive integral tables—sometimes running to dozens of pages. Mathematical software such as *Maple* and *Mathematica* also produces antiderivatives quickly and easily. As with all power tools, however, using tables and mathematical software requires care, and their results require interpretation. We illustrated some of the possibilities and pitfalls in Section 5.5.

The integral as a limit of approximating sums Like the derivative, the integral is defined rigorously as a limit of approximating sums. Section 5.6 described the procedure. Although symbolically messy, approximating sums make good graphical sense. Several varieties were introduced, compared, and calculated.

Approximating sums and applications of the integral In Section 5.6 we defined the integral as a limit of approximating sums. In Section 5.7 we focused on approximating sums themselves—their definition, properties, meanings, and uses.

Thinking of the integral as a limit of sums suggests a variety of applications. One application of the integral concerns rates and amounts:

If $f(t)$ tells the rate *at which a quantity varies at time t, then $\int_a^b f$ tells the* amount *by which the same quantity changes over the interval from $t = a$ to $t = b$.*

We applied this idea using speed data for a car to estimate the distance traveled.

REVIEW EXERCISES

1. Let f be the function shown graphically below.

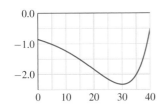

(a) Which of the following is the best estimate of $\int_0^{40} f(x)\,dx$: -65, -20, 0, 30, 60? Justify your answer.

(b) Show that the average value of f over the interval $[10, 30]$ is less than -1.

2. The graph of a function h is shown below.

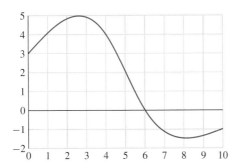

List, from smallest to largest, (i) $h'(5)$, (ii) the average value of h over the interval $[0, 10]$, (iii) the average rate of change

of h over the interval $[0, 10]$, (iv) $\int_0^{10} h(x)\,dx$, (v) $\int_0^5 h(x)\,dx$, (vi) $\int_6^{10} h(x)\,dx$, and (vii) $\int_5^5 h(x)\,dx$.

3. Sketch the graph of a continuous function f such that $\int_{-2}^3 f(t)\,dt = -10$ and

(a) $\int_{-2}^3 |f(t)|\,dt = 10$.

(b) $\int_{-2}^3 |f(t)|\,dt \neq 10$.

4. Show that $0.4 \leq \int_0^1 \sin(e^x)\,dx \leq 1$.

5. Show that $\pi/2 < \int_0^\pi \cos(\sin x)\,dx \leq \pi$.
[HINT: $\cos 1 > \cos(\pi/3) = 1/2$.]

6. (a) Show that $\pi(1 + 1/e) \leq \int_0^{2\pi} e^{\sin x}\,dx \leq \pi(1 + e)$.
   [HINT: If $0 \leq x \leq \pi$, $0 \leq \sin x \leq 1$. If $\pi \leq x \leq 2\pi$, $-1 \leq \sin x \leq 0$.]

(b) Use the result in part (a) to find numbers A and B such that $A \leq \int_0^{50\pi} e^{\sin x}\,dx \leq B$.

7. (a) Let $y = f(x)$ be the line through the points $(0, \arctan 0)$ and $(1, \arctan 1)$. Explain why $f(x) \leq \arctan x$ if $0 \leq x \leq 1$. [HINT: Consider the concavity of the arctangent function.]

(b) Find an equation of the line tangent to the curve $y = \arctan x$ at $x = 0$.

(c) Show that $\pi/8 \leq \int_0^1 \arctan x\,dx \leq 1/2$.

8. (a) Suppose that $f(x) \geq 0$ if $0 \leq x \leq 1$. Show that $\int_0^1 x^k f(x)\, dx \leq \int_0^1 f(x)\, dx$ for all integers $k \geq 0$.

(b) Does the result in part (a) remain valid if the assumption "$f(x) \geq 0$ if $0 \leq x \leq 1$" is not made? Explain.

9. (a) Show that $\int_0^{\pi/2} \sin^2 x\, dx + \int_0^{\pi/2} \cos^2 x\, dx = \pi/2$.

(b) Explain why $\int_0^{\pi/2} \sin^2 x\, dx = \int_0^{\pi/2} \cos^2 x\, dx$.

[HINT: $\cos x = -\sin(x - \pi/2) = \sin(\pi/2 - x)$.]

(c) Use parts (a) and (b) to show that $\int_0^{\pi/2} \sin^2 x\, dx = \pi/4$.

10. (a) Show that $\int_0^{\pi/2} \sqrt{1 + \cos(2x)}\, dx = \sqrt{2} \int_0^{\pi/2} \cos x\, dx$.

[HINT: $1 + \cos(2x) = 2\cos^2 x$ for all x.]

(b) Does $\int_{\pi/2}^{\pi} \sqrt{1 + \cos(2x)}\, dx = \sqrt{2} \int_{\pi/2}^{\pi} \cos x\, dx$? Justify your answer.

11. Suppose that f is a continuous function and that $-2 \leq f(x) \leq 5$ if $1 \leq x \leq 3$. Explain why the average value of f over $[1, 3]$ must be between -2 and 5.

12. Show that the expression $\int_a^b (f(x) - c)^2\, dx$ assumes its minimum value when c is the average value of f over the interval $[a, b]$.

13. Suppose that f and g are continuous on $[a, b]$ and $\int_a^b f(x)\, dx \leq \int_a^b g(x)\, dx$.

(a) Must $f(x) \leq g(x)$ for every x in $[a, b]$? If so, explain why. If not, give a counterexample.

(b) Must there be a number c such that $a \leq c \leq b$ and $f(c) \leq g(c)$? If so, explain why. If not, give a counter-example.

14. Let $F(x) = \int_a^x f(t)\, dt$ and $G(x) = \int_b^x f(t)\, dt$, where a and b are constants, and f is a continuous function. Explain why $G(x) = F(x) + C$, where C is a constant.

Some antiderivatives—trigonometric ones, especially—can be written in more than one way. In Exercises 15–17, verify the formula using the trigonometric identities $\sin^2 x + \cos^2 x = 1$, $\sin(2x) = 2\cos x \sin x$, *and* $\cos(2x) = 2\cos^2 x - 1$.

15. $\int 2\sin x \cos x\, dx = \sin^2 x + C = -\cos^2 x + C$

16. $\int \cos(2x)\, dx = \frac{1}{2}\sin(2x) + C = \cos x \sin x + C$

17. $\int \cos^2 x\, dx = \frac{x}{2} + \frac{1}{2}\cos x \sin x + C = \frac{x}{2} + \frac{1}{4}\sin(2x) + C$

18. (a) Use the trigonometric identity $2\sin^2 x = 1 - \cos(2x)$ to find $\int \sin^2 x\, dx$.

(b) Explain why $\int_0^{2\pi} \sqrt{1 - \cos(2x)}\, dx \neq \sqrt{2} \int_0^{2\pi} \sin x\, dx$.

Some antiderivatives are difficult to find from scratch but may be checked easily by differentiation. In Exercises 19–25, verify the antiderivative formula by differentiation.

19. $\int \ln x\, dx = x \ln x - x + C$

20. $\int \arctan x\, dx = x \arctan x - \frac{1}{2}\ln(1 + x^2) + C$

21. $\int \frac{dx}{x^2 + a^2} = \frac{1}{a}\arctan\left(\frac{x}{a}\right) + C$

22. $\int \tan x\, dx = \ln|\sec x| + C$

[HINT: Recall that $\left(\ln|x|\right)' = 1/x$ for all $x \neq 0$.]

23. $\int \sec x\, dx = \ln|\sec x + \tan x| + C$

24. $\int \frac{dx}{1 - x^2} = \frac{1}{2}\ln\left|\frac{1 + x}{1 - x}\right| + C$

25. $\int \arcsin x\, dx = x \arcsin x + \sqrt{1 - x^2} + C$

In Exercises 26–45, evaluate the antiderivative. Check your answer by differentiation.

26. $\int (3x^5 + 4x^{-2})\, dx$

27. $\int \frac{dx}{3x}$

28. $\int \frac{dx}{4\sqrt{1 - x^2}}$

29. $\int \frac{3}{x^2 + 1}\, dx$

30. $\int 3e^{4x}\, dx$

31. $\int \left(2\sin(3x) - 4\cos(5x)\right) dx$

32. $\int 4\sec^2(3x)\, dx$

33. $\int \left(1 + \sqrt{x}\right)^2 dx$

34. $\int (x + 1)^2 \sqrt[3]{x}\, dx$

35. $\int \frac{(3 - x)^2}{x}\, dx$

36. $\int \frac{(1 - x)^3}{\sqrt{x}}\, dx$

37. $\int e^x(1 - e^x)\, dx$

38. $\int \frac{dx}{\sqrt{2x - x^2}}$. [HINT: $2x - x^2 = 1 - (x - 1)^2$.]

39. $\int x \sin(x^2)\, dx$

40. $\int \dfrac{dx}{2x+1}$

41. $\int x^3 e^{x^4} \, dx$

42. $\int \dfrac{\arctan x}{1+x^2} \, dx$

43. $\int \dfrac{(\ln x)^3}{x} \, dx$

44. $\int \dfrac{dx}{1+9x^2}$

45. $\int \ln|\cos x| \tan x \, dx$

46. Evaluate $\displaystyle\int_0^1 x\sqrt{1-x^4}\,dx$. [HINT: Use the substitution $u = x^2$ to relate the given integral to one that can be evaluated geometrically.]

47. Use the substitution $u = 1 - x$ to show that

$$\int_0^1 x^n(1-x)^m \, dx = \int_0^1 x^m(1-x)^n \, dx.$$

48. Find the flaw in the following "proof" that $I = \displaystyle\int_{-1}^1 \dfrac{dx}{x^2+1} = 0$:

$$I = \int_{-1}^1 \frac{dx}{x^2+1} = \int_{-1}^1 \frac{x^{-2}}{1+x^{-2}} \, dx = -\int_{-1}^1 \frac{du}{1+u^2} = -I$$

so $I = 0$.

49. Find the flaw in the following "proof" that $\displaystyle\int_0^\pi \sqrt{1-\sin x}\,dx = 0$:

$$\int \sqrt{1-\sin x}\,dx = \int \frac{\sqrt{1-\sin^2 x}}{\sqrt{1+\sin x}}\,dx$$

$$= \int \frac{\cos x}{\sqrt{1+\sin x}}\,dx$$

$$\rightarrow \int \frac{du}{\sqrt{u}}$$

$$= 2\sqrt{u} + C \rightarrow 2\sqrt{1+\sin x} + C.$$

Thus,

$$\int_0^\pi \sqrt{1-\sin x}\,dx = 2\sqrt{1+\sin \pi} - 2\sqrt{1+\sin 0} = 0.$$

50. The rate at which the world's oil is being consumed is increasing. Suppose that the rate (measured in billions of barrels per year) is given by the function $r(t)$, where t is measured in years and $t = 0$ is January 1, 1990.

    (a) Write a definite integral that represents the total quantity of oil used between the start of 1990 and the start of 1995.

(b) Suppose that $r(t) = 32e^{0.05t}$. Find an approximate value for the definite integral from part (a) using a right sum with $n = 5$ equal subintervals.

(c) Interpret each of the five terms in the sum from part (b) in terms of oil consumption.

(d) Evaluate the definite integral from part (a) exactly using the FTC.

51. Suppose that the rate of change of the temperature of a cup of coffee t minutes after it is placed in a room is $T'(t) = -3.5e^{-0.05t}\,°$C/minute. If the temperature of the coffee is initially $90\,°$C, what is the temperature of the coffee 5 minutes later?

52. Let $I = \displaystyle\int_0^5 \sqrt[3]{2x}\,dx$. Use sigma notation to write the left, right, and midpoint approximations to I. (Assume that $N = 10$ equal subintervals are used.)

53. Let $I = \displaystyle\int_0^5 \sqrt{3x}\,dx$. Use sigma notation to write the left, right, and midpoint approximations to I. (Assume that N equal subintervals are used.)

54. **(a)** Write the sum

$$\frac{5}{2}(2.3)^2 + \frac{5}{6}(2.8)^2 + \frac{5}{12}(3.3)^2 + \frac{5}{20}(3.8)^2$$

using sigma notation.
[HINT: $2 = 1\cdot 2, 6 = 2\cdot 3, 12 = 3\cdot 4$, and $20 = 4\cdot 5$.]

(b) Is the sum in part (a) a Riemann sum approximation to $\displaystyle\int_0^4 x^2\,dx$? Explain.

In Exercises 55–58, rewrite the expression in terms of G, where $G(x) = \displaystyle\int_1^x g(t)\,dt$.

55. $\displaystyle\int_1^3 g(u)\,du$

56. $\displaystyle\int_0^1 g(x)\,dx$

57. $\displaystyle\int_{-2}^2 g(t)\,dt$

58. $\displaystyle\int_2^4 g(t)\,dt$

59. Suppose that f is a function such that $\displaystyle\int_0^3 f(x)\,dx = -1$.

    (a) Suppose that f is an *even* function. Explain why

$$\int_{-3}^3 f(x)\,dx = -2.$$

(b) If f is an *odd* function, what is the value of $\displaystyle\int_{-3}^3 f(x)\,dx$? Explain.

60. Show that $0 \le \displaystyle\int_0^\pi x\sin x\,dx \le \pi^2/2$.
[HINT: $0 \le \sin x \le 1$ if $0 \le x \le \pi$.]

In Exercises 61–64, $h(x) = \int_0^x g(t)\,dt$, where g is the function graphed below.

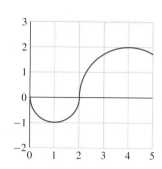

61. Rank the eight numbers -1, 3, 5, $h(1)$, $h(4)$, $\int_0^2 g(u)\,du$, $\int_1^1 g(x)\,dx$, and $\int_5^3 g(t)\,dt$ in increasing order.

62. Evaluate $h'(1)$.

63. Is $h''(3) < 0$? Justify your answer.

64. How many roots does h have in the interval $[0, 5]$? Justify your answer.

65. Suppose that $\int_4^6 h(w)\,dw = 15$. Evaluate $\int_4^6 (2h(z) + 3)\,dz$.

66. Suppose that $\int_1^5 (3f(x) + 2)\,dx = 68$. Evaluate $\int_1^5 f(z)\,dz$.

In Exercises 67–70, suppose that $\int_{-3}^1 h(t)\,dt = -2$ and that $\int_{-1}^1 h(u)\,du = 4$.

67. Evaluate $\int_1^{-3} h(w)\,dw$.

68. What is the average value of h over the interval $[-1, 1]$?

69. Evaluate $\int_{-3}^{-1} h(z)\,dz$.

70. Is it possible that h is an odd function? Justify your answer.

In Exercises 71–74, H is the function $H(w) = \int_{-2}^w xe^x\,dx$.

71. Indicate whether each of the following values is positive, negative, or zero.
  (a) $H(-3)$
  (b) $H(-2)$
  (c) $H(0)$
  (d) $H(2)$

72. Show that $H(1) \geq -3e^{-1}$.

73. Evaluate $H'(1)$.

74. On which subintervals of $(-\infty, \infty)$, if any, is the graph of H concave up?

75. Show that $\lim_{n\to\infty} \left(\dfrac{3}{n}\right) \sum_{k=1}^n \left(1 + \dfrac{3k}{n}\right)^2 = 21$.

76. Is $\left(\dfrac{3}{5}\right) \cdot \left(-1 + 0 + \dfrac{1}{2} + 1 + 2\right)$ a Riemann sum approximation to $\int_{-1}^2 x\,dx$? Justify your answer.

77. Let $G(x) = \int_x^2 \cos(\pi t^2/4)\,dt$. Evaluate $G'(2)$.

78. Suppose that $\int_1^e \dfrac{g(\ln x)}{x}\,dx = 42$. Find values of the parameters K, a, and b so that $K\int_a^b g(u)\,du = 21$.

79. Is $\dfrac{2}{7} \sum_{k=0}^6 \sqrt{3 + 2k/7}$ a Riemann sum approximation of the integral $\int_3^5 \sqrt{x}\,dx$? Justify your answer.

80. What is the *exact* value of $\lim_{n\to\infty} \dfrac{5}{n} \sum_{k=1}^n \sqrt{4 + 5k/n}$? Justify your answer.

81. Suppose that $\int_1^4 h(w)\,dw = 17$. Evaluate $\int_4^1 (2h(z) - 5)\,dz$.

82. Suppose that $\int_{-2}^3 (f(x) + 1)\,dx = 0$. Evaluate $\int_{-2}^3 (f(x) - x)\,dx$.

In Exercises 83–90, $S(z) = \int_1^z \sin(t^2/2)\,dt$.

83. Rank the six numbers -0.5, 0.5, $S(0)$, $S(1)$, $S(2)$, and $S(3)$ in increasing order.

84. Is $S(z)$ decreasing at $z = 2$? Justify your answer.

85. Is $S(z)$ concave up at $z = 3$? Justify your answer.

86. Evaluate $S''(4)$.

87. How many roots does S have in the interval $[0, 5]$? Justify your answer.

88. Where does S attain its minimum value over the interval $[0, 5]$?

89. Where does S attain its maximum value over the interval $[0, 5]$?

90. It can be shown that $S(4) \approx 0.74380$ and $S(7) \approx 0.60917$. Use this information to estimate the average value of $\sin(t^2/2)$ over the interval $[4, 7]$.

In Exercises 91–96, evaluate the integral (if possible) assuming that f is an even function such that $\int_1^2 f(t)\,dt = 3$ and $\int_1^3 f(t)\,dt = 7$. If the value of an integral cannot be determined from the information given about f, indicate this.

91. $\int_2^1 f(t)\,dt$

92. $\int_2^2 f(t)\,dt$

93. $\int_{-1}^1 f(t)\,dt$

94. $\int_{-2}^{-1} f(t)\,dt$

95. $\int_{2}^{3} f(t)\,dt$

96. $\int_{1}^{3} |f(t)|\,dt$

97. What is the average value of the function $g(z) = |1+z|$ over the interval $[-2, 3]$?

98. Suppose that $h(t) = \sqrt{\ln(1 + e^{\sin t})}$. Evaluate $\int_{0}^{\pi} h'(x)\,dx$.

99. Explain why $\displaystyle\lim_{n\to\infty} \frac{3}{n} \sum_{j=1}^{n} \cos\left(1 + \frac{3j}{n}\right) = \sin 4 - \sin 1$.

   [HINT: What is the definition of $\int_{a}^{b} f(x)\,dx$?]

In Exercises 100–103, $F(x) = \int_{0}^{x} f(t)\,dt$, where f is a function such that f, f', and f'' are continuous on the interval $(-\infty, \infty)$ and

   (i) f is positive on the interval $(-\infty, 1)$ and negative on the interval $(1, \infty)$.

   (ii) f is decreasing on the interval $(-\infty, 2)$ and increasing on the interval $(2, \infty)$.

   (iii) f is concave up on the interval $(-\infty, 5)$ and concave down on the interval $(5, \infty)$.

100. Is $F(9) > F(6)$? Justify your answer.

101. Is $\int_{7}^{9} f'(x)\,dx > 0$? Justify your answer.

102. Let $F(x) = \int_{0}^{x} f(t)\,dt$. At which value of x in the interval $[0, 10]$ does $F(x)$ attain its maximum value? Justify your answer.

103. Is $F(x)$ concave up at $x = 3$? Justify your answer.

104. Suppose that $\int_{0}^{5} f(3x)\,dx = 12$. Find values of the constants K, a, and b such that

$$K \int_{a}^{b} f(u)\,du = 3.$$

105. Use the fact that $3 - 2x - x^2 = 4 - (x+1)^2$ and the substitution $u = (x+1)/2$ to show that $\displaystyle\int_{0}^{1} \frac{dx}{\sqrt{3 - 2x - x^2}} = \frac{\pi}{3}$.

106. Suppose that $\int_{1}^{e} f(u)\,du = 7$. Find values of the constants K, a, and b such that $K \int_{a}^{b} e^x f(e^x)\,dx = 7$.

107. Suppose that $f(t) = \sqrt{1 + t^4}$. Show that $\dfrac{d}{dx}\left(\int_{0}^{x} f(t)\,dt\right) \neq \int_{0}^{x} f'(t)\,dt$.

108. Is there a function g with the property $\int_{0}^{x} g(t)\,dt = x^2 e^x$? Justify your answer.

109. Explain why there is not a function h with the property $\int_{0}^{x} h(t)\,dt = x^2 - e^x$.

110. Show that $\dfrac{d}{dx}\left(\int_{1}^{x^3} \sin(t^2)\,dt\right) = 3x^2 \sin(x^6)$.
   [HINT: Use the chain rule.]

111. Is $f(0) \cdot 1 + f(2) \cdot 2 + f(4) \cdot 3 + f(6) \cdot 4$ a Riemann sum approximation to $\int_{0}^{10} f(x)\,dx$? Justify your answer.
   [HINT: Draw a picture.]

112. Suppose that f is a continuous function such that $\int_{-1}^{1} f(x)\,dx = 7$. Use this fact to evaluate $\int_{0}^{\pi} f(\cos x) \sin x\,dx$. Show your work.

In Exercises 113–116, $g(x) = \int_{0}^{x} \dfrac{e^{-t}}{1 + t^2}\,dt$.

113. Is $g(3) > 2$? Justify your answer.

114. Is $g(-1) > 1$? Justify your answer.

115. Is g increasing at $x = 1$? Justify your answer.

116. Is g concave up at $x = 1$? Justify your answer.

117. Suppose that $\int_{0}^{8} f(t)\,dt = 4$. Evaluate $\int_{0}^{2} x^2 f(x^3)\,dx$.

118. Explain why a trapezoid approximating sum is *not* a Riemann sum approximation.

119. Evaluate $\int_{-1}^{2} |z|\,dz$.

120. Find an antiderivative of $h(x) = \cos^3 x \sin x$.

Mean Value Theorems and Integrals

Here are two similarly named theorems:

> **THEOREM 6 (Mean value theorem for derivatives)** Let f be continuous on the closed interval $[a, b]$ and differentiable on the open interval (a, b). Then there is a number c in (a, b) for which
> $$f'(c) = \frac{f(b) - f(a)}{b - a}.$$

> **THEOREM 7 (Mean value theorem for integrals)** Let f be continuous on the closed interval $[a, b]$. Then there is a number c in $[a, b]$ for which
> $$f(c) = \frac{\int_a^b f(x)\, dx}{b - a}.$$

Theorems 6 and 7 look (and are) closely related to each other, but the first is about derivatives and the second about integrals. In this interlude we'll call these theorems the MVTD and the MVTI, respectively. The names suggest both similarities and differences in the theorems; we explore both.

What the theorems say We met Theorem 6 in Section 4.9 (there we called it just the mean value theorem, or MVT). As the name MVTD suggests, Theorem 6 is about *derivatives* and *averages*: it says that for some c between a and b, the derivative $f'(c)$ is equal to the average rate of change of f over the interval $[a, b]$. Graphically, the MVTD says that, for at least one c between a and b, the slope of the tangent line at $x = c$ is equal to the slope of the secant line over the interval $[a, b]$; see the familiar picture in Figure 1(a).

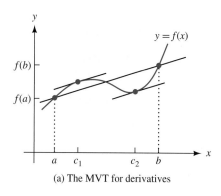

(a) The MVT for derivatives

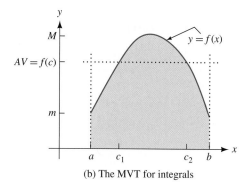

(b) The MVT for integrals

FIGURE 1

Two mean value theorems

Figure 1(b) illustrates the MVTI. It shows *two* values of c in the interval $[a, b]$ that satisfy the equation in Theorem 7. For each such c, the value $f(c)$ is the *average value* (AV) of the function f over the interval $[a, b]$. (We defined average value in Section 5.1 to be the quantity on the right side of Theorem 7.) The MVTI can thus be restated as follows:

If f is continuous on [a, b], then there is at least one c in [a, b] for which the value *f(c) equals the* average value *of f on [a, b].*

This should sound reasonable; unless f is constant we'd expect $f(x) > AV$ for some inputs x and $f(x) < AV$ for others inputs. (If f is constant then $f(x) = AV$ for *all* x, so the theorem's claim is clearly true in this case.)

The following problems explore further the MVTI and its connection to the MVTD.

PROBLEM 1   (MVTD and FTC imply MVTI) One way to prove the MVTI is to use the fundamental theorem of calculus (FTC) and the MVTD. To do so, let F be an antiderivative for f. Then we know from the FTC that $\int_a^b f(x)\,dx = F(b) - F(a)$. Use this fact and the MVTD to prove the MVTI.

PROBLEM 2   (MVTI and FTC imply MVTD) Given the FTC, the MVTD (in a slightly weaker form) can also be proved using the MVTI, as follows: Let f be differentiable with continuous derivative f' on $[a, b]$. By the FTC, we have $f(b) - f(a) = \int_a^b f'(x)\,dx$. Now apply the MVTI to the function $g = f'$ on $[a, b]$ to conclude that the MVTD holds.

PROBLEM 3   (Proving the MVTI without the FTC) Using the FTC to prove the MVTI may be a bit of overkill, especially since the MVTD is sometimes used to prove the FTC. Fortunately, it's not hard to avoid the FTC in proving the MVTI, as follows.

We observed in Section 5.1 that if m and M are, respectively, the minimum and maximum values of f on $[a, b]$, then

$$m(b - a) \le \int_a^b f(x)\,dx \le M(b - a).$$

This implies (give details) that the average value of f lies *between* m and M. Use this fact and the intermediate value theorem to prove the MVTI. (You might also mention the extreme value theorem.)

PROBLEM 4   In each part following find a value of c that satisfies the MVTI for the given f and $[a, b]$.

(a) $f(x) = x$ on $[a, b] = [0, 1]$

(b) $f(x) = x^2$ on $[a, b] = [0, 1]$

(c) $f(x) = x^3$ on $[a, b] = [-1, 1]$

(d) $f(x) = e^x$ on $[a, b] = [0, 1]$

PROBLEM 5   Let $f(x) = mx + n$ be any *linear* function and $[a, b]$ any interval. Show that the MVTI holds for f on $[a, b]$ with c equal to the *midpoint* of $[a, b]$.

NUMERICAL INTEGRATION

6.1 APPROXIMATING INTEGRALS NUMERICALLY

In Chapter 5 we defined the definite integral $\int_a^b f(x)\, dx$ as a limit of certain approximating sums. Since the fundamental theorem of calculus permits many integrals to be calculated by antidifferentiation—without any explicit reference to approximating sums—we might feel tempted to forget approximating sums entirely. That would be a serious mistake. As we'll see in this chapter, approximating sums have down-to-earth, practical value; they lead easily to good approximations for definite integrals, even ones whose "exact" values are hard or impossible to find.

In this chapter we'll see how to use approximating sums of various kinds to estimate integrals. We'll compare how well different types of sums work for this purpose, and we'll even offer "accuracy guarantees" for many of the estimates we produce. We'll also apply similar ideas to find approximate solutions to differential equations. The techniques of this chapter are called **numerical methods** (as opposed to symbolic methods) because, as we'll see, they involve direct numerical computations more than symbolic manipulation.

A sampler of sums

In Sections 5.6 and 5.7 we described several types of approximating sums for integrals (left, right, midpoint, and trapezoid sums) and studied some of their basic properties. For a given integral $I = \int_a^b f(x)\, dx$, each such sum involves the following steps:

- **Partition:** Chop $[a, b]$ into n equal subintervals, each of length $\Delta x = (b - a)/n$, with successive endpoints $x_0 < x_1 < x_2 < \cdots < x_n$.
- **Evaluate:** Choose in each subinterval an input c_i at which to evaluate f (left endpoint, right endpoint, midpoint, etc.).
- **Sum:** Add up the contributions $f(c_i)\, \Delta x$ (one for each subinterval) to calculate the approximating sum.

We'll often write L_n, R_n, T_n, and M_n for left, right, trapezoid, and midpoint sums, each with n equal subdivisions. Here are general formulas for all of them:

$$L_n = (f(x_0) + f(x_1) + \cdots + f(x_{n-1})) \, \Delta x = \sum_{i=0}^{n-1} f(x_i) \, \Delta x$$

$$R_n = (f(x_1) + f(x_1) + \cdots + f(x_n)) \, \Delta x = \sum_{i=1}^{n} f(x_i) \, \Delta x$$

$$T_n = \left(\frac{f(x_0)}{2} + f(x_1) + \cdots + f(x_{n-1}) + \frac{f(x_n)}{2} \right) \, \Delta x = \sum_{i=1}^{n} \frac{1}{2}(f(x_{i-1}) + f(x_i)) \, \Delta x$$

$$M_n = (f(m_1) + f(m_1) + \cdots + f(m_n)) \, \Delta x = \sum_{i=1}^{n} f(m_i) \, \Delta x.$$

In the last line, $m_i = (x_{i-1} + x_i)/2$, the midpoint of the ith subdivision.

EXAMPLE 1 Let $I = \int_0^1 \sin x \, dx$. Write out the approximating sums R_{100} and M_{100}, both without and with sigma notation.

Solution Chopping the interval $[0, 100]$ into 100 equal-length subintervals gives the partition points $x_0 = 0$, $x_1 = 0.01$, $x_2 = 0.02$, ..., $x_{99} = 0.99$, $x_{100} = 1.00$, and so the right sum has the form

$$R_{100} = \big(\sin(0.01) + \sin(0.02) + \cdots + \sin(1.00)\big) \cdot \frac{1}{100}.$$

For M_{100} we evaluate f at the *midpoint* of each subinterval:

$$M_{100} = \big(\sin(0.005) + \sin(0.015) + \cdots + \sin(0.085) + \sin(0.095)\big) \cdot \frac{1}{100}.$$

To write these sums in sigma notation, we first express the right endpoints and the midpoints using an index variable i. The ith right endpoint has the form $x_i = 0.01i$ (for i from 1 to 100), whereas the ith midpoint can be written as $m_i = 0.005 + 0.01i$ (for i from 0 to 99). In sigma notation, the sums have the form

$$R_{100} = \sum_{i=1}^{100} \sin(0.01\,i) \cdot \frac{1}{100}; \qquad M_{100} = \sum_{i=0}^{99} \sin(0.005 + 0.01i) \cdot \frac{1}{100}.$$

Evaluating such monstrous sums—no matter the form—is no job fit for a human. ∎

Help from technology Computers and calculators have no trouble calculating sums like those in Example 1. Exactly *how* one uses technology to calculate such sums depends, of course, on the machine at hand. On the TI–86 calculator, for instance, commands of the form

```
sum(seq(sin(0.01*i),i,1,100))/100
sum(seq(sin(0.005+0.01*i),i,0,99))/100
```

evaluate the sums in Example 1. The results (to five decimal places) are

$$R_{100} = 0.46390 \quad \text{and} \quad M_{100} = 0.45970.$$

With the computer package *Maple*, to give another example, one might give commands something like the following to calculate the approximating sums L_{20}, R_{20}, T_{20}, and M_{20} for the integral $\int_0^1 \sin x \, dx$:

```
left(sin(x), x=0..1, 20)          right(sin(x), x=0..1, 20)
trapezoid(sin(x), x=0..1, 20)     midpoint(sin(x), x=0..1, 20)
```

In this book we'll sometimes use *Maple*-style commands like those above—but only as illustrations, not necessarily as literal commands to be typed. It's the ideas that count—and exactly the *same* ideas apply with any form of technology.

EXAMPLE 2 Calculate several approximating sums for $\int_0^1 \sin x \, dx$. Compare results to the "exact" answer.

Solution The fundamental theorem of calculus (FTC) gives the exact answer (we rounded the numerical version to five decimal places):

$$\int_0^1 \sin x \, dx = -\cos x \Big]_0^1 = -\cos 1 + \cos 0 \approx 0.45970.$$

For comparison, here are results rounded to five decimal places:

Sum	Input	Output
L_{20}	`left(sin(x), x=0..1, 20)`	0.43857
R_{20}	`right(sin(x), x=0..1, 20)`	0.48064
T_{20}	`trapezoid(sin(x), x=0..1, 20)`	0.45960
M_{20}	`midpoint(sin(x), x=0..1, 20)`	0.45975

Notice that all four results are in the general ballpark, but the trapezoid and midpoint sums are much better estimates than the other two. Understanding how well various sums approximate integrals and why some do better than others is an important theme of this chapter. ∎

Real examples Example 2 shows nicely how (and why) approximating sums work, but in an important sense it's artificial: If we *know* the exact value of an integral, then there's little practical value in approximating it. The best use of approximating sums is in approximating integrals we *can't* evaluate exactly.

Why not? Because (for deep reasons) the antiderivative of $\sin(x^2)$ is not an elementary function.

EXAMPLE 3 Use approximating sums to estimate $I = \int_0^1 \sin(x^2) \, dx$. (The integral I can't be evaluated by antidifferentiation.) ◄ Can we say anything for *certain* about the exact value of I?

Solution With technology we can use fairly large values of n. Here are some results rounded to five decimals:

$$L_{100} = 0.30607; \quad R_{100} = 0.31448; \quad T_{100} = 0.31028; \quad M_{100} = 0.31026.$$

The four sums are close to each other, and so we can reasonably hope that they're also close to the true value of I. Without further information, we'd probably guess (but it's only a guess) that I lies somewhere near 0.310. We'll explain soon how to find a small interval (which contains 0.310) that's *guaranteed* to contain I, but it's important to realize that we may never know I exactly. ∎

Errors and error bounds

Many quantities of interest, such as the integral in Example 3 or the *true* fraction of voters who prefer Smith to Jones in an election, are difficult or impossible to compute exactly; they *must* be estimated. But an approximation alone has little value. We'd also like some measure of how accurate an approximation is. A credible opinion poll, for instance, reports not only that 58% of voters surveyed prefer Smith but also that the true statistic *probably* falls between, say, 55% and 61%.

We'd like some similar measure of accuracy for approximating sums. Our wish is about to be granted—and in a much stronger way than opinion polls permit. Remarkably enough, we can often guarantee that approximating sums are accurate to within a small tolerance. We treat this theme informally in the rest of this section and in more detail in the next section.

Errors committed by approximating sums   For a given integral $I = \int_a^b f(x)\,dx$ and a large value of n, we expect the approximating sums L_n, R_n, T_n, and M_n to be "close" to the exact value of I. But *how* close? Within 0.01? Within 0.000000000001? Can we be sure? Surprisingly often, we can.

The word "error" means several (mostly disagreeable) things in everyday speech. In mathematics, "error" has a simpler (and more agreeable) meaning: it's the *difference* between an exact value and an approximation. For an approximating sum T_n, for example, we call the difference $I - T_n$ the **error committed by** T_n. In the same spirit, $|I - T_n|$ is called the **absolute error** (or the **magnitude** of the error) committed by T_n. In Example 2, for instance, the error committed by M_{20} is (with numbers rounded to five decimals)

$$I - M_{20} = -\cos 1 + \cos 0 - 0.45975 = -0.00005.$$

The error is *negative* because M_{20} slightly *overestimates* I. The magnitude of the error, 0.00005, is *small* because M_{20} approximates I so closely. The following table collects a variety of error results:

\multicolumn{5}{c}{**Errors committed in approximating $I = \int_0^1 \sin(x)\,dx$**}
n	$I - L_n$	$I - R_n$	$I - T_n$	$I - M_n$
4	0.107580	−0.102787	0.002396	−0.001200
8	0.053190	−0.051993	0.000598	−0.000300
16	0.026445	−0.026146	0.000149	−0.000075
32	0.013185	−0.013111	0.000036	−0.000018

Reading down any column shows how errors tend to decrease in magnitude as n increases; reading across any row shows how one method compares with another.

Bounding errors   We can't expect (except in some contrived cases) to compute *exactly* the error a given approximation commits—doing so would be tantamount to finding the unknown quantity exactly. But it is often reasonable to ask for an **error bound**—a guarantee that the error committed by an approximation does not exceed some given amount.

Error bounds are usually stated as inequalities, sometimes involving absolute values. Here are two examples:

$$0 \le I - L_{10} \le 0.01; \qquad |I - M_n| \le \frac{32}{n^2}.$$

Notice:

- **Worst-case scenarios**   Both inequalities guarantee that the error is *no worse than* the quantity on the right; the *actual* error committed may be much less.
- **An underestimate**   The first (double) inequality means that L_{10} lies within 0.01 of, but *underestimates*, the exact integral I.

- **Size, not sign** The second inequality says that, for given n, the absolute error committed by M_n doesn't exceed $32/n^2$. We bound the *absolute* error when we care (or know) more about the size than the sign of the error.

- **Improvement as n increases** Finding and using inequalities like the second one above is the subject of the next section. But we can notice right now that the inequality depends on n: As n increases, the right side decreases toward zero, and so the absolute error committed by M_n also decreases toward zero. ◄

This stands to reason; we expect more work to give a better result.

Is more always better? Roundoff errors Common sense suggests that for a given integral I, L_{100} commits less error than L_{10}, T_{100} outperforms T_{10}, and so on—for more work we expect better accuracy. But more isn't *always* better. In the real world, approximating sums like L_{10000} and $T_{1000000}$ are less useful than we might expect. One drawback to using huge values of n is that adding many numbers can take a lot of time, even for a fast computer. A more serious problem is that small roundoff errors (which *always* occur in decimal arithmetic) can accumulate in repetitive calculations. In the worst cases roundoff errors can make a calculation completely meaningless.

A full treatment of roundoff errors is outside the scope of a calculus text. We'll control roundoff errors (though we'll never completely escape them) by avoiding unnecessarily large-scale computations.

Trapping the elusive integral

One simple method of controlling the error committed by approximating sums is to "trap" the (unknown) exact value I *between* two approximating sums. The following example shows the basic idea.

EXAMPLE 4 Let $I = \int_0^1 \sin(x^2)\,dx$. In Example 3 we calculated the approximating sums

$$L_{100} = 0.30607; \quad R_{100} = 0.31448; \quad T_{100} = 0.31028; \quad M_{100} = 0.31026.$$

Use these numbers to "trap" I.

Solution The first key observation is that the integrand $\sin(x^2)$ is *increasing* on the interval $[0, 1]$. A useful conclusion follows:

Every left sum L_n underestimates I; every right sum R_n overestimates I.

Figure 1 illustrates the first part of our claim.

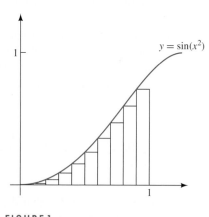

FIGURE 1
Left sums underestimate $\int_0^1 \sin(x^2)dx$

It follows that I lies somewhere *between* L_{100} and R_{100}:

$$L_{100} = 0.30607 \leq I \leq 0.31449 = R_{100}.$$

This is well worth knowing—especially because we don't know I exactly.

We could pull L_n and R_n even closer together by using larger values of n. But we can do even better, with less work, by using the trapezoid and midpoint estimates T_{100} and M_{100}. Figure 1 shows that the integrand is *concave up* on the interval $[0, 1]$. As a result the trapezoid rule T_n *overestimates* the integral for every n. ➡ It's less obvious, but also true, that M_n *underestimates* the integral; thus,

Do you see why the shape of the graph ensures this?

$$M_{100} = 0.31026 \leq I \leq 0.31028 = T_{100}.$$

This is an even "tighter" trap than we had from left and right sums.   ■

Trapping with left and right sums: monotone functions   The main trapping idea in the preceding example was that if the integrand f happens to be *increasing* on $[a, b]$, then every left sum *underestimates* $\int_a^b f(x)\,dx$, while every right sum *overestimates*. The reverse occurs if f happens to *decrease* on $[a, b]$. The following simple but useful theorem covers both possibilities: ➡

Recall: "monotone" means "increasing or decreasing but not both."

THEOREM 1 (Trapping the integral of a monotone function)   Suppose that f is monotone on $[a, b]$, and let $I = \int_a^b f(x)\,dx$. For any positive integers n and m, either

$$L_n \leq I \leq R_m \quad \text{or} \quad R_m \leq I \leq L_n.$$

In either case, I lies *between* L_n and R_m.

(Often we'll use $n = m$, but that's not necessary; inequalities like $L_4 \leq I \leq R_7$ are sometimes useful.)

We can't expect a simple theorem to solve *all* our problems—after all, many integrals don't involve monotone functions. Still, we can often get useful information from the theorem—sometimes with only a little extra work. We illustrate the idea with two examples.

EXAMPLE 5   (Exploiting symmetry)   Use left and right sums to find good upper and lower bounds for $I = \int_{-1}^1 e^{x^2}\,dx$.

Solution   The integrand $f(x) = e^{x^2}$ is *not* monotone on $[-1, 1]$; it decreases on $[-1, 0]$ and increases on $[0, 1]$. But f is an *even* function and so its graph is symmetric about the y-axis. Thus, we can write

$$I = \int_{-1}^1 e^{x^2}\,dx = 2\int_0^1 e^{x^2}\,dx = 2I_0.$$

Now $I_0 = \int_0^1 e^{x^2}\,dx$ *does* satisfy the hypothesis of Theorem 1, and so we'll use technology to find some left and right sums for I_0. Following are several, rounded to five decimal places:

Sum	Input	Output
L_{10}	`left(exp(x^2), x=0..1, 10)`	1.38126
R_{10}	`right(exp(x^2), x=0..1, 10)`	1.55309
L_{100}	`left(exp(x^2), x=0..1, 100)`	1.45412
R_{100}	`right(exp(x^2), x=0..1, 100)`	1.47130

The first two rows and Theorem 1 imply that $1.38126 \le I_0 \le 1.55309$. The last two rows say something stronger: $1.45412 \le I_0 \le 1.47130$. Thus, we've trapped I_0 in a small interval. To trap our *real* prey, $I = 2I_0$, we multiply the preceding inequality by 2:

$$1.45412 \le I_0 \le 1.47130 \implies 2.90824 \le 2I_0 = I \le 2.94260.$$

(Multiplying the weaker inequality $1.38126 \le I_0 \le 1.55309$ by 2 gives the weaker result $2.76252 \le I \le 3.10618$.) ∎

EXAMPLE 6 **(Divide and conquer)** Use left and right sums to trap $I = \int_0^2 \sin(x^2)\, dx$.

This is easy to check using derivatives or by plotting.

Solution As in Example 5, the integrand is not monotone—it increases for $0 \le x \le \sqrt{\pi/2} \approx 1.2533$ and decreases for $\sqrt{\pi/2} \le x \le 2$. ◄ Thus, we can split the desired integral into two pieces, each of which *does* satisfy the hypothesis of Theorem 1:

$$I = I_1 + I_2 = \int_0^{\sqrt{\pi/2}} \sin(x^2)\, dx + \int_{\sqrt{\pi/2}}^2 \sin(x^2)\, dx.$$

Now we apply Theorem 1, first to I_1 and then to I_2. For I_1 we find

$$L_{100} = 0.54301 \quad \text{and} \quad R_{100} = 0.55554, \quad \text{so} \quad 0.54301 \le I_1 \le 0.55554.$$

For I_2,

$$L_{100} = 0.26205 \quad \text{and} \quad R_{100} = 0.24893, \quad \text{so} \quad 0.24893 \le I_2 \le 0.26205.$$

Having trapped I_1 and I_2, we can add the preceding inequalities to trap I:

$$0.54301 + 0.24893 = 0.79194 \le I_1 + I_2 \le 0.81759 = 0.55554 + 0.26205.$$ ∎

Midpoint and trapezoid traps As we said in Example 4, some integrals can be trapped between trapezoid and midpoint sums; *concavity* is what matters.

EXAMPLE 7 Use M_4 and T_4 to approximate $I = \int_0^1 x^2\, dx$. How much error does each sum commit?

Solution By the FTC, the exact value is $I = 1/3$. Routine calculations (by hand or machine) give

$$M_4 = \frac{63}{192}; \quad T_4 = \frac{66}{192}; \quad I = \frac{1}{3} = \frac{64}{192}.$$

(We calculated with fractions for ease of comparison.) The numbers show that T_4 *overestimates* I (by 2/192) and M_4 *underestimates* I (by 1/192). Figure 2 illustrates these results:

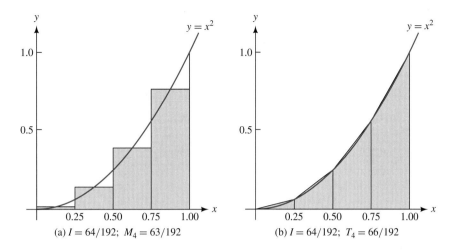

(a) $I = 64/192$; $M_4 = 63/192$ (b) $I = 64/192$; $T_4 = 66/192$

FIGURE 2
Midpoint and trapezoid estimates to $\int_0^1 x^2\, dx$

Figure 2(b) shows that (and why) T_4 overestimates I (though not by much). Upward concavity of the integrand means that the f-graph sags *below* the secant lines that define the trapezoid estimate. It's less obvious (but visible if you look *really* closely at Figure 2(a)) that M_4 underestimates I.   ■

The midpoint rule and tangent lines   Thinking of the midpoint rule in a slightly different way will help explain why it underestimates the integral in Example 7. So far we've thought of M_n as a sum of *rectangular* areas, obtained by replacing the graph of f over a subinterval with a horizontal line determined at the midpoint m_i, as in Figure 3(a). An alternative, but equivalent, approach to M_n often reveals the *direction* of error more clearly. The key idea is as follows:

> *Replace the graph of f over the ith subinterval with the tangent line (rather than a horizontal line) at $x = m_i$.*

Figure 3(b) illustrates this view of the midpoint rule:

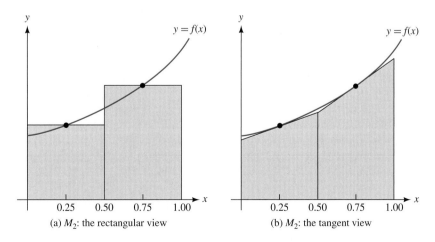

(a) M_2: the rectangular view   (b) M_2: the tangent view

FIGURE 3
Two views of the midpoint rule

From this point of view, M_n becomes a sum of *trapezoidal* areas; the tops (or bottoms if the function is negative) of the trapezoids are determined by *tangent lines*. For the trapezoid rule T_n, by contrast, the tops (or bottoms) are *secant* lines.

Why do both the rectangular and tangent approaches to M_n produce the same answer? The reason has nothing special to do with tangent lines. The fortunate fact is that the area of *any* trapezoid (with parallel vertical sides, as shown) is its base times its *height at the midpoint*. Thus, the shaded areas in Figure 3 are indeed equal.

The "tangent view" of M_n explains clearly why *concavity* determines the sign of the error committed by M_n. Upward concavity of the integrand means that the f-graph lies *above* the tangent line on each subinterval. The following theorem collects what we know:

> **THEOREM 2** If f is concave *up* on $[a, b]$, then T_n overestimates I and M_n underestimates I. If f is concave *down*, the reverse holds. In both cases, I lies between T_n and M_n.

Splitting the difference: averaging different sums

Theorem 1 says that for a monotone integrand the exact integral I lies between L_n and R_n. This suggests an obvious strategy: split the difference. Indeed, trapezoid sums can be thought of as doing exactly that:

We discussed this in Section 5.7.

$$\text{For all } n, \ T_n = \frac{L_n + R_n}{2} \text{ is the average of } L_n \text{ and } R_n. \ \blacktriangleleft$$

We can approach trapezoid and midpoint estimates in a similar way. Theorem 2 guarantees for certain integrals that T_n and M_n trap I. Therefore, it's reasonable to "compromise" between trapezoid and midpoint sums. This time, however, we'll split the difference unevenly because, as a rough rule, T_n commits about *twice* as much approximation error as M_n. (Example 7 illustrates this; so do the numbers in Example 2.) Therefore, we use, instead of the ordinary average of T_n and M_n, the number one-third of the way from M_n to T_n (called a **weighted average**). We state the result as a definition:

> **DEFINITION (Simpson's rule)** For any positive integer n, the quantity
> $$S_{2n} = \frac{2}{3}M_n + \frac{1}{3}T_n$$
> is the **Simpson's rule approximation** with $2n$ subdivisions.

Notice the subscript on the left: It's $2n$, not n, because the subdivision endpoints x_i together with the midpoints m_i partition the interval $[a, b]$ into $2n$ subdivisions. Notice also the unequal **weights** attached to M_n and T_n; the more accurate approximation M_n gets twice as much weight as the less accurate T_n.

Simpson's rule is remarkably accurate and therefore well known both to humans and to many calculators and computers. Using the data in Example 2, for example, we calculate (to five decimal places)

$$S_{40} = \frac{2}{3}M_{20} + \frac{1}{3}T_{20} = \frac{2}{3}0.45975 + \frac{1}{3}0.45960 = 0.45970,$$

which agrees with the exact value of I to all five decimal places shown. In fact, calculating with more decimal places shows that S_{40} agrees with I to eight or nine decimal places. (We explore Simpson's rule further in an Interlude at the end of this chapter.)

BASIC EXERCISES

1. Let $I = \int_1^4 \frac{dx}{\sqrt{x}}$.

   (a) Use the FTC to evaluate I exactly.

   (b) Write out the approximating sums L_3, R_3, T_3, and M_3.

   (c) Compute the approximation errors $|I - L_3|$, $|I - R_3|$, $|I - T_3|$, and $|I - M_3|$.

   (d) What is the approximation error made by S_6?

2. Let $I = \int_0^1 x^3 \, dx$.

   (a) Use the FTC to evaluate I exactly.

   (b) Write out the approximating sums L_4, R_4, T_4, and M_4.

   (c) Compute the approximation errors made by L_4, R_4, T_4, M_4, and S_8.

3. Let $I = \int_0^1 \sin(x^2) \, dx$. Explain why the inequalities $L_7 \le I \le R_4$ are valid.

4. Let $I = \int_1^3 \cos(1/x) \, dx$.

   (a) Explain why the inequalities $L_7 \le I \le R_{13}$ are valid.

   (b) Explain why the inequalities $T_{11} \le I \le M_7$ are valid.

5. Let $I = \int_0^2 e^{x^2} \, dx$.

   (a) Compute L_{10}, R_{10}, T_{10}, M_{10}, and S_{20}.

   (b) Use L_{10} and R_{10} to find a bound on the approximation error made by T_{10}.

   (c) Use M_{10} and T_{10} to find a bound on the approximation error made by S_{20}.

6. Let $I = \int_1^3 e^{-x^2} \, dx$.

   (a) Compute L_{20}, R_{20}, T_{20}, M_{20}, and S_{40}.

   (b) Does L_{20} underestimate I? Justify your answer.

   (c) Does M_{20} underestimate I? Justify your answer.

7. Let $I = \int_0^1 f(x) \, dx$, where the function f has the values shown in the table:

x	0.00	0.25	0.50	0.75	1.00
$f(x)$	1.307	1.096	1.018	1.173	1.435

   (a) Compute L_4, R_4, T_4, M_2, and S_4.

   (b) A plot of the data shows that it reasonable to assume that f is concave up on the interval $[0, 1]$. Use this assumption and your results from part (a) to find upper and lower bounds on I. Justify your choices.

8. Let $I = \int_{-1}^2 f(x) \, dx$, where f has the values shown in the table.

x	−1.00	−0.25	0.50	1.25	2.00
$f(x)$	0.0000	2.6522	4.8755	6.8328	8.6790

   (a) Compute L_4, R_4, T_4, M_2, and S_4.

   (b) A plot of the data shows that it reasonable to assume that f is increasing and concave down on the interval $[0, 1]$. Use this assumption and your results from part (a) to find upper and lower bounds on I. Justify your choices.

9. Let $I = \int_0^1 \sin x \, dx$.

   (a) Use the table in Example 2 to compute the approximation errors committed by each of the approximating sums L_{20}, M_{20}, R_{20}, and T_{20}.

   (b) Why does L_{20} underestimate I?

   (c) Why does T_{20} underestimate I?

10. Let $I = \int_0^1 \sin x \, dx$. How does the table on page 375 show that, for all the methods, the approximation errors tend to decrease in magnitude as n increases?

11. Let $I = \int_0^1 \sin x \, dx$. How does the table on page 375 show that M_n overestimates I for all of the values of n shown?

12. Let I and I_0 be as in Example 5.

   (a) Compute the approximating sum T_{100} for I_0.

   (b) Use part (a) to compute an estimate of I.

13. Let $I = \int_0^1 \sin(x^2) \, dx$. Use the information in Example 3 to compute S_{200}.

14. Let $I = \int_5^7 \cos x \, dx$.

   (a) Write out the approximating sum S_4.

   (b) Compute the approximation error made by S_4.

Let $I = \int_1^5 f(x) \, dx$. In Exercises 15–18, give an example (graphical or symbolic) of a function f such that the given inequality is true for all $n \ge 1$. In each case, identify the property (or properties) of f that guarantee that the inequality is true.

15. $L_n > I$

16. $R_n < I$

17. $T_n < I$

18. $M_n > I$

19. **(a)** Let $I = \int_a^b f(x) \, dx$ and suppose that f is positive and concave up on the interval $[a, b]$. Use words and pictures to explain why T_n overestimates I for all $n \ge 1$.

   (b) Let I and f be as in part (a), but suppose that f is negative on some or all of the interval $[a, b]$. Does T_n still overestimate I for all $n \ge 1$? Justify your answer.

20. Let $I = \int_{-2}^5 f(x) \, dx$, where f is an increasing function over the interval $[-2, 5]$. Furthermore, suppose that $L_{10} = 9.4132$ and $R_{10} = 9.5768$.

   (a) Explain why $|I - L_{10}| \le R_{10} - L_{10}$.

   (b) Evaluate T_{10}.

   (c) Explain why $|I - T_{10}| \le 0.0818$. [HINT: $0.0818 = (R_{10} - L_{10})/2$.]

21. Let $I = \int_a^b f(x) \, dx$. Show that $T_n = (L_n + R_n)/2$.

22. Suppose that f is monotone on $[a, b]$. Explain why
$$|I - T_n| \leq \frac{|L_n - R_n|}{2}.$$

23. Suppose that f does not have any inflection points in $[a, b]$. Explain why $|I - S_{2n}| \leq |M_n - T_n|$.

24. Let $I = \int_a^b f(x)\,dx$ and suppose that f is decreasing on $[a, b]$. Rank L_n, R_n, and T_n in increasing order.

25. Let $I = \int_a^b f(x)\,dx$ and suppose that f is increasing on $[a, b]$. Explain why $L_n \leq M_n \leq R_n$ for all $n \geq 1$.

26. Let $I = \int_a^b f(x)\,dx$, where f is the function shown below. Rank the values I, L_{30}, R_{30}, T_{30}, and M_{30} in increasing order. Justify your answer.

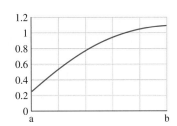

27. Suppose that f is positive, increasing, and concave down on the interval $[1, 7]$. Rank the following estimates of $\int_1^7 f(x)\,dx$ in increasing order: L_{100}, M_{100}, R_{100}, S_{200}, T_{100}.

28. Let f be the function shown below. Estimates of $\int_a^b f(x)\,dx$ computed using the left, right, midpoint, and trapezoid rules (all with the same number of equal subintervals) are 8.52974, 9.71090, 9.74890, and 11.04407. Which rule produced which estimate? Justify your answer.

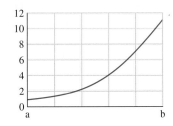

Let $I = \int_a^b f(x)\,dx$, where f is positive and increasing over the interval $[a, b]$. In Exercises 29–34, indicate whether, for all $n \geq 1$, the statement must be true, cannot be true, or may be true. Justify your answers.

29. $L_n < I$ 30. $M_n < I$

31. $L_n < R_n$ 32. $L_n < M_n$

33. $L_n < T_n$ 34. $T_n < M_n$

Let $I = \int_a^b f(x)\,dx$, where f is positive and concave up over the interval $[a, b]$. In Exercises 35–40, indicate whether, for all $n \geq 1$, the statement must be true, cannot be true, or may be true. Justify your answers.

35. $R_n \leq I$ 36. $M_n \leq I$

37. $L_n \leq R_n$ 38. $R_n \leq M_n$

39. $R_n \leq T_n$ 40. $T_n \leq M_n$

Let $I = \int_a^b f(x)\,dx$, where f is positive, decreasing, and concave down over the interval $[a, b]$. In Exercises 41–46, indicate whether, for all $n \geq 1$, the statement must be true, cannot be true, or may be true. Justify your answers.

41. $R_n \leq I$ 42. $T_n \leq I$

43. $L_n \leq R_n$ 44. $L_n \leq T_n$

45. $R_n \leq M_n$ 46. $T_n \leq M_n$

Let $I = \int_0^5 f(x)\,dx$, and suppose that the following inequalities are true if $0 \leq x \leq 5$:

    (i) $f(x) \geq 0$;

    (ii) $1 \leq f'(x) \leq 3$;

    (iii) $2 \leq f''(x) \leq 7$.

In Exercises 47–50, indicate whether the statement must be true, may be true, or cannot be true. Justify your answers.

47. $I - L_{10} > 0$ 48. $I - T_{10} > 0.00005$

49. $M_{10} < R_{10}$ 50. $S_{20} < M_{10}$

FURTHER EXERCISES

51. Explain how the "divide and conquer" strategy can be used with midpoint and trapezoid approximating sums to trap $I = \int_0^2 \sin(x^2)\,dx$.

52. Let $I = \int_0^4 xe^{-x}\,dx$.

    (a) Explain how to use left and right approximating sums to bound the value of I.

    (b) Explain how to use midpoint and trapezoid approximating sums to bound the value of I.

53. Suppose that f is a linear function and that $I = \int_a^b f(x)\,dx$.

    (a) Explain why $T_n = I$ for every $n \geq 1$.

    (b) Use words and pictures to show that the area of any trapezoid with parallel vertical sides is its base times its height at the midpoint.

    (c) Use part (a) to show that $M_n = T_n$ for every $n \geq 1$.

54. Show that $R_n = L_n + \big(f(b) - f(a)\big) \cdot \Delta x$, where $\Delta x = (b - a)/n$.

55. Show that $T_n = L_n + \frac{1}{2}\big(f(b) - f(a)\big) \cdot \Delta x$, where $\Delta x = (b - a)/n$.

56. Let $I = \int_a^b f(x)\,dx$, where f is increasing and concave up on the interval $[a, b]$. Explain why $|I - L_n| \leq |I - R_n|$.

57. Let $I = \int_a^b f(x)\,dx$, where f is increasing and concave up on the interval $[a, b]$.

   (a) Explain why $|I - M_n| \leq |I - L_n|$.

   (b) What can be said about the relationship between $|I - M_n|$ and $|I - R_n|$? Justify your answer.

58. Suppose that $f'(x) > 0$ and $f''(x) < 0$ if $a \leq x \leq b$. Use a picture to show that T_n is a more accurate estimate of $I = \int_a^b f(x)\,dx$ than L_n (i.e., $|I - T_n| < |I - L_n|$).

59. Suppose that $f'(x) > 0$ and $f''(x) > 0$ if $a \leq x \leq b$. Is R_n a more accurate estimate of $I = \int_a^b f(x)\,dx$ than T_n? Justify your answer.

60. Let $I = \int_a^b F(x)\,dx$, where F is an antiderivative of the function f in Exercise 26. Does T_{100} underestimate or overestimate the value of I? Explain.

In Exercises 61–64, $I = \int_a^b f(x)\,dx$ where f' is the function shown below.

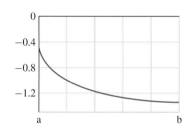

61. Rank the values I, L_n, R_n, M_n, and T_n in increasing order. Justify your answer.

62. Suppose that $f(x) \geq 0$ if $a \leq x \leq b$. Does L_n overestimate $\int_a^b \dfrac{dx}{1 + f(x)}$? Justify your answer.

63. Does R_n overestimate $\int_a^b \sqrt{1 + (f'(x))^2}\,dx$? Justify your answer.

64. Does M_n overestimate $\int_a^b e^{-f(x)}\,dx$? Justify your answer.

In Exercises 65–68, f is an antiderivative of the function shown below.

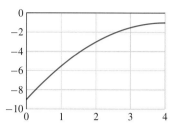

65. Let $I = \int_0^4 f(x)\,dx$. Rank the values of the approximating sums L_n, R_n, M_n, T_n, and S_{2n} in increasing order. Justify your answer.

66. Suppose that $f(0) = 0$ and that $I = \int_0^4 x f(x)\,dx$. Rank the values L_n, R_n, and I in increasing order.

67. Suppose that $f(0) = 0$ and that $I = \int_0^4 (f(x))^2\,dx$. Rank the values L_n, R_n, and I in increasing order.

68. Suppose that $I = \int_0^4 e^{f(x)}\,dx$. Rank the values M_n, T_n, and I in increasing order.

Let $I = \int_a^b f(x)\,dx$ and suppose that f'' is the function shown below. In Exercises 69–74, indicate whether the inequality must *be true,* cannot *be true, or* may *be true. Justify your answers.*

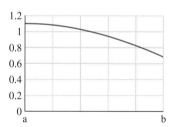

69. $I \leq M_{100}$ **70.** $I \leq L_{200}$

71. $L_{100} \leq R_{100}$ **72.** $T_{200} \leq M_{200}$

73. $M_{50} \leq L_{50}$ **74.** $M_{100} \leq S_{200}$

6.2 ERROR BOUNDS FOR APPROXIMATING SUMS

Any estimate to an unknown quantity is most useful when it comes with an **error bound**—an ironclad guarantee that the estimate is in error by no more than some predetermined, computable amount. We develop several such guarantees in this section.

In Section 6.1 we described one approach to accuracy guarantees: "trapping" the desired integral between a known *underestimate* and a known *overestimate*. The trapping method is simple, but it works only for rather special integrals—those whose integrands are either monotone or don't change their direction of concavity on the interval of integration. Unfortunately, many useful integrals fail both these tests.

EXAMPLE 1   Estimate $I = \int_0^3 \sin(x^2)\,dx$.

Solution   As we said in Section 6.1, the integrand $\sin(x^2)$ has no elementary antiderivative formula, and so the fundamental theorem of calculus doesn't help us find I. The trapping methods of Section 6.1 don't apply either because the integrand is neither monotone nor of constant concavity on $[0, 3]$. Still, it's easy to calculate approximating sums using technology. Here are several (rounded to five decimals):

$$L_{50} = 0.75956; \quad R_{50} = 0.78428; \quad T_{50} = 0.77192; \quad M_{50} = 0.77439.$$

These nice numbers all lie in the same vicinity, and so it seems safe to guess that I lies *somewhere* nearby. But how good is each of these estimates? How much better accuracy could we expect from, say, doubling the number of subdivisions? The error bound formulas of this section let us answer such questions. For this integral I, for instance, these formulas produce the following **error bounds**: ◄

We'll give more details later; the general form of error bounds is the main point here.

$$|I - L_{50}| \leq 0.54 \quad \text{and} \quad |I - M_{50}| \leq 0.02.$$

We can rewrite these results as double inequalities if we prefer. For the first, we have

$$0.21956 = L_{50} - 0.54 \leq I \leq L_{50} + 0.54 = 1.29956.$$

That's a large margin of error; we'll do better with the trapezoid and midpoint rules.

This means that I, whatever its *exact* value may be, lies somewhere between 0.21956 and 1.29956. ◄

These inequalities illustrate several points:

- **Which does better?**   The second inequality is much "better" than the first—in practice, midpoint sums usually commit less error than left sums.

- **Ironclad promises**   Error bounds are *guarantees*, not just guesses. The second inequality, for example, means that I *must* lie within 0.02 of M_{50}, that is, somewhere in the interval $[0.75439, 0.79439]$.

- **Pessimistic views**   Error bounds represent worst-case scenarios; in practice, approximating sums often commit less error than their pessimistic error bounds predict.   ■

Sources of error

Different types of approximating sums commit errors for different reasons. Understanding the reasons is the key to controlling errors.

Left- and right-rule errors, constant functions, and first derivatives   Left and right sums are conceptually almost identical; left- and right-rule errors have the same source. ◄ The left and right rules "pretend," in effect, that f is *constant* on each subinterval. In particular:

For simplicity, we'll concentrate mainly on left sums, but all the important ideas apply to right sums too.

If f is constant on $[a, b]$, then the left and right rules commit zero *error in estimating $I = \int_a^b f$.*

The first derivative tells whether this situation applies: f is *constant* on an interval if and only if $f' = 0$ on that interval.

Thus, constant functions represent the best possible case for left and right sums. But most functions of interest are not constant; the more f deviates from being constant, the worse the left and right rules behave. In graphical terms, the steeper the graph of f, the more error L_n and R_n commit. (The same principle holds whether f rises or falls; the only difference is whether L_n underestimates or overestimates.)

The quantity $|f'(x)|$ measures how steeply the f-graph rises or falls at x. If $|f'|$ is large on $[a, b]$, then f is far from constant, and so we expect L_n and R_n to produce poor approximations. The following example illustrates how $|f'|$ affects left-rule errors for three linear functions.

EXAMPLE 2 How well does L_4 approximate each of the three integrals $\int_0^1 x \, dx$, $\int_0^1 2x \, dx$, and $\int_0^1 (4 - 4x) \, dx$? How are the results linked to first derivatives?

Solution Figure 1 shows how well or poorly L_4 works for each integral. Shaded areas represent L_4 *errors*.

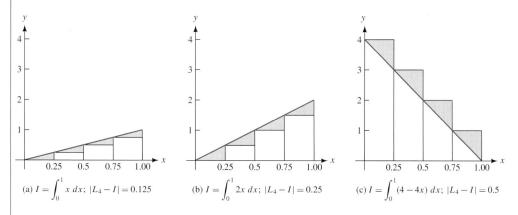

(a) $I = \displaystyle\int_0^1 x \, dx$; $|L_4 - I| = 0.125$ (b) $I = \displaystyle\int_0^1 2x \, dx$; $|L_4 - I| = 0.25$ (c) $I = \displaystyle\int_0^1 (4 - 4x) \, dx$; $|L_4 - I| = 0.5$

FIGURE 1

Comparing left-rule errors

Exact and approximate values of all the integrals are easily found using the FTC, technology, or hand calculation. The following table shows the results.

L_4 errors: how steepness matters			
Integral	$\int_0^1 x \, dx$	$\int_0^1 2x \, dx$	$\int_0^1 (4 - 4x) \, dx$
Exact value	0.500	1.000	2.000
L_4	0.375	0.750	2.500
$I - L_4$	0.125	0.250	−0.500
$|I - L_4|$	0.125	0.250	−0.500

Figure 1 shows, and the table confirms, that approximation errors increase in magnitude as graphs get steeper. Derivatives tell the same story: For the three given integrands, $|f'| = 1$, $|f'| = 2$, and $|f'| = |-4| = 4$, respectively. We won't be surprised, therefore, to see the first derivative f' appear in error bound formulas for left and right sums. ∎

Trapezoid- and right-rule errors, linear functions, and second derivatives   Midpoint and trapezoid sums usually do a better job than left and right sums in approximating an integral $\int_a^b f(x)\,dx$ for a simple but important reason:

> *If f is a linear function, then the trapezoid and midpoint rules commit zero error in estimating $I = \int_a^b f$.*

(This is geometrically obvious for trapezoid sums; it's not hard to see for midpoint sums as well.) This means that midpoint- and trapezoid-rule errors result not from the steepness of the f-graph but from its *concavity*:

> *The more concave (upward or downward) the graph of f, the more error M_n and T_n commit.*

If f is linear, then $f' = k$ (a constant) and $f'' = 0$. In general, the magnitude of the second derivative, $|f''|$, measures the concavity or "bulginess" of the f-graph, which in turn generates trapezoid and midpoint errors. It's no surprise, therefore, that the *second* derivative appears in the error bound theorem for trapezoid and midpoint sums.

EXAMPLE 3   Let $f(x) = 1 - x^4$, and consider $I = \int_0^1 f(x)\,dx$. How is the error committed by T_4 related to the concavity of f and to f''?

Solution   Figure 2 shows I (the area under the curve), T_4 (the shaded area), and the error committed (the difference).

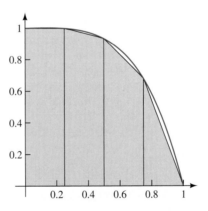

FIGURE 2
Trapezoid rule errors and concavity:
Approximating $\int_0^1 (1 - x^4)\,dx$

The figure shows that the errors committed over successive subdivisions *increase* as we move from left to right. This happens because $|f''(x)| = 4x^3$ also increases as we move from left to right along the interval of integration. ∎

Error bound theorems

With the examples as background, we're ready for more general results. The following theorem formally presents all of the promised error bounds.

> **THEOREM 3 (Error bounds for approximating sums)** Let $I = \int_a^b f(x)\,dx$, and let L_n, R_n, T_n, and M_n denote left, right, trapezoid, and midpoint sums for I, each with n equal subdivisions.
>
> - **Left- and right-rule errors** Let K_1 be a constant such that $|f'(x)| \le K_1$ for all x in $[a, b]$. Then
>
> $$|I - L_n| \le \frac{K_1(b-a)^2}{2n}; \qquad |I - R_n| \le \frac{K_1(b-a)^2}{2n}.$$
>
> - **Midpoint- and trapezoid-rule errors** Let K_2 be a constant such that $|f''(x)| \le K_2$ for all x in $[a, b]$. Then
>
> $$|I - T_n| \le \frac{K_2(b-a)^3}{12n^2}; \qquad |I - M_n| \le \frac{K_2(b-a)^3}{24n^2}.$$

Some of the ingredients of the error bounds—the numbers 2, 12, and 24, the different powers of $(b-a)$ and n, the constants K_1 and K_2—may look mysterious. We'll explain them all soon. For the moment, we focus on the general form of the error bounds.

- **Finding K_1 and K_2: estimate** The constants K_1 and K_2 can be *any* upper bounds for $|f'|$ and $|f''|$ on $[a, b]$. Finding such bounds can be difficult, and so we usually resort to ballpark estimates, often from graphs. (We illustrate the process in examples to follow.) Doing so may seem suspect, but it's perfectly legal. *Over*estimating K_1 and K_2 is harmless—the error bounds "work" even if K_1 and K_2 are larger than necessary. ➡ The only penalty for such sloppiness is slightly weaker error bounds.

  But not if K_1 and K_2 are too small!

- **Inequalities** All of the formulas are inequalities. The actual errors (on the left sides) are often less than the right sides predict. ➡

  But never more!

- **The size, not the sign, of the error** Each inequality bounds the *absolute* error committed by a rule; the theorem doesn't distinguish overestimates from underestimates. (That distinction is sometimes clear from pictures or other information.)

- **Dependence on n** In every formula, n appears in the denominator; naturally enough, increasing n leads to decreasing error bounds. Notice, too, that for any *particular* integral I, the right side of each error bound formula depends *only* on n—all other symbols are numerical constants. In other words, the inequalities have the form

$$|I - L_n| \le \frac{C_1}{n}; \quad |I - R_n| \le \frac{C_1}{n}; \quad |I - T_n| \le \frac{C_2}{n^2}; \quad |I - M_n| \le \frac{C_3}{n^2},$$

where C_1, C_2, and C_3 are numerical constants. The last two denominators involve n^2, not n; this big advantage reflects the better performance of the trapezoid and midpoint rules.

Using error bounds

We illustrate by example what the theorem says and why it's useful. Throughout, we round numerical results to five decimals.

EXAMPLE 4 **(A linear integrand)** How much error do L_4 and L_8 commit in estimating $I = \int_0^1 10x\, dx$? What does the error bound theorem predict?

Solution The FTC gives $I = 5$ (exactly!); a calculator or computer gives

$$L_4 = 3.75 \quad \text{and} \quad L_8 = 4.375.$$

Thus, the *actual* errors committed are

$$|I - L_4| = |5 - 3.75| = 1.25; \qquad |I - L_8| = |5 - 4.375| = 0.625.$$

What does Theorem 3 say? Because $f(x) = 10x$, we have $f' = 10$, and so we can take $K_1 = 10$. Now the theorem says

$$|I - L_4| \leq \frac{10 \cdot 1^2}{2 \cdot 4} = 1.25; \qquad |I - L_8| \leq \frac{10 \cdot 1^2}{2 \cdot 8} = 0.625.$$

In short, our worst fears are realized: L_4 and L_8 behave as badly as the theorem allows. Figure 3 illustrates this sad state of affairs.

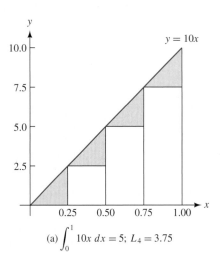

$$(a) \int_0^1 10x\, dx = 5; \ L_4 = 3.75$$

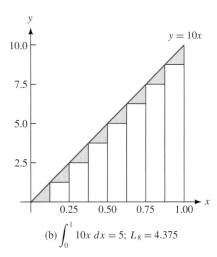

$$(b) \int_0^1 10x\, dx = 5; \ L_8 = 4.375$$

FIGURE 3
Errors committed by L_4 and L_8

EXAMPLE 5 **(A quadratic integrand)** How much error do L_4 and L_8 commit in estimating $I = \int_0^1 x^2\, dx$. What do the error bounds predict?

Solution Easy calculations give $I = 1/3$, $L_4 = 0.21875$, and $L_8 = 0.27344$, so the errors committed are

$$|I - L_4| \approx 0.11458; \qquad |I - L_8| \approx 0.05990.$$

The first error is about twice the second—doubling n roughly halved the error committed.

To decide what Theorem 3 says here we first need a value for K_1. Because $f'(x) = 2x$, the inequality

$$|f'(x)| = |2x| \leq 2$$

holds for all x in $[0, 1]$, the interval of integration. Thus, we can use $K_1 = 2$, and

Theorem 3 gives

$$|I - L_4| \leq \frac{2 \cdot 1^2}{2 \cdot 4} = 0.25; \qquad |I - L_8| \leq \frac{2 \cdot 1^2}{2 \cdot 8} = 0.125.$$

We're OK—the actual errors committed are safely less than the theorem's guarantees. Notice, too, that doubling n cuts the error bound in half—just as we saw for the actual errors. ➡ ■

Not all integrals behave this predictably.

Getting real   Examples 4 and 5 are artificial in the sense that we could easily find *exact* values for the integrals. The following example is more realistic.

EXAMPLE 6   Use M_{10}, T_{10}, and L_{10} to approximate $I = \int_0^1 \sin(x^2)\,dx$. What do the error bound theorems predict?

Solution   Here are the numbers:

$$M_{10} = 0.30982; \qquad T_{10} = 0.31117; \qquad L_{10} = 0.26910.$$

To find values for K_1 and K_2, we find first and second derivatives and plot the results. A little symbolic work with $f(x) = \sin(x^2)$ gives

$$f'(x) = 2x\cos(x^2) \quad \text{and} \quad f''(x) = -4x^2\sin(x^2) + 2\cos(x^2).$$

It's challenging with symbolic methods to maximize $|f'(x)|$ and $|f''(x)|$ for x in $[0, 1]$, but it's easy (with technology) to *plot* f' and f''. The results are shown in Figure 4.

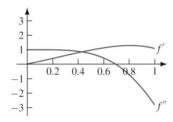

FIGURE 4
Estimating K_1 and K_2 graphically

The graphs suggest reasonable values: $K_1 = 1.5$ and $K_2 = 3$ will do. (Because we're bounding $|f'|$ and $|f''|$, negative values of the functions must be considered.)
  Only arithmetic remains:

$$|I - M_{10}| \leq \frac{K_2(b-a)^3}{24n^2} = \frac{3.0 \cdot 1^3}{2400} = 0.00125;$$

$$|I - T_{10}| \leq \frac{K_2(b-a)^3}{12n^2} = \frac{3.0 \cdot 1^3}{1200} = 0.00250;$$

$$|I - L_{10}| \leq \frac{K_1(b-a)^2}{2n} = \frac{1.5 \cdot 1^2}{20} = 0.075.$$

The errors committed by M_n, T_n, and L_n don't exceed 0.00125, 0.00250, and 0.075, respectively. ■

Choosing n in advance   How many subdivisions does a given method need to approximate an integral with specified accuracy? The error-bound theorem lets us choose n in advance.

EXAMPLE 7   How large an n does each method require to approximate $I = \int_0^1 \sin(x^2)\, dx$ with error guaranteed to be less than 0.0001? (We used the same integral in Example 6, and so the same values of K_1 and K_2 apply.)

Solution   We set each error bound less than or equal to 0.0001 and solve for n:

$$\text{For } M_n: \qquad \frac{K_2(b-a)^3}{24n^2} = \frac{3}{24n^2} \le 0.0001 \iff n \ge 35.35534;$$

$$\text{For } T_n: \qquad \frac{K_2(b-a)^3}{12n^2} = \frac{3}{12n^2} \le 0.0001 \iff n \ge 50;$$

$$\text{For } L_n: \qquad \frac{K_1(b-a)^2}{2n} = \frac{1.5}{2} \le 0.0001 \iff n \ge 7500.$$

Thus, to *guarantee* accuracy within 0.0001, the midpoint, trapezoid, and left rules require at least 36, 50, and 7500 subdivisions, respectively. Notice the striking difference in efficiency: The left rule requires almost 250 times as much work as the midpoint rule to guarantee the same accuracy! ∎

Why the theorem holds

We won't prove Theorem 3 in detail, but we'll outline the idea of a rigorous proof for the L_n bound and briefly indicate the analogous arguments for the other rules.

Left-rule error bounds   Let $I = \int_a^b f(x)\, dx$, and let K_1 be any upper bound for $|f'|$ on $[a, b]$. The error any approximating sum commits in approximating I is the total of the errors committed over each subinterval. (In the best cases such errors tend to cancel out; in the worst cases, they add up.)

How much error, at worst, can L_n commit over *one* subinterval $[x_{i-1}, x_i]$ of width $\Delta x = (b-a)/n$? Figure 5 illustrates the situation.

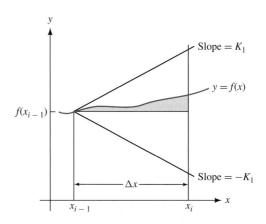

FIGURE 5
Left-rule error over one subdivision

Here's what the figure shows:

- **What the left rule pretends**   The left rule "pretends" that f is constant on $[x_{i-1}, x_i]$. The shaded area represents the error that results from this pretense.

- **What K_1 means graphically**   The derivative inequality $|f'(x)| < K_1$ means that the f-graph over $[x_{i-1}, x_i]$ must stay in the wedge between the lines with slopes $\pm K_1$. (To "escape" this wedge, f would have to increase or decrease faster than K_1 allows.) ➡

- **When the worst happens**   The worst possible error—positive or negative—occurs if the f-graph is either the top line or the bottom line of the wedge. A careful look at the picture shows that these errors correspond to the areas of two right triangles, one above and one below the horizontal line $y = f(x_{i-1})$. Each of these triangles has area

A rigorous proof of this fact requires the mean value theorem.

$$\frac{1}{2} \text{ base} \cdot \text{height} = \frac{1}{2} \Delta x \cdot (K_1 \Delta x) = \frac{1}{2} K_1 \Delta x^2 = \frac{K_1 (b-a)^2}{2n^2}.$$

The last quantity is the greatest possible error L_n can commit over *one* subdivision. Over all n subdivisions, therefore, the total maximum error is no more than n times as much—just what the theorem says.

Trapezoid- and midpoint-rule bounds   Again, let $I = \int_a^b f(x)\,dx$, and suppose that $|f''(x)| \leq K_2$ for all x in $[a, b]$. Among all functions that satisfy this inequality, the quadratic function $q(x) = K_2 x^2 / 2$ represents an extreme because $q''(x) = K_2$ for all x—in other words, $|q''|$ is always as large as the inequality permits.

How much error, at worst, can M_n and T_n commit over one subinterval? We answer by computing directly with the "worst offender" function q. To simplify the algebra, we take $[0, h]$ as our subinterval.

The first step is to compare I_h (the exact integral of q over $[0, h]$) with M_h and T_h (the midpoint and trapezoid estimates over the same interval):

$$I_h = \int_0^h q(x)\,dx = \int_0^h \frac{K_2 x^2}{2}\,dx = \frac{K_2 x^3}{6}\Big]_0^h = \frac{K_2 h^3}{6};$$

$$M_h = q(h/2) \cdot h = \frac{K_2 (h/2)^2}{2} \cdot h = \frac{K_2 h^3}{8};$$

$$T_h = \frac{q(0) + q(h)}{2} \cdot h = \frac{K_2 h^3}{4}.$$

The *errors* committed over $[0, h]$ are therefore

$$I_h - M_h = \frac{K_2 h^3}{6} - \frac{K_2 h^3}{8} = \frac{K_2 h^3}{24};$$

$$I_h - T_h = \frac{K_2 h^3}{6} - \frac{K_2 h^3}{4} = -\frac{K_2 h^3}{12}.$$

Notice that the trapezoid and midpoint errors have *opposite* signs as we've seen occur in examples. The mysterious constants 12 and 24 from Theorem 3 have also finally appeared.

Thus, we've found the worst-case error committed by each method over *one* subinterval. Multiplying the results by n gives the worst-case error over all n subintervals. Replacing h with $(b - a)/n$ gives the error bound formulas of Theorem 3.

BASIC EXERCISES

1. Let $\int_0^3 \sin(x^2)\,dx$.

   (a) Use Theorem 3 to show that $|I - M_{50}| \leq 0.02$ as claimed in Example 1.

   (b) Explain why $0.754 \leq I \leq 0.795$.

2. Explain the concept of an error bound in your own words. Be sure that your explanation discusses why error bounds are "pessimistic" estimates of approximation errors.

3. Consider the three integrands in Example 2. Show that the magnitude of the approximation error committed by L_4 is proportional to $|f'|$ for each integrand.

4. Let $I = \int_1^b f(x)\,dx$, where f is a linear function.

(a) Show that $T_n = I$ for every $n \geq 1$.

(b) Show that $M_n = I$ for every $n \geq 1$.

5. Let $I = \int_0^2 5x\,dx$.

(a) What does Theorem 3 say about the error committed by R_8? (See Example 4.)

(b) What does Theorem 3 say about the error committed by T_8?

6. Let $I = \int_0^4 3x^2\,dx$.

(a) What does Theorem 3 say about the error committed by L_5? (See Example 5.)

(b) What does Theorem 3 say about the error committed by M_5?

7. Let I be as in Example 6.

(a) Explain why 2 is an acceptable value for K_1.

(b) Is 2 an acceptable value for K_2? Justify your answer.

8. Suppose that $-4 \leq f'(x) \leq 3$ if $1 \leq x \leq 2$. Explain why Theorem 3 does *not* guarantee that $|I - L_n| \leq 3/(2n)$.

9. Let $I = \int_1^4 \dfrac{dx}{\sqrt{x}}$.

(a) Use the FTC to evaluate I exactly.

(b) Compute the actual approximation errors $|I - L_3|$ and $|I - R_3|$. Are these errors consistent with the error bound formulas in Theorem 3? Justify your answer.

(c) Compute the actual approximation errors committed by T_3 and M_3. Are these errors consistent with the error bound formulas in Theorem 3? Justify your answer.

(d) Use Theorem 3 to find a value of n that guarantees that $|I - M_n| \leq 0.005$ (i.e., that M_n has two-decimal-place accuracy).

10. Let $I = \int_0^1 x^3\,dx$.

(a) Use the FTC to evaluate I exactly.

(b) Compute the actual approximation errors $|I - L_4|$ and $|I - R_4|$. Are these errors consistent with the error bound formulas in Theorem 3? Justify your answer.

(c) Compute the actual approximation errors committed by T_4 and M_4. Are these errors consistent with the error bound formulas in Theorem 3? Justify your answer.

(d) Use the error bounds in Theorem 3 to find a value of n that guarantees that $|I - M_n| \leq 0.005$ (i.e., that M_n has two-decimal-place accuracy).

In Exercises 11–14, find a value of n for which Theorem 3 guarantees that L_n approximates the value of the integral within ± 0.005. Justify your answers.

11. $\int_0^3 e^{-x^2}\,dx$

12. $\int_0^2 \sin(x^2)\,dx$

13. $\int_0^1 (1+x^2)^{-1}\,dx$

14. $\int_1^{10} \sin(1/x)\,dx$

15–18. For each of the integrals in Exercises 11–14, find a value of n for which Theorem 3 guarantees that M_n approximates the value of the integral within ± 0.005. Justify your answers.

Let $I = \int_0^{10} f(x)\,dx$. In Exercises 19–22, give an example of a function f with the given property.

19. L_{200} underestimates I, and the magnitude of the approximation error made is as large as Theorem 3 allows.

20. L_{200} overestimates I, and the magnitude of the approximation error made is as large as Theorem 3 allows.

21. M_{200} underestimates I, and the magnitude of the approximation error made is as large as Theorem 3 allows.

22. M_{200} overestimates I, and the magnitude of the approximation error made is as large as Theorem 3 allows.

FURTHER EXERCISES

23. (a) According to Theorem 3, how much smaller is the error bound when the number of subdivisions used in a left sum estimate of an integral is doubled?

(b) Repeat part (a) for a midpoint sum estimate.

(c) What do parts (a) and (b) suggest about the practical utility of these two numerical integration methods?

24. (a) Compare the error bound for L_n to the error bound for L_{10n}. Approximately how many additional decimal places of guaranteed accuracy are gained by using 10 times as many subintervals?

(b) Repeat part (a) for M_n and M_{10n}.

(c) What do parts (a) and (b) suggest about the practical utility of these two numerical integration methods?

25. Let $I = \int_a^b f(x)\,dx$. If $|I - L_{100}| \leq 5 \times 10^{-4}$, does Theorem 3 guarantee that $|I - L_{1000}| \leq 5 \times 10^{-5}$? Justify your answer.

26. Let $I = \int_a^b f(x)\,dx$. If $|I - T_{100}| \leq 5 \times 10^{-4}$, does Theorem 3 guarantee that $|I - M_{100}| \leq 2.5 \times 10^{-5}$? Justify your answer.

27. Let $I = \int_1^{11} f(x)\,dx$, where $f(x) = \cos(1/x)$.

(a) What is the smallest value of n for which Theorem 3 guarantees that $|I - L_n| \le 0.005$?

(b) What is the smallest value of n for which Theorem 3 guarantees that $\left| \int_1^6 f(x)\,dx - L_n \right| \le 0.004$?

(c) What is the smallest value of n for which Theorem 3 guarantees that $\left| \int_6^{11} f(x)\,dx - L_n \right| \le 0.001$?

(d) Show that the estimates computed in parts (b) and (c) can be combined to produce an estimate of I that is guaranteed to be within 0.005 of the true value.

(e) The values in parts (a) and (d) are both estimates of I with two-decimal-place accuracy. How does the amount of computational effort necessary to compute the estimate in part (d) compare with that needed to compute the estimate in part (a)? (The number of values of f used is a reasonable measure of computational effort.)

28. Let $I = \int_0^5 f(x)\,dx$, where $f(x) = e^{-x^2}$.

(a) What is the smallest value of n for which Theorem 3 guarantees that $|I - R_n| \le 0.01$?

(b) What is the smallest value of n for which Theorem 3 guarantees that $\left| \int_0^2 f(x)\,dx - R_n \right| \le 0.005$?

(c) What is the smallest value of n for which Theorem 3 guarantees that $\left| \int_2^5 f(x)\,dx - R_n \right| \le 0.005$?

(d) Show that the estimates computed in parts (b) and (c) can be combined to produce an estimate of I that is guaranteed to approximate I within 0.01 of its true value.

(e) The estimates in parts (a) and (d) both estimate I within 0.01. How does the amount of computational effort necessary to compute the estimate in part (d) compare with that needed to compute the estimate in part (a)? (The number of values of f used is a reasonable measure of computational effort.)

29. Let $I = \int_0^2 \sqrt{4 - x^2}\,dx$.

(a) Explain why $I = \pi$. [HINT: Draw a picture.]

(b) Compute $|I - L_{10}|$.

(c) Why doesn't Theorem 3 provide a useful bound on the magnitude of the approximation error?

In Exercises 30–32, f is an antiderivative of the function shown below.

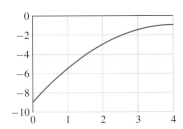

30. Find a value of n that guarantees that $\left| \int_0^4 f(x)\,dx - L_n \right| \le 0.0001$.

31. Suppose that F is an antiderivative of f. Find a value of n such that $\left| \int_1^2 F(x)\,dx - M_n \right| \le 0.01$.

32. Suppose that F is an antiderivative of f and that $f(2) = 0$. Find a value of n such that
$$\left| L_n - \int_2^4 F(x)\,dx \right| \le 0.01.$$

33. Suppose that $f''(x) = \dfrac{e^x \cos x}{1 + x^2}$. Find an integer n such that T_n approximates $\int_0^5 f(x)\,dx$ within 0.001.

34. Let $I = \int_{-1}^2 f(x)\,dx$, where f is a function with the following properties:
(i) f is increasing on the interval $[-1, 2]$;
(ii) f'' is the function shown;

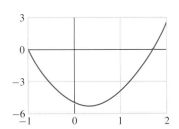

(iii) f has the values shown in the table.

x	-1.00	-0.25	0.50	1.25	2.00
$f(x)$	2.0000	26.522	48.755	68.328	86.790

(a) Compute R_4 and an upper bound on the value of $|I - R_4|$.

(b) Compute T_4 and an upper bound on the value of $|I - T_4|$.

Let $I = \int_{-4}^6 h(x)\,dx = 9$, where h is a function with the following properties:
(i) $-7 \le h(x) \le 3$ if $-4 \le x \le 6$,
(ii) $-5 \le h'(x) - 2$ if $-4 \le x \le 6$, and
(iii) $0 \le h''(x) \le 4$ if $-4 \le x \le 6$.

In Exercises 35–38, indicate whether the statement must *be true,* cannot *be true, or* may *be true. Justify your answers.*

35. $L_{100} = 11.75$

36. $M_{100} = 8.98$

37. $T_{100} = 8.99$

38. $I - T_{100} \le 5 \times 10^{-6}$

39. Let $I = \int_a^b f(x)\,dx$ and suppose that f is monotone on the interval $[a, b]$. Show that $|I - L_n| \le |f(b) - f(a)| \cdot \dfrac{b - a}{n}$.

40. Let $I = \int_a^b f(x)\, dx$ and suppose that f is monotone on the interval $[a, b]$. Show that $|I - M_n| \le |f(b) - f(a)| \cdot \dfrac{b-a}{n}$.

41. Let $I = \int_a^b f(x)\, dx$ and suppose that f is monotone on the interval $[a, b]$. Show that $|I - T_n| \le |f(b) - f(a)| \cdot \dfrac{b-a}{2n}$.

EULER'S METHOD: SOLVING DEs NUMERICALLY

The problem of solving differential equations (DEs) resembles the problem of integrating functions. In each case, symbolic (or "exact") methods are sometimes—but not always—available. When symbolic methods fail, graphical and numerical methods are often the best recourse. For DEs as for integrals, the key idea is to take small steps toward a destination, adding up small contributions as we go.

In Section 4.1 we showed how to solve a first-order initial value problem (IVP)

$$y' = f(t, y), \qquad y(a) = b$$

graphically, using its slope field:

> *Starting at (a, b), move through the slope field, "going with the flow." The result is a solution curve (i.e., the graph of a solution function) for the IVP.*

The numerical version of this idea is known as **Euler's method**. It tells exactly *how* to move step by step through a slope field to approximate a solution curve. We illustrate with another look at an example from Section 4.1.

Time $t = 0$ can denote any convenient reference time.

EXAMPLE 1   Let $y(t)$ denote the temperature (in degrees Fahrenheit) of a cup of coffee at time t (in minutes). ◄ Room temperature is $70°$F; the coffee starts at $190°$F. The coffee's temperature is described by the IVP

$$y' = -0.1(y - 70); \qquad y(0) = 190.$$

How hot is the coffee after 10 minutes? After 20 minutes?

The bulleted starting point represents the initial condition.

Solution   Figure 1 shows the slope field and one plausible solution curve. ◄

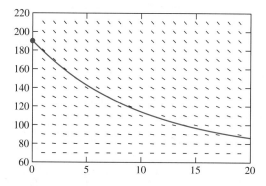

FIGURE 1
**Graphically solving the IVP $y' = -0.1(y - 70)$;
$y(0) = 190$**

The solution curve shows that the temperature at 10 minutes is a little over $110°$F. After 20 minutes, the temperature is about $84°$F.   ∎

Improving on the graphical approach   Drawing a solution curve by "going with the flow" is simple and natural. Any two humans, given the slope field and initial condition shown in Figure 1, would draw similar solution curves and thus arrive at similar estimates for $y(10)$ and $y(20)$. This graphical approach, for all its simplicity, has two main flaws:

(i) it gives approximate answers (as do all graphical methods); and

(ii) "drawing a curve" isn't precisely defined as a mathematical process.

Of these flaws, (ii) is worse than (i). Approximate answers are fully respectable, and often they're the best available. To have any idea how *accurate* estimates are, however, we must first describe precisely, in mathematical terms, how we *produce* the estimates. A precise description is also essential if we want (as we do now) to use technology to produce and improve our estimates.

Let's describe—in terms a computer can understand—one sensible way to move through the slope field of Example 1, starting from $(0, 190)$ and ending at $t = 10$. We'll take 10 *straight-line steps, each of horizontal length 1*. The slope of each linear step is determined by the slope field at the *beginning* of the step. To make a "curve," we join successive linear steps. Figure 2 shows the result, with a dot at each successive **Euler point**.

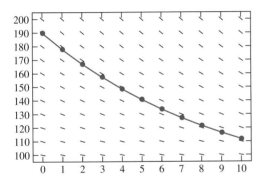

Each step corresponds to a 1-second time interval.
Notice that the "curve" is really 10 straight-line segments.

FIGURE 2
Ten Euler steps through a slope field

Here's how we found the successive Euler points:

- **Step 1**   The first linear step starts at the point $(0, 190)$ and ends one unit to the right, at the point $(1, 178)$. The *slope* of this line segment is given by the DE, as follows: When $t = 0$ and $y = 190$, the DE $y' = -0.1(y - 70)$ says

$$y'(0) = -0.1\big(y(0) - 70\big) = -0.1(190 - 70) = -12.$$

Thus, the segment has slope -12, and our estimate is

$$y(1) \approx y(0) + y'(0) \cdot 1 = 190 - 12 \cdot 1 = 178.$$

- **Step 2**   The second step begins at $(1, 178)$ and moves one unit to the right. As before, its slope is determined by the DE, but with the *updated* y-value. With $t = 1$ and $y = 178$, we get

$$y'(1) = -0.1(178 - 70) = -10.8,$$

and so

$$y(2) \approx y(1) + y'(1) \cdot 1 \approx 178 - 10.8 \cdot 1 = 167.2.$$

Each step of the type just described is an **Euler step** with **step size** 1. It takes 10 Euler steps to reach $t = 10$. In practice, of course, such repetitive calculations are usually exported to machines. The following table (with most entries rounded to two decimals) summarizes our numerical work:

From $t = 0$ to $t = 10$ in ten Euler steps										
Step	**0**	**1**	**2**	**3**	**4**	**5**	**6**	**7**	**8**	**9**	**10**
t	0	1	2	3	4	5	6	7	8	9	10
$y'(t)$	−19.00	−10.80	−9.72	−8.75	−7.87	−7.09	−6.34	−5.74	−5.17	−4.65	−4.18
$y(t)$	190.00	178.00	167.20	157.48	148.73	140.86	133.77	127.40	121.66	116.49	111.84

We estimate, therefore, that the coffee is at about 112°F after 10 minutes.

How good is Euler's method?

Euler's method and the left-rule method are closely related. Both methods find approximate answers, and the errors each method commits come from the same source: The left rule "pretends" that the integrand function is constant over short intervals, while Euler's method "pretends" that the *slope of a solution curve* remains constant over short intervals. Neither of these conditions is likely to hold *exactly*, but in many cases these quantities are "close enough" to being constant that the resulting error stays small.

In some cases, including that of Example 1, we can tell *exactly* how "good" our Euler's method results are. It turns out (as an easy calculation shows) that the function $y(t) = 70 + 120e^{-0.1t}$ is a solution of the IVP $y' = -0.1(y - 70)$; $y(0) = 190$. In particular,

$$y(10) = 70 + 120e^{-1} \approx 114.15.$$

Note that this result is significantly different from what Euler's method predicted (see the preceding table). Plotting both the Euler "curve" and the exact solution curve together (Figure 3) shows how errors in Euler's method accumulate over the time interval.

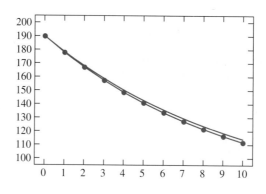

FIGURE 3
Exact vs. Euler's method solutions

The two "curves" aren't terribly different, but they spread apart slightly as time goes on.

Shorter steps, better accuracy We've seen for integrals that the left rule's accuracy improves as the number of subdivisions increases (or as the step size decreases). Euler's method is similar: More (and smaller) steps usually give more accurate results. But at a price—more steps mean more work.

What happens in the coffee situation if we use 100 steps of size 0.1 rather than 10 steps of size 1? Carrying out 100 steps by hand would be tedious, but computers don't complain. Following are some sample results; *exact* values of y appear for comparison in the last row.

From $t = 0$ to $t = 10$ in 100 Euler steps									
step	0	1	2	3	10	20	30	90	100
t	0	0.1	0.2	0.3	1.0	2.0	3.0	9.0	10.0
y'	−12.00	−11.89	−11.76	−11.64	−10.85	−9.82	−8.88	−4.86	−4.39
y_{Euler}	190.00	188.80	187.61	186.43	178.53	168.15	158.76	118.57	113.92
y_{exact}	190.00	188.81	187.62	186.45	178.58	168.25	158.90	118.79	114.15

The Euler estimates are better this time.

We might think of using even smaller step sizes, with thousands or millions of Euler steps. As with numerical integration, however, this strategy is computationally foolish. For one thing, roundoff error eventually becomes significant. For another, more efficient numerical strategies are available. We won't pursue such "smart" strategies except to drop the name of one of the most popular, the **Runge–Kutta method**. This method is related to Simpson's rule in the same way as Euler's method is related to the left rule. ➡

We met Simpson's rule in Section 6.1.

Euler's method in general

Suppose that we're given a first-order DE $y' = f(t, y)$ and an initial condition $y(a) = y_0$ and that we estimate $y(b)$ using Euler's method with n steps on the interval $a \le t \le b$. What exactly does Euler's method give? The answer depends on our point of view.

- **The graphical view:** Euler's method says how to move step by step through the DE's slope field starting at (a, y_0) and ending when $t = b$. Each Euler step has *horizontal* length (or **step size**) $\Delta t = (b - a)/n$. Each step goes in the direction of the slope field at the *beginning point* of the step. Joining successive **Euler points** with line segments (connecting the dots, in other words) produces a piecewise–linear curve. ➡ This curve is the graph of an approximate solution function $Y(t)$ for the DE. (We'll reserve the lowercase $y(t)$ for the exact solution.)

Mathspeak for a "curve" built from linear pieces.

- **The numerical view:** Euler's method produces a list

$$Y(t_0), \ Y(t_1), \ Y(t_2), \ Y(t_3), \ \ldots, \ Y(t_n)$$

of numerical values of an approximate solution function $Y(t)$, for equally spaced inputs

$$a = t_0 < t_1 < t_2 < t_3 < \cdots < t_n = b;$$

each t_i is Δt units greater than its predecessor. Plotting these values and joining successive points with line segments produces the graph just mentioned.

No formulas Euler's method uses—and produces—only *numerical* information. Thus, we can use Euler's method to produce a graph or a table of values but not a *formula* for a solution function. (The same principle holds if the left rule is applied to an integral $\int_a^b f(x)\,dx$.)

The numerical nature of Euler's method is a mixed bag. To its credit, Euler's method works with almost *any* DE, no matter how complicated. On the debit side, Euler's method produces *only* approximate results—even for simple DEs that can easily be solved exactly.

A general recipe for Euler's method Here is a general description—rather than an example—of how Euler's method works.

- **Step 1:** The starting point $Y(t_0) = y_0$ comes from the initial condition. The DE $y' = f(t, y)$ says that a solution curve through (t_0, y_0) has slope $m_0 = f(t_0, y_0)$. Thus, the first Euler step has slope m_0, and its rise (or fall) is $m_0 \cdot \Delta t$ vertical units. That is,

$$Y(t_1) = Y(t_0) + m_0 \Delta t,$$

where $t_1 = t_0 + \Delta t$.

- **Step 2:** The second Euler step starts at $\big(t_1, Y(t_1)\big)$, where the first step ended. At $\big(t_1, Y(t_1)\big)$, says the DE, the solution curve has slope $m_1 = f\big(t_1, Y(t_1)\big)$. Thus, the second Euler step rises (or falls) $m_1 \cdot \Delta t$ vertical units, and so

$$Y(t_2) = Y(t_1) + m_1 \Delta t.$$

- **Continue:** The pattern should now be clear. At each Euler step we do two things: (1) use the DE to "update" the slope m_i; and (2) move with slope m_i from $\big(t_i, Y(t_i)\big)$ to $\big(t_{i+1}, Y(t_{i+1})\big)$. After n steps, we arrive at $t_n = b$.

Reduced entirely to symbols, Euler's recipe for assembling an approximate solution $Y(t)$ looks like this:

$$
\begin{aligned}
Y(t_0) &= y_0 \\
Y(t_1) &= Y(t_0) + f\big(t_0, Y(t_0)\big) \cdot \Delta t \\
Y(t_2) &= Y(t_1) + f\big(t_1, Y(t_1)\big) \cdot \Delta t \\
&\ \ \vdots \\
Y(t_{i+1}) &= Y(t_i) + f\big(t_i, Y(t_i)\big) \cdot \Delta t \\
&\ \ \vdots \\
Y(t_n) &= Y(t_{n-1}) + f\big(t_{n-1}, Y(t_{n-1})\big) \cdot \Delta t
\end{aligned}
$$

Simple but powerful. It is striking that Euler's method—the idea of which could hardly be simpler—should be named after one of history's greatest mathematicians. One lesson may be the surprising power of simple ideas. Another lesson may be that powerful ideas, properly understood, *become* simple.

Leonhard Euler (1707–1783), born in Switzerland, is among the most influential and prolific mathematicians in history. His contributions to calculus are especially notable: Combining key ideas of Newton and Leibniz, Euler established the field of mathematical analysis. To gauge Euler's importance to our field, consider just a few of the now familiar symbols and notations he introduced: $f(x)$, π, e, i, $\sum$, and Δx.

Euler's method in action: modeling logistic growth

The same flies and more discussion of logistic growth appear in an Interlude in Chapter 4, but what follows is self-contained.

To show Euler's method in action we apply it to model a population of fruit flies living in a controlled environment. ◆ We assume that the fly population grows **logistically**. More precisely:

> *The population's growth rate is proportional both to the population itself and to the difference between the environment's carrying capacity and the population.*

This condition is nicely expressed symbolically as a DE:

$$P' = kP(C - P),$$

where P represents the population, P' the population's growth rate, and C the carrying capacity of the environment; k is a constant of proportionality. ➜

P and P' vary with time; k and C are constants.

EXAMPLE 2 Through empirical measurements, researchers determine that their captive fly population, initially 1000 members strong, satisfies the logistic IVP

$$P' = 0.00000556\,P(10000 - P); \qquad P(0) = 1000.$$

(Here t measures time in days since the original measurement, $P = P(t)$ is the population at time t, and $P' = P'(t)$ is the growth rate in flies per day at time t.)

What will the population be in 10 days? How will the population evolve in the long run?

Solution To estimate $P(10)$, we take 10 (1-day) Euler steps. The numbers are clumsy, so we get technological help.

- **Step 1:** When $t = 0$, $P = 1000$, and so

$$P'(0) = 0.00000556 \cdot 1000 \cdot 9000 \approx 50.04 \text{ flies per day.}$$

After 1 day, therefore, we estimate a population of
$P(1) \approx 1000 + 50.04 = 1050.04$ flies.

- **Step 2:** When $t = 1$, $P \approx 1050.04$, and so

$$P'(1) \approx 0.00000556 \cdot 1050.04 \cdot (10000 - 1050.04) \approx 52.25 \text{ flies per day.}$$

Therefore, we estimate

$$P(2) \approx P(1) + P'(1) \cdot 1 \approx 1050.04 + 52.25 = 1102.29 \text{ flies.}$$

The idea should now be clear, so we'll let a computer do the rest. Here are the results with most numbers rounded to two decimal places:

Flies: From $t = 0$ to $t = 10$ in 10 Euler steps						
step	0	1	2	3	4	5
t	0	1	2	3	4	5
$P'(t)$	50.04	52.25	54.53	56.88	59.29	61.77
$P(t)$	1000.00	1050.04	1102.29	1156.82	1213.70	1272.99

step	6	7	8	9	10
t	6	7	8	9	10
$P'(t)$	64.31	66.91	69.56	72.27	75.02
$P(t)$	1334.76	1399.07	1465.97	1535.53	1607.80

Beware... The tabulated estimate $P(10) \approx 1607.80$ should be viewed cautiously:

- **Whole flies** The decimal numbers look impressive, but flies don't come in fractions. Our best guess in context is 1608 flies on day 10. (Such estimation comes with the territory. Real populations don't adhere *precisely* to mathematical models.)

- **Too big or too small?** The tabulated figure 1607.80 for day 10 *underestimates* the number of flies that would exist if the population behaved *exactly* as the DE predicts. The reason is related to step size: Taking 1-day Euler steps amounts to pretending that the growth rate remains constant over 1-day periods. In reality, the growth rate *increases* along with the population. (With symbolic techniques [coming in Chapter 7] it can be shown that an *exact* solution to the DE satisfies $P(10) = 1638.09$.)

• **Smaller steps**   Using a smaller step size would update the growth rate more often than daily and so might better estimate the true population on day 10. With 100 Euler steps, each 0.1 days long, the computer gives

$$P(0.1) \approx 1005.00, \quad P(0.2) \approx 1010.03, \quad \ldots \quad P(10) \approx 1621.43.$$

The last estimate is *still* below what the DE model predicts, but it's better than before.

In the long run   With machine help, it's easy to use Euler's method to project the fly population far into the future—if one remembers, of course, that errors accumulate over time. Figure 4 shows our results for a 150-day period using a step size of 0.1 day. ◄

The graph looks smooth because it's made from 1500 dots.

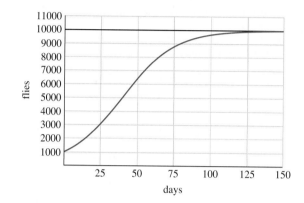

FIGURE 4
Flies in the long run

The graph's general shape looks reasonable. The population rises slowly at first, speeds up as the population increases, and then tapers off as the population nears the environment's carrying capacity. ∎

BASIC EXERCISES

1. Consider the IVP in Example 1. Show how the estimate $y(3) \approx 157.48$ shown in the table on page 396 was computed.

2. Consider the IVP in Example 1. Show how the estimate $y(8) \approx 121.66$ shown in the table on page 396 was computed.

3. Show that $y(t) = 70 + 120e^{-0.1t}$ is the exact solution of the IVP in Example 1.

4. Consider the IVP in Example 1.
  (a) Use the DE to show that $y''(t) > 0$ for all $t > 0$. [HINT: $y'' = -0.1y'$.]
  (b) Use part (a) to explain why Euler's method underestimates the exact solution at each step.

5. Consider the DE in Example 2 but suppose that $P(0) = 6000$.
  (a) Use the DE to show that $P(t)$ is concave down for all $t \geq 0$.
  (b) Will Euler's method underestimate $P(100)$? Justify your answer.

6. Consider the DE in Example 2 but suppose that $P(0) = 0$.
  (a) Carry out three Euler steps by hand; use step size 1 day.
  (b) Estimate the population on day 200. Explain what's going on in biological terms.

7. Consider the DE $y' = f(t, y)$. Suppose that the slope of the line segment joining the successive Euler points $(t_i, Y(t_i))$ and $(t_{i+1}, Y(t_{i+1}))$ is m_i.
  (a) What, if anything, does this imply about $f(t_i, Y(t_i))$?
  (b) What, if anything, does this imply about $f(t_{i+1}, Y(t_{i+1}))$?

8. Consider the IVP $y' = 3$, $y(0) = 2$.
  (a) Find the exact solution $y(t)$ to this IVP. [HINT: What functions have constant derivatives?]
  (b) Use Euler's method, with step size 1, to estimate $y(1)$, $y(2)$, $y(3)$, $y(4)$, and $y(5)$. How accurate are these estimates?
  (c) What accounts for the accuracy of Euler's method in this case?

FURTHER EXERCISES

9. The number e is sometimes defined as $y(1)$, where $y(t)$ is the solution to the IVP $y' = y$, $y(0) = 1$.

  (a) Use Euler's method with one subdivision to estimate e. Why is the answer an underestimate?

  (b) Repeat part (a) with four subdivisions.

  (c) Imagine that Euler's method with 10,000 steps is used to estimate $y(1)$. Will the result underestimate e? Justify your answer.

  (d) Show that Euler's method with n steps produces the estimate $y(1) \approx \left(1 + \dfrac{1}{n}\right)^n$.

10. Consider the IVP $y' = 1 + t - y$, $y(0) = 0$.

  (a) Use Euler's method with five steps of size 0.4 to estimate $y(2)$.

  (b) Use Euler's method with 10 steps of size 0.2 to estimate $y(2)$.

  (c) Show that the exact solution of the IVP is $y(t) = t$.

  (d) How does the exact value of $y(2)$ compare with the estimates you computed in parts (a) and (b)? Explain how you could have predicted this from the slope field for the DE.

11. Consider the IVP $y' = 1 + t - y$, $y(0) = 1$.

  (a) Use Euler's method with five steps of size 0.4 to estimate $y(2)$.

  (b) Use Euler's method with 10 steps of size 0.2 to estimate $y(2)$.

  (c) Show that the exact solution of the IVP is $y(t) = t + e^{-t}$.

  (d) How does the exact value of $y(2)$ compare with the estimates you computed in parts (a) and (b)? Explain how you could have predicted this from the slope field for the DE.

12. Consider the IVP $y' = 1 + t - y$, $y(0) = -1$.

  (a) Use Euler's method with five steps of size 0.4 to estimate $y(2)$.

  (b) Use Euler's method with 10 steps of size 0.2 to estimate $y(2)$.

  (c) Show that the exact solution of the IVP is $y(t) = t - e^{-t}$.

  (d) How does the exact value of $y(2)$ compare with the estimates you computed in parts (a) and (b)? Explain how you could have predicted this from the slope field for the DE.

13. Consider the IVP $y' = y - 2$, $y(0) = 1$.

  (a) Use Euler's method with 10 steps of size 0.1 to estimate $y(1)$.

  (b) Use Euler's method with 20 steps of size 0.05 to estimate $y(1)$.

  (c) Show that the exact solution of the IVP is $y(t) = 2 - e^t$.

  (d) How does the exact value of $y(1)$ compare with the estimates you computed in parts (a) and (b)? Explain how

you could have predicted this from the slope field for the DE.

14. Consider the IVP $y' = xy$, $y(0) = 1$.

  (a) Use Euler's method with one step of size 1 to estimate $y(1)$.

  (b) Use Euler's method with two steps of size 0.5 to estimate $y(1)$.

  (c) Use Euler's method with four steps of size 0.25 to estimate $y(1)$.

  (d) Use Euler's method with eight steps of size 0.125 to estimate $y(1)$.

  (e) Show that the exact solution of the IVP is $y(x) = e^{x^2/2}$.

  (f) Show that the results in parts (a)–(d) suggest that the error made by Euler's method is proportional to $1/n$, where n is the number of steps.

15. Consider the IVP $y' = y^2$, $y(0) = 1$.

  (a) Use Euler's method with four steps of size 0.2 to estimate $y(0.8)$.

  (b) Show that the exact solution of the IVP is $y(t) = (1 - t)^{-1}$.

  (c) Explain why Euler's method doesn't provide a good estimate of $y(0.8)$.

16. Consider the IVP $y' = \sin t$, $y(0) = 0$.

  (a) Show that $y(t) = 1 - \cos t$ is the exact solution of this IVP.

  (b) Carry out Euler's method with four steps to estimate $y(1)$. Do all steps by hand.

  (c) Let $y(t)$ be the *exact* solution function, and let $Y(t)$ be the *approximate* solution function constructed by Euler's method. Plot $y(t)$ and $Y(t)$ on the same axes for $0 \le t \le 1$.

  (d) Explain why $y(t) - Y(t) \ge 0$ for any n.

17. Consider the IVP $y' = e^{-t}$, $y(0) = 0$.

  (a) Show that $y(t) = 1 - e^{-t}$ is the exact solution of this IVP.

  (b) Estimate $y(1)$ by carrying out Euler's method with five steps. Do all steps by hand.

  (c) Let $y(t)$ be the *exact* solution function, and let $Y(t)$ be the *approximate* solution function constructed by Euler's method. Plot $y(t)$ and $Y(t)$ on the same axes for $0 \le t \le 1$.

  (d) Explain why $y(t) - Y(t) \le 0$ for any n.

18. Consider the IVP $y' = f(t)$, $y(t_0) = 0$.

  (a) Show that the exact solution of this IVP is $y(t) = \displaystyle\int_{t_0}^{t} f(x)\,dx$.

  (b) Explain why the Euler's method estimate of $y(t)$ computed using n steps is equal to the left-sum estimate, L_n, of $\displaystyle\int_{t_0}^{t} f(x)\,dx$.

Simpson's Rule

On beyond trapezoid and midpoint sums

Recall the following properties of left, right, trapezoid, and midpoint approximations to a definite integral $I = \int_a^b f(x)\,dx$:

- **Constant functions** The **left and right rules** find I exactly (i.e., commit zero error) if f is a *constant* function. The error bounds

$$|I - L_n| \le \frac{K_1(b-a)^2}{2n} \quad \text{and} \quad |I - R_n| \le \frac{K_1(b-a)^2}{2n}$$

reflect this property: If f is any *constant* function, then K_1 (an upper bound for $|f'|$ on $[a, b]$) is zero.

- **Linear functions** The **trapezoid and midpoint rules** find I exactly (i.e., commit zero error) if f is a *linear* function (and therefore for constant functions, too). Again, the error bounds

$$|I - T_n| \le \frac{K_2(b-a)^3}{12n^2} \quad \text{and} \quad |I - M_n| \le \frac{K_2(b-a)^3}{24n^2}$$

reflect this property: If f is any *linear* function, then K_2 (an upper bound for $|f''|$ on $[a, b]$) is zero.

- **Quadratic and cubic functions** In this unit we explore **Simpson's rule**, also known as the **parabolic rule**. As we'll discover, Simpson's rule represents two *steps* in the right direction: Simpson sums (which we denote by S_n) commit zero error if f is a *cubic* function. (Therefore, the error is zero for constant, linear, and quadratic functions, too.) Once again, an error bound formula reflects this property. Here is the formula for Simpson's rule: ◄

We won't prove this formula; its general form is the main point.

$$|I - S_n| \le \frac{K_4(b-a)^5}{180n^4}.$$

Here, the constant K_4 is an upper bound for $|f^{(\mathrm{iv})}(x)|$, the magnitude of the *fourth* derivative of f for x in $[a, b]$. It's easy to see that $K_4 = 0$ for any for constant, linear, quadratic, or cubic function.

The definition Simpson's method of approximating $I = \int_a^b f$ starts with a partition $a = x_0 < x_1 < x_2 < \cdots < x_n = b$ of $[a, b]$ into an *even* number, $n = 2m$, of subdivisions, each of length $\Delta x = (b-a)/n$. ◄ Simpson's rule is defined as follows:

We'll see soon why n must be even.

DEFINITION Simpson's approximation to I is the sum

$$S_n = \frac{\Delta x}{3}\big(f(x_0) + 4f(x_1) + 2f(x_2) + 4f(x_3) + 2f(x_4) + \cdots + 4f(x_{n-1}) + f(x_n)\big).$$

In sigma notation,

$$S_n = S_{2m} = \frac{\Delta x}{3}\sum_{i=1}^{m}\big(f(x_{2i-2}) + 4f(x_{2i-1}) + f(x_{2i})\big).$$

The definition raises obvious questions: What accounts for the coefficient pattern $1, 4, 2, 4, \ldots, 1$ in the summation, and for the division by 3? How is all this related to quadratic functions? We'll explore these questions in what follows.

Simpson sums as weighted averages In Section 6.1 we described Simpson's rule as a **weighted average** of the trapezoid and midpoint sums: if $n = 2m$, then

$$S_n = \frac{T_m + 2M_m}{3}.$$

PROBLEM 1 Let $I = \int_0^1 \sin x \, dx$. By the definition above,

$$S_6 = \frac{1}{18} \left(\sin(0) + 4\sin\left(\tfrac{1}{6}\right) + 2\sin\left(\tfrac{2}{6}\right) + 4\sin\left(\tfrac{3}{6}\right) + 2\sin\left(\tfrac{4}{6}\right) + 4\sin\left(\tfrac{5}{6}\right) + \sin(1) \right).$$

Write out T_3 and M_3 in the same form and check directly that $S_6 = (T_3 + 2M_3)/3$. Then find numerical values for all three approximating sums. How close is each to the exact answer?

Simpson's rule and parabolic approximation A geometric approach to Simpson's rule involves *parabolas*—in effect, Simpson's rule pretends that an integrand f is *quadratic* on successive pairs of subintervals. Figure 1 illustrates the idea; the shaded area represents S_6. ➡

Notice that S_6 involves three parabola arcs, not six.

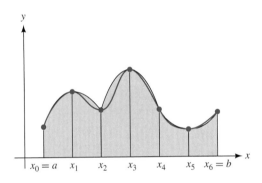

The area under the (blue) graph is approximated by three parabola-topped regions.

FIGURE 1
Simpson's rule: The idea

Understanding Simpson's rule geometrically is a three-step process:

1. Partition $[a, b]$ into an even number n of equal subdivisions, each of width $\Delta x = (b - a)/n$. Consider successive *pairs* of subdivisions.

2. Each pair of subdivisions involves *three* equally-spaced partition points: x_{2i-2}, x_{2i-1}, and x_{2i}. Over each such double subinterval, imagine replacing the integrand f with a *quadratic* function q—the one whose graph (a parabola) passes through the three points $\left(x_{2i-2}, f(x_{2i-2})\right)$, $\left(x_{2i-1}, f(x_{2i-1})\right)$, and $\left(x_{2i}, f(x_{2i})\right)$. For algebraic reasons, only *one* such function q exists. If the three points happen to be collinear, then q turns out to be a linear function.

3. The area under (or over, if $f < 0$) each small parabolic arc turns out to be

$$\text{area} = \frac{\Delta x}{3} \left(f(x_{2i-2}) + 4f(x_{2i-1}) + f(x_{2i}) \right);$$

summing all $n/2$ such contributions gives S_n as in the definition. (Notice the sudden appearance of the once mysterious coefficients 4 and 3.)

The claim about areas in the last step isn't obviously true, but it's not terribly difficult to defend. The following problem explores a special case.

PROBLEM 2 Let f be a function, and suppose that $f(-1) = 23$, $f(0) = 41$, and $f(1) = 11$. Find a quadratic function $q(x) = a + bx + cx^2$ with these same values. Then show by direct calculation that the area under the q-graph from $x = -1$ to $x = 1$ satisfies

$$\text{area} = \frac{1}{3}(f(-1) + 4f(0) + f(1)).$$

In general, one considers any three points of the form $(x_{2i-2}, f(x_{2i-2}))$, $(x_{2i-1}, f(x_{2i-1}))$, and $(x_{2i}, f(x_{2i}))$ on the graph of f, with equally-spaced x-coordinates. Then one finds a quadratic function q whose graph passes through these points. Integrating q over the double subinterval $[x_{2i-2}, x_{2i}]$ turns out, when the dust settles, to give

$$\frac{\Delta x}{3}(f(x_{2i-2}) + 4f(x_{2i-1}) + f(x_{2i})),$$

as desired.

Zero error on quadratics

Simpson's rule is—by design—error free for constant, linear, and quadratic integrands. The following problems give further evidence of this.

PROBLEM 3 In each part, show that S_2 evaluates the integral exactly.

$$\text{(a)} \int_a^b 1\,dx \qquad \text{(b)} \int_a^b x\,dx \qquad \text{(c)} \int_a^b x^2\,dx.$$

Do each part by calculating both the exact integral and S_2. (For S_2, use the partition $x_0 = a$, $x_1 = (a+b)/2$, $x_3 = b$.)

Problem 3 implies that Simpson's rule (even with $n = 2$) commits zero error for *every* integrand of the form $Ax^2 + Bx + C$ because

$$\int_a^b (Ax^2 + Bx + C)\,dx = A\int_a^b x^2\,dx + B\int_a^b x\,dx + C\int_a^b 1\,dx,$$

and S_2 evaluates each summand on the right side exactly.

One free degree: zero error for cubics By a stroke of good luck, Simpson's rule commits zero error even for *cubic* integrands. The following example shows that this is true for $f(x) = x^3$; arguing as in the preceding paragraph shows that the same result holds for *every* cubic function.

EXAMPLE 1 Let $I = \int_a^b x^3\,dx$; show that S_2 commits *zero* error.

Solution Direct calculation (with $x_0 = a$, $x_1 = (a+b)/2$, and $x_2 = b$) gives

$$I = \int_a^b x^3\,dx = \frac{b^4 - a^4}{4}; \qquad S_2 = \frac{b-a}{6}\left(a^3 + 4\left(\frac{a+b}{2}\right)^3 + b^3\right).$$

Straightforward algebra (see the following problem) shows that $I = S_2$. ■

PROBLEM 4 Expand and simplify S_2 in Example 1 to show that $S_2 = I$.

An error bound It's no surprise, given Example 1, that the error committed by Simpson's rule involves the *fourth derivative* $f^{(4)}$ on $[a, b]$; the fourth derivative detects how much f differs from being cubic. ➡

For a cubic function, the first four derivatives are all zero.

THEOREM 4 (Error bound for Simpson's rule)
Let $I = \int_a^b f(x)\,dx$, and let K_4 be any upper bound for $|f^{(4)}|$ on $[a, b]$. Then

$$|I - S_n| \le \frac{K_4(b-a)^5}{180n^4}.$$

Notice:

- As with K_1 and K_2 in earlier theorems, K_4 may be *any* upper bound for $|f^{(4)}|$. As before, rough estimates are best found graphically.
- Notice that n^4 appears in the denominator; this high power of n accounts for the remarkable efficiency of Simpson's rule.

PROBLEM 5 Earlier we calculated that if $I = \int_0^1 \sin x\,dx$, then $S_6 = 0.45969967$. What does the error formula predict? How much *actual* error does S_6 commit?

Finally, we estimate an unknown quantity.

EXAMPLE 2 Use S_{10} to estimate $I = \int_0^1 \sin(x^2)\,dx$. How much error, at worst, might S_{10} commit? How many subdivisions would be needed to approximate I with error less than 10^{-8}?

Solution Any convenient technology gives $S_{10} = 0.310260$ (rounded to six decimals). A slightly tedious hand calculation (it's much easier with symbol-manipulating software) finds the fourth derivative of $f(x) = \sin(x^2)$:

$$f^{(iv)}(x) = 16\,\sin(x^2)x^4 - 48\,\cos(x^2)x^2 - 12\,\sin(x^2).$$

Bounding the fourth derivative is best done graphically, as in Figure 2:

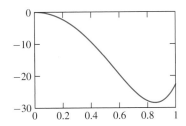

FIGURE 2
Bounding $|f^{(iv)}(x)|$ on $[0, 1]$

The graph suggests that $K_4 = 30$ looks safe. ← The rest is arithmetic. By the theorem,

$$|I - S_{10}| \leq \frac{30 \cdot (1-0)^5}{180 \cdot 10^4} \approx 0.000017.$$

The error, if any, enters around the fifth decimal place, and so 0.3103 is a meaningful estimate for I.

To ensure error less than 10^{-8}, we need n so that

$$|I - S_n| \leq \frac{30 \cdot (1-0)^5}{180 \cdot n^4} < 10^{-8}.$$

Solving the last inequality for n gives

$$\frac{1}{6 \cdot n^4} < 10^{-8} \iff n^4 > \frac{10^8}{6} \iff n > \frac{100}{\sqrt[4]{6}} \approx 63.9,$$

and we conclude that any even $n \geq 64$ will do. ∎

FURTHER EXERCISES

1. Let $I = \int_0^2 e^{x^2}\, dx$.

   (a) Compute S_4 (use technology).

   (b) Bound the approximation error made by S_4.

   (c) Use the error bound formula for Simpson's rule to find a value of n so that S_n is guaranteed to approximate I within 10^{-5}.

2. Let $I = \int_0^1 \cos(100x)\, dx$.

   (a) Compute the value of I exactly, using the FTC.

   (b) Compute S_{10}. Show that the *actual* approximation error is less than 0.05.

   (c) Find an upper bound on the magnitude of the approximation error. Why is this upper bound so enormous?

   (The moral of this problem is that the error bound of Theorem 4 sometimes *vastly* overstates the actual error committed.)

3. Let $I = \int_5^7 \cos x\, dx$.

   (a) Compute S_4.

   (b) What does the error bound formula say about the error made if I is approximated by S_4? How much error does S_4 actually commit?

4. Let $I = \int_{-1}^2 f(x)\, dx$, where f is a function such that $1 \leq f^{(4)}(x) \leq 8$ if $-1 \leq x \leq 2$ and f has the values shown in the table.

x	−1.00	−0.25	0.50	1.25	2.00
$f(x)$	2.000	26.522	48.755	68.328	86.790

   Compute S_4 and an upper bound on the value of $|I - S_4|$.

5. Compute an estimate of $\int_0^2 xe^{-x^3}\, dx$ that is guaranteed to be correct within ± 0.001. Explain carefully why your estimate has the desired accuracy.

Gaussian Quadrature: Approximating Integrals Efficiently

Calculations have costs. Even fast computers require time and other resources to complete calculations. In estimating a definite integral $\int_a^b f(x)\,dx$, for example, it may be "expensive" to find many values of the integrand f. In such cases we want an approximate method that uses relatively few evaluations of the integrand and makes the best possible use of whatever values are computed. The method called **Gaussian quadrature** works efficiently in this sense.

Special integrals—for now   To avoid symbolic clutter, we'll focus mainly on integrals of the form

$$\int_{-1}^{1} f(x)\,dx,$$

for which the interval of integration is $[-1, 1]$. At the end we'll show how to extend the ideas developed here to definite integrals defined on *any* interval.

Quadrature formulas   All of the approximate integration methods (left, right, midpoint, and trapezoid rules) we've studied have similar symbolic forms. Here are several examples for an integrand f on $[-1, 1]$: ➡

L, M, and T stand for left, midpoint, and trapezoid.

$$L_2 = 1\,f(-1) + 1\,f(0)$$
$$M_4 = \frac{1}{2}\,f\left(-\tfrac{3}{4}\right) + \frac{1}{2}\,f\left(-\tfrac{1}{4}\right) + \frac{1}{2}\,f\left(\tfrac{1}{4}\right) + \frac{1}{2}\,f\left(\tfrac{3}{4}\right)$$
$$T_3 = \frac{1}{3}\,f(-1) + \frac{2}{3}\,f\left(-\tfrac{1}{3}\right) + \frac{2}{3}\,f\left(\tfrac{1}{3}\right) + \frac{1}{3}\,f(1)$$

All of these rules have the same general form:

$$w_1 f(x_1) + w_2 f(x_2) + \cdots + w_n f(x_n), \tag{1}$$

where $x_1, x_2, \ldots, x_n$ are specific input points (called the **nodes**) in the interval of integration (in this case, that's $[-1, 1]$) and the $w_1, w_2, \ldots, w_n$ are positive numbers (called the **weights**).

Any expression of the form (1) is called a **quadrature formula** on the interval $[-1, 1]$. All three of the expressions M_2, L_4, and T_3 above are examples of quadrature formulas. For a given integrand function f, a quadrature formula combines values of f at the nodes with the weights to produce a *number* that (with any luck) should approximate the definite integral $\int_{-1}^{1} f(x)\,dx$. Thus, a *good* quadrature formula is one that approximates such integrals closely for many integrand functions f.

> **EXAMPLE 1**   Let $f(x) = \cos x$. What do the quadrature formulas M_2, L_4, and T_3 give? How do these numbers compare with the "true value" of $\int_{-1}^{1} \cos x\,dx$? Which quadrature formula is "best"?
>
> **Solution**   Applying the quadrature formulas to $f(x) = \cos x$ gives
>
> $$L_2 = 1\,\cos(-1) + 1\,\cos(0) \approx 1.5403;$$
> $$M_4 = \frac{1}{2}\,\cos\left(-\tfrac{3}{4}\right) + \frac{1}{2}\,\cos\left(-\tfrac{1}{4}\right) + \frac{1}{2}\,\cos\left(\tfrac{1}{4}\right) + \frac{1}{2}\,\cos\left(\tfrac{3}{4}\right) \approx 1.7006;$$
> $$T_3 = \frac{1}{3}\,\cos(-1) + \frac{2}{3}\,\cos\left(-\tfrac{1}{3}\right) + \frac{2}{3}\,\cos\left(\tfrac{1}{3}\right) + \frac{1}{3}\,\cos(1) \approx 1.6201.$$

The exact value, by comparison, is $\int_{-1}^{1} \cos x \, dx = \sin(1) - \sin(-1) \approx 1.6829$. In this case, at least, M_4 comes closest to the exact value and so might be considered "best" among the three quadrature formulas. On the other hand, M_4 used *four* values of the cosine, whereas L_2 used only two, and thus the better approximation came at a higher price. ■

PROBLEM 6 In each of the following parts apply the quadrature formulas L_2, M_4, and T_3 to the given function f on $[-1, 1]$. Compare results to the "true value" of $\int_{-1}^{1} f(x) \, dx$. Try to explain any "surprises" that appear.

(a) $f(x) = e^x$

(b) $f(x) = x$

(c) $f(x) = 1$

(d) $f(x) = \sin x$

PROBLEM 7 Suppose that f is an *odd* function (i.e., $f(-x) = -f(x)$). Explain why (a) $\int_{-1}^{1} f(x) \, dx = 0$; and (b) all three quadrature formulas L_2, M_4, and T_3 give 0 (the exact answer) when applied to f.

PROBLEM 8 Suppose that f is a *constant* function (i.e., $f(x) = k$ for some real number k). Explain why all three quadrature formulas L_2, M_4, and T_3 give the *exact* value of $\int_{-1}^{1} f(x) \, dx$.

PROBLEM 9 Suppose that f is a *linear* function (i.e., $f(x) = ax + b$ for some real numbers a and b). Which (if any) of the three quadrature formulas L_2, M_4, and T_3 give the *exact* value of $\int_{-1}^{1} f(x) \, dx$? Why?

PROBLEM 10 Write out the quadrature formulas M_5 and T_5 on $[-1, 1]$. What estimate does each formula produce for $\int_{-1}^{1} \cos x \, dx$?

PROBLEM 11 Consider any quadrature formula

$$Q = w_1 f(x_1) + w_2 f(x_2) + \cdots + w_n f(x_n)$$

on $[-1, 1]$. For any function g, we'll write $Q(g)$ for the result of applying Q to g. Let g and h be any functions defined on $[-1, 1]$, and let C be any constant. Show:

(a) $Q(g + h) = Q(g) + Q(h)$

(b) $Q(Cg) = C \, Q(g)$.

(The point of the problem is that quadrature formulas behave like definite integrals in the sense that both operations "respect" sums and constant multiples.)

Gaussian quadrature formulas In all three quadrature formulas, L_2, M_4, and T_3 (and in such close relatives as L_5, M_{237}, and T_{79}), the nodes are evenly spaced—all nodes are endpoints or midpoints of *equal-length* subdivisions of the interval $[-1, 1]$.

Choosing nodes this way seems natural, but it's not the only possibility. In the early 1800s the German mathematician Carl Friedrich Gauss found that placing the nodes elsewhere can give even better results. Here are two of Gauss's quadrature formulas:

$$GQ_2 = 1 f\left(-1/\sqrt{3}\right) + 1 f\left(1/\sqrt{3}\right) \approx f(-0.5774) + f(0.5774);$$

$$GQ_3 = \frac{5}{9} f\left(-\sqrt{3/5}\right) + \frac{8}{9} f(0) + \frac{5}{9} f\left(\sqrt{3/5}\right)$$

$$\approx 0.5556 f(-0.7746) + 0.8889 f(0) + 0.5556 f(0.7746).$$

Here, GQ stands for **Gaussian quadrature**, and the subscript reflects the number of nodes. With $f(x) = \cos x$, for example, the rules give

$$GQ_2 = \cos\left(-1/\sqrt{3}\right) + \cos\left(1/\sqrt{3}\right) \approx 1.6758;$$

$$GQ_3 = \frac{5}{9}\cos\left(-\sqrt{3/5}\right) + \frac{8}{9}\cos(0) + \frac{5}{9}\cos\left(\sqrt{3/5}\right) \approx 1.6830.$$

These results agree to two or three decimal places with the exact value of $\int_{-1}^{1} \cos x\, dx$—a remarkable result given that only two or three values of the integrand were used.

We'll explore in a moment how formulas GQ_2 and GQ_3 can be found, but first let's see how well they work.

PROBLEM 12 In each of the following parts, apply the Gaussian quadrature formulas GQ_2 and GQ_3 to the given function f on $[-1, 1]$. Compare results to the "true value" of $\int_{-1}^{1} f(x)\, dx$. (Do these by hand to be sure you see the patterns.)

(a) $f(x) = 1$

(b) $f(x) = x$

(c) $f(x) = x^2$

(d) $f(x) = x^3$

(e) $f(x) = x^4$

(f) $f(x) = x^5$

What the problems show Results of Problem 12 are worth emphasizing:

> *Formula GQ_2 gives the exact value of the integral $\int_{-1}^{1} f(x)\, dx$ if $f(x) = 1$, $f(x) = x$, $f(x) = x^2$, and $f(x) = x^3$. Formula GQ_2 gives the exact value of $\int_{-1}^{1} x^k\, dx$ for $k = 0, 1, \ldots, 5$.*

Because quadrature formulas "respect" sums and constant multiples (see Problem 11), we can say more:

> **FACT** Formula GQ_2 gives the *exact* value of $\int_{-1}^{1} f(x)\, dx$ if $f(x)$ is any polynomial of degree *three* or lower; GQ_3 gives the exact value if $f(x)$ is any polynomial of degree *five* or lower.

PROBLEM 13 In each of the following parts, apply GQ_2 and GQ_3 to the given function f on $[-1, 1]$. Compare results to the "true value" of $\int_{-1}^{1} f(x)\, dx$.

(a) $f(x) = e^x$

(b) $f(x) = \ln(x + 2)$

(c) $f(x) = x^6$

Finding Gauss's weights and nodes Quadrature formulas GQ_2 and GQ_3 seem to work very well, but how were the weights and nodes in formulas GQ_2 and GQ_3 found? Problem 12 suggests how Gauss might have reasoned:

- **For GQ_2:** Choose weights w_1 and w_2 and nodes x_1 and x_2 to ensure that the formula gives exact answers for integrands 1, x, x^2, and x^3.

- **For GQ_3:** Choose weights w_1, w_2, and w_3 and nodes x_1, x_2, and x_3 to ensure that the formula gives exact answers for integrands 1, x, x^2, x^3, x^4, and x^5.

We work out the details in the next two problems.

PROBLEM 14 **(Finding the formula GQ_2)**   Formula GQ_2 has the general form

$$GQ_2 = w_1 f(x_1) + w_2 f(x_2).$$

Find numerical values for w_1, x_1, w_2, and x_2 by working through the following steps. (Explain anything that needs explanation.)

(a) Assume (it's reasonable to do so) that the two weights are equal and that the nodes are symmetrically placed with respect to the origin. This leads (explain how) to the simpler form $GQ_2 = wf(-n) + wf(n)$, where w and n are positive numbers still to be found.

(b) Use the fact that GQ_2 gives the exact integral for $f(x) = 1$ to show that $w = 1$, and hence that $GQ_2 = f(n) + f(-n)$, for some n still to be found.

(c) Use the fact that GQ_2 gives the exact integral for $f(x) = x^2$ to show that $n = 1/\sqrt{3}$, and hence that $GQ_2 = f(-1/\sqrt{3}) + f(1/\sqrt{3})$, as above.

(d) Check that $GQ_2 = f(-1/\sqrt{3}) + f(1/\sqrt{3})$ gives the exact integral for both $f(x) = x$ and $f(x) = x^3$, as desired.

PROBLEM 15 **(Finding the formula GQ_3)** Because GQ_3 has three nodes and three weights, it has the general form

$$GQ_3 = w_1 f(x_1) + w_2 f(x_2) + w_3 f(x_3).$$

Find numerical values for the weights and nodes by working through the following steps. (Explain anything that needs explanation.)

(a) It's reasonable to assume (or guess) that the nodes are symmetrically placed with respect to the origin; this means that $x_2 = 0$ and that $x_3 = -x_1$. It's also reasonable to suppose (again for reasons of symmetry) that the leftmost and rightmost nodes have the same weights. This leads (explain how) to the simpler form

$$GQ_3 = w_1 f(x_1) + w_2 f(0) + w_1 f(-x_1),$$

where w_1, w_2, and x_1 are still to be found.

(b) Use the fact that GQ_3 gives the exact integral for $f(x) = 1$ to show that $2w_1 + w_2 = 2$.

(c) Use the fact that GQ_3 gives the exact integral for $f(x) = x^2$ to show that $2w_1 x_1^2 = 2/3$.

(d) Use the fact that GQ_3 gives the exact integral for $f(x) = x^4$ to show that $2w_1 x_1^4 = 2/5$.

(e) Use results from earlier parts to show that $x_1 = \sqrt{3/5}$ and $w_1 = 5/9$.

(f) Use parts (b) and (e) to show that $w_2 = 8/9$. Conclude that GQ_3 has the desired form.

Integration over any interval   These ideas would be of little value if they applied *only* to integrals over the interval $[-1, 1]$. The following example suggests how to use a linear change of variable to apply the Gaussian quadrature idea to an integral over *any* finite interval.

EXAMPLE 2 Use Gaussian quadrature to approximate the integral $I = \int_2^5 \cos x \, dx$. (The exact value is $\sin 5 - \sin 2 \approx -1.8682$.)

Solution First, we find a linear function $u(x) = Ax + B$ so that $u(2) = -1$ and $u(5) = 1$. A little algebra shows that $u(x) = 2x/3 - 7/3$ does the trick. Second, we make the u-substitution $u = 2x/3 - 7/3$ in the integral I (note that $x = 3u/2 + 7/2$ and $dx = 3\,du/2$) to get the new integral

$$I = \int_2^5 \cos x \, dx = \int_{-1}^1 \frac{3}{2} \cos\left(\frac{3}{2}u + \frac{7}{2}\right) du.$$

Finally, we apply GC_2 and GC_3 to the *new* integrand function (on the right) to get

$$GC_2(g) = -1.8201 \quad \text{and} \quad GC_3(g) = -1.8692,$$

both of which compare respectably to the exact integral. ■

PROBLEM 16 In each part, mimic the preceding example to apply GC_3 to the given integral. Compare results to the exact integral values.

(a) $\int_0^1 e^x \, dx$

(b) $\int_1^5 \frac{1}{x} \, dx$

7

USING THE INTEGRAL

7.1 MEASUREMENT AND THE DEFINITE INTEGRAL; ARC LENGTH

Introduction

Many important applications of calculus involve measuring something, such as the *area* of a plane region, the *volume* of a solid object, the *length* of a curve from one point to another, the *net distance* a moving object travels over an interval of time, the *work* done against gravity in raising a satellite into orbit, or the *present value* of an income stream. ◄

We'll define quantities such as work and present value as we go along.

In the simplest cases such quantities can be measured using basic formulas. For example, measuring the area of a rectangular region, the length of a straight line segment, or the distance covered by an object moving at constant speed requires no big ideas or clever techniques from calculus. In practice, though, most regions are not rectangular, most curves are not straight, and most speeds are not constant. In these more usual and more interesting situations, the definite integral turns out to be an essential tool.

In Chapter 5 we defined integrals informally in terms of signed area.

The link between integrals and areas is nothing new—we've exploited it from the beginning. ◄ But if definite integrals measured *only* area, they wouldn't deserve the fuss we make over them. In fact, as we'll see, measurement problems of *many* kinds boil down to definite integrals. The last step—evaluating a definite integral—is often relatively easy. We can use the FTC if an antiderivative is readily available; if not, we can use the numerical methods developed in Chapter 6. For many measurement problems, the main step is choosing *which* definite integral to evaluate—both the integrand and the limits of integration may take some thought to decide upon. Because our main interest in this chapter is usually in setting up integrals, we'll sometimes skip (or leave to a machine) the details of routine calculations.

Technical note: assuming good behavior To avoid unhelpful distractions, we assume that all the integrals $\int_a^b f(x)\,dx$ we meet in this chapter make good mathematical sense. To guarantee this, it's enough to assume that every integrand f is *continuous* on $[a, b]$. We will assume this from now on unless something explicit is said to the contrary. (In fact, discontinuous integrands *do* sometimes arise in practical applications. Even in such cases, however, the basic ideas of this section often apply, although perhaps in slightly modified forms.)

412

Two views of the definite integral

In applying the integral in varied settings, it's useful to remember two different but closely related interpretations of a definite integral $\int_a^b f(x)\,dx$.

*   **A limit of approximating sums**   The integral is defined formally as a limit of approximating sums. Chapter 6 discusses and compares several kinds of approximating sums: left, right, trapezoid, midpoint, and so on. Using right sums, for instance, we can write

$$\int_a^b f(x)\,dx = \lim_{n\to\infty} \sum_{i=1}^n f(x_i)\Delta x,$$

    where the inputs x_i are the right endpoints of n equal-length subintervals of $[a, b]$. From this point of view, the integral "adds up" many contributions, each of the form $f(x_i)\Delta x$.

*   **Accumulated change in an antiderivative**   The FTC says that

$$\int_a^b f(x)\,dx = F(b) - F(a),$$

    where the function F on the right can be *any* antiderivative of f on $[a, b]$. ➧ The difference $F(b) - F(a)$ represents, in a natural way, the **accumulated change** (or net change) in F over the interval $[a, b]$. In other words, to find the net change in F over $[a, b]$, integrate f—the *rate function* associated with F—over $[a, b]$.

Remember: Antiderivatives are not unique. If F is one antiderivative of f, then so is F + k, where k is any constant.

These two approaches to the integral are mathematically equivalent. The FTC guarantees that both methods give the same "answer." Having two different ways to think about the integral makes it more versatile in applications. Which viewpoint is "better" depends on the situation. The following example illustrates both viewpoints.

EXAMPLE 1   The function $v(t) = 10 + 20t - 10t^2$ gives a car's eastward velocity, in miles per hour at time t hours, for $0 \le t \le 3$. Calculate $I = \int_0^3 v(t)\,dt$; interpret I both as a limit of sums and as accumulated change in an antiderivative.

Solution   Calculating I symbolically is easy:

$$I = \int_0^3 v(t)\,dt = 10t + 10t^2 - \frac{10t^3}{3}\bigg]_0^3 = 30.$$

What does the answer mean? Figure 1 gives two different views.

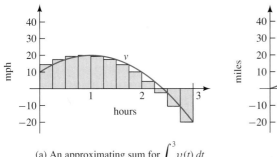

(a) An approximating sum for $\int_0^3 v(t)\,dt$

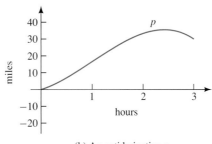

(b) An antiderivative p

FIGURE 1
Two views of one answer

Figure 1(a) suggests the limit-of-sums view of I. It shows R_{12}, the right approximating sum for I with 12 subdivisions. The signed area of each rectangle approximates the *eastward* distance covered by the car over that subinterval. The three rightmost rectangles contribute *negative* eastward distance (i.e., *westward* distance). Therefore, the right sum R_{12} approximates the *net* eastward distance covered over the entire interval $0 \le t \le 3$. As the limit of such approximating sums, I represents the *exact* net eastward distance covered over $[0, 3]$—precisely 30 miles. Numerical evidence supports this idea:

$$R_{12} \approx 25.938; \quad R_{50} \approx 29.082; \quad R_{200} \approx 29.774; \quad \ldots \quad I = 30.$$

Figure 1(b) illustrates the accumulated-change view of I. Because the function v describes the car's eastward *velocity*, an antiderivative function p describes the car's east–west *position* given appropriate units and points of reference. Since $v(t) = 10 + 20t - 10t^2$, we can choose the antiderivative function $p(t) = 10t + 10t^2 - 10t^3/3$. Then $p(0) = 0$, so the car starts at position 0 and $p(t)$ represents the car's position at time t measured in miles east of the starting point. The graph of v in Figure 1(b) shows a net movement of 30 miles east—the accumulated change in *position* over the 3-hour interval. ■

Measuring areas in the plane

"Signed" means that areas under the x-axis count as negative.

Recall the familiar interpretation of the integral as *signed* area: ◄ For any continuous function f on $[a, b]$,

$$\int_a^b f(x)\, dx = \text{signed area bounded by the } f\text{-graph for } a \le x \le b.$$

Keeping track of positive and negative contributions to signed area takes a little care. Here's a simple example to bring the issue back to mind.

EXAMPLE 2   Easy calculations give

$$I_1 = \int_0^\pi \sin x \, dx = 2 \quad \text{and} \quad I_2 = \int_0^{2\pi} \sin x \, dx = 0.$$

What do these results say about ordinary and signed areas?

Solution   Because $\sin x \ge 0$ on $[0, \pi]$, the integral I_1 gives the ordinary area—2 square units—under one arch of the sine curve. That $I_2 = 0$ means that the *signed* area over $[0, 2\pi]$ is zero. This makes good geometric sense: The symmetry of the sine graph means that areas above and below the x-axis exactly cancel each other out. ■

Areas defined by more than one curve   The connection between areas and integrals can be used, with only minor changes, to measure areas defined by more than one function graph. The shaded part of Figure 2 is one such region; it is bounded above and below by function graphs and on the left and right by vertical lines.

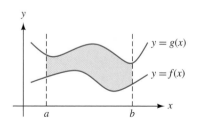

FIGURE 2
A region defined by two curves

Notice that the shaded area in Figure 2 is the *difference* between the areas associated with $\int_a^b g(x)\,dx$ and $\int_a^b f(x)\,dx$. Therefore,

$$\text{shaded area} = \int_a^b g(x)\,dx - \int_a^b f(x)\,dx = \int_a^b \left(g(x) - f(x)\right) dx. \qquad (1)$$

EXAMPLE 3   Measure the shaded area in Figure 3.

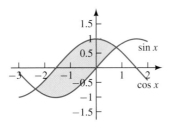

FIGURE 3
Area between sine and cosine curves

Solution   The sine and cosine curves intersect at $x = -3\pi/4$ and $x = \pi/4$, and thus Formula 1 gives

$$\text{area} = \int_{-3\pi/4}^{\pi/4} (\cos x - \sin x)\,dx = 2\sqrt{2} \approx 2.82843.$$

The numerical result looks graphically reasonable, but attentive readers may wonder: Does Formula 1 apply even when (as here) one or both curves dip below the x-axis. The answer, fortunately, is yes: Formula (1) requires only that $g(x) \geq f(x)$ when $a \leq x \leq b$.   *So that $g(x) - f(x) \geq 0$.* ∎

Other regions; integrating in x or in y   In practice, plane regions of interest may be bounded by various types of curves—not necessarily graphs of the standard $y = f(x)$ style. Figure 4 illustrates two common types. ➡

More complicated regions may be combinations of these two types.

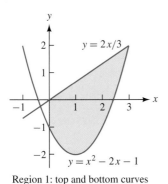

Region 1: top and bottom curves

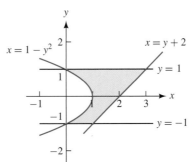

Region 2: left and right curves

FIGURE 4
Two types of plane regions

The following formulas apply to finding areas such as those in Figure 4.

> **FACT**   Let f and g be continuous functions.
>
> - **Integrating in x**   Let R be the region bounded above by $y = g(x)$, below by $y = f(x)$, on the left by $x = a$, and on the right by $x = b$. Then R has area
>
> $$\int_a^b \left(g(x) - f(x)\right) dx. \tag{2}$$
>
> - **Integrating in y**   Let R be the region bounded on the right by $x = g(y)$, on the left by $x = f(y)$, below by $y = c$, and above by $y = d$. Then R has area
>
> $$\int_c^d \left(g(y) - f(y)\right) dy. \tag{3}$$

Depending on the region, one or the other formula may be simpler; some areas can be calculated by integration either in x or in y. We illustrate the possibilities by example.

EXAMPLE 4   Measure the two shaded areas in Figure 4.

Solution   The two graphs in Figure 4(a) intersect at $x = 3$ and $x = -1/3$, as Figure 4 suggests—and the quadratic formula confirms. (The intersection at $x = -1/3$ is outside Region 1 and therefore doesn't appear in the calculation.) Now Formula (2) gives

$$\text{area} = \int_0^3 \left(\tfrac{2}{3}x - (x^2 - 2x - 1)\right) dx = 6.$$

Integration in x seems impractical for Region 2, because defining upper and lower curves is troublesome. But Formula (3) applies conveniently:

$$\text{area} = \int_{-1}^1 \left(y + 2 - (1 - y^2)\right) dy = \frac{8}{3}. \qquad \blacksquare$$

EXAMPLE 5   Use *both* Formulas (2) and (3) to find the area of the region bounded by the curves $x = 0$, $y = 2$, and $y = e^x$, shown shaded in Figure 5.

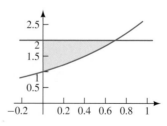

FIGURE 5
Curves $y = e^x$ and $y = 2$

Solution   The two curves intersect where $e^x = 2$ or $x = \ln 2$. Integrating in x gives

$$\text{area} = \int_0^{\ln 2} (2 - e^x) \, dx = 2\ln 2 - 1 \approx 0.3863.$$

To find the same area in terms of y, we rewrite the equation $y = e^x$ in the equivalent form $x = \ln y$. This curve is the *right* boundary; $x = 0$ is the left boundary. Integrating in y now gives

$$\text{area} = \int_1^2 (\ln y - 0) \, dy = 2 \ln 2 - 1,$$

just as above.   ∎

Arc length

Arc length offers a good example of using integrals to measure geometric quantities *other* than area; arc length integrals also show the role of approximating sums.

EXAMPLE 6   How long is the curve $y = \sin x$ from $x = 0$ to $x = \pi$? (First try to estimate the length—draw your own graph.)

Solution   Straight line segments are easy to measure using the distance formula. In the spirit of approximating sums, a natural strategy suggests itself for measuring *curves*:

Approximate the curve with straight line segments and add up their lengths.

Figure 6 shows a four-segment piecewise-linear approximation to the sine curve.

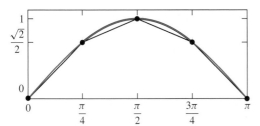

The endpoints of each line segment are shown with black dots.

FIGURE 6
Estimating the length of one arch

Knowing the coordinates of the segments' endpoints, we can readily (if tediously) calculate the total length of the four-segment chain; it's about 3.79. With more and shorter line segments, we expect increasingly better approximations to the curve's true length L. In other words, we expect L to be the *limit* of polygonal lengths as the number of subdivisions tends to infinity. This language suggests—rightly—the presence of a definite integral, and, indeed, we'll derive one in a moment. But first let's see the limiting principle in action numerically, by tabulating lengths (computed electronically, of course) of several piecewise-linear approximations to the elusive L. Each length corresponds to subdividing the x-interval $[0, \pi]$ into n equal subintervals.

Approximating the length of a curve					
Number of segments	4	8	16	31	100
Polygonal length	3.7901	3.8125	3.8183	3.8197	3.8201

For now, 3.8201 is our best guess.   ∎

Arc length in general   A general arc length problem reads as follows:

Find the length of the f-graph from x = a to x = b.

The preceding example suggests a general strategy:

Approximate with piecewise–linear arcs and then take a limit.

How does this work in practice? How is an integral involved? *Which* integral is involved? To describe the situation more precisely, we need a more general picture.

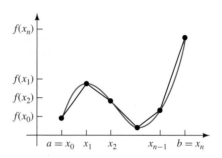

FIGURE 7

Approximating a curve C with a polygonal arc C_n

Figure 7 has three main ingredients:

1. C, the graph of f from $x = a$ to $x = b$;

2. A partition of $[a, b]$ into n subintervals with endpoints $x_0, x_1, \ldots, x_n$;

3. A **polygonal arc** C_n made from n line segments joining the points $(x_0, f(x_0))$, $(x_1, f(x_1)), \ldots, (x_n, f(x_n))$ on C.

Now we can put the idea succinctly: Length of $C = \lim_{n \to \infty} (\text{length of } C_n)$.

How long is C_n? A useful estimate   To measure C_n we add the lengths of its n segments. The ith segment joins $(x_{i-1}, f(x_{i-1}))$ to $(x_i, f(x_i))$, and so its length is

$$\sqrt{\left(f(x_i) - f(x_{i-1})\right)^2 + (x_i - x_{i-1})^2}.$$

Thus,

$$\text{total length of } C_n = \sum_{i=1}^{n} \sqrt{\left(f(x_i) - f(x_{i-1})\right)^2 + (x_i - x_{i-1})^2}. \tag{4}$$

We'll rewrite Equation 4 as an approximating sum for an integral. The first step is an algebraic trick—factoring out $(x_i - x_{i-1})$:

$$\text{length of } C_n = \sum_{i=1}^{n} \sqrt{\left(f(x_i) - f(x_{i-1})\right)^2 + (x_i - x_{i-1})^2}$$

$$= \sum_{i=1}^{n} \sqrt{\left(\frac{f(x_i) - f(x_{i-1})}{x_i - x_{i-1}}\right)^2 + 1} \cdot (x_i - x_{i-1}).$$

This clumsy expression can be simplified. By the definition of derivative, the approximation

$$\frac{f(x_i) - f(x_{i-1})}{x_i - x_{i-1}} \approx f'(x_i)$$

holds if $x_i - x_{i-1}$ is small. Putting the pieces together and writing Δx_i for $(x_i - x_{i-1})$, we get

$$\text{Length of } C_n = \sum_{i=1}^{n} \sqrt{\left(\frac{f(x_i) - f(x_{i-1})}{x_i - x_{i-1}}\right)^2 + 1} \cdot (x_i - x_{i-1})$$

$$\approx \sum_{i=1}^{n} \sqrt{(f'(x_i))^2 + 1} \cdot \Delta x_i.$$

The final line rewards our work: It is an approximating sum for an integral. We state the conclusion as a Fact:

FACT (Arc length by integration) The integral

$$\int_a^b \sqrt{(f'(x))^2 + 1} \, dx$$

gives the length of the f-graph from $x = a$ to $x = b$.

EXAMPLE 7 Write the length of the sine curve from $x = 0$ to $x = \pi$ as an integral. Estimate its value.

Solution By the Fact, the length is

$$I = \int_0^\pi \sqrt{(\cos x)^2 + 1} \, dx.$$

The FTC won't help us evaluate I because there's no convenient antiderivative. But we can estimate I by using (say) the midpoint rule with 100 subdivisions: $I \approx M_{100} \approx 3.8202$, which is close to our earlier estimate. ∎

Do curves have length? Do piecewise-linear approximations successfully approximate a curve's arc length, as we claimed?

With modern computer graphics, the answer seems obvious. Almost every "curve" in this book, for instance, is really a collection of short line segments—shorter than the eye can perceive individually—strung end to end. In this case, appearances don't deceive: All but the very worst-behaved curves *can* be approximated in the sense just discussed.

Curves do have length.

EXAMPLE 8 Interpret the integral $I = \int_0^1 \sqrt{1 + 4x^2} \, dx$ in two different ways: (i) as the *length* of an appropriate curve and (ii) as the *area* of something. Describe these quantities and evaluate them numerically.

Solution The integral I fits the arc length "template" if $f'(x) = 2x$. Clearly, $f(x) = x^2$ has this property. Therefore, I gives the length of the curve $y = x^2$ from $x = 0$ to $x = 1$. ➡

Any definite integral measures the signed area bounded by its integrand. In this case, therefore, I measures the area under the curve $y = \sqrt{1 + 4x^2}$ from $x = 0$ to $x = 1$.

Important note: The curve we're measuring is not the integrand.

How large, numerically, is I? Antidifferentiation is algebraically messy, and so we'll find M_{20}, the midpoint estimate with 20 subdivisions. Reported to three decimal places:

$$I \approx M_{20} \approx 1.479.$$

The respective units might be inches and square inches.

Thus, both quantities (length and area) have the same approximate numerical value: 1.479 units of *length* and 1.479 units of *area*, respectively. ◄

Figure 8 illustrates both interpretations of I.

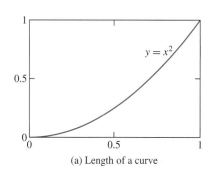

(a) Length of a curve

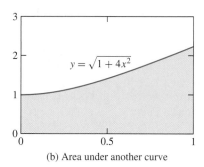

(b) Area under another curve

FIGURE 8

$\displaystyle\int_0^1 \sqrt{1+4x^2}\, dx \approx 1.479$: **One value, two meanings**

BASIC EXERCISES

1. Let R be the region between the curves $y = x$ and $y = x^2$.

  (a) Draw a sketch of the region R.

  (b) Use five left-rule rectangles to approximate the area of R. (Give a numerical answer.)

  (c) Draw a picture that illustrates the sum you computed in part (b). Use this picture to decide whether your answer in part (b) overestimates or underestimates the area of R.

  (d) Find the area of R *exactly* using the FTC.

2. Let A be the region bounded by the curves $y = \sin x$ and $y = \cos x$ and the lines $x = 1$ and $x = 3$.

  (a) Draw a sketch of the region R.

  (b) Use five midpoint-rule rectangles to approximate the area of R. (Give a numerical answer.)

  (c) Explain why M_n overestimates the area of R for every $n \geq 1$.

  (d) Find the area of R *exactly* using the FTC.

3. In this exercise, integration in x is used to find the area of Region 2 in Figure 4.

  (a) Find the area of the region bounded below by the x-axis, above by the line $y = 1$, on the right by the line $x = 1$, and on the left by the curve $x = 1 - y^2$.

  (b) Find the area of the region bounded above by the x-axis, below by the line $y = -1$, on the right by the line $x = 1$, and on the left by the curve $x = 1 - y^2$.

  (c) Find the area of the region bounded by the lines $x = 1$, $x = y + 2$, and $y = 1$.

  (d) Use the results of parts (a), (b), and (c) to find the area of Region 2.

4. This exercise provides yet another approach to finding the area of Region 2 in Figure 4.

  (a) Find the area of the region bounded by the line $x = 0$ and the curve $x = 1 - y^2$.

  (b) Find the area of the region bounded by the lines $x = y + 2$, $y = -1$, and $x = 3$.

  (c) Find the area of the region bounded by the lines $x = 0$, $x = 3$, $y = 1$, and $y = -1$.

  (d) Use the results of parts (a), (b), and (c) to find the area of Region 2.

5. Let $I = \displaystyle\int_0^1 \sqrt{1+4x^2}\, dx$. (See Example 8.)

  (a) Use a trapezoid sum with $n = 20$ subintervals to estimate I.

  (b) Explain why $M_{20} < I < T_{20}$.

6. Write an integral I that gives the length of the curve $y = x^2 + 1$ from $x = 0$ to $x = 1$.

  (a) Use a midpoint sum with $n = 20$ equal subintervals to estimate I.

  (b) Evaluate I exactly. [HINT: Use the table of integrals.]

7. Write an integral I that gives the length of the curve $y = 2 + \sin x$ from $x = 0$ to $x = \pi$. Use a midpoint sum with $n = 20$ subintervals to estimate I.

8. Compute the length of the line $y = mx + b$ from $x = A$ to $x = B$. Does the answer agree with the usual distance formula?

9. A moral of Example 8 is that a given integral can be interpreted in various ways. In each of the following parts, interpret the numerical value of the integral in the stated context using appropriate units of measurement.

(a) $\int_0^3 f(t)\,dt$, where the function $f(t) = 5t^2 - 20t + 50$ tells a certain car's speed, in miles per hour, t hours after midnight

(b) $\int_0^3 f(t)\,dt$, where the function $f(t) = 5t^2 - 20t + 50$ tells a certain car's speed, in feet per second, t seconds after midnight

(c) $\int_0^3 f(t)\,dt$, where the function $f(t) = 5t^2 - 20t + 50$ tells a certain car's acceleration, in feet per minute per minute, t minutes after midnight

10. Interpret the integral $\int_1^3 f(t)\,dt$ in each of the following situations and comment on what it means in each case if the integral is negative. For instance, if $f(t)$ is the eastward velocity of a car in miles per hour at time t hours, then $\int_1^3 f(t)\,dt$ gives the car's net eastward distance traveled between $t = 1$ and $t = 3$. If the integral is negative, the car's position at time 3 is west of its position at time 1.

(a) $f(t)$ is an object's upward velocity, in feet per second, at time t seconds.

(b) $f(t)$ is an object's upward acceleration, in feet per second per second, at time t seconds

(c) $f(t)$ is the net rate of flow of water into a tank, in liters per minute, at time t minutes.

(d) $f(t)$ is the slope at $x = t$ of the graph of $y = g(x)$.

11. Values of a function f are given in the following table:

t	0.9	1.0	1.1	1.2	1.3	1.4	1.5	1.6
$f(t)$	0.974	0.909	0.808	0.675	0.516	0.335	0.141	−0.058

t	1.7	1.8	1.9	2.0	2.1
$f(t)$	−0.256	−0.443	−0.612	−0.757	−0.872

(a) Use the table to estimate values of $f'(t)$ for $t = 1.0, 1.1, \ldots, 2.0$.

(b) Use the result of part (a) to approximate $\int_1^2 f'(t)\,dt$. Compare your result with $f(2) - f(1)$.

(c) Use the table to estimate $\int_1^2 f(t)\,dt$. What does the answer mean geometrically?

(d) Use the table to estimate $\int_1^2 \sqrt{f'(t)^2 + 1}$. What does the answer mean geometrically?

(e) In fact, $f(t) = \sin(2t)$. Use this information to find exact values of the integrals in parts (b) and (c).

12. Let $I = \int_0^1 \sqrt{1 + x}\,dx$.

(a) Calculate I exactly by antidifferentiation.

(b) I can be thought of as the area under some curve. Decide which curve and draw the region in question.

(c) I can also be thought of as the length of a graph $y = f(x)$, $x = 0$ to $x = 1$. Find a function f with this property.

FURTHER EXERCISES

In Exercises 13–24, sketch the region bounded by the given curves and find the area of this region by evaluating integrals of the form $\int_a^b \big(f(x) - g(x)\big)\,dx$.

13. $y = x^4$, $y = 1$

14. $y = x$, $y = x^3$

15. $y = x^2$, $y = x^3$

16. $y = x^2 - 1$, $y = x + 1$

17. $y = \sqrt{x}$, $y = 0$, $x = 4$

18. $y = \sqrt{x}$, $y = x^2$

19. $y = 9(4x + 5)^{-1}$, $y = 2 - x$

20. $y = 9(4x^2 + 5)^{-1}$, $y = 2 - x^2$

21. $y = e^x$, $y = 0$, $x = 0$, $x = 1$

22. $y = 2^x$, $y = 5^x$, $x = -1$, $x = 1$

23. $x = y^2 - 4$, $y = 2 - x$

24. $x = 9 - y^2$, $y = -3 - x$

25–36. Find the area of each region in Exercises 13–24 by evaluating integrals of the form $\int_a^b \big(f(y) - g(y)\big)\,dy$.

37. Suppose that a 12-in.-diameter pizza is cut into "thirds" by making vertical cuts 4 inches on either side of the center. How much bigger is the center piece than the other two pieces?

38. Let $g(x) = f(x) + C$, where C is a constant, L_f be the length of the curve $y = f(x)$ from $x = a$ to $x = b$, and L_g be the length of the curve $y = g(x)$ from $x = a$ to $x = b$.

(a) Use the arc length formula to show that $L_f = L_g$.

(b) Give a geometric explanation of the result in part (a).

In Exercises 39–53, find the length of the curve $y = f(x)$ from $x = a$ to $x = b$.

39. $f(x) = \sqrt{1 - x^2}$, $a = -1/2, b = \sqrt{3}/2$

40. $f(x) = \sqrt{4 - x^2}$, $a = 0, b = 1$

41. $f(x) = x^2$ $a = 1, b = 2$

42. $f(x) = e^x$, $a = 0, b = 1$

43. $f(x) = x^{3/2}$, $a = 1, b = 4$

44. $f(x) = x^{2/3}$, $a = 0, b = 8$

45. $f(x) = 4x^{5/4}/5$, $a = 1, b = 16$

46. $f(x) = \dfrac{\ln x}{2} - \dfrac{1}{4x^2}, \quad a = 1/4, b = 4$

47. $f(x) = \dfrac{x^2}{2} - \dfrac{\ln x}{4}, \quad a = 3, b = 9$

48. $f(x) = \dfrac{x^3}{6} + \dfrac{1}{2x}, \quad a = 1/2, b = 2$

49. $f(x) = \dfrac{x^3}{3} + \dfrac{1}{4x}, \quad a = 1/2, b = 2$

50. $f(x) = \dfrac{x^4}{4} + \dfrac{1}{8x^2}, \quad a = 1, b = 3$

51. $f(x) = \dfrac{1}{2x^2} + \dfrac{x^4}{16}, \quad a = 1, b = 3$

52. $f(x) = \dfrac{x^5}{5} + \dfrac{1}{12x^3}, \quad a = 2, b = 4$

53. $f(x) = \cosh x, \quad a = 1, b = 4$

54. Let L_1 be the length of the curve $y = f(x)$ from $x = 0$ to $x = a$, L_2 be the length of the curve $y = 2f(x)$ from $x = 0$ to $x = a$, and L_3 be the length of the curve $y = f(2x)$ from $x = 0$ to $x = a/2$. Rank the values of L_1, L_2, and L_3 in increasing order.

55. Let I be the length of the curve $y = f(x)$ from $x = a$ to $x = b$. Suppose that the function f is increasing and concave down on the interval $[a, b]$. Show that L_n, the left-sum approximation computed with n equal subintervals, overestimates the value of the arc length integral.

56. Let J be the length of the curve $y = f(x)$ from $x = 0$ to $x = 4$, where f is an antiderivative of the function shown.

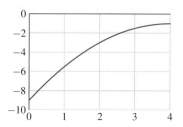

(a) Estimate J using a left-sum approximation with $n = 4$ equal subintervals.

(b) Is $L_4 < J$? Justify your answer.

57. At time $t = 0$, a bug starts crawling along the path $y = \frac{1}{3}(x^2 + 2)^{3/2}$ with constant speed $v = 6$ ft/min. If the bug starts its journey at the point $(1, \sqrt{3})$, where is it 5 minutes later?

58. A snail crawls along the curve $y = \sqrt{x^3}$ at a speed of 3 ft/h. How long does it take the snail to travel from the point $(1, 1)$ to the point $(4, 8)$? (Assume that both x and y are measured in feet.)

59. Suppose that g is a function with the property $g'(t) \geq 1$ if $a \leq t \leq b$. Let $f(x) = \displaystyle\int_a^x \sqrt{\left(g'(t)\right)^2 - 1}\, dt$. Show that the length of the curve $y = f(x)$ from $x = a$ to $x = b$ is $g(b) - g(a)$.

60. Suppose that $h(x) = \left(f(x) + g(x)\right)/2$, where f and g are functions with the property that $f'(x) \cdot g'(x) = -1$ if $a \leq x \leq b$. Show that the length of the curve $y = h(x)$ from $x = a$ to $x = b$ is $L = \dfrac{1}{2}\displaystyle\int_a^b \left| f'(x) - g'(x) \right| dx$.
[HINT: For any real number a, $\sqrt{a^2} = |a|$.]

7.2 FINDING VOLUMES BY INTEGRATION

Slicing the loaf

Imagine a solid object—a long loaf of French bread, say—lying along the x-axis in the xy-plane, between $x = a$ and $x = b$. The usual way to slice such a loaf is with cuts perpendicular to the x-axis. Mathematically speaking, the "knives" that cut such slices are vertical planes, each of the form $x = k$ for some constant k. Figure 1 shows one such cut viewed from overhead.

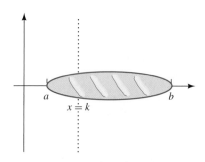

FIGURE 1
Slicing a loaf: an overhead view

The loaf's total volume is, of course, the sum of the volumes of all the slices. This obvious fact is surprisingly useful mathematically. We'll use it to compute volumes by integration for various solid objects.

The cross-sectional area function   If the loaf is irregular, different slices have different volumes even if all have the same thickness. This difference arises because slices from the middle of the loaf have larger **cross-sectional area** (where the butter goes) and, hence, more volume than slices from the narrow ends.

For any value of x between a and b, let $A(x)$ denote the cross-sectional area revealed by a knife cut at x (i.e., the area of the intersection of the loaf with the "vertical" plane at x, perpendicular to the x-axis). For an irregular loaf, $A(x)$ varies with x, rising and falling with the loaf's thickness.

The *exact* volume of one slice of bread is hard to compute because the cross section $A(x)$ varies with x. If it happens that $A(x)$ is constant, then the following simple but important fact applies:

> *If a slice with thickness Δx has* constant *cross-sectional area A, then the slice has volume $A \cdot \Delta x$.*

In a small x-interval (i.e., for a very thin slice), the cross-sectional area $A(x)$ is *nearly* constant, and so we can approximate the volume of a slice of thickness Δx as follows:

$$\text{Volume of one slice} \approx A(x)\Delta x,$$

where $A(x)$ is the cross-sectional area at any convenient value of x within the slice.

Reassembling Riemann's loaf: estimating volume   To estimate the volume of the loaf, let's slice it into n slabs of equal thickness $\Delta x = (b-a)/n$ with cuts at $x = x_0 = a$, $x = x_1$, $x = x_2, \ldots, x = x_n = b$. (Cuts are crumbless in this ideal world.) Figure 2 gives an overhead view with $n = 4$: ➡

Cut on the dotted lines.

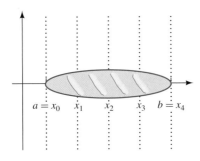

FIGURE 2
Four slices from the loaf

Using the cross-sectional area $A(x_i)$ for the ith slab leads to an estimate for *total* volume:

$$\text{Total volume} \approx A(x_1)\Delta x + A(x_2)\Delta x + \cdots + A(x_n)\Delta x. \tag{1}$$

Here is the crucial observation: The right side of Equation (1) is an approximating sum (a right sum) for the integral $\int_a^b A(x)\,dx$ of the cross-sectional area function.

Now we play the standard limit game: As n tends to infinity, the slab thickness Δx tends to zero, and so the volume estimate tends to the true volume of the loaf. Meanwhile, the approximating sums tend to the true value of the integral. We state our conclusion as a general theorem:

> **THEOREM 1**   A solid lies with its base on the xy-plane between the vertical planes $x = a$ and $x = b$. For all x in $[a, b]$, let $A(x)$ denote the area of the cross section at x perpendicular to the x-axis. If $A(x)$ is a continuous function, then
>
> $$\text{Volume} = \int_a^b A(x)\,dx.$$

Observe:

- **Why continuous?**   Requiring that $A(x)$ be a continuous function of x ensures that the integral exists. For simple, smooth, solid objects, A *is* continuous; more complicated objects can often be handled by (mentally) breaking them up into simpler objects to which the formula of Theorem 1 applies.
- **Which axis?**   Although the theorem mentions the x-axis, which is usually shown "horizontal," there is nothing sacred about these choices. For objects such as trees and Egyptian pyramids, the convenient "slicing axis" may well be vertical. In context this causes little trouble.

The simplest case: constant cross sections   For solids with *constant* cross sections, called **prisms** (sticks of butter, glasses of milk, and some mass-produced bread loaves are all prisms), the theorem merely restates elementary volume formulas.

EXAMPLE 1   A rectangular parallelepiped (or "brick") has edge lengths L cm, W cm, and H cm, so its volume is LWH cm^3. Does Theorem 1 agree?

Solution   We certainly hope so.

If we place the brick anywhere in the xy-plane with one edge (let's use the edge of length L) parallel to the x-axis, then the brick lies between $x = a$ and $x = a + L$. All cross sections perpendicular to the x-axis are rectangles with area WH cm^2, and so $A(x) = WH$ for all x with $a \le x \le a + L$; the theorem gives

$$\text{Volume} = \int_a^{a+L} A(x)\,dx = \int_a^{a+L} WH\,dx = LWH$$

as we expect. Notice, too, that there's nothing special about the L-edge; we could have aligned *any* edge parallel to the x-axis.   ∎

No formula? Think numerically   Using the theorem directly requires an explicit formula for the area function $A(x)$—we need something to integrate. Finding such formulas isn't difficult for simple solids, but for complicated or irregular solids it may be hard or impossible. In this case, however, we can still use numerical methods to estimate the volume integral.

EXAMPLE 2   Hoping to estimate the volume of wood in a 20-meter log, Ranger Rick uses a tape measure to gauge the log's girth (circumference) at 5-meter intervals, starting from the big end. Here are Rick's results:

Measuring a natural log				
distance from end (m)	0	5	10	15	20
circumference (m)	4.0	3.6	3.2	2.6	2.0

What does Rick do next?

Solution   Recalling Theorem 1, Rick defines $A(x)$ to be the tree's cross-sectional area (in square meters) x meters from the big end. Because tree cross sections are roughly circular, a section with circumference C has area $C^2/(4\pi)$, and Rick tabulates the following approximate values of A:

The log's cross-sectional areas					
x	0	5	10	15	20
$A(x)$	1.27	1.03	0.81	0.54	0.32

Finally, Rick uses these values to calculate T_4, the trapezoid-rule estimate with four subdivisions for the integral in Theorem 1: ➧

See the definition of T_4 in Section 6.1.

$$\int_0^{20} A(x)\,dx \approx T_5 = (1.27 + 2 \cdot 1.03 + 2 \cdot 0.81 + 2 \cdot 0.54 + 0.32)\frac{20}{8} \approx 15.88,$$

or just under 16 cubic meters.   ■

Solids of revolution

Real-world solid objects come in all shapes and sizes. Vastly overrepresented in calculus books, however, are **solids of revolution**—solids formed by "revolving" some region in the xy-plane about an axis (often the x-axis). ➧

Solids of revolution could be made on a lathe or a potter's wheel.

Solids of revolution are convenient in calculus for one main reason: their cross sections perpendicular to the axis of rotation are *circular*. Thin slices of such solids are either circular **disks** or, for hollow solids, **washers** (disks with holes in the middle). This property simplifies calculating cross-sectional areas—and explains these solids' popularity.

Figure 3(a) shows one such solid; Figure 3(b) shows five disk-shaped slices that approximate the solid. (We calculate the solid's volume in Example 3.)

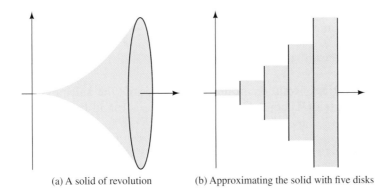

(a) A solid of revolution   (b) Approximating the solid with five disks

FIGURE 3
Disks approximating a solid

EXAMPLE 3   Find the volume of the solid formed by revolving the shaded region shown in Figure 4 about the x-axis.

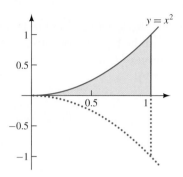

FIGURE 4
Rotate the shaded region about the x-axis

Solution   For any x in $[0, 1]$, the cross section at x is a circle with radius $y = x^2$. Thus,

$$A(x) = \pi(x^2)^2 = \pi x^4.$$

By Theorem 1,

$$\text{Volume} = \int_0^1 A(x)\,dx = \int_0^1 \pi x^4\,dx = \frac{\pi}{5} \approx 0.628.$$ ∎

A similar computation applies whenever the region under the graph of a positive function $y = f(x)$, from $x = a$ to $x = b$, is revolved around the x-axis.

EXAMPLE 4   Find the volume of the sphere of radius r.

Solution   The sphere is the result of revolving the area under the curve $f(x) = \sqrt{r^2 - x^2}$, from $x = -r$ to $x = r$, about the x-axis. The cross section at x, therefore, is a circle with radius $f(x)$. Hence,

$$A(x) = \pi\big(f(x)\big)^2 = \pi(r^2 - x^2).$$

Therefore,

$$\text{Volume} = \pi \int_{-r}^{r} (r^2 - x^2)\,dx = \pi\left(r^2 x - \frac{x^3}{3}\right)\Bigg]_{-r}^{r} = \frac{4\pi r^3}{3},$$

which is the well-known volume formula.   ∎

Other axes of rotation   Rotation around the x-axis is not the only possibility. Whatever the axis, the principle of Theorem 1 applies: Volume is found by integrating the cross-sectional area.

EXAMPLE 5   Find the volume of the solid formed by rotating the shaded region in Figure 4 about the axis $y = -1$.

Solution   The solid formed in this way is a *hollow* "horn." For any x in $[0, 1]$ the cross section at x is a **washer** with inner radius 1 and outer radius $1 + x^2$. The area $A(x)$ of such a washer is

$$A(x) = \pi\,(\text{outer radius})^2 - \pi\,(\text{inner radius})^2$$
$$= \pi(1 + x^2)^2 - \pi 1^2 = \pi(2x^2 + x^4).$$

By Theorem 1,

$$\text{Volume} = \int_0^1 A(x)\,dx = \pi \int_0^1 (2x^2 + x^4)\,dx = \frac{13\pi}{15} \approx 2.723.$$ ∎

EXAMPLE 6   Find the volume of the dish-shaped solid formed by rotating the shaded region in Figure 4 about the y-axis.

Solution   We slice the solid *horizontally* (with slices perpendicular to the y-axis). The slice at height y is a washer with cross-sectional area $A(y)$; it has outer radius determined by the line $x = 1$ and inner radius determined by the curve $y = x^2$. To express $A(y)$ as a function of y, it's convenient to rewrite the curve equation $y = x^2$ in the form $x = \sqrt{y}$. Now the washer at height y has area

$$A(y) = \pi \, (\text{outer radius})^2 - \pi \, (\text{inner radius})^2 = \pi \, (1 - y),$$

and so the volume in question is (by an easy calculation)

$$\int_0^1 A(y)\, dy = \pi \int_0^1 (1 - y)\, dy = \frac{\pi}{2}.$$ ■

Noncircular cross sections

For solids with noncircular cross sections, $A(x)$ may be harder to find; otherwise, the idea is exactly the same.

EXAMPLE 7   A pyramid of height 1 has square horizontal cross sections; at the bottom, the edge length is 1. Find its volume.

Solution   Let the x-axis run vertically from $x = 0$ to $x = 1$. Because the edge length decreases *linearly* from 1 (at the base) to 0 (at the apex), the edge length $\ell(x)$ at height x is $1 - x$. Thus, the cross-sectional area at height x is

$$A(x) = \ell(x)^2 = (1 - x)^2 = 1 - 2x + x^2,$$

and the volume is

$$\text{Volume} = \int_0^1 A(x)\, dx = \int_0^1 (1 - 2x + x^2)\, dx = \frac{1}{3}.$$

(Note that the pyramid has 1/3 the volume of the rectangular prism with the same base.) ■

Prisms and ''hats'': a general volume formula   It is well known that if a **prism** has base area A and height h, then the volume is just the product Ah. This simple formula applies, as we saw in Example 1, because, for prisms, the cross-sectional area function is *constant*.

A similar but not identical volume formula applies to solids we'll call **hats**. ➤ These are solids that start from a base and "taper linearly to a point." In Figure 5, for example, the solid's base is a rectangle; the intermediate slices (smaller but similar rectangles) taper linearly to a point.

The name isn't official, but it suggests the shape.

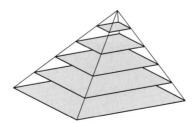

FIGURE 5
Five slices from a "hat" solid

A number of well-known solids can be thought of as hats; they include square pyramids (such as in Example 7), tetrahedra (triangular pyramids), and circular cones. (We do *not* require that the "point" of a hat be "centered" above the base, though this does occur for Egyptian pyramids and for ordinary funnel-shaped cones.)

How is the volume of a hat with base area A and height h related to the volume of a *prism* with the same base area and height? The answer is surprisingly simple:

> **FACT (Hat sizes)** Suppose that a hat solid has base area A and height h measured perpendicular to the base. Then the hat's volume is
>
> $$\text{Volume} = \frac{1}{3} Ah = \frac{1}{3} \text{ base} \times \text{height}.$$

Figure 5 shows how the area decreases with "altitude."

The Fact holds because of a special property of slices parallel to the base. If $A(x)$ denotes the area of the slice at height x, then $A(x)$ decreases from $A(0) = A$ (at the bottom) to $A(h) = 0$ (at the top). ◄ The special geometry of hats lets us be more specific:

The area of the slice at height x is $A(x) = \left(\dfrac{h-x}{h}\right)^2 \cdot A$.

Figure 5 suggests why this particular formula holds (notice especially the power 2). In the figure, the base and all four slices shown are similar rectangles. These rectangles have edge lengths that decrease *linearly* to zero as x (the height) runs from 0 to h. If, say, the base rectangle has dimensions a and b (so $A(0) = ab$), then the rectangle at height x has dimensions $a(h-x)/h$ and $b(h-x)/x$, ◄ and thus its area is $A(x) = ab(h-x)^2/h^2 = A(h-x)^2/h^2$. A similar argument applies to *any* hat-shaped solid.

The multiplier $(h-x)/h$ represents a "fraction" of the total height.

Now Theorem 1 lets us calculate the hat's volume:

$$\text{Volume} = \int_0^h A(x)\,dx = \int_0^h A\frac{(h-x)^2}{h^2}\,dx = \frac{1}{3}Ah.$$

(The last step is a straightforward calculation.)

BASIC EXERCISES

1. (See Example 2.) Ranger Rick's apprentice, Nick, is a city kid unfamiliar with forest lore. Nick used Rick's circumference data and the trapezoid rule—but he assumed that cross sections of the tree were squares. What's Nick's volume estimate?

2. Seeing his calculator battery dying, Rick decides to use only two of his measurements, at $x = 5$ and $x = 15$, to estimate the volume using the midpoint rule. What is his estimate now?

3. Yet another way to estimate the volume of a log is to regard the log as part of a circular cone (the entire tree trunk) with radius decreasing linearly from a maximum at the bottom to zero at the top. Rick's log has circumference 4.0 m at the bottom and 2.0 m at height 20 m. Use these measurements to estimate the log's volume yet again.

4. Let V be the volume of a circular cone with base radius r and height h.

   (a) Use the Fact about hat sizes on page 428 to explain the formula $V = \pi r^2 h/3$.

   (b) Find the formula for V in part (a) by evaluating an integral of the form $V = \int_0^h A(y)\,dy$.

5. Let V be the volume of a square pyramid with base side s and height h.

   (a) Use the Fact about hat sizes on page 428 to find a formula for V.

   (b) Find the formula for V in part (a) by evaluating an integral of the form $V = \int_0^h A(y)\,dy$.

6. (a) Imagine a circular cone with base radius r and height h. Now lop off the top part of the cone with a horizontal cut at height $h/2$. What fraction of the cone's *volume* did you cut off?

   (b) Imagine a square pyramid with base side s and height h. Now lop off the top part of the pyramid with a horizontal cut at height $h/2$. What fraction of the pyramid's *volume* did you cut off?

(c) Consider *any* hat solid with base area A and height h. What proportion of the total volume is contained in the top half of the solid (i.e., the part with "altitude" $h/2$ or greater)?

7. Imagine a circular cone with base radius r and height h. At what vertical height would you have to cut off the cone's top to get two pieces of equal volume?

8. Imagine a square pyramid with base side s and height h. At what vertical height would you have to cut off the pyramid's top to get two pieces of equal volume?

9. Let V be the volume of the solid obtained by rotating about the x-axis the region R bounded above by the curve $y = f(x)$, below by the x-axis, on the left by the vertical line $x = a$, and on the right by the vertical line $x = b$. Furthermore, suppose that $a = x_0 < x_1 < \ldots x_{n-1} < x_n = b$ and that $x_{j-1} \le t_j \le x_j$ if $0 \le j \le n$.

(a) Explain why $V \approx \pi \sum_{j=1}^{n} \left(f(t_j) \right)^2 (x_j - x_{j-1})$.

(b) Suppose that f is decreasing over the interval $[a, b]$ and that $t_j = x_j$. Does the approximating sum in part (a) underestimate V? Justify your answer.

(c) Suppose that $f(x) = x^2$, $a = 0$, and $b = 2$. Explain why
$$\lim_{n \to \infty} \pi \sum_{j=1}^{n} \left(f(t_j) \right)^2 (x_j - x_{j-1}) = \frac{32\pi}{5}.$$

10. Let V be the volume of the solid obtained by rotating about the y-axis the region R bounded on the right by the curve $x = g(y)$, on the left by the y-axis, on the bottom by the horizontal line $y = a$, and on the top by the horizontal line $y = b$.

Furthermore, suppose that $a = y_0 < y_1 < \ldots y_{n-1} < y_n = b$, and that $y_{j-1} \le t_j \le y_j$ if $0 \le j \le n$.

(a) Explain why $V \approx \pi \sum_{j=1}^{n} \left(g(t_j) \right)^2 (y_j - y_{j-1})$.

(b) Suppose that g is decreasing over the interval $[a, b]$ and that $t_j = x_{j-1}$. Does the approximating sum in part (a) overestimate V? Justify your answer.

(c) Suppose that $g(y) = \sqrt{y}$, $a = 0$, and $b = 4$. Explain why
$$\lim_{n \to \infty} \pi \sum_{j=1}^{n} \left(g(t_j) \right)^2 (x_j - x_{j-1}) = 8\pi.$$

In Exercises 11–14, sketch the region in the first quadrant bounded by the given curves and find the volume of the solid that is formed when this region is revolved around the x-axis.

11. $y = x^3$, $y = 0$, $x = 8$

12. $y = x^4$, $y = 1$, $x = 0$

13. $y = x + 6$, $y = x^3$, $x = 0$

14. $y = x^2$, $y = x^3$

In Exercises 15–18, sketch the region bounded by the given curves and find the volume of the solid of revolution formed when the region is revolved around the y-axis.

15. $y = \sqrt{x}$, $y = 0$, $x = 4$

16. $y = \sqrt{x}$, $y = x^2$

17. $y = x^2$, $y = 2^x$, $x = 0$

18. $y = e^x$, $y = 0$, $x = 0$, $x = 1$

FURTHER EXERCISES

19. The base of a solid object is the circle $x^2 + y^2 = 4$; cross sections of the object perpendicular to the x-axis are squares with one side in the xy-plane. What is the volume of the object?

20. The base of a solid object is the region bounded by the parabola $y = x^2/2$ and the line $y = 2$; cross sections of the object perpendicular to the y-axis are semicircles. What is the volume of the object?

21. The base of a solid object is the region bounded by the parabola $y^2 = 3x$ and the line $x = 3$; cross sections of the object perpendicular to the x-axis are equilateral triangles. What is the volume of the object?

22. The base of a certain solid is the region enclosed by $y = 1/x$, $y = 0$, $x = 1$, and $x = 4$. Every cross section of the solid taken perpendicular to the x-axis is an isosceles right triangle with its hypotenuse across the base. Find the volume of the solid.

23. Consider a right circular cylinder with radius r and height h.

(a) Compute the volume of the cylinder using circular cross sections.

(b) Compute the volume of the cylinder using rectangular cross sections.

24. Suppose that a $45°$ wedge is cut out of a circular cylinder with radius r. (One edge is horizontal and the other is inclined to the horizontal at an angle of $45°$.)

(a) Find the volume of the wedge using cross sections that are isosceles right triangles.

(b) Find the volume of the wedge using cross sections that are rectangles.

25. A drinking glass has circular cross sections. The glass has height 5 inches, bottom diameter 2 inches, and top diameter 3 inches. How much liquid can the glass hold?

26. A pie dish has base diameter 8 in., top diameter 10 in., and height 2 in. What is the volume of the dish?

27. Assume that the Earth is a sphere with circumference $C \approx 24{,}900$ mi.

(a) Find the volume of the Earth north of latitude $45°$.

(b) Find the volume of the Earth between the equator and latitude $45°$.

28. A spherical balloon with radius 3 in. is partially filled with water. If the water in the balloon is 4 in. deep at the deepest point, how much water is released when the balloon hits the wall and breaks?

29. A cylindrical gasoline tank with radius 4 ft and length 25 ft is buried on its side under a service station (i.e., its flat ends are perpendicular to the ground surface). If the gasoline in the tank is 6 ft deep, what is the volume of gasoline in the tank?

30. Suppose that a hemispherical bowl of radius r, initially full of a liquid, is tilted $45°$. How much liquid remains in the bowl?

31. Cross sections of a table leg are squares. The following table gives the width (in inches) of cross sections at several heights (also in inches). Use numerical integration (e.g., the midpoint rule) to estimate the volume of the table leg.

Height	0	5	10	15	20	25	30
Width	1.6	1.4	1.2	1.0	0.8	0.6	0.4

32. Suppose that the volume of water required to fill a hollow object to a depth of h inches is $V(h) = 1.5h + \sin h$ cubic inches. What is the cross-sectional area of the object 1 inch above its base?

33. A fuel-oil tank is 10 ft long and has flat ends that are perpendicular to the ground surface. Cross sections parallel to the flat ends have the shape of the ellipse $x^2/9 + y^2/36 = 1$. If the fuel oil in the tank is 9 ft deep, what is the volume of the fuel oil in the tank?

34. Suppose that two circular cylinders of radius R intersect in such a way that their axes meet at right angles. Find the volume of the space that is inside both cylinders. [HINT: Draw a sketch. The space that is inside both cylinders has an axis perpendicular to the axes of the cylinders. Furthermore, any cross section of the space perpendicular to this axis is a square.]

35. Let R be a rectangle with vertices at the points $(a, 0)$, (a, h), $(b, 0)$, and (b, h), where $h > 0$ and $b > a$.
(a) What is the area of R?
(b) What is the volume of the solid obtained by rotating R about the x-axis?
(c) What is the volume of the solid obtained by rotating R about the line $y = -c$, where $c \geq 0$?
(d) What is the volume of the solid obtained by rotating R about the line $y = c$, where $c \geq h$?

36. Let R denote the region bounded by the curves $y = x^2$, $y = 6 - x$, and $y = 0$. Write an integral for the volume of the solid that is obtained when R is rotated about the line
(a) $x = -2$
(b) $y = -1$

In Exercises 37–40, sketch the region bounded by the given curves and find the volume of the solid of revolution formed when the region is revolved around the line $y = a$.

37. $y = \sqrt{x}, \quad y = 0, \quad x = 1, \quad a = 1$

38. $y = x^2 - 1, \quad y = x + 1, \quad a = -1$

39. $x = y^2 - 4, \quad y = 2 - x, \quad a = 2$

40. $y = \sqrt{x}, \quad y = x^2, \quad a = -2$

41. Find the volume of the solid formed when the region bounded by the curves $y = \arctan x$, $y = 0$, and $x = 1$ is revolved around the y-axis.

42. Let V be the volume of the solid formed when the region bounded by the curves $y = \arctan x$, $y = 0$, $x = -3$, and $x = -1$ is revolved around the x-axis.
(a) Express V as a definite integral.
(b) If the integral in part (a) is estimated using a left sum with $n = 10$ subintervals, does the result underestimate V? Justify your answer.

*Exercises 43–55 provide an introduction to the **method of cylindrical shells**: the volume of the region formed when the area below the curve $y = f(x) \geq 0$ from $x = a$ to $x = b$ is rotated around the y-axis is $V = 2\pi \int_a^b x f(x)\, dx$.*

43. Let R be a rectangle with vertices at the points $(a, 0)$, (a, h), $(b, 0)$, and (b, h), where $0 \leq a < b$ and $h > 0$. If R is rotated about the y-axis, the resulting solid is a **cylindrical shell**. Find a formula for the volume of this solid of revolution in terms of a, b, and h.

44. Let f be a function that is continuous and nonnegative on the interval $[a, b]$ and R be the region bounded above by the curve $y = f(x)$, below by the x-axis, on the left by the vertical line $x = a$, and on the right by the vertical line $x = b$. Furthermore, suppose that $0 \leq a = x_0 < x_1 < \ldots x_{n-1} < x_n = b$, $m_j = (x_{j-1} + x_j)/2$, and that V is the volume of the solid obtained by rotating R about the y-axis.
(a) Explain why $V \approx \pi \sum_{j=1}^{n} f(m_j)\left((x_j)^2 - (x_{j-1})^2\right)$.
(b) Show that $f(m_j)\left((x_j)^2 - (x_{j-1})^2\right) = 2m_j f(m_j)(x_j - x_{j-1})$. [HINT: $u^2 - v^2 = (u + v)(u - v)$.]
(c) Use parts (a) and (b) to justify the formula $V = 2\pi \int_a^b x f(x)\, dx$.

45. Find the volume of the solid of revolution formed when the region bounded by the curve $y = \sin x$ between $x = 0$ and $x = \pi$ is rotated around the y-axis.

46. Find the volume of a hemisphere with radius R by rotating the region enclosed by the curve $y = \sqrt{R^2 - x^2}$, the x-axis, and the y-axis around the y-axis.

47. Find the volume of the cone formed by rotating the region enclosed by the line $y = 1 - x/2$, the x-axis, and the y-axis around the y-axis.

48. Find the volume of the solid of revolution formed when the region bounded by the curve $y = e^{-x^2}$ and the lines $x = 0$ and $x = 1$ is rotated around the y-axis.

49. Find the volume of the solid of revolution formed when the region bounded by the curve $y = \ln x$ and the lines $x = 1$ and $x = e$ is rotated around the y-axis.

50. Suppose that a circular hole of radius 2 cm is bored through the center of a sphere of radius 3 cm. What is the volume of the portion of the sphere that remains?

51. Suppose that a circular hole of radius a is bored through the center of a cone of height h and radius r. What is the volume of the portion of the cone that remains?

52. Let V be the volume of the solid of revolution formed when the region bounded by the curve $y = x^{1/3}$ and the lines $y = 0$, $x = 0$, and $x = 8$ is revolved around the y-axis.
  (a) Compute V using the method of disks and washers.
  (b) Compute V using the method of cylindrical shells.

53. Let V be the volume of the solid of revolution formed when the region bounded by the curve $y = x\sqrt{1 - x^2}$ and the lines $y = 0$, $x = 0$, and $x = 1$ is revolved around the y-axis.
  (a) Use the method of disks and washers to show that
$$V = \int_0^{1/2} \frac{\pi}{2}\left(1 + \sqrt{1 - 4y^2}\right) dy - \int_0^{1/2} \frac{\pi}{2}\left(1 - \sqrt{1 - 4y^2}\right) dy.$$

(b) Use the method of cylindrical shells to show that
$$V = \int_0^1 2\pi x^2 \sqrt{1 - x^2}\, dx.$$

(c) Show that $V = \pi^2/8$.

54. The rate at which a viscous fluid (e.g., water or blood) moves through a tube is not uniform. The fluid moves fastest near the center and slowest along the wall.

Suppose that a fluid is flowing in a circular pipe of radius R centimeters and that the velocity of the fluid r centimeters from the center of the pipe is $v(r)$ centimeters per second. Explain why the rate (in cubic centimeters per second) at which the liquid is flowing through the pipe is the value of the integral
$$2\pi \int_0^R r v(r)\, dr.$$

55. Suppose that a city is circular and that the population density decreases linearly to zero from the center to the rim.
  (a) Let K be the population density (in people per square mile) at the center of the city, and let R be the radius of the city. Explain why the population density at a distance r from the center of the city is $p(r) = K(1 - r/R)$ people/mi^2.
  (b) What is the population of the city? [HINT: Use an integral.]

7.3 WORK

How much work does a rocket do in lifting a satellite into orbit? How much work does a car do on the 934-mile drive from Perham, Minnesota to Durham, North Carolina? How much work does a horse do in plowing the back 40? Is it more work to raise a 60-lb bucket from a 60-ft well or a 50-lb bucket from a 70-ft well?

Those are all good questions, with thoroughly practical consequences. Any kind of work requires fuel (hydrogen, gasoline, hay, cheeseburgers, . . .), and it's important to know how much is needed "on board." **Work** is the common ingredient (and the common word) in each situation mentioned. We need a mathematically useful definition.

The meaning of work

The everyday meaning of "work" is all too familiar; the word covers everything from completing tax returns to digging ditches. Physicists have a narrower, more precise definition. (To a physicist, digging ditches is a lot of work; filing tax returns is almost none.) Work is done when a *force acts through a distance*. Without movement, *no* work is done; for instance, a house does no work by holding up its roof. Even Atlas, the Titan forced to bear the world on his shoulders, accomplished no work (except, perhaps, in *lifting* the globe)—fatiguing though his labors must have been.

Simplifying assumptions: Are they reasonable?   Throughout this section, we assume the following:

A force (either constant or variable) acts along a straight line; movement occurs along the same line.

These assumptions are convenient, but are they reasonable? After all, few real-life physical phenomena are that simple. Forces may act at *angles* to the direction of motion, and motion may occur along *curves*. Moving cars, for instance, encounter many forces—gravity, air drag, rolling resistance, crosswinds—all of which vary over the course of a trip. Can anything useful survive our simplifying assumptions?

In fact, much *does* survive. Some interesting physical phenomena conform to our "straight-line" assumptions. Complicated combinations of forces, moreover, can often be understood as *sums* of simpler forces, each acting as we have assumed.

Work done by a constant force   In the simplest instance, the force is constant along its line of action, and the "obvious" definition applies:

$$\text{Work} = \text{force} \times \text{distance}.$$

The *units* of work reflect this definition. In the United States, lifting a 30-pound toddler 3 feet to hip height takes

$$30 \text{ pounds} \cdot 3 \text{ feet} = 90 \text{ foot-pounds}$$

One pound is about 4.44 newtons.

of work. A common unit of force in the metric system is the **newton**(N)—about 0.225 lb. ◄ Thus, lifting the same child to the same hip in Canada takes

$$133.44 \text{ newtons} \cdot 0.914 \text{ m} \approx 121.964 \text{ newton-meters}$$

James P. Joule was a 19th-century British physicist.

of work. (Newton meters are known in physics as **Joules**.) ◄

Other units of work, such as the inch-ounce and the mile-pound are possible, but all have the same "dimension": units of *distance* times units of *force*.

Work done by a variable force   In typical physical situations work is performed by a force that *varies* along its line of action. Even hoisting a toddler can be construed this way: On the way up, the child moves away from the center of the Earth and so becomes "lighter." For practical purposes this effect is usually small enough to ignore. (But see Example 4, just ahead.) In more interesting situations, the force varies more dramatically:

- **Springs**   The farther a spring is stretched from its "natural length," the greater the force with which it pulls back. According to **Hooke's law**, an ideal spring's **restoring force** $F(x)$ is proportional to x, the distance the spring is stretched. That is,

$$F(x) = kx$$

The numerical value of k depends on the units of measurement and on the "stiffness" of the spring.

for some constant k, called the **spring constant**. ◄

- **Gravity**   An object's weight decreases as it moves away from the Earth. **Newton's law of universal gravitation** sharpens this commonplace observation:

  The force of Earth's gravity on an object is inversely proportional to the square of the distance from the object to the center of the Earth.

  Thus, the force F required to lift an object (i.e., to counteract gravity) is $F(x) = k/x^2$, where x is the object's distance from the center of the Earth. The value of the proportionality constant k depends on the object and on the units used to measure force and distance.

- **Buckets and ropes**   The force required to draw water from a well *decreases* as the bucket rises—the higher the bucket, the less rope to be pulled up. (We pursue this observation in Example 5.)

- **Pumping iron**   One way to "press" a heavy barbell is to exert a *constant* upward force (equal to the barbell's weight) all the way from shoulder level to the top of one's reach. This strategy is mathematically simple but not physiologically ideal. The mechanics of the human body make it easier to exert more upward force at some levels of arm extension than at others. Therefore, to the extent that the rules permit, a human weight lifter exerts a *varying* upward force through a barbell's travel. (But note: Whatever the weight lifter's strategy, the *work* of lifting is the same: the barbell's weight times the distance lifted.)

Handling such examples requires a definition that permits *variable* forces.

DEFINITION (Work) Let F be a continuous function. If the force $F(x)$ acts along an axis from $x = a$ to $x = b$, then the work done by the force F is

$$\int_a^b F(x)\,dx.$$

After a simple example we'll discuss why this definition is reasonable.

EXAMPLE 1 Stretching a spring 1 ft beyond its natural length requires 10 lb of force. How much work is done in stretching the spring from rest to 1 ft? From rest to a ft? How does the answer depend on a?

Solution By Hooke's law, $F(x) = kx$. The given conditions say that $F(1) = 10$, and so $k = 10$. By the preceding definition,

$$\text{Work to stretch 1 ft} = \int_0^1 10x\,dx = 10\frac{x^2}{2}\bigg]_0^1 = 5 \text{ ft-lb};$$

$$\text{Work to stretch } a \text{ ft} = \int_0^a 10x\,dx = 10\frac{x^2}{2}\bigg]_0^a = 5a^2 \text{ ft-lb}.$$

The last answer reveals an interesting relationship between force and work:

The force required to stretch a spring a feet from rest is proportional to a; the work done in the process is proportional to a^2. ■

Work as an integral

Why is it "right" to define work as an integral? Does the integral really represent a physically meaningful quantity, one that deserves the name "work"? Any sensible definition of work done by a *variable* force should, at least, agree with the simpler definition of work done by a *constant* force. Ours does: If $F(x) = k$, then

$$\text{Work} = \int_a^b F(x)\,dx = \int_a^b k\,dx = k(b - a) = \text{force} \times \text{distance}.$$

So far so good. But why is this definition right for variable forces? We'll give two arguments.

Adding small contributions: the integral as a sum The work integral $\int_a^b F(x)\,dx$ can be thought of as a limit of approximating sums. Specifically, let's slice the interval $[a, b]$ into n equal subintervals, each of length Δx, and let x_i be the right endpoint of the ith subinterval. ➡ Then

Left endpoints or midpoints would have done as well.

$$\int_a^b F(x)\,dx = \lim_{n \to \infty} \sum_{i=1}^n F(x_i)\,\Delta x.$$

Figure 1 should look familiar from earlier work:

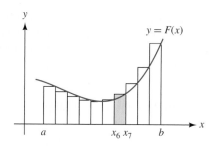

FIGURE 1

Approximating work: a right sum

The sum $\sum_{i=1}^{n} F(x_i)\,\Delta x$ is the key quantity. Each summand $F(x_i)\,\Delta x$ represents the work done by the *constant* force $F(x_i)$ over the subinterval $[x_{i-1}, x_i]$. (The shaded area in Figure 1 represents the work done over the subinterval $[x_6, x_7]$.) The sum adds up these contributions; the entire polygonal area represents this total. Here is the punch line:

> *Each approximating sum is an estimate of the total work done over the interval* $[a, b]$.

Now as $n \to \infty$, the sums tend (by definition of the integral) to $\int_a^b F(x)\,dx$, and so the integral plausibly measures the "true" work done.

Increments and elements   As further evidence for the integral definition of work, we offer the following mnemonic argument using "increments" and "elements." Let x be any point in the interval $[a, b]$. Consider the tiny interval $[x, x + \Delta x]$, with **increment of distance** Δx. Because F remains essentially constant over any tiny interval, the work ΔW done over the interval $[x, x + \Delta x]$ satisfies

$$\Delta W \approx F(x)\,\Delta x.$$

Letting Δx tend to zero now gives

$$dW = F(x)\,dx;$$

dW is called the **element of work.** "Adding up" these elements by integration gives the total work done from $x = a$ to $x = b$:

$$\text{Total work} = W = \int dW = \int_a^b F(x)\,dx.$$

(The equation $W = \int dW$ should be understood informally; it means that the integral "adds up small contributions.")

Calculating work: Miscellaneous examples

The definition of work as an integral is simple enough. Deciding *what* to integrate, and over what *interval*, is usually the hardest part of applying the definition. We illustrate the process with several examples.

EXAMPLE 2   (Working against gravity)   Earth's gravity affects even very distant objects; it never dies out completely. How, then, is it possible to put objects into orbit? Calculus gives the answer.

Communications satellites weigh around 4000 newtons and orbit the Earth at altitudes of about 25,000 kms. How much work is done against Earth's gravity in lifting such a satellite into orbit?

Solution By Newton's law of gravitation, the (variable) force F needed to counteract Earth's gravity is

$$F(x) = \frac{k}{x^2},$$

where k is a constant and x denotes the distance from Earth's center. That the satellite weighs 4000 N on Earth's surface—about 6000 km from Earth's center—means (allowing for approximation)

$$4000 = \frac{k}{6000^2} \implies k = 1.44 \times 10^{11} \implies F(x) = \frac{1.44 \times 10^{11}}{x^2},$$

where force is measured in newtons. Therefore, the work done (in newton-kilometers) in moving from $x = 6000$ to $x = 31,000$ ➡ is found by an easy integral calculation:

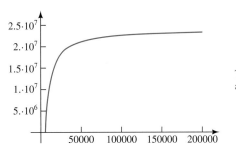

$x = 31,000$ *corresponds to an* "*altitude*" *of 25,000 km above Earth's surface.*

$$\int_{6000}^{31000} \frac{1.44 \times 10^{11}}{x^2}\, dx = -\frac{1.4 \times 10^{11}}{x}\Bigg]_{6000}^{31000} \approx 19,350,000 \text{ N-km}.$$

(In more conventional physical units, this is about 1.9×10^{10} Joules.) That's heavy lifting to be sure, but not impossibly heavy. A loaded Boeing 747 airplane accomplishes about the same work against gravity to reach cruising altitude. ■

EXAMPLE 3 (Flying to infinity) How much *additional* work is required to lift the satellite another 75,000 km to an altitude of 100,000 km? Is there *any* upper limit on altitude?

Solution The next 75,000 km of altitude come almost for free. The work done in raising the satellite from $x = 31,000$ to $x = 106,000$ km is a relatively trifling

$$\int_{31000}^{106000} \frac{1.44 \times 10^{11}}{x^2}\, dx \approx 3,290,000 \text{ N-km}.$$

which is only about 17% of the work required to gain the first 25,000 km of altitude. A computation similar to those above gives the work $W(a)$ done in lifting the satellite from Earth's surface to *any* altitude a:

$$W(a) = \int_{6000}^{a} \frac{1.44 \times 10^{11}}{x^2}\, dx = 2.4 \times 10^7 \left(1 - \frac{6000}{a}\right).$$

A close look at the last expression reveals a surprising conclusion that has important consequences for space flight:

> *Lifting the satellite to* any *altitude a, no matter how great, requires no more than* 2.4×10^7 *N-km of work.*

Plotting W against a (Figure 2) shows that W is bounded above—and suggests why space flight is possible.

The graph has a horizontal asymptote at height 2.4×10^7.

FIGURE 2
Work (N-km) as a function of altitude (km)

■

EXAMPLE 4   Gravity's force diminishes—very slightly—when a 30-lb toddler is lifted 3 ft to hip height. We said earlier that this effect is negligible. Is it?

Solution   If we *ignore* diminishing gravity, the work done is simply 30 lb times 3 ft, or 90 ft-lb.

Now let's take Newton's law into account. The computations resemble those of Example 2 but with different numbers. We'll need these facts:

- The Earth's radius is about 21 million ft, and so the toddler's vertical travel is from $x = 21,000,000$ to $x = 21,000,003$ ft.
- At $x = 21,000,000$, $F(x) = k/x^2 = 30$, and so $k \approx 1.323 \times 10^{16}$.

All ingredients are now ready; the total work is found by integration:

$$\int_{21000000}^{21000003} \frac{1.323 \times 10^{16}}{x^2} \, dx = -\frac{1.323 \times 10^{16}}{x} \Bigg]_{21000000}^{21000003} \approx 89.99998714.$$

Thus, the "work savings" from diminishing gravity is therefore only 0.00001286 ft-lb—about the work done when a medium-sized raindrop falls one foot.   ■

EXAMPLE 5   Is it more work to raise a 70-lb bucket of water from a 50-ft well or a 50-lb bucket from a 70-ft well? Rope weighs 0.25 pounds per foot.

Solution   Let x denote "height," that is, the distance (in feet) from either bucket to the bottom of its well. As the first bucket travels from $x = 0$ to $x = 50$, the force required to raise it varies with x. At height x, this force must counteract both the weight of the bucket and the weight of $50 - x$ ft of rope. Thus, the first bucket's force function is

$$F(x) = 70 + 0.25 \cdot (50 - x) = 82.5 - 0.25x,$$

and the work done is

$$\int_0^{50} F(x) \, dx = \int_0^{50} (82.5 - 0.25x) \, dx = 3812.5 \text{ ft-lb.}$$

The second bucket's computation is almost identical. This time,

$$F(x) = 50 + 0.25 \cdot (70 - x) = 67.5 - 0.25x,$$

and the work done is

$$\int_0^{70} F(x) \, dx = \int_0^{70} (67.5 - 0.25x) \, dx = 4112.5 \text{ ft-lb.}$$

Note, finally, that the answer might have been guessed. The same amount of work (350 ft-lb) is done on each *bucket*, but drawing up the longer rope takes more work.   ■

EXAMPLE 6   (Pumping water)   A conical tank with radius r feet and depth h feet is full of water; the cone "points" downward. How much work is needed to pump all the water out the top of the tank? (Water weighs 62.4 lb per cubic foot.)

Solution   Here it's hard to visualize a force acting through a distance, and so we take a slightly different tack: We "slice the tank" horizontally into thin circular layers, calculate the work done on each layer, and "add up" our results using an integral. Figure 3 shows a side view of the tank and a typical thin layer.

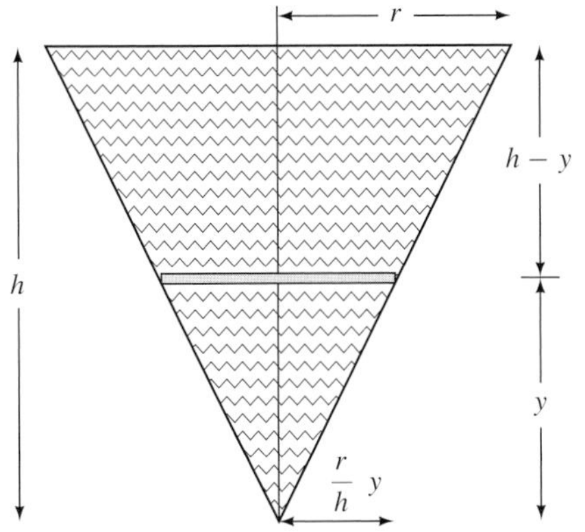

The "layer" at height y is shaded. Use similar triangles to find its radius.

FIGURE 3
Slicing a conical tank: a side view

Let y be the height (in feet) above the tank's bottom; then the tank itself lies between $y = 0$ and $y = h$. The layer at height y (shown shaded) is essentially a disk of radius ry/h with thickness Δy, and thus its approximate *volume* is

$$\text{area} \times \text{thickness} = \pi \left(\frac{r}{h} y\right)^2 \cdot \Delta y = \pi \frac{r^2}{h^2} y^2 \, \Delta y \quad \text{cubic feet.}$$

Thus, its approximate *weight* is

$$62.4 \cdot \pi \frac{r^2}{h^2} y^2 \, \Delta y \text{ lb.}$$

This layer is lifted $h - y$ feet in pumping, and so the work done on this layer is approximately

$$\Delta W = (h - y) \cdot 62.4\pi \frac{r^2}{h^2} y^2 \, \Delta y \text{ ft-lb.}$$

This means that the **element of work** in this situation is

$$dW = 62.4\pi \frac{r^2}{h^2} y^2 (h - y) \, dy;$$

the total work done (in foot-pounds) is the integral

$$\int dW = 62.4 \, \pi \frac{r^2}{h^2} \int_0^h y^2 (h - y) \, dy = 62.4 \frac{\pi}{12} r^2 h^2.$$

Notice, finally, that the *entire* tank of water weighs $62.4\pi r^2 h/3$ lb, so the work just calculated is the same as that required to lift the entire tank of water a distance of $h/4$ ft—one fourth the height of the cone. ■

BASIC EXERCISES

1. A certain spring has a natural length of 18 in; a force of 10 lb is enough to compress it to a length of 16 in.

(a) What is the value of the spring constant k?

(b) How much work is done in compressing the spring from 16 to 12 in.?

2. It takes $40x$ lb of force to keep a certain spring stretched x feet from rest.

(a) Find the work done in stretching the spring 2 ft, starting from rest.

(b) Find the work done in stretching the same spring s ft from rest. (The answer depends on s.)

(c) Farmer Ole's aging plow horse, Sven, can do only 10,000 ft-lb of work on his daily oats ration. Ole hitches Sven to the spring described above. How far can Sven pull?

3. It takes kx pounds of force to keep a spring stretched x feet from rest.

   (a) How much work is done in stretching the spring 10 ft from rest? (The answer depends on k.)

   (b) How much work is done in stretching the spring from $x = a$ ft to $x = a + 10$ ft? (The answer depends on both k and a.)

4. Redo Example 2, but assume that the orbit is 30,000 km from Earth's center.

5. The cable attached to an elevator car weighs 5 lb/ft. If 200 ft of cable must be wound onto a pulley to lift the car from the bottom floor to the top floor of a building, how much work is done in lifting just the cable?

6. A cylindrical tank with radius 5 ft and height 10 ft is half filled with water. How much work must be done to pump all the water over the top rim of of the tank?

7. Is it more work to raise a 60-lb bucket from a 60-ft well or a 50-lb bucket from a 70-ft well? Assume that the rope weighs 0.25 lb/ft.

FURTHER EXERCISES

8. In some situations, the force required to stretch a "spring" is not linear. Consider a simple model of a bow and arrow in which the force required to draw the arrow a distance x is $F(x) = ax + bx^2$.

   (a) If a force of 15 pounds is required to draw the arrow one foot and a force of 50 pounds is required to draw the arrow two feet, find the constants a and b.

   (b) How much work is required to draw the arrow 1 foot?

   (c) How much work is required to draw the arrow 2 feet?

9. A hemispherical tank with radius r is filled with a fluid of density ρ. How much work must be done to pump all the fluid over the top rim of the tank?

10. Suppose that a tank in the shape of the frustum of a cone is filled with a fluid of density ρ kilograms per cubic meter. If the radius of the top of the tank is 2 m, the radius of the bottom of the tank is 1 m, and height of the tank is 1 m, how much work must be done to pump all the fluid over the top rim of the tank?

11. A cylindrical gasoline tank with radius 4 ft and length 15 ft is buried under a service station. The top of the tank is 10 ft underground; its flat ends are perpendicular to the ground surface. Find the amount of work needed to pump all the gasoline in the tank to a nozzle that is 3 ft above the ground. (Gasoline weighs 42 lb/ft³.) [HINT: Each horizontal cross section of the tank is a rectangle, one side of which has length 15 ft.]

12. A bucket that weighs 80 lb when filled with water is lifted from the bottom of a well that is 75 ft deep. However, the bucket has a hole in it, and so it weighs only 40 lb when it reaches the top of the well. Suppose that water leaks from the bucket at a constant rate and that the rope weighs 0.65 lb/ft. Find the work required to lift the bucket from the bottom of the well to the top.

13. A certain spring exerts $4x$ pounds of force when stretched x feet from rest. One end is fixed to the ceiling. A chain that is 10 ft long and weighs 2 lb/ft hangs from the other end. The end of the chain just brushes the floor. Find the work done in pulling down on the chain a distance of 2 ft. (Don't ignore the weight of the chain!)

14. An object weighing k pounds slides without friction along the graph of $y = f(x)$. (The x and y units are feet.) A horizontal force $F(x)$ is applied to move the object from $x = a$ to $x = b$. Physical intuition says that at x, the horizontal force $F(x)$ required is proportional to the slope of the graph: $F(x) = kf'(x)$.

   (a) Find the work done if $k = 10, a = 0, b = 1$, and $f(x) = x$.

   (b) Find the work done if $k = 10, a = 0, b = 1$, and $f(x) = x^3$.

   (c) Find the work done if $k = 10, a = 0, b = 1$, and $f(x) = x^n$, for n any positive integer.

   (d) Explain the relations among your answers to parts (a)–(c).

15. Suppose that an object moves along the x-axis and that the force on the object varies with x. Explain how the work done by the force F as the object moves from $x = a$ to $x = b$ can be determined from a graph of $F(x)$.

16. The force between two molecules can be modeled by a function of the form $A/x^7 + B/x^{13}$, where A and B are constants and x is the distance between the molecules. Find the work done if the distance between the molecules changes from $x = a$ to $x = b$.

17. The cross-sectional area of a tank at height y above its base is $A(y)$; the top of the tank corresponds to $y = h$.

   (a) What does the expression $\int_0^h A(y)\,dy$ represent in this context?

   (b) Suppose that the tank is filled with water to depth $y = a$. Find an expression for the work required to pump all of the water out the top of the tank.

18. A conical tank with radius r and depth h feet is full of water; the cone "points" upward.

   (a) Suppose that all of the water is pumped out the top of the tank. Explain why the work necessary is *greater* than in Example 6.

   (b) Compute the work needed to pump all of the water out the top of the tank.

7.4 SEPARATING VARIABLES: SOLVING DEs SYMBOLICALLY

The idea — an example and remarks

In earlier sections we approached DEs and IVPs graphically (with slope fields) and numerically (with Euler's method). Now we solve DEs *symbolically*, by **separating variables**. Before describing the method in general terms, we illustrate how it works—when it works—with another look at cooling coffee.

EXAMPLE 1 Let $y(t)$ denote the temperature (in degrees Fahrenheit) of a cup of coffee at time t (in minutes). Room temperature is $70°$F; the coffee starts at $190°$F. The coffee's temperature is described by the IVP

$$y' = -0.1(y - 70); \quad y(0) = 190.$$

How hot is the coffee after 10 min? How hot is the coffee at *any* time t? (We addressed similar questions in Example 1 of Section 6.3, page 394.)

Solution First we rewrite the DE using the Leibniz notation for y':

$$\frac{dy}{dt} = -0.1(y - 70).$$

Next comes the key step: *Separate the y's and dy's from the t's and dt's.* The result is

$$\frac{dy}{y - 70} = -0.1\, dt;$$

antidifferentiating both sides gives

$$\int \frac{dy}{y - 70} = \int -0.1\, dt.$$

Working out the antiderivatives is straightforward; the result is

$$\ln|y - 70| + C_1 = -0.1t + C_2,$$

where C_1 and C_2 are constants. Because both C_1 and C_2 can take *any* value, it's convenient (and legal) to combine the two constants into one:

$$\ln|y - 70| = -0.1t + C.$$

To solve for y, we exponentiate both sides:

$$\ln|y - 70| = -0.1t + C \implies |y - 70| = e^{-0.1t + C} \implies y = \pm e^C e^{-0.1t} + 70.$$

One more simplification helps reduce clutter. Because C is an arbitrary constant, so is $K = \pm e^C$; thus, y has the form

$$y = Ke^{-0.1t} + 70.$$

Finally, we use the initial condition to assign K a numerical value:

$$y(0) = 190 \implies 190 = Ke^0 + 70 \implies K = 120.$$

The problem is now completely solved. The function $y(t) = 120e^{-0.1t} + 70$ solves the IVP *exactly* for any time t. ➡ At $t = 10$, for example,

$$y(10) = 120e^{-1} + 70 \approx 114.146°\text{F}. \quad \blacksquare$$

A plot of $y(t)$ appears on page 394.

A legal separation? It's easy to check by differentiation that $y(t) = 120e^{-0.1t} + 70$ is a solution of the preceding IVP. But does the method really make sense? Can we really "separate" dy from dt as we just did?

That's a good question; dy/dt normally denotes *one* quantity, not the quotient of two separate quantities. The bottom line is that, although care with symbols is always wise, things work out all right in this case. The mathematical reason, roughly speaking, is that the antiderivative notations $\int f(t)\,dt$ and $\int g(y)\,dy$ are designed to ensure that dt and dy can be handled separately if appropriate care is taken.

Our best defense of the separation method, however, is that it works—and can be verified to work by an easy direct check. The same important principle applies to solving DEs and to finding antiderivatives: *Finding* answers may be difficult, but *checking* answers is (relatively) easy.

Symbolic solutions: from one, many   Solving the preceding DE by symbolic antidifferentiation was easy. Solving more complicated DEs symbolically can be *much* harder, if not impossible. Symbolic methods—when they work—have two important advantages over numerical and graphical approaches:

1. They produce *exact* solutions, not approximations.

2. They handle DEs that include parameters (symbols that can stand for any constant).

The second property means that solving *one* DE symbolically can amount, in effect, to solving whole families of DEs. The next example illustrates this advantage.

Time $t = 0$ can denote any convenient reference time.

EXAMPLE 2   Let $y(t)$ denote the temperature (in degrees Fahrenheit) of a cup of coffee at time t (in minutes). ◄ Room temperature is T_r °F; at time $t = 0$ the coffee is at T_0 °F. The coffee's temperature is described by the IVP

$$y' = -0.1(y - T_r); \quad y(0) = T_0.$$

How hot is the coffee after 10 minutes? How hot is it at *any* time t?

Solution   (Notice first that neither slope fields nor Euler's method would be any use here because both require numerical data.)

This example and the last are virtual clones. So are their solutions; we need only replace 70 with T_r, 190 with T_0, and 0 with t_0 wherever they appear in the preceding solution. Here's the result:

$$y(t) = (T_0 - T_r)e^{-0.1t} + T_r.$$

The beauty of this solution is that it applies for *all* values of the parameters T_0 and T_r. Figure 1(a) shows several solution curves, all with $T_r = 70$ but with various values of T_0. Figure 1(b) shows another set of solution curves, all with $T_0 = 190$ but with various values of T_r.

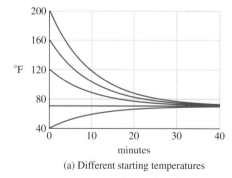

(a) Different starting temperatures

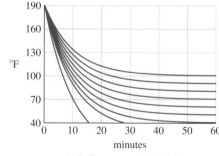

(b) Different room temperatures

FIGURE 1

Cups of coffee; how they cool

The method in general: when it works, when it doesn't

Separable and inseparable DEs   Separating variables can "work" only with separable DEs—those in which the t and y variables *can* be separated. Following are several **separable** DEs, each with its "separated" form:

$$\frac{dy}{dt} = ty \quad \Longrightarrow \quad \frac{dy}{y} = t\,dt;$$

$$\frac{dy}{dt} = \frac{\sin t}{y} \quad \Longrightarrow \quad y\,dy = \sin t\,dt;$$

$$\frac{dy}{dt} = \frac{\sin(t^2)}{y} \quad \Longrightarrow \quad y\,dy = \sin(t^2)\,dt.$$

Separable DEs often have, or can be written in, one of the forms

$$\frac{dy}{dt} = f(t)g(y) \quad \text{or} \quad \frac{dy}{dt} = \frac{f(t)}{g(y)}$$

and thus either

$$\frac{dy}{g(y)} = f(t)\,dt \quad \text{or} \quad g(y)\,dy = f(t)\,dt.$$

(This is the case in all three examples above.)

In some DEs (called **inseparable**), the t and y variables *cannot* be separated. Here are two simple-looking examples:

$$\frac{dy}{dt} = t + y; \quad \frac{dy}{dt} = \sin(t + y).$$

Both DEs *can* be solved—but not by separating variables.

Antiderivative problems   Even if t and y can be separated, the antiderivative problems

$$\int \frac{dy}{g(y)} = \int f(t)\,dt \quad \text{or} \quad \int g(y)\,dy = \int f(t)\,dt$$

remain. Again, trouble may loom: Some elementary functions cannot be antidifferentiated in elementary symbolic form. One of the separable DEs seen earlier poses just this problem; the right-hand antiderivative

$$\int y\,dy = \int \sin(t^2)\,dt$$

has no elementary solution.

The big picture and two morals   The separation of variables method applies only to separable DEs. In practice, DEs take many other symbolic forms; fortunately, other symbolic methods exist to solve them. For example, many practically useful DEs have the form

$$y' = a(t)y + b(t);$$

they're called **first-order linear DEs.** General methods exist for solving first-order (and higher-order) linear DEs. (We describe some in an Interlude in Chapter 8.) If $a(t)$ and $b(t)$ are suitably simple functions, solutions can be found explicitly in terms of elementary functions.

For example, the DE $y' = y + t$ is not separable, but using other methods one can produce the (easily checked) solution $y = Ce^t - t - 1$; C can be any constant.

There are two main morals: First, if separating variables fails, other methods may succeed. Second, solving DEs symbolically, like finding antiderivatives symbolically, is art as well as science; there is no fail-safe route to a solution.

Rumors, separable DEs, and logistic growth

In earlier work we used the logistic DE

$$\frac{dP}{dt} = kP(C - P) = kP(t)\big(C - P(t)\big)$$

to model populations that grow logistically. Both constants k and C can be interpreted in biological terms: C represents the environment's long-run carrying capacity, and k measures the population's reproduction rate.

Biological populations come first to mind, but the logistic DE works elsewhere, too. Translating the logistic DE into words gives the key property of *any* quantity that grows logistically toward a long-run value C:

> *The quantity P grows at a rate that is proportional* both *to P itself* and *to $C - P$, the "room available" for further growth.*

Solving the logistic DE   For later use, let's solve the logistic DE in general form. Separating variables sets up the antiderivative problem:

$$\frac{dP}{dt} = kP(C - P) \implies \int \frac{dP}{P(C - P)} = \int k\, dt.$$

See Formula 23.

The right side of the last equation is easy to integrate. The left side is stickier, but it can be found in an integral table ◄ by using technology or by symbolic methods we'll study in the next chapter. In any event, the answer works out to this:

$$\int \frac{dP}{P(C - P)} = \frac{1}{C} \ln \left| \frac{P}{C - P} \right|.$$

We can drop the absolute value signs if we assume—as we will—that $0 < P < C$.

Putting everything together and isolating P solves the logistic DE. The steps are a bit messy:

$$\frac{dP}{dt} = kP(C - P) \implies \int \frac{dP}{P(C - P)} = \int k\, dt$$

$$\implies \ln \frac{P}{C - P} = C(kt + D)$$

(D is a constant of integration)

$$\implies \frac{P}{C - P} = e^{Ckt} e^{CD} \qquad \text{(exponentiate both sides)}$$

$$\implies \frac{C - P}{P} = e^{-Ckt} e^{-CD} \qquad \text{(invert both sides)}$$

$$\implies \frac{C}{P} = e^{-Ckt} e^{-CD} + 1$$

$$\implies P = \frac{C}{de^{-Ckt} + 1}. \qquad (d = e^{-CD})$$

(Because D is an arbitrary constant, so is $d = e^{-CD}$.) The result is prettier than the calculation:

> **FACT**   For any positive constants C, d, and k, the function
> $$P(t) = \frac{C}{de^{-Ckt} + 1}$$
> solves the logistic differential equation $P' = kP(C - P)$.

To apply this result, we need only choose appropriate values for the constants.

Rumors and logistic growth

A rumor spreads as "tellers" pass it on to "hearers." (All hearers immediately become tellers.) The rumor spreads slowly at first, when tellers are few. It spreads faster when both tellers and hearers are plentiful, slows down again as hearers become scarce, and stops when everyone knows the rumor. The teller population, in other words, grows as shown in Figure 2.

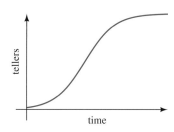

FIGURE 2
Logistic rumor-spreading

As the shape of the graph suggests, logistic growth is one plausible model for rumor-spreading. ➡

It's not the only possible model, as we'll see.

EXAMPLE 3   Riverdale High has 1000 students. On day 0, Archie, Jughead, Betty, Veronica, and 16 of their friends start a rumor that spreads logistically. A day later, 50 students know it. What happens over the next 10 days? When is the rumor spreading fastest?

Solution   Let $P(t)$ denote the number of tellers (people who know the rumor) after t days. Then $P(t)$ satisfies the logistic DE and two additional conditions:
$$\frac{dP}{dt} = kP(1000 - P); \quad P(0) = 20; \quad P(1) = 50.$$

From the preceding Fact (with $C = 1000$) we get
$$P(t) = \frac{1000}{de^{-1000kt} + 1}$$

for appropriate constants d and k. The other given conditions let us evaluate d and k:
$$P(0) = \frac{1000}{d + 1} = 20 \quad \Longrightarrow \quad d = 49;$$

$$P(1) = \frac{1000}{49e^{-1000k} + 1} = 50 \quad \Longrightarrow \quad k \approx 0.000947.$$

We've found our solution function:

$$P(t) = \frac{1000}{49e^{-0.947t} + 1}.$$

Figure 3 shows the graph.

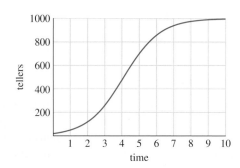

FIGURE 3
A rumor at Riverdale

By 10 days, almost everyone knows the rumor. The rumor seems to spread fastest, moreover, when $P = 500$, just after day 4, where the graph is steepest. (With elementary calculus we can find this time exactly.) ◂

The answer is $t \approx 4.11$—see the exercises.

EXAMPLE 4   Chagrined that bad news travels so fast, Archie, Jughead, Betty, and Veronica start the next rumor all by themselves. It travels through Riverdale with the same "transmission coefficient" $k = 0.000947$. What happens this time?

Solution   Except for the new initial condition, the situation is exactly as in the preceding example. Thus,

$$P(0) = \frac{1000}{d+1} = 4 \quad \Longrightarrow \quad d = 249 \quad \Longrightarrow \quad P(t) = \frac{1000}{249e^{-0.947t} + 1}.$$

Figure 4 shows the new rumor graph; the old one appears for comparison.

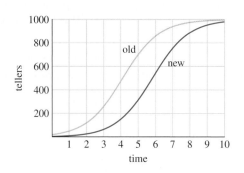

FIGURE 4
Two rumors at Riverdale

Yesterday's news: non-logistic rumors

Not all rumors spread logistically. One possible flaw in the logistic model $P' = kP(C - P)$ concerns the "transmission coefficient" k. In the logistic model, k remains constant over time. In practice, however, k probably *shrinks* over time; as a rumor ages, people care less and less.

A more realistic rumor model might somehow reflect this "staling" process. One approach is to replace the constant k in the logistic DE with a *decreasing function of t*. Here's one possibility. If

$$k(t) = \frac{K}{1+t},$$

where K is any positive constant, then (as we'd hope) $k(t) \to 0$ as $t \to \infty$. Then, the new differential equation would have the form

$$\frac{dP}{dt} = k(t)P(t)\big(C - P(t)\big) = \frac{K}{1+t}P(C - P).$$

EXAMPLE 5 Still another Riverdale rumor spreads according to the preceding DE. When the principal hears the rumor at time 0, 100 students know the rumor, and it's spreading at the rate of 100 students per day. What happens as time goes on?

Solution As before, let $P(t)$ denote the number of students who know the rumor at time t. The new DE, like the old, is separable:

$$\frac{dP}{dt} = \frac{K}{1+t}P(1000 - P) \Longrightarrow \int \frac{dP}{P(1000-P)} = \int \frac{K}{1+t}\,dt.$$

Both sides can be antidifferentiated and the resulting equation solved (with some effort) for P. When the dust settles the result is

$$P(t) = \frac{1000}{d(1+t)^{-1000K} + 1}.$$

The additional information lets us find the constants K and d. From $P(0) = 100$ it follows that $d = 9$. To find K we use the DE itself:

$$P'(0) = 100 = \frac{K}{1+0}100(1000 - 100) \Longrightarrow K = \frac{1}{900}.$$

All the numbers are now in place. Here's the formula:

$$P(t) = \frac{1000}{9(1+t)^{-10/9} + 1}.$$

Figure 5 shows what happens over a 100-day interval.

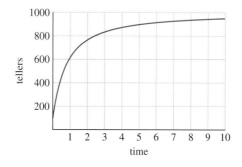

FIGURE 5

Stale news: a non-logistic rumor

Stale news travels slowly.   ■

BASIC EXERCISES

Exercises 1–4 are based on the five curves shown in Figure 1(a); label them C_1 through C_5, from top to bottom. Each curve describes the temperature "evolution" of a different cup of coffee.

1. (a) What values of T_0 correspond to each curve?

   (b) What does this mean in the context of Example 2?

2. Curve C_2 is the solution the IVP $y' = -0.1(y - 70)$, $y(0) = 160$. What IVP corresponds to curve C_1?

3. What IVP corresponds to curve C_4?

4. Another cup of coffee is $100°$F after 10 minutes. Find the temperature of the coffee at time $t = 0$.

Exercises 5–8 are based on the nine curves shown in Figure 1(b); label them C_1 through C_9, from top to bottom. Each curve describes the temperature "evolution" of a different cup of coffee.

5. (a) What values of T_r correspond to C_1, C_4, and C_9?

   (b) What does this mean in the context of Example 2?

6. Curve C_4 is the solution the IVP $y' = -0.1(y - 70)$, $y(0) = 190$. What IVP corresponds to curve C_1?

7. What IVP corresponds to curve C_9?

8. Another cup of coffee, initially at $190°$F, is $100°$F after 20 minutes. Find the room temperature.

Exercises 9–13 explore the IVP $y' = -0.1(y - T_r)$, $y(0) = T_0$ from Example 2.

9. Show, by differentiation that $y(t) = (T_0 - T_r)e^{-0.1t} + T_r$ is the solution of the IVP.

10. Suppose that $T_0 = 200$ and $y(10) = 100$. Find T_r.

11. Suppose that $T_r = 80$ and $y(10) = 120$. Find T_0.

12. Suppose that $y(10) = 100$ and $y(20) = 80$.

   (a) How hot was the coffee at time $t = 0$?

   (b) What's the room temperature?

13. (a) It's true that $\lim_{t \to \infty} e^{-0.1t} = 0$. Use this fact to find $\lim_{t \to \infty} y(t)$.

   (b) Interpret part (a) in coffee terms.

14. In Example 3, we said that $k \approx 0.00947$. Verify this claim by finding the *exact* value of k.

15. In Example 3, we said that the rumor spreads fastest when the population is 500. Show that this is true by showing that the maximum value of P' occurs when $P = C/2$.

16. In Example 3, we said that the rumor spreads fastest just after day 4. Verify this claim by finding the value of t for which $P(t) = 500$.

In Exercises 17–22, find the solution of the IVP.

17. $y' = y^2$, $\quad y(1) = 1$

18. $y' = 2e^{-y}$, $\quad y(0) = 1$

19. $y' = x^2 y$, $\quad y(0) = -2$

20. $y' = \dfrac{x}{1 + y^2}$, $\quad y(2) = 3$

21. $y' - xy^2 = 0$, $\quad y(1) = 1$

22. $y' = x/y$, $\quad y(2) = 3$

FURTHER EXERCISES

23. Consider the DE $y' = xy$.

   (a) Find functions f and g such that this DE can be written as $f(y)y' = g(x)$.

   (b) Find functions F and G such that $F' = f$ and $G' = g$.

   (c) Explain why $F(y) = G(x) + C$ implicitly defines a solution y of the DE.

   (d) Use part (c) to show that $y(x) = Ce^{x^2/2}$ is a solution of the DE.

24. Consider the logistic DE $P'(t) = kP(t)(C - P(t))$, where k and C are positive constants. Show that $\lim_{t \to \infty} P(t) = C$. What does this mean in biological terms?

25. Here are plots of four variants on the logistic-rumor graph of Example 4. In each case, 50 people know the rumor on day 0.

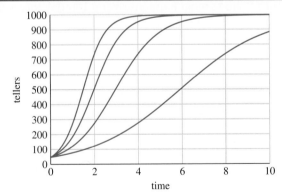

   (a) Label the curves P_1 through P_4 from top to bottom. Each curve's equation is of the form $P(t) = 1000/(19e^{-kt} + 1)$ for some value of k. Use the graph to estimate a value of k for each curve P_1 through P_4.

(b) In this context, the parameter k can be thought of as the rumor's "transmission coefficient." It measures how likely a given teller is to pass the rumor on to a given hearer. Alternatively, k can be thought of as a measure of how "interesting" the rumor may be. Which rumor curve corresponds to the hottest rumor? Which corresponds to the dullest rumor? Justify your answers.

In Exercises 26–31, find a solution of the DE.

26. $y' = \sqrt{1 - y^2}$

27. $y' = 1 + y^2$

28. $y' = 1 - y$

29. $y' = 4 - y^2$

30. $y' = y \ln|y|$

31. $y' = x^2 \cos y$

32. Find an equation of a curve $y = f(x)$ that passes through the point $(0, 1)$ and has the property that the length of any segment of the graph is the same as the area under that segment of the graph (ignore units).

(a) Show that f must satisfy the equation $\sqrt{1 + \left(f'(x)\right)^2} = f(x)$.

(b) Use the fact that $\displaystyle\int \frac{du}{\sqrt{u^2 - 1}} = \cosh^{-1} x$ to find the solution function $f(x)$.

7.5 PRESENT VALUE

Money grows

Anyone, given the choice, would prefer a dollar today to a dollar tomorrow. This makes excellent sense. A dollar tomorrow is worth less than a dollar today because today's dollar, prudently invested, can earn some interest by tomorrow.

But *how much* less is a dollar tomorrow worth than a dollar today? What is an advertised $1 million lottery prize really worth if it's paid in 20 yearly $50,000 installments? Given the choice, should I collect my salary in annual, monthly, or weekly installments, or does it matter? How much should I deposit now to cover 4 years of college tuition starting in 10 years?

All these questions concern what economists call **present value**, that is, the value *now* of a future payment or "stream" of payments. Present-value calculations using integral calculus are vital in economic decision-making, such as in choosing among competing investment options.

The present value of one future payment   The first and most important example illustrates the simplest case: the present value of *one* future payment. Notice especially that exponential functions are involved and that present value depends on an interest rate.

EXAMPLE 1   A savings bond will pay $1000 on its maturity date 10 years from today. What is it worth now if 5% compound interest is available? What if 10% interest were available?

Solution   Let $V(t)$ denote the bond's value t years from now. We're given that $V(10) = 1000$; we want to find $V(0)$.

To start, recall this important fact: If interest is continuously compounded, the value V grows *exponentially*. In other words, $V(t)$ has the form

$$V(t) = V(0)e^{rt},$$

where r is the annual interest rate. Remember why? Here's a quick refresher. Growth at continuously compounded interest rate r means, in differential equation language, that for all t,

$$V'(t) = rV(t).$$

It's easy to check that for any constant C, the function $V(t) = Ce^{rt}$ solves this DE. Because $V(0) = Ce^0 = C$, it follows that $V(0)$ is the "right" value for C.

Knowing that $V(10) = 1000$, we can solve for $V(0)$. For any interest rate r,

$$V(10) = V(0)e^{10r} = 1000 \implies V(0) = 1000e^{-10r}.$$

Thus, if $r = 0.05$, $V(0) = 1000e^{-0.5} \approx 606.53$; if $r = 0.1$, $V(0) = 1000e^{-1.0} \approx 367.88$. These calculations mean that the **present value** of a single \$1000 payment 10 years in the future is about \$607 based on 5% annual interest, or about \$368 with 10% interest. ∎

The preceding example prompts three general remarks:

• **A negative exponential** The present value PV of one future payment depends on three things: P, the size of the payment; t, the time until the payment is made; and r, the interest rate. We found that

$$PV = Pe^{-rt}.$$

The general form of this expression will reappear soon. The negative exponent reflects the fact that the present value of a future payment is normally *less* than the future payment itself.

• **Backward in time** The present value of a future payment P can be thought of as the amount to be deposited now, at interest rate r, to accrue to value P after time t. Equivalently, one can think of following an investment's value *backward* in time from value P at time t to value PV at time zero.

• **Simplifying assumptions** Present-value calculations are simplest if we assume that (1) the interest rate r remains constant over time, and (2) interest is compounded continuously. (The first assumption is just for convenience; the second lets us use calculus.) We make both of these assumptions in most of what follows.

The discrete case: present value of several future payments Many real life cases involve several payments occurring at future times. The *total* present value of a sequence of payments is found, naturally, by addition.

EXAMPLE 2 A lottery jackpot, although advertised as \$1 million, is paid in 20 annual installments of \$50,000. What's the present value of the prize given 6% annual interest? What if 8% interest were available?

Solution By the preceding formula, the present value of one \$50,000 payment k years in the future is $50,000e^{-0.06k}$. Thus, the total present value of all 20 payments is

$$50,000 \left(e^{-0.06 \times 0} + e^{-0.06 \times 1} + e^{-0.06 \times 2} + \cdots + e^{-0.06 \times 19}\right) \approx \$599,983.$$

That's much less than the advertised million. At 8% interest, the computation is similar; the present value is

$$50,000 \left(e^{-0.08 \times 0} + e^{-0.08 \times 1} + e^{-0.08 \times 2} + \cdots + e^{-0.08 \times 19}\right) \approx \$519,033,$$

still farther from the advertised million-dollar payoff. ∎

Several payments: the general form Example 2 illustrated the general picture. If r is the interest rate and payments $P_1, P_2, \ldots, P_n$ are made at times $t_1, t_2, \ldots, t_n$, then the present value of all n payments is found by summation:

$$PV = \sum_{i=1}^{n} P_i e^{-rt_i}. \tag{1}$$

The continuous case: Present value of an income stream

In real life, financial transactions occur at specific moments—the beginning of a month, the end of a pay period, 11:59 P.M. on April 15, and so on. Nevertheless, economists often picture income as an unbroken "stream," flowing continuously in time (hence, for instance, the term "cash flow"). The next example—in which all three graphs are drawn as continuous, unbroken curves—illustrates this point of view. Notice that the graphs show the *rates* of income flow as functions of time.

EXAMPLE 3 The graphs in Figure 1 show rates of daily income, over a 360-day "year," for three idealized workers: a tax consultant, an ice-cream vendor, and a factory worker. (Which is which? Day 0 is January 1; U.S. tax returns are due April 15.) How much does each worker earn over the whole year? Who's best off? Why?

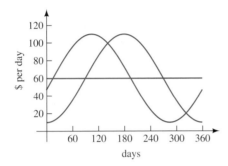

FIGURE 1
Three income streams

Solution First, which graph is which? Judging from their peaks (around April 15 = day 105 and July 1 = day 180), we'd guess that the two sinusoidal graphs represent the tax consultant and the ice-cream vendor. ➤ The straight-line graph therefore represents the factory worker's constant $60 daily wage. Label the graphs T, I, and F.

Our fable occurs north of the equator. How do the graphs show this?

The factory worker's total annual income is easiest to calculate. At $60 per day for a 360-day year, the factory worker earns $21,600—the area under the F-graph from $t = 0$ to $t = 360$. The total incomes of the other two workers can also be found as areas under the T- and the I-graphs, respectively. A close look reveals a mild surprise. All three graphs enclose the *same* total area; all three workers earn the same total annual income: $21,600.

Nevertheless, the workers are not (quite) equally well off. The tax consultant is luckiest, because T receives more income early in the year than I and F do. For now that's all we can say. Soon we'll *quantify* T's advantage by calculating the present value of each income stream at $t = 0$. ∎

Fair assumptions? We made several simplifying assumptions in Examples 1–3: that interest rates remain constant over time, that interest is compounded continuously, that daily earnings follow simple patterns, and so on. None of these assumptions need be literally true. For instance, few factory workers are paid by the day, including weekends. Ice-cream sales rise and fall with temperature, not just with the time of year.

Do such simplifications render our results useless? Certainly not. *All* economic decisions require some combination of calculation and guesswork. To some extent, the future is always unknowable. In practice, making sensible simplifying assumptions is inevitable—and it's standard operating procedure.

Why treat income as a stream? Treating income as a continuous stream lends itself nicely to calculus; functions, derivatives, and integrals all arise naturally. From this point of view, an income stream is a function p of time: At time t, $p(t)$ is the (instantaneous) *rate* at which income "flows" to the recipient. It follows, then, that the integral $\int_a^b p(t)\,dt$ is the total income received over the time interval $a \le t \le b$. (That's why we looked at areas under the curves in Example 3.)

Present value as an integral

Building on the preceding examples, let's define the present value of a continuous income stream. Three ingredients are involved:

- A time scale t; $t = 0$ corresponds to the *present time.*
- An interest rate r, usually taken as a constant.

The income stream starts at time a and ends at time b.

- A continuous income-stream function $p(t)$ defined for $a \le t \le b$. ◄ At time t, $p(t)$ tells the rate of income flow with respect to time.

Finally, an important caution about compatible units of measurement:

The quantities t, r, and p must use the same time unit.

For example, if t measures days, then r is the *daily* interest rate and $p(t)$ is measured in money units per *day*.

> **DEFINITION** Let t, r, and p be as before. The **present value of the income stream** p is defined by the integral
>
> $$PV = \int_a^b p(t)e^{-rt}\,dt.$$

First we'll use the definition and then defend it as sensible. In the meantime, notice the similarity to Equation (1), page 448.

EXAMPLE 4 Given a 5% annual interest rate, find the present value (at the beginning of the 360-day year) of the income stream of each of the three workers T, I, and F in Example 3. How different are the results if 10% interest is available?

Solution Let t denote time in days; then $t = 0$ is the present time, and the income flow lasts from $t = 0$ to $t = 360$. Because t involves days, r should measure the daily, not yearly, interest rate. Here, therefore, $r = 0.05/360$.

The three stream functions—we'll call them p_T, p_I, and p_F—are shown graphically on page 449. To find present values we need formulas for p_T, p_I, and p_F. One formula can be "read" easily from the graphs: $p_F(t) = 60$. The other two graphs look periodic; with some effort we could produce their formulas from trigonometric ingredients. Here we just state them:

$$p_T(t) = 50\cos\left(\pi \cdot \frac{t - 105}{180}\right) + 60; \quad p_I(t) = 50\cos\left(\pi \cdot \frac{t - 180}{180}\right) + 60.$$

The rest of the solution consists of straightforward (though messy) calculation. Here are the results. With $r = 0.05/360$,

$$PV_F = \int_0^{360} 60 e^{-0.05t/360} \, dt \approx 21068.89;$$

$$PV_T = \int_0^{360} \left(50\cos(\pi(t-105)/180) + 60 \right) e^{-0.05t/360} \, dt \approx 21203.55;$$

$$PV_I = \int_0^{360} \left(50\cos(\pi(t-180)/180) + 60 \right) e^{-0.05t/360} \, dt \approx 21067.78.$$

As predicted, T comes out on top, about \$135 ahead of the others.

At a 10% annual interest rate ($r = 0.1/360$) the tax consultant's advantage is a little greater:

$$PV_F = \int_0^{360} 60 e^{-0.1t/360} \, dt \approx 20555.12;$$

$$PV_T = \int_0^{360} \left(50\cos(\pi(t-105)/180) + 60 \right) e^{-0.1t/360} \, dt \approx 20817.26;$$

$$PV_I = \int_0^{360} \left(50\cos(\pi(t-180)/180) + 60 \right) e^{-0.1t/360} \, dt \approx 20550.78.$$ ∎

Why the integral definition makes sense   The typographical similarity between

$$\sum_{i=1}^{n} P_i e^{-rt_i} \quad \text{and} \quad \int_a^b p(t) e^{-rt} \, dt$$

(the formulas for present value in the discrete and continuous cases, respectively) is no accident. Roughly speaking, the integral is the continuous version of the discrete sum.

To relate the integral and the sum more precisely, we approximate a continuous income stream with a discrete sequence of payments. Suppose, then, that for $a \le t \le b$, $p(t)$ describes a continuous income stream, and let $a = t_0 < t_1 < t_2 < \cdots < t_n = b$ partition the interval $[a, b]$ into n short subintervals. We approximate the income stream p with a finite sequence of payments $P_1, P_2, \ldots, P_n$ occurring at times $t_1, t_2, \ldots, t_n$, respectively. ➤

Payments P_i occur at the end of each subinterval.

How large should each payment P_i be? Over a (short) subinterval $[t_{i-1}, t_i]$, the function p doesn't change much (because p is continuous), and so the estimate $p(t) \approx p(t_i)$ is reasonable. ➤ Thus, the *total income* received over the time interval $[t_{i-1}, t_i]$ is approximately $p(t_i)\,\Delta t_i$. In our (approximate) discrete version, therefore, it's natural to use $P_i = p(t_i)\,\Delta t_i$. In summary, we approximate the continuous stream p, defined for $a \le t \le b$, with n separate payments; at time t_i, the payment is $P_i = p(t_i)\Delta t_i$. As n tends to infinity, we expect our approximation to improve.

We pretend, in effect, that $p(t)$ is constant *on each subinterval.*

We saw earlier that the total present value of these n payments, given interest rate r, is

$$\sum_{i=1}^{n} P_i e^{-rt_i} = \sum_{i=1}^{n} p_i e^{-rt_i} \Delta t_i.$$

The key point is that the right side is an approximating sum for $\int_a^b p(t) e^{-rt} \, dt$—just the integral we've been waiting for. Therefore, as n tends to infinity, the approximation tends both to the desired integral and to the desired present value.

BASIC EXERCISES

1. Al and Bob, both 42 years old, hope to retire as millionaires. To that end, they'll deposit money now to accrue to $1 million by the time they're 65.

   (a) How much should Al deposit now if 6% interest is available?

   (b) How much should Al deposit now if 8% interest is available?

   (c) Bob has $100,000 on hand. What interest rate will he need to meet his retirement goal?

2. Christine, now 20, wants to retire at 65 with a $1 million nest egg. How much should she deposit now at 6%? At 8%? How much should she invest in speculative junk bonds that pay 20%?

3. **Real interest rates.** Inflation, the rise of prices over time, reduces the buying power of future money. In effect, inflation imposes negative interest on a deposit. Economists "control for inflation" by defining the real interest rate as

   real interest rate = nominal rate − inflation rate.

   Predicting future inflation necessarily involves guesswork; so, therefore, must the real interest rate. Historically, real interest rates have tended to hover in the range 2–4%.

   Anne and Betty, now 42 years old, hope to retire as *real* millionaires, that is, as millionaires after inflation. They can accomplish this by depositing enough money now to accrue—at real interest rate *r*—to $1 million by the time they're 65.

   (a) How much should Anne deposit now if 2% real interest is available?

   (b) How much should Anne deposit now if 4% real interest is available?

   (c) Betty has $200,000 on hand. What real interest rate will she need to meet her retirement goal?

4. Christopher, now 20, wants to retire at 65 with a $1 million real nest egg. How much should he deposit now at 2% real interest? At 4%? How much should he invest in speculative junk bonds that pay 15% real interest?

5. A child born in 1993 will typically start college in 2011 and graduate in 2015. Nobody really knows what college tuitions will be then. However, St. Lena College offers an "early decision" guarantee for the class of 2015: Enroll as an infant and lock in yearly tuition payments of $40,000, $42,000, $44,000, and $46,000 payable in 2011, 2012, 2013, and 2014. How much should new parents deposit in 1993 to cover these future payments if 6% annual interest is available? What if 8% interest is available?

6. Consider again the college tuition scenario in Exercise 5. This time we take a continuous point of view. Over the 4 years 2011–2015, total tuition paid is $172,000. Suppose, now, that this money is paid continuously at the *constant* rate of $43,000 per year. How much difference does the continuous point of view make? To decide, answer the same questions as in Exercise 5.

7. Consider carefully the two wavy graphs in Example 3. We claimed that their formulas are

   $$p_T(t) = 50\cos\left(\pi\,\frac{t-105}{180}\right) + 60;$$

   $$p_I(t) = 50\cos\left(\pi\,\frac{t-180}{180}\right) + 60.$$

   (a) Which graph goes with which formula? Explain briefly, in words, how you know.

   (b) The functions $\cos t$, p_T, and p_I are all periodic functions; that is, they repeat themselves on time intervals of a fixed length. How long is the period of each of these functions?

   (c) The formulas for p_T and p_I are "built" from $\cos t$ using various constants: 50, 60, 105, 180, and so on. Briefly discuss the effect each constant has on the corresponding graph.

FURTHER EXERCISES

8. Find the present value of the income stream in parts (a)–(e). In each case, treat income as a continuous stream. (Thus, for instance, think of yearly income of $12,000 as flowing in at the *constant rate* of $12,000 per year.)

   (a) A yearly income of $12,000 beginning 10 years in the future and lasting for 10 years. Assume annual interest rate $r = 0.06$.

   (b) A yearly income of $12,000 beginning 10 years in the future and lasting for 10 years. Assume annual interest rate $r = 0.08$.

   (c) A yearly income of $12,000 beginning 10 years in the future and lasting for 10 years. Assume annual interest rate $r = 0$. (This investor doesn't trust banks; he "deposits" the money under the floorboards.)

   (d) At time *t* years, with $10 \le t \le 20$, income flows at the rate $p(t) = 12,000e^{0.04t}$. Assume annual interest rate $r = 0.08$. [NOTE: In practice, many income streams do increase over time—for example, to correct for inflation. This one rises at a 4% instantaneous rate.]

   (e) Repeat part (d), but assume an annual interest rate of $r = 0.1$.

9. Suppose that an investment that pays a continuous return of $5000 per year for 8 years is offered for sale at $30,000. If the current interest rate is 6%, should you invest?

10. Symbolic calculation of the integrals in Example 4 is a rather messy affair. Instead, use the midpoint rule with 50 subdivisions to estimate the first three integrals there.

11. Symbolic calculation of the integrals in Example 4 is messy, but it can be done. One way is to use this antiderivative formula:

$$\int e^{ax} \cos(bx)\, dx = \frac{e^{ax}}{a^2+b^2}\left(a\cos(bx)+b\sin(bx)\right)$$

(a) Verify by differentiation that the antiderivative formula really holds.

(b) Use the integral formula to show symbolically (as claimed in Example 4) that

$$\int_0^{360}\left(50\cos\left(\pi\cdot\frac{t-180}{180}\right)+60\right)e^{-0.1t/360}\, dt \approx 20{,}550.78.$$

12. Suppose that an acquaintance offers to sell you an investment that will produce income at a continuous rate of $1000/(1+t)$ dollars per year for 10 years. The asking price is $2000. If 10-year bank certificates of deposit that pay 5% per year compounded continuously are available, should you purchase this investment from your acquaintance?

13. Suppose that an income stream p is continuously invested at the same continuously compounded interest rate r that is currently available. Show that the total income accrued between $t=0$ and $t=T$ is

$$e^{rT}\, PV = e^{rT}\int_0^T p(t)e^{-rt}\, dt.$$

Mass and Center of Mass

Density and mass   In many situations it's possible to consider a quantity, such as mass, as being "distributed" along the x-axis from $x = a$ to $x = b$. For example, imagine a straight piece of wire laid along the x-axis from $x = a$ to $x = b$. If the wire has constant thickness and is made of some homogeneous material such as pure copper, then the wire's mass is *evenly* distributed along the interval $[a, b]$ and the wire's **density** is constant. In this case the wire's total mass is simply the product of the density and the length. If, say, the wire has constant density 0.3 grams/centimeter (g/cm) and lies between $x = 0$ and $x = 5$ cm, then the wire's mass is $0.3 \times 5 = 1.5$ g.

If the wire has *varying* thickness or is made of some inhomogeneous material, then the wire's mass is *unevenly* distributed along the interval $[a, b]$. In this case the wire's density is *not* constant—the density is large where the wire is "fat" or "heavy" and small where the wire is "thin" or "light." In this case it's convenient to think of a **density function**, which is often denoted by $\rho(x)$. (Numerical values of $\rho(x)$ depend on the units of measurement.)

In general, the **linear density** of a distribution is *defined* as a derivative: If $Q(x)$ is the total quantity (e.g., the total mass of wire) over the interval $[a, x]$, then the density function $\rho(x)$ is defined by $\rho(x) = Q'(x)$, the rate of change of Q at x. (Note that $Q(a) = 0$ by our definition.) In the case of a wire, for instance, the statement $\rho(2) = 0.345$ could mean (depending on the units) that at $x = 2$ cm, the wire's density is 0.345 g/cm.

Now the FTC says that

$$\int_a^b \rho(x)\,dx = Q(b) - Q(a) = Q(b);$$

the right side is the total quantity in question from $x = a$ to $x = b$. Here is the formal result:

> **FACT (Measuring mass)**   Suppose that a mass distribution from $x = a$ to $x = b$ has density function $\rho(x)$. Then the total mass over the interval $[a, b]$ is given by the integral
> $$\int_a^b \rho(x)\,dx.$$

EXAMPLE 1   A wire laid along the x-axis from $x = 0$ to $x = 2$ has density function $\rho(x) = x^2$. Find (and interpret) the total mass.

Solution   By the preceding Fact, the total mass is

$$\int_0^2 \rho(x)\,dx = \int_0^2 x^2\,dx = \frac{8}{3}.$$

What the answer means depends on the units. If length (along the x-axis) is measured in centimeters and mass is measured in grams, then the density $\rho(x)$ is in grams per centimeter and the wire's total mass is 8/3 g.   ■

PROBLEM 1   A wire laid along the x-axis from $x = 0$ to $x = 10$ has density function $\rho(x) = 2 + \sin x$. Find the total mass if x is measured in feet and $\rho(x)$ in ounces per foot. What are the units of the answer?

Beyond wires   The wire of Example 1 can be thought of as essentially a one-dimensional object. In other cases a mass distribution along the x-axis may arise from a three-dimensional figure.

EXAMPLE 2 Consider the horn-shaped solid of revolution formed by revolving the shaded region shown in Figure 1 about the x-axis. (We saw in Section 7.2 how to find the solid's volume by integration; see Example 3, page 425.)

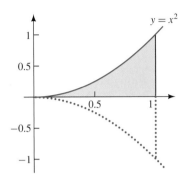

FIGURE 1
Revolve the shaded region about the x-axis

Now imagine that the solid is made of some homogeneous material (such as plastic) that weighs 1 g/cm^3. Use the distribution idea discussed here to find the solid's mass (and hence also its volume).

Solution We can think of the solid as determining a distribution of mass along the x-axis from $x = 0$ to $x = 1$. The horn's outward-flaring shape means that the mass distribution is not constant along the x-axis. If $M(x)$ denotes the mass of the part of the solid over the interval $[0, x]$ and $V(x)$ denotes the volume of the same object, then $M(x)$ and $V(x)$ increase faster and faster as x moves from 0 to 1.

The linear density $\rho(x)$ of this mass distribution is the derivative $M'(x)$. In this case mass and volume have the same numerical value, and so $M'(x) = V'(x)$. We observed in Section 7.2 that $V'(x)$ is the solid's cross-sectional area (perpendicular to the x-axis) at x. (This is essentially the content of Theorem 1, page 424.) In the present case, the cross section at x is a disk with radius x^2, and thus we see that the linear density is $M'(x) = \pi x^4$. Integrating the density function gives the answer we want:

$$\int_0^1 \pi x^4 \, dx = \frac{\pi}{5}.$$

The numerical answer (the same as we found in Section 7.2) can be interpreted either as mass (in grams) or as volume in cubic centimeters. ■

PROBLEM 2 Mimic the preceding example to find the mass of a cone of height h cm and radius r cm. Assume that the cone is made of some material that weighs 1 g/cm^3.

Center of mass Given a distribution of mass along the x-axis (as considered above), how can we find its **center of mass**—the point at which the distribution balances? For a *uniform* mass distribution from $x = a$ to $x = b$ there's no problem: The system balances at the midpoint $x = (a + b)/2$. For a general distribution, the center of mass, like mass itself, is found by integration.

To get started, imagine several **point masses**, say $m_1, m_2, \ldots, m_n$ arranged along the x-axis at positions $x_1, x_2, \ldots, x_n$, as in Figure 2:

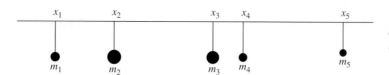

Think of the masses as lead weights hanging from threads.

FIGURE 2
Five (unequal) point masses along an axis

(A point mass is a mass that can be thought of as located at one point, as can the hanging masses in Figure 2.) For a fixed value of x, say $x = p$, the quantity

$$m_1(x_1 - p) + m_2(x_2 - p) + \cdots m_n(x_n - p)$$

is called the **moment** of the system about $x = p$. It's a fact of physics that the system *balances* at p if this moment is zero. (This means, in particular, that $(x_i - p)$ must be positive for some i and negative for others—hardly surprising to anyone who's ridden a seesaw.) To find the center of mass (denoted by $\bar{x}$) one sets the moment about p equal to zero and solves for p.

PROBLEM 3 A mass distribution consists of two point masses: $m_1 = 5$ at $x_1 = -1$ and $m_2 = 5$ at $x_2 = 1$. Find the center of mass by setting the moment about p equal to zero and solving for p.

PROBLEM 4 Another mass distribution consists of two point masses: $m_1 = 5$ at $x_1 = -1$ and $m_2 = 10$ at $x_2 = 1$. Where is the center of mass now?

PROBLEM 5 Yet another mass distribution consists of three point masses: $m_1 = 1$ at $x_1 = 0$, $m_2 = 2$ at $x_2 = 2$, and $m_3 = 3$ at $x_3 = 3$. Find the center of mass.

Continuous distributions The situation for a continuous mass distribution is similar to that just described. If the mass distribution has density $\rho(x)$ from $x = a$ to $x = b$, then the system's **moment** about $x = p$ is an *integral* rather than a sum:

$$\text{moment} = \int_a^b (x - p)\rho(x)\,dx.$$

As before, the **center of mass** is the number $\bar{x}$ for which

$$\text{moment} = \int_a^b (x - \bar{x})\rho(x)\,dx = 0.$$

PROBLEM 6 Solve the preceding equation to show that the center of mass $\bar{x}$ satisfies the equation

$$\bar{x} = \frac{\int_a^b x\rho(x)\,dx}{\int_a^b \rho(x)\,dx}.$$

(Notice that, by the discussion above, the denominator is just the total mass of the system.)

PROBLEM 7 Find the center of mass of the wire in Example 1.

PROBLEM 8 Find the center of mass of a circular cylinder (made of a homogeneous material) of radius r and "height" h. [HINT: Visualize the cylinder as running from $x = 0$ to $x = h$. Then, as discussed in Example 2, the density function $\rho(x)$ can be taken as the cross-sectional area function—which is constant in this case.]

PROBLEM 9   Find the center of mass of a cone (made of a homogeneous material) of radius r and "height" h. [HINT: Visualize the cone as running from $y = 0$ to $y = h$ with the big end up. Then the density function $\rho(y)$ can be taken as the cross-sectional area function. (Integrate in y, not in x.)]

PROBLEM 10   In Example 6 of Section 7.3 (page 436), we found the work needed to pump all the water out the top of a conical tank of radius r feet and height h feet. Show that this is the same amount of work as is needed to raise the entire mass of the cone from the center of gravity of the cone to the top of the cone.

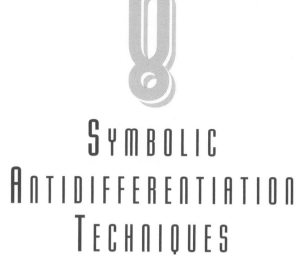

SYMBOLIC ANTIDIFFERENTIATION TECHNIQUES

8.1 INTEGRATION BY PARTS

Introduction

In this chapter we return to an old project—finding, for a given elementary function f, a new elementary function F for which $F' = f$. (Recall: An *elementary* function is one built from the standard basic "elements"—power functions, exponential and logarithmic functions, trigonometric functions, and so on.) ◄

Roughly speaking, an elementary function is one that has a "formula."

As we said in Chapter 5, finding elementary *antiderivatives* is, as a rule, harder than finding *derivatives*. Sometimes it is even impossible; some elementary functions do not *have* elementary antiderivatives. The function $f(x) = \sin(x^2)$ is one simple-looking example. (That's one reason we took so much trouble in Chapter 6 to approximate integrals like $\int_0^1 \sin(x^2)\,dx$.)

Many important elementary functions *do* have elementary antiderivatives, however, and in this chapter we describe several methods for finding them. If we combine the symbolic methods discussed here with a moderate-sized integral table, we can handle many of the integrals that arise in practical applications. With additional help from modern mathematical software, we can do better still.

Why bother? Elementary antiderivative formulas are obviously useful; they let us calculate definite integrals exactly, for one thing. But why do we take the trouble to *calculate* antiderivatives? Why not just "ask" an integral table or mathematical software for an answer? One answer is that, as sentient beings, we'd like to know *something* about where tabulated or computer-generated answers come from. Another answer is that looking for antiderivatives, with or without technology, reveals important—and sometimes surprising—relations among the key functions of calculus.

EXAMPLE 1 Find $\int \left(\dfrac{1}{x^2} + \dfrac{1}{x^2+1} + \dfrac{2x}{x^2+1} \right) dx$. Are there any surprises?

Solution An answer is easy to find (and to check by differentiation):

$$\int \left(\frac{1}{x^2} + \frac{1}{x^2+1} + \frac{2x}{x^2+1} \right) dx = -\frac{1}{x} + \arctan x + \ln(x^2+1) + C.$$

The calculation isn't especially interesting, but notice the somewhat surprising *form* of the result. All three summands of the integrand are simple rational functions, but the antiderivative has three quite different pieces—a power of x, an arctangent, and a logarithm. One goal of this chapter is to explain how such similar parents can produce such different offspring. ∎

Integration by parts and the product rule

We saw in Section 5.6 that the method of substitution is a "reversed" version of the *chain rule* for derivatives. In a similar sense, the method of **integration by parts** reverses the *product rule* for derivatives. The product rule says that, for differentiable functions u and v,

$$(u \cdot v)' = u \cdot v' + v \cdot u'.$$

The product rule can also be understood as a statement about *anti*derivatives; it says that uv is an antiderivative of $uv' + vu'$. In symbolic language, the product rule says

$$u(x) \cdot v(x) = \int \big(u(x) \cdot v'(x) + v(x) \cdot u'(x) \big) \, dx.$$

Rearranging terms gives an equivalent equation:

$$\int u(x) \cdot v'(x) \, dx = u(x) \cdot v(x) - \int v(x) \cdot u'(x) \, dx.$$

This last identity is worth a closer look. Notice especially how derivatives and antiderivatives are swapped inside the two integral signs. ➤ This simple but useful idea appears often in mathematics. In calculus, this equation is known as the **integration-by-parts formula**. As we'll see, this formula sometimes lets us trade a difficult integration problem for an easier one.

We trade $u \cdot v'$ for $v \cdot u'$.

If we write

$$u = u(x), \quad v = v(x), \quad du = u'(x) \, dx, \quad \text{and} \quad dv = v'(x) \, dx,$$

the formula looks simpler still. We state the result as a theorem; it applies both to definite and to indefinite integrals. ➤

The FTC explains why the formula applies to definite integrals.

> **THEOREM 1 (Integration by parts)** If u and v are differentiable functions, then
>
> - $\int u \, dv = uv - \int v \, du;$
> - $\int_a^b u \, dv = uv \Big]_a^b - \int_a^b v \, du.$

First we illustrate how the method works under ideal conditions. (Later we'll see what can go wrong . . . .)

EXAMPLE 2 Find $\int x \, e^x \, dx$ and $\int_0^1 x \, e^x \, dx$.

Solution Setting $u = x$ and $dv = e^x \, dx$ fits the theorem's template:

$$\int x \, e^x \, dx = \int u \, dv.$$

To use the formula, we need values for du and v. The first is simple: Since $u = x$, $du = dx$. Finding a suitable v is sometimes a hard problem in its own right, but in this case it's easy: If $v = e^x$, then it's easy to check that $dv = e^x\,dx$, as desired. Now the integration-by-parts formula kicks in:

$$\int x\,e^x\,dx = \int u\,dv = uv - \int v\,du = x\,e^x - \int e^x\,dx = x\,e^x - e^x + C.$$

Checking antiderivatives produced by u-substitution involves the chain rule.

Is the answer correct? Checking the answer by differentiation involves—as we might have guessed—the product rule: ◄

$$\left(x\,e^x - e^x + C \right)' = e^x + x\,e^x - e^x = x\,e^x.$$

Thus, our "candidate" antiderivative behaves as it should.

Evaluating the *definite* integral is now a routine calculation:

$$\int_0^1 x\,e^x\,dx = x\,e^x \Big]_0^1 - \int_0^1 e^x\,dx = x\,e^x - e^x \Big]_0^1 = 1. \qquad ■$$

EXAMPLE 3 Find $\displaystyle\int x^2 \ln x\,dx.$

Solution If we set $u = \ln x$, then

$$dv = x^2 dx, \quad du = \frac{1}{x}\,dx, \quad v = \frac{x^3}{3},$$

and everything works nicely:

$$\int x^2 \ln x\,dx = \int u\,dv$$

$$= uv - \int v\,du$$

$$= \frac{x^3}{3}\ln x - \int \frac{x^2}{3}\,dx$$

$$= \frac{x^3}{3}\ln x - \frac{x^3}{9} + C.$$

The answer is easy to check by differentiation. ■

Tricks of the trade: Wise and foolish choices

How to choose u and dv successfully isn't always obvious. Setting $u = \ln x$ and $dv = x^2\,dx$ in Example 3, for instance, was natural enough, but not the only possibility. We might have tried, say, $u = x^2$ and $dv = \ln x\,dx$. These latter choices are not incorrect, but they turn out to be unhelpful. ◄

Try them to see why they're unhelpful.

Sometimes, surprising choices of u and dv turn out to work.

EXAMPLE 4 Find $\displaystyle\int \ln x\,dx.$

Solution At first glance the problem seems not to fit the mold at all. Yet it does fit. If we set $u = \ln x$, then

$$dv = dx, \quad du = \frac{1}{x}\,dx, \quad v = x;$$

fitting things into the formula gives

$$\int \ln x\,dx = x\,\ln x - \int 1\,dx = x\,\ln x - x + C.$$

The answer is easy to check by differentiation, using the product rule. ■

When things go wrong Integration by parts trades one integral expression, $\int u\, dv$, for another, $uv - \int v\, du$. The bargain is worth making only if the second expression is simpler or more cooperative than the first. In practice, such satisfaction cannot be guaranteed. Indeed, choosing u and dv unwisely can make things *worse*, not better.

> **EXAMPLE 5** (**A step in the wrong direction**) What's wrong with choosing $u = e^x$ and $dv = x\, dx$ for the indefinite integral $\int x\, e^x\, dx$?
>
> **Solution** Our choices aren't *wrong*, but they don't help with the problem at hand. Setting
>
> $$u = e^x, \quad dv = x\, dx, \quad du = e^x\, dx, \quad \text{and} \quad v = \frac{x^2}{2}$$
>
> gives
>
> $$\int x\, e^x\, dx = uv - \int v\, du = e^x \frac{x^2}{2} - \int \frac{x^2}{2} e^x\, dx.$$
>
> The new integral (on the right) looks harder than the old one (on the left). This time we struck out. ∎

Once u and dv are chosen, we need corresponding expressions for du and v. Finding du from u is *always* straightforward because it involves only differentiation. But finding v from dv is an antidifferentiation problem in its own right; it may be hard or even impossible.

> **EXAMPLE 6** Find $\int 2x\, \exp(x^2)\, dx$. ➡
>
> *Recall:* $\exp(u)$ *is another notation for* e^u.
>
> **Solution** The integrand is a product and so looks like a candidate for integration by parts. Let's try $u = 2x$ and $dv = \exp(x^2)\, dx$. It's clear that $du = 2\, dx$, but what's v? The answer: Nothing useful—$\exp(x^2)$ *has* no elementary antiderivative. We can't even write v down, let alone use it to solve our problem.
>
> Did we choose u and dv unwisely? Or is integration by parts the wrong tool for this job? In this case the latter alternative holds. In fact, the ordinary u-substitution $u = x^2$, $du = 2x\, dx$ works nicely:
>
> $$\int 2x\, \exp(x^2)\, dx = \int e^u\, du = \exp(x^2) + C.$$
>
> There's a moral here: Try easy methods first—especially substitution. ∎

When things go right Mishaps are certainly possible, but for surprisingly many antiderivatives—even ones that don't appear to be products—integration by parts is effective. The trick is to choose u and dv successfully, where "successfully" means two things:

(i) dv can be antidifferentiated to give v;

(ii) $\int v\, du$ is simpler or more familiar than $\int u\, dv$.

Not *every* choice of u and dv succeeds, but if one choice fails, another may work. Intuition for recognizing good and bad choices comes with practice.

LIATE: A mnemonic for choosing *u* and *dv* Whether an attempt at integration by parts succeeds or fails usually depends on the choices of u and dv.

In his article "A Technique for Integration by Parts" (*American Mathematical Monthly* 90 (1983), pp. 210–211), Herbert E. Kasube proposes the LIATE rule for choosing u. (Choosing u is enough; given u, dv must be "everything else.")

He observes that most integrands are built from functions of five types: logarithmic (L), inverse (I) trigonometric, algebraic (A), trigonometric (T), and exponential (E). The LIATE rule says the following:

Choose u in the order LIATE.

In other words, u should be a function of the *first* available type in the list L, I, A, T, E. In Example 1, for instance, the choice $u = x$ was of type A (algebraic) because neither logarithmic (L) nor inverse trigonometric (I) functions were present.

Not every indefinite integral succumbs to integration by parts. For those that do, the LIATE method usually seems to work.

Mixing methods   Some problems require more than one antidifferentiation method.

What does LIATE say here?

EXAMPLE 7   Find $\int \arctan x \, dx$. ◄

Solution   Let $u = \arctan x$ and $dv = dx$. Then $du = (1 + x^2)^{-1} dx$, $v = x$, and integration by parts gives

$$\int \arctan x \, dx = uv - \int v \, du = x \arctan x - \int \frac{x}{1 + x^2} \, dx.$$

We're nearly finished—the last integral yields to the ordinary substitutions $u = 1 + x^2$ and $du = 2x \, dx$:

$$x \arctan x - \int \frac{x}{1 + x^2} \, dx = x \arctan x - \frac{1}{2} \int \frac{du}{u}$$

$$= x \arctan x - \frac{1}{2} \ln |u| + C$$

$$= x \arctan x - \frac{1}{2} \ln(1 + x^2) + C. \qquad \blacksquare$$

Beyond the basics

There are several useful variations on the basic integration-by-parts theme; two of them appear in the following examples.

EXAMPLE 8   (Do it again)   Find $\int x^2 e^x \, dx$.

Solution   If $u = x^2$, then $dv = e^x \, dx$, $du = 2x \, dx$, $v = e^x$, and

$$\int x^2 e^x \, dx = x^2 e^x - 2 \int x e^x \, dx.$$

With $u = x$, $dv = e^x \, dx$.

Did we get anywhere? Yes! We've already seen that the last integral can be handled by *another* integration by parts. ◄ We won't redo the problem. Here's the result:

$$\int x^2 e^x \, dx = x^2 e^x - 2 \int x e^x \, dx = x^2 e^x - 2 \, (x e^x - e^x) + C. \qquad \blacksquare$$

EXAMPLE 9   (Integrate twice; then solve)   Find $\int e^x \sin x \, dx$.

Solution   If we set $u = \sin x$, then

$$dv = e^x \, dx, \quad du = \cos x \, dx, \quad v = e^x,$$

and we get

$$I = \int e^x \sin x \, dx = e^x \sin x - \int e^x \cos x \, dx.$$

Not much progress yet; let's try the method *again* on the last integral. If $u = \cos x$, then

$$dv = e^x \, dx, \quad du = -\sin x \, dx, \quad \text{and} \quad v = e^x,$$

and so

$$I = e^x \sin x - \int e^x \cos x \, dx = e^x \sin x - \left(e^x \cos x + \int e^x \sin x \, dx \right).$$

(Watch the signs carefully.) The original integral I has reappeared! It might seem that we're chasing our tail, but in fact we can *solve* the last equation for I:

$$I = e^x \sin x - (e^x \cos x + I) \Longrightarrow 2I = e^x \sin x - e^x \cos x$$

$$\Longrightarrow I = \frac{e^x}{2} (\sin x - \cos x) + C.$$

The result is easy to check by differentiation. ∎

Reality check The "wraparound" trick in Example 9 seems almost too good to be true. Let's check the result graphically. Figure 1 shows graphs of $f(x) = e^x \sin x$ and $F(x) = e^x (\sin x - \cos x)/2$, which supposedly is an antiderivative:

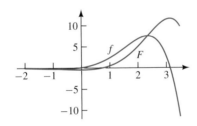

FIGURE 1

$$f(x) = e^x \sin x \text{ and } F(x) = \frac{e^x}{2} (\sin x - \cos x)$$

The graphs look reasonable—f appears to be the derivative of F. In particular:

- **Asymptotes** The integrand $f(x)$ tends quickly to zero as $x \to -\infty$. Properly, therefore, the antiderivative F looks nearly horizontal toward the left of the picture.

- **Wild swings in the long run** As $x \to \infty$, $f(x)$ changes sign at each multiple of π, oscillating between larger and larger positive and negative values. The antiderivative F therefore also oscillates dramatically, with peaks and valleys at multiples of π. Figure 1 just begins to suggest this behavior; a larger window would show it more clearly.

Reduction formulas: Stepping down to a solution

The equation

$$\int x^n e^x \, dx = x^n e^x - n \int x^{n-1} e^x \, dx, \tag{1}$$

which holds for all integers $n > 0$, is called a **reduction formula**. (The name makes sense because the power of x is "reduced" by one on the right-hand side.) It expresses the left-hand indefinite integral in terms of another, slightly "reduced" integral. We'll first use the formula and then explain why it holds.

EXAMPLE 10   Use reduction formula (1) to find $\int x^3 e^x \, dx$.

Solution   The formula says, for $n = 3$, that

$$\int x^3 e^x \, dx = x^3 e^x - 3 \int x^2 e^x \, dx.$$

In Example 8 we found a value for the last integral; plugging that result into the formula gives the answer:

$$\int x^3 e^x \, dx = x^3 e^x - 3 \int x^2 e^x \, dx$$

$$= x^3 e^x - 3\left(x^2 e^x - 2xe^x + 2e^x + C\right)$$

$$= x^3 e^x - 3x^2 e^x + 6xe^x - 6e^x - 3C.$$

Because C is an arbitrary constant, $-3C$ is arbitrary too, and we write simply

$$\int x^3 e^x \, dx = x^3 e^x - 3x^2 e^x + 6xe^x - 6e^x + C.$$

A routine check by differentiation shows that the answer is right.   ∎

Why the formula holds   Example 10 shows that the reduction Formula (1) works, at least for $n = 3$. But *why* must it hold? How might we discover it? Integration by parts gives some answers. If we set

$$u = x^n, \quad dv = e^x \, dx, \quad du = nx^{n-1} \, dx, \quad \text{and} \quad v = e^x,$$

then integration by parts yields what we want:

$$\int x^n e^x \, dx = x^n e^x - n \int x^{n-1} e^x \, dx.$$

Integral tables contain many reduction formulas; most can be found using integration by parts.

BASIC EXERCISES

1. **(a)** Show that $\dfrac{d}{dx} x \ln x = 1 + \ln x$.

   (b) Explain why part (a) implies that $\int \ln x \, dx = x \ln x - x + C$.

2. **(a)** Show that $\dfrac{d}{dx} x^4 \ln x = x^3 + 4x^3 \ln x$.

   (b) Explain why part (a) implies that $\int x^3 \ln x \, dx = \dfrac{x^4 \ln x}{4} - \dfrac{x^4}{16} + C$.

   (c) Find $\int x^3 \ln x \, dx$ using integration by parts with $u = \ln x$ and $dv = x^3 \, dx$.

3. **(a)** Show that $\dfrac{d}{dx} x e^{-x} = e^{-x} - xe^{-x}$.

   (b) Explain why part (a) implies that $\int xe^{-x} \, dx = -e^{-x} - xe^{-x} + C$.

   (c) Find $\int xe^{-x} \, dx$ using integration by parts with $u = x$ and $dv = e^{-x} \, dx$.

4. **(a)** Show that $\dfrac{d}{dx} xe^{2x} = e^{2x} + 2xe^{2x}$.

   (b) Use part (a) to show that $\int xe^{2x} \, dx = \dfrac{xe^{2x}}{2} - \dfrac{e^{2x}}{4} + C$.

   (c) Find $\int xe^{2x} \, dx$ using integration by parts with $u = x$ and $dv = e^{2x} \, dx$.

5. **(a)** Show that $\dfrac{d}{dx} x \arctan(2x) = \arctan(2x) + \dfrac{2x}{1 + 4x^2}$.

   (b) Use part (a) to find $\int \arctan(2x) \, dx$.

   (c) Find $\int \arctan(2x) \, dx$ using integration by parts with $u = \arctan(2x)$ and $dv = dx$.

6. **(a)** Show that $\dfrac{d}{dx} x \arcsin x = \arcsin x + \dfrac{x}{\sqrt{1 - x^2}}$.

   (b) Use part (a) to find $\int \arcsin x \, dx$.

   (c) Find $\int \arcsin x \, dx$ using integration by parts with $u = \arcsin x$ and $dv = dx$.

7. (a) Show that $\dfrac{d}{dx}e^{2x}\sin x = 2e^{2x}\sin x + e^{2x}\cos x$.

   (b) Find $\dfrac{d}{dx}e^{2x}\cos x$.

   (c) Use parts (a) and (b) to find $\displaystyle\int e^{2x}\sin x\,dx$.

   (d) Use integration by parts twice — first with $u=\sin x$, then with $u=\cos x$ — to find $\displaystyle\int e^{2x}\sin x\,dx$.

8. (a) Show that $\dfrac{d}{dx}e^{x}\cos(2x) = e^{x}\cos(2x) - e^{x}\sin(2x)$.

   (b) Find $\dfrac{d}{dx}e^{x}\sin(2x)$.

   (c) Use parts (a) and (b) to find $\displaystyle\int e^{x}\cos(2x)\,dx$.

   (d) Use integration by parts to find $\displaystyle\int e^{x}\cos(2x)\,dx$.

In Exercises 9–14, use integration by parts with the suggested u and dv to find the antiderivative. Check your answers by differentiation.

9. $\displaystyle\int x\sin(3x)\,dx \qquad u=x, \quad dv=\sin(3x)\,dx$

10. $\displaystyle\int \frac{\ln x}{x^2}\,dx \qquad u=\ln x, \quad dv=x^{-2}\,dx$

11. $\displaystyle\int x\sec^2 x\,dx \qquad u=x, \quad dv=\sec^2 x\,dx$

12. $\displaystyle\int x\sec x\tan x\,dx \qquad u=x, \quad dv=\sec x\tan x\,dx$

13. $\displaystyle\int(\ln x)^2\,dx \qquad u=\ln x, \quad dv=\ln x\,dx$

14. $\displaystyle\int \sqrt{x}\ln x\,dx \qquad u=\sqrt{x}, \quad dv=\ln x\,dx$

In Exercises 15–20, evaluate the definite integral using integration by parts. Check your answers by comparing them with a midpoint rule estimate computed using n = 20.

15. $\displaystyle\int_0^1 xe^{-x}\,dx$

16. $\displaystyle\int_0^{\pi} x\cos(2x)\,dx$

17. $\displaystyle\int_1^e x\ln x\,dx$

18. $\displaystyle\int_{\pi/4}^{\pi/2} x\csc^2 x\,dx$

19. $\displaystyle\int_{-1}^{\sqrt{2}/2} x^2\arctan x\,dx$

20. $\displaystyle\int_1^4 e^{3x}\cos(2x)\,dx$

21. (a) Show that $\displaystyle\int x^2\cos x\,dx = x^2\sin x - 2\int x\sin x\,dx$.

   (b) Find $\displaystyle\int x\sin x\,dx$.

   (c) Use parts (a) and (b) to show that $\displaystyle\int x^2\cos x\,dx = x^2\sin x + 2x\cos x - 2\sin x$.

22. (a) Show that $\displaystyle\int x^2\sin x\,dx = -x^2\cos x + 2\int x\cos x\,dx$.

   (b) Find $\displaystyle\int x\cos x\,dx$.

   (c) Use parts (a) and (b) to show that $\displaystyle\int x^2\sin x\,dx = -x^2\cos x + 2x\sin x + 2\cos x$.

23. Use the reduction formula

$$\int(\ln x)^n\,dx = x(\ln x)^n - n\int(\ln x)^{n-1}\,dx, \quad n\geq 1$$

to find $\displaystyle\int(\ln x)^3\,dx$.

24. Use the reduction formula

$$\int x(\ln x)^n\,dx = \frac{1}{2}x^2(\ln x)^n - \frac{n}{2}\int x(\ln x)^{n-1}\,dx, \quad n\geq 1$$

to find $\displaystyle\int x(\ln x)^2\,dx$.

FURTHER EXERCISES

25. Show that if $r\neq -1$ is a constant,

$$\int x^r\ln x\,dx = \frac{x^{r+1}}{r+1}\left(\ln x - \frac{1}{r+1}\right) + C.$$

26. Derive the reduction formula

$$\int x^n e^x\,dx = x^n e^x - n\int x^{n-1}e^x\,dx.$$

27. Derive the reduction formula in Exercise 23.

28. Derive the reduction formula in Exercise 24.

29. (a) Use the substitution $u=x^3$ to find $\displaystyle\int x^2\ln(x^3)\,dx$.

   (b) Use integration by parts to find $\displaystyle\int x^2\ln(x^3)\,dx$. [HINT: $\ln(x^3) = 3\ln x$.]

30. (a) Use the substitution $u=a+bx$ to find $\displaystyle\int x\sqrt{a+bx}\,dx$.

   (b) Use integration by parts with $u=x$ and $dv=\sqrt{a+bx}\,dx$ to find $\displaystyle\int x\sqrt{a+bx}\,dx$.

31. Use the substitution $u=x^2$ and integration by parts to find $\displaystyle\int x^3 e^{x^2}\,dx$.

32. Use the substitution $u=-x^3$ and integration by parts to find $\displaystyle\int x^5 e^{-x^3}\,dx$.

33. (a) Show that $\displaystyle\int \sin^2 x\,dx = -\sin x\cos x + \int\cos^2 x\,dx$.

   (b) Use part (a) and the trigonometric identity $\sin^2 x + \cos^2 x = 1$ to show that

$$\int\sin^2 x\,dx = \frac{1}{2}(x - \sin x\cos x).$$

34. Use integration by parts and the trigonometric identity $\sin^2 x + \cos^2 x = 1$ to find $\displaystyle\int\cos^2 x\,dx$.

35. Find $\displaystyle\int x\cos^2 x\,dx$ using integration by parts. [HINT: $\cos^2 x = \frac{1}{2}(1+\cos(2x))$.]

36. Find $\displaystyle\int x\sin^2 x\,dx$. [HINT: $\sin^2 x = \frac{1}{2}(1-\cos(2x))$.]

In Exercises 37–58, find the antiderivative. Check your results by differentiation.

37. $\int x^2 \cos x \, dx$

38. $\int x \sin x \cos x \, dx$

39. $\int x \csc x \cot x \, dx$

40. $\int \arccos x \, dx$

41. $\int \sqrt{x} \ln \left(\sqrt[3]{x}\right) dx$

42. $\int x e^x \sin x \, dx$

43. $\int \arctan(1/x) \, dx$

44. $\int x^3 (\ln x)^2 \, dx$

45. $\int e^{\sqrt{x}} \, dx$

46. $\int x^5 \sin \left(x^3\right) dx$

47. $\int \sin (\ln x) \, dx$

48. $\int \sqrt{x} e^{-\sqrt{x}} \, dx$

49. $\int \sin \left(\sqrt{x}\right) dx$

50. $\int \dfrac{\arctan \left(\sqrt{x}\right)}{\sqrt{x}} \, dx$

51. $\int \sqrt{x} \arctan \left(\sqrt{x}\right) dx$

52. $\int x^2 \arcsin x \, dx$

53. $\int \cos x \ln(\sin x) \, dx$

54. $\int x^5 e^{-x^2} \, dx$

55. $\int x \sinh x \, dx$

56. $\int x \cosh(2x) \, dx$

57. $\int x^2 \sinh x \, dx$

58. $\int \cosh(\sqrt{x}) \, dx$

59. Let $I_n = \displaystyle\int_1^e (\ln x)^n \, dx$, where $n \geq 1$ is an integer.

  (a) Show that $I_1 = 1$.

  (b) Use the reduction formula in Exercise 23 to show that $I_2 = e - 2I_1 = e - 2$.

  (c) Show that $I_3 = e - 3I_2 = 6 - 2e$.

  (d) Evaluate I_5.

60. Let $I_n = \displaystyle\int_0^1 x^n e^x \, dx$, where $n \geq 0$ is an integer.

(a) Show that $I_0 = e - 1$.

(b) Use the reduction formula in Exercise 26 to show that $I_1 = e - I_0 = 1$.

(c) Show that $I_2 = e - 2I_1 = e - 2$.

(d) Evaluate I_5.

61. Find a function f such that

$$\int f(x) \sin x \, dx = -f(x) \cos x + \int x^3 \cos x \, dx.$$

62. Suppose that f is a continuous function, that $f(0) = 2$, and that $\displaystyle\int_0^\pi f(x) \sin x \, dx + \int_0^\pi f''(x) \sin x \, dx = 6$. Find $f(\pi)$.

63. Suppose that f is a continuous function, that $\displaystyle\int_{-\pi/2}^{3\pi/2} f'(x) \, dx = 1$, and that $\displaystyle\int_{-\pi/2}^{3\pi/2} f''(x) \cos x \, dx = 4$. Evaluate $\displaystyle\int_{-\pi/2}^{3\pi/2} f(x) \cos x \, dx$.

64. Show that v in the integration by parts formula can be *any* antiderivative of dv.

65. (a) Explain why $f(b) - f(a) = \displaystyle\int_a^b f'(x) \, dx$.

  (b) Use integration by parts and part (a) to show that
  $$f(b) = f(a) + f'(a)(b - a) - \int_a^b (x - b) f''(x) \, dx.$$
  [HINT: Let $dv = dx$ and $v = x - b$.]

  (c) Show that $f(b) = f(a) + f'(a)(b - a) + \dfrac{1}{2} f''(a)(b - a)^2 + \dfrac{1}{2} \displaystyle\int_a^b (x - b)^2 f'''(x) \, dx$.

  (d) Let $p_3(x)$ be the third-degree Taylor polynomial for $f(x)$ based at $x = a$. Show that
  $$f(b) - p_3(b) = -\frac{1}{6} \int_a^b (x - b)^3 f^{(4)}(x) \, dx.$$

8.2 PARTIAL FRACTIONS

We studied rational functions in Section 1.3; see also Appendix B.

A **rational function** is any function that can be written as a *ratio* of two polynomials. ← All three of the following expressions define rational functions:

$$\frac{2}{1 - x^2}; \qquad \frac{2 + 5x + 3x^2 + 3x^3}{x(1 + x^2)(x + 2)}; \qquad \frac{x^3}{1 + x^2}.$$

This section is about the problem of antidifferentiating rational functions. Notice first that there *is* a problem. Polynomials themselves are easy to antidifferentiate, but *quotients* of polynomials are another matter entirely. Derivatives of quotients can be sticky to compute; *anti*derivatives can be even worse.

The **method of partial fractions** is a systematic technique for antidifferentiating rational functions. The basic idea is to "divide and conquer": We rewrite the given rational function as a sum of simpler rational functions and then antidifferentiate the terms separately.

First examples: The method in action

Before describing the method in full theoretical regalia, we use several examples to introduce the general discussion that follows.

EXAMPLE 1 Find $\int \dfrac{2}{1-x^2}\,dx$.

Solution The problem looks hard at first glance. The integrand doesn't match any of our standard basic forms, and neither u-substitution nor integration by parts looks promising. ➡

Some clever substitution might succeed, but probably with more labor than necessary.

The new trick is to rewrite the integrand as a sum of two *simpler* terms, as follows:

$$\frac{2}{1-x^2} = \frac{2}{(1+x)(1-x)} = \frac{1}{1+x} + \frac{1}{1-x}. \tag{1}$$

The last two summands are the **partial fractions** (i.e., parts of the original fraction) for which the method is named.

We'll explain soon how we *found* Equation (1). But notice now that, once written down, the equation is easy to *check*. Notice also that Equation (1) makes the antiderivative problem much easier. Each summand on the right is easy to antidifferentiate, and adding the results completes the problem. Here are the details: ➡

Watch the minus signs.

$$\int \frac{2}{1-x^2}\,dx = \int \frac{dx}{1+x} + \int \frac{dx}{1-x} = \ln|1+x| - \ln|1-x| + C.$$

Differentiation shows that the answer is right:

$$\left(\ln|1+x| - \ln|1-x| + C\right)' = \frac{1}{1+x} - \frac{-1}{1-x} = \frac{2}{1-x^2}$$

as claimed.   ∎

EXAMPLE 2 Find $\int \dfrac{2+5x+3x^2+3x^3}{x(1+x^2)(x+2)}\,dx$.

Solution It's easy to *check* (but harder to guess) that

$$\frac{2+5x+3x^2+3x^3}{x(1+x^2)(x+2)} = \frac{1}{x} + \frac{1}{1+x^2} + \frac{2}{x+2}.$$

With the integrand subdivided into partial fractions, the problem is now easy to conquer:

$$\int \frac{2+5x+3x^2+3x^3}{x(1+x^2)(x+2)}\,dx = \int \frac{dx}{x} + \int \frac{dx}{1+x^2} + \int \frac{2}{x+2}\,dx$$

$$= \ln|x| + \arctan x + 2\ln|x+2| + C.$$

Notice the curious pattern, which is one we'll see repeatedly: Antidifferentiating a rational function produces a combination of logs and arctangents.   ∎

EXAMPLE 3 Find $\int \dfrac{x^3}{1+x^2}\,dx$.

Solution In this case the denominator cannot be factored. But the numerator has degree three and the denominator only two, and so we can long divide the numerator by the denominator. Here's the result:

$$\frac{x^3}{1+x^2} = x - \frac{x}{1+x^2}.$$

Again we've rewritten the integrand as the sum of simpler terms, each of which is relatively easy to antidifferentiate: ➡

For the second integral, substitute $u = 1+x^2$.

$$\int x\,dx - \int \frac{x}{1+x^2}\,dx = \frac{x^2}{2} - \frac{1}{2}\ln|1+x^2| + C.$$ ∎

Authors have been known to rig problems.

Good news: The method works—in principle   We've seen that not *every* elementary function has an elementary antiderivative, but must every *rational* function have one? True, we found a nice antiderivative in each of the preceding examples, but maybe that was just luck. ◂

In fact, every rational function *does* have an elementary antiderivative, which involves *only* the ingredients seen in the preceding examples: logarithms, arctangents, and rational functions. A fully rigorous proof of this fact would take us somewhat beyond our mathematical depth, but we remark that partial fractions play a key role.

Knowing that an antiderivative formula exists is one thing; finding one explicitly is quite another. The hardest problem is usually algebraic—rewriting the given rational function as a sum of partial fractions. Once this is done, antidifferentiating the separate pieces is usually relatively easy. Because the method itself is our main interest, we'll arrange to keep the algebra simple.

Rational numbers, rational functions, and partial fractions

Rational *functions* are like rational *numbers* in that both are quotients of simpler objects. Many of the standard ideas, operations, and vocabulary that apply to ordinary fractions apply to rational functions too.

Proper vs. improper   A positive rational number p/q is called **improper** if $p \geq q$. Any improper fraction can be written as the sum of an integer and a proper fraction. For example,

$$\frac{29}{6} = 4 + \frac{5}{6};$$

the numbers 4 and 5 can be found by long dividing 29 by 6.

Rational *functions* behave similarly to rational *numbers*. A rational function $r(x) = p(x)/q(x)$ is **proper** if p has lower degree than q; otherwise, r is **improper**. (The integrand in Example 3 is improper.) The analogy with rational numbers continues:

> *Any improper rational function can be written as the sum of a polynomial and a proper rational function.*

For instance,

$$\frac{x}{x+1} = 1 - \frac{1}{x+1}; \qquad \frac{x^2}{x+1} = x - 1 + \frac{1}{x+1}.$$

As with ordinary fractions, the right sides of these equations can be found by long division.

Miss Manners would approve.   Rational functions, no matter how "improper," commit no breach of morals or manners. Learning such quaint jargon is, depending on one's point of view, either part of the fun of mathematics or an annoying obstacle.

Partial fractions—of numbers and of polynomials   Any rational *number* m/n can be written as a sum or difference of fractions of a special type: fractions whose denominators are the prime factors of n or powers thereof. For example, 6 has prime factors 2 and 3, and

$$\frac{5}{6} = \frac{5}{2 \cdot 3} = \frac{1}{2} + \frac{1}{3}; \qquad \frac{5}{12} = \frac{5}{2^2 \cdot 3} = \frac{3}{4} - \frac{1}{3}.$$

The idea for rational functions is similar. Any rational function $p(x)/q(x)$ can be written as the sum of **partial fractions**—that is, other rational functions whose denominators are either **irreducible factors** of $q(x)$ or powers thereof. (An **irreducible** polynomial is one that cannot be written as a product of lower-order factors.) ➤ We used this idea when we wrote

$$\frac{p(x)}{q(x)} = \frac{2 + 5x + 3x^2 + 3x^3}{x(1 + x^2)(x + 2)} = \frac{1}{x} + \frac{1}{1 + x^2} + \frac{2}{x + 2}.$$

Irreducible factors are analogous to prime numbers: neither can be factored.

Notice especially the denominators on the right: Each one is an irreducible factor of $q(x)$.

Two basic forms   All this jargon may sound formidably abstract. Indeed, the theory of polynomials and rational functions *is* a formidable subject, but we need to examine just the tip of that iceberg. Only the following facts will matter to us:

- **Factoring:**   Every polynomial can be written as a product of *linear* and *quadratic* factors.

- **Recognizing irreducible factors:**   Linear polynomials are automatically irreducible, but a quadratic polynomial $q(x) = ax^2 + bx + c$ may or may not have linear factors, depending on whether or not q has roots. The quadratic formula tells which is the case: If $b^2 - 4ac < 0$, then q has no roots and so can't be factored. If $b^2 - 4ac \geq 0$, then q has roots and can be factored.

- **Proper behavior:**   Any *proper* rational function can be written as a sum of *proper* partial fractions.

Two main types   The situation boils down to the fact that any proper rational function can be written as a sum of partial fractions of just two basic types:

$$\text{Type I:} \quad \frac{A}{(ax + b)^n}; \quad \text{Type II:} \quad \frac{Ax + B}{(ax^2 + bx + c)^n}.$$

(In each case n is a positive integer; a, b, c, A, and B are real constants.) Antidifferentiating *any* rational function, therefore, reduces to antidifferentiating these two basic types.

Type I antiderivatives   Partial fractions of the first type are always easy to antidifferentiate. The substitution $u = ax + b$ succeeds; here are the results:

$$\int \frac{A}{(ax + b)^n}\, dx = \begin{cases} \dfrac{A}{a} \ln |ax + b| & \text{if } n = 1; \\[2ex] \dfrac{A}{a(1 - n)(ax + b)^{n-1}} & \text{if } n > 1. \end{cases}$$

The general formula may look forbidding, but in most actual cases antidifferentiation is easy.

EXAMPLE 4   Find $\displaystyle\int \frac{6}{2x + 1}\, dx$ and $\displaystyle\int \frac{6}{(2x + 1)^5}\, dx.$

Solution   For both integrals, substituting $u = 2x + 1$ and $du = 2\, dx$ does the trick:

$$\int \frac{6}{2x + 1}\, dx = \int \frac{3}{u}\, du = 3 \ln |2x + 1| + C;$$

$$\int \frac{6}{(2x + 1)^5}\, dx = \int \frac{3}{u^5}\, du = -\frac{3}{4(2x + 1)^4} + C.$$ ∎

Type II antiderivatives   Antidifferentiating partial fractions of the form

$$\frac{Ax + B}{(ax^2 + bx + c)^n}$$

can be quite messy, especially if $n > 1$. In that case, reduction formulas can be used to reduce the exponent n.

For $n = 1$, **completing the square** in the denominator simplifies things considerably, as the next example illustrates.

EXAMPLE 5   Find $\displaystyle\int \frac{dx}{x^2 + 2x + 2}$ and $\displaystyle\int \frac{2x + 3}{x^2 + 2x + 2}\, dx$.

Solution   Because $b^2 - 4ac < 0$, the denominator is irreducible; factoring is impossible. So let's complete the square:

$$\frac{1}{x^2 + 2x + 2} = \frac{1}{x^2 + 2x + 1 + 1} = \frac{1}{(x + 1)^2 + 1}.$$

Now we substitute $u = x + 1$ and $du = dx$ in both integrals. The first integral becomes

$$\int \frac{dx}{(x + 1)^2 + 1} = \int \frac{du}{u^2 + 1} = \arctan(x + 1) + C.$$

The same substitution in the second integral gives

$$\int \frac{2x + 3}{x^2 + 2x + 2}\, dx = \int \frac{2(u - 1) + 3}{u^2 + 1}\, du \qquad (u = x + 1 \text{ implies } x = u - 1)$$

$$= \int \frac{2u + 1}{u^2 + 1}\, du$$

$$= \int \frac{2u}{u^2 + 1}\, du + \int \frac{du}{u^2 + 1}$$

$$= \ln|u^2 + 1| + \arctan u + C$$

$$= \ln|x^2 + 2x + 2| + \arctan(x + 1) + C,$$

which completes the problem.   ■

Reality check: A graphical interlude

We've been throwing plenty of symbols around, but does it all make sense? We've seen, for instance, that antidifferentiating rational functions seems to spawn logarithms and arctangents. When the dust cleared in Example 2, for instance, we concluded that

$$\int \frac{2 + 5x + 3x^2 + 3x^3}{x(1 + x^2)(x + 2)}\, dx = \ln|x| + \arctan x + 2\ln|x + 2| + C.$$

A look at graphs helps explain the presence of these ingredients. Figure 1 shows graphs of the integrand f and an antiderivative F.

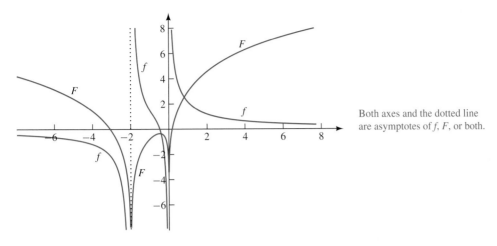

FIGURE 1
A rational function f and an antiderivative F

Both axes and the dotted line are asymptotes of f, F, or both.

The graphs are complicated (like the functions themselves), but they show the right things:

- **Two vertical asymptotes:** The integrand f has vertical asymptotes at $x = -2$ and $x = 0$; they correspond to the factors $(x + 2)$ and x in the denominator. The antiderivative F also has two vertical asymptotes, at the same places. Near these values of x, the F-graph resembles a logarithmic function. The remaining factor in the denominator, $(1 + x^2)$, is never zero, and so it contributes no vertical asymptotes either to f or to F.

- **In the long run:** The general shape of the F-graph looks roughly logarithmic as $x \to \pm\infty$. As $x \to \infty$, for instance, the F-graph is increasing and concave down just like a logarithmic function. Here's an intuitive explanation for this "log-like" behavior. Expanding the denominator of f gives

$$f(x) = \frac{2 + 5x + 3x^2 + 3x^3}{x(1 + x^2)(x + 2)} = \frac{2 + 5x + 3x^2 + 3x^3}{2x + x^2 + 2x^3 + x^4}.$$

For large $|x|$ the highest powers in the numerator and denominator "dominate," and so

$$f(x) = \frac{2 + 5x + 3x^2 + 3x^3}{2x + x^2 + 2x^3 + x^4} \approx \frac{3x^3}{x^4} = \frac{3}{x}.$$

Thus, for large positive x we have

$$F(x) \approx \int \frac{3}{x}\, dx = 3\ln|x| + C,$$

which corresponds to the log-like shape of F to the right of Figure 1.

Last, not least: How to find the partial fractions

We've seen *why* we'd want to write a rational function as a sum of partial fractions but not yet *how* to do so. How, for instance, did we arrive at the equation

$$\frac{2 + 5x + 3x^2 + 3x^3}{x(1 + x^2)(x + 2)} = \frac{1}{x} + \frac{1}{1 + x^2} + \frac{2}{x + 2}$$

in Example 2?

Writing a rational function $p(x)/q(x)$ as a sum of partial fractions takes four steps:

- **Make it proper:** If $p(x)/q(x)$ is improper, use long division to rewrite it as the sum of a polynomial and a proper rational function.
- **Factor q:** Write the denominator $q(x)$ as a product of linear and irreducible quadratic factors; some may be repeated.
- **Write as a sum:** Write $p(x)/q(x)$ as a sum of partial fractions with **undetermined coefficients** (represented by letters) in their numerators.
- **Find the coefficients:** Solve one or more algebraic equations to find numerical values for the coefficients.

The first two steps are straightforward; the last two need elaboration.

Writing the integrand as a sum As earlier examples show, each partial fraction summand comes from some factor of the denominator of the original rational function. For instance, we expect a partial fraction sum of the general form

$$\frac{2 + 5x + 3x^2 + 3x^3}{x(1 + x^2)(x + 2)} = \frac{A}{x} + \frac{Bx + C}{1 + x^2} + \frac{D}{x + 2}$$

for the rational function on the left; note that one partial fraction corresponds to each irreducible factor of the denominator. Because the middle summand has a *quadratic* denominator, its numerator can have degree up to 1 but no higher. The other summands, which have *linear* denominators, can have only constant numerators. (Otherwise, a summand would be improper.)

Repeated factors The general form of an answer is slightly different if the denominator has a **repeated factor**—that is, a factor that appears to a power higher than 1. Each such factor contributes several partial fractions—one for each power of the factor—to the general form. For example, the partial fraction version of

$$\frac{1}{x^3(x^2 + 1)^2(x + 1)}$$

has the general form

$$\frac{A}{x} + \frac{B}{x^2} + \frac{C}{x^3} + \frac{Dx + E}{x^2 + 1} + \frac{Fx + G}{(x^2 + 1)^2} + \frac{H}{x + 1}.$$

(These "extra" summands are necessary to guarantee that a partial fraction form exists.)

Finding values for the coefficients Finding numerical values for the unknown constants $A, B, C, \ldots$ is a problem in *algebra*. (The calculus part comes later.) Examples of several techniques follow.

Math jargon—impress your friends.

EXAMPLE 6 (**The simplest case: only linear factors**) Derive the **partial fraction decomposition** ◄

$$\frac{2}{(1 + x)(1 - x)} = \frac{1}{1 + x} + \frac{1}{1 - x}.$$

(We used it in Example 1, page 467.)

Solution The denominator's factors show that we need constants A and B such that

$$\frac{2}{(1 + x)(1 - x)} = \frac{A}{1 + x} + \frac{B}{1 - x}.$$

Multiplying both sides by $(1+x)(1-x)$ clears out the denominators; we get

$$2 = A(1-x) + B(1+x). \qquad (2)$$

Now it's easy to find A and B. One way is to multiply out the right side and collect like powers of x:

$$2 = A(1-x) + B(1+x) = A - Ax + B + Bx = (A+B) + (B-A)x.$$

Equating powers of x on the far left and far right shows that $A + B = 2$ and $B - A = 0$, and so $A = B = 1$ as claimed earlier.

Another, perhaps simpler, route to the same result is to plug cleverly chosen values of x into Equation (2). Doing so may make the appropriate values of A and B obvious. If, say, $x = 1$, then Equation (2) becomes simply $2 = 2B$, and so $B = 1$. Similarly, plugging $x = -1$ into Equation (2) yields $2 = 2A$, or $A = 1$.   ■

E X A M P L E 7   (One irreducible quadratic)   It's not hard to *check* that

$$\frac{2 + 5x + 3x^2 + 3x^3}{x(1+x^2)(x+2)} = \frac{1}{x} + \frac{1}{1+x^2} + \frac{2}{x+2}. \qquad (3)$$

Explain how to *find* this equation.

Solution   The form of the denominator on the left means that we want constants A, B, C, and D such that

$$\frac{2 + 5x + 3x^2 + 3x^3}{x(1+x^2)(x+2)} = \frac{A}{x} + \frac{Bx+C}{1+x^2} + \frac{D}{x+2}.$$

We'll find these constants by solving an appropriate set of equations.

Multiplying both sides by the denominator on the left and collecting powers of x gives the new equation

$$2 + 5x + 3x^2 + 3x^3 = A(1+x^2)(x+2) + (Bx+C)x(x+2) + Dx(1+x^2)$$
$$= 2A + (A + 2C + D)x + (2A + 2B + C)x^2 + (A + B + D)x^3.$$

Equating coefficients of powers of x in the first and the last expressions gives *four* equations in *four* unknowns:

$$2 = 2A; \quad 5 = A + 2C + D; \quad 3 = 2A + 2B + C; \quad 3 = A + B + D.$$

Solving these equations simultaneously takes some algebraic work (by hand or by calculator), but the process is straightforward. The results are $A = 1$, $B = 0$, $C = 1$, and $D = 2$; replacing these symbols with their values produces Equation (3).   ■

B A S I C   E X E R C I S E S

1. **(a)** Verify, by adding the terms on the right, that

   $$\frac{2}{1-x^2} = \frac{1}{1+x} + \frac{1}{1-x}.$$

   (b) Use part (a) to show that $\displaystyle\int \frac{dx}{1-x^2} = \frac{1}{2}\ln\left|\frac{1+x}{1-x}\right| + C$, where C is a constant.

2. Verify, by adding the terms on the right, that

   $$\frac{2 + 5x + 3x^2 + 3x^3}{x(1+x^2)(x+2)} = \frac{1}{x} + \frac{1}{1+x^2} + \frac{2}{x+2}.$$

3. Use the algebraic identity $x^3 = x(1+x^2) - x$ to find the partial fraction decomposition of $x^3/(1+x^2)$.

4. **(a)** Use the algebraic identity $x = (x+1) - 1$ to find the partial fraction decomposition of $x/(1+x)$.

   (b) Find the partial fraction decomposition of $x^2/(1+x)$.

5. **(a)** Verify, by adding the terms on the right, that

   $$\frac{5x+7}{(x+1)(x+2)} = \frac{2}{x+1} + \frac{3}{x+2}.$$

   (b) Find $\displaystyle\int \frac{5x+7}{(x+1)(x+2)}\, dx$.

6. Suppose that $p(x) = \dfrac{4x^2 + 3x + 2}{x^2(x+1)}$.

   (a) Can constants A and B be chosen so that $p(x) = \dfrac{A}{x} + \dfrac{B}{x+1}$? Justify your answer.

   (b) Can constants A and B be chosen so that $p(x) = \dfrac{A}{x^2} + \dfrac{B}{x+1}$? Justify your answer.

   (c) Find constants A and B so that $p(x) = \dfrac{A}{x} + \dfrac{B}{x^2} + \dfrac{C}{x+1}$.

   (d) Find $\displaystyle\int p(x)\,dx$.

7. Let $f(x) = \dfrac{x^2 + 1}{(x-1)^2}$.

   (a) Explain why f is a rational function.

   (b) Explain why f is not a proper rational function.

   (c) Write f as the sum of a polynomial and a proper rational function.

   (d) Use part (c) to find an antiderivative of f.

8. Let $g(x) = \dfrac{(x+1)^3}{x^2 + x + 1}$.

   (a) Explain why g is a rational function.

   (b) Is g is a proper rational function? Justify your answer.

   (c) Write g as the sum of a polynomial and a proper rational function.

   (d) Use part (c) to find an antiderivative of g.

9. Suppose that $q(x) = (x-2)(x^2 + 2x + 3)$ and that $f(x) = p(x)/q(x)$ is a proper rational function. Is $\dfrac{A}{x-2} + \dfrac{Bx+C}{x^2 + 2x + 3}$ the partial fraction decomposition of f? Justify your answer.

10. Suppose that $q(x) = (x+1)(x^2 - 3x + 4)$ and that $g(x) = p(x)/q(x)$ is a proper rational function. Find the form of the partial fraction decomposition of g. (Your answer will involve unknown constant coefficients.)

11. Suppose that $q(x) = (x+1)(x^2 - 2x + 1)$ and that $g(x) = p(x)/q(x)$ is a proper rational function. Is $\dfrac{A}{x+1} + \dfrac{Bx+C}{x^2 - 2x + 1}$ the partial fraction decomposition of g? Justify your answer.

12. Suppose that $q(x) = (x+1)(x^2 + 2x + 1)$ and that $g(x) = p(x)/q(x)$ is a proper rational function. Is $\dfrac{A}{x+1} + \dfrac{Bx+C}{x^2 + 2x + 1}$ the partial fraction decomposition of g? Justify your answer.

13. Suppose that $q(x) = (x+1)^2(x^2 + 1)$ and that $h(x) = p(x)/q(x)$ is a proper rational function. Is $\dfrac{A}{(x+1)^2} + \dfrac{Bx+C}{x^2 + 1}$ the partial fraction decomposition of h? Justify your answer.

14. Suppose that $q(x) = (x+1)\left(x^2 + 1\right)^3$ and that $k(x) = p(x)/q(x)$ is a proper rational function. Is $\dfrac{A}{x+1} + \dfrac{Bx+C}{(x^2+1)^3}$ the partial fraction decomposition of k? Justify your answer.

15. Consider the partial fraction decomposition equation
$$\frac{x^2 + 3x - 1}{x(x+1)(x-2)} = \frac{A}{x} + \frac{B}{x+1} + \frac{C}{x-2}.$$

   (a) Multiply both sides of this equation by x and then substitute $x = 0$ to show that $A = 1/2$.

   (b) Multiply both sides of this equation by $x+1$ and then substitute $x = -1$ to show that $B = 1$.

   (c) Multiply both sides of this equation by $x-2$ and then substitute $x = 2$ to show that $C = 3/2$.

   (d) Find $\displaystyle\int \frac{x^2 + 3x - 1}{x(x+1)(x-2)}\,dx$.

16. (a) Use the technique illustrated in parts (a)–(c) of Exercise 15 to find values of the constants A, B, and C such that
$$\frac{6}{(x-2)\,(x^2 - 1)} = \frac{A}{x-2} + \frac{B}{x-1} + \frac{C}{x+1}.$$

   (b) Find $\displaystyle\int \frac{6}{(x-2)\,(x^2 - 1)}\,dx$.

17. Find $\displaystyle\int \frac{x^2 + 2x + 5}{(x-1)(x+1)(x+2)}\,dx$.

18. Find $\displaystyle\int \frac{x^2 + 2x + 5}{x(x^2 - 4)}\,dx$.

19. Consider the partial fraction decomposition equation
$$\frac{x^2 - 1}{x\,(x^2 + 4)} = \frac{A}{x} + \frac{Bx+C}{x^2 + 4}.$$

   (a) Multiply both sides of this equation by x and then substitute $x = 0$ to show that $A = -1/4$.

   (b) Substitute $x = 1$ into this equation to show that $B + C = 5/4$.

   (c) Substitute $x = -1$ into this equation to show that $B - C = 5/4$.

   (d) Use parts (b) and (c) to show that $B = 5/4$ and $C = 0$.

   (e) Find $\displaystyle\int \frac{x^2 - 1}{x\,(x^2 + 4)}\,dx$.

20. Find $\displaystyle\int \frac{3x^2 + 7x + 5}{(x+1)(x^2 + 2x + 2)}\,dx$.

21. Is there a quadratic polynomial $q(x)$ such that $4/(1-x) + 5\ln(3+x) + C$ is an antiderivative of $q(x)/((1-x)^2(3+x))$? If so, find it. If not, explain why no such polynomial q exists.

22. Is there a quadratic polynomial $q(x)$ such that $4/(1-x) + 5\ln(3+x) + C$ is an antiderivative of $q(x)/((1-x^2)(3+x))$? If so, find it. If not, explain why no such polynomial q exists.

23. Is there a quadratic polynomial $q(x)$ such that $2/(1-x) + \arcsin x + \ln(3+x) + C$ is an antiderivative of $q(x)/((1-x)^2(3+x))$? If so, find it. If not, explain why no such polynomial q exists.

24. Is there a quadratic polynomial $q(x)$ such that $4/(1-x)^2 + 5\ln|x+3| + C$ is $\displaystyle\int \frac{q(x)}{(1-x)^2(x+3)}\,dx$? If so find it. If not, explain why no such polynomial q exists.

FURTHER EXERCISES

25. Consider the partial fraction decomposition

$$\frac{x^2}{(x+1)^3} = \frac{A}{x+1} + \frac{B}{(x+1)^2} + \frac{C}{(x+1)^3}.$$

(a) Multiply both sides of the equation by $(x+1)^3$ and then substitute $x = -1$ to show that $C = 1$.

(b) Substitute $x = 0$ into the equation to show that $A + B = -1$.

(c) Substitute $x = 1$ into the equation to show that $2A + B = 0$.

(d) Use parts (b) and (c) to show that $A = 1$ and $B = -2$.

(e) Find $\displaystyle\int \frac{x^2}{(x+1)^3}\, dx$.

26. Find $\displaystyle\int \frac{2-x}{x^2(x+2)}\, dx$.

27. Let $p(x) = (x+1)(x-2)$.

(a) Find the partial fraction decomposition of $1/p(x)$.

(b) Find the partial fraction decomposition of $x/p(x)$.

(c) Use parts (a) and (b) to find $\displaystyle\int \frac{4-3x}{(x+1)(x-2)}\, dx$.

28. Let $p(x) = (x-2)(x+3)$.

(a) Find the partial fraction decomposition of $1/p(x)$.

(b) Find the partial fraction decomposition of $x/p(x)$.

(c) Find $\displaystyle\int \frac{4x+5}{(x-2)(x+3)}\, dx$.

29. Let $p(x) = (x-1)(x^2+1)$.

(a) Find the partial fraction decomposition of $1/p(x)$.

(b) Find the partial fraction decomposition of $x/p(x)$.

(c) Find the partial fraction decomposition of $x^2/p(x)$.

(d) Use parts (a)–(c) to find $\displaystyle\int \frac{2+3x+4x^2}{(x-1)(x^2+1)}\, dx$.

30. Let $p(x) = (x+2)(x^2+2x+5)$.

(a) Find the partial fraction decomposition of $1/p(x)$.

(b) Find the partial fraction decomposition of $x/p(x)$.

(c) Find the partial fraction decomposition of $x^2/p(x)$.

(d) Find $\displaystyle\int \frac{3-4x+5x^2}{(x+2)(x^2+2x+5)}\, dx$.

31. Evaluate $\displaystyle\int_0^1 \frac{dx}{(x-2)(x^2+1)}$.

32. Evaluate $\displaystyle\int_0^2 \frac{dx}{x^3+1}$. [HINT: $x^3+1 = (x+1)(x^2-x+1)$.]

In Exercises 33–40, find the antiderivative.

33. $\displaystyle\int \frac{2x+1}{(x-2)(x+3)}\, dx$

34. $\displaystyle\int \frac{x+1}{(x-1)(x+2)}\, dx$

35. $\displaystyle\int \frac{5x^2+3x-2}{x^3+2x^2}\, dx$

36. $\displaystyle\int \frac{4x^2-3x+2}{x\,(2x-1)^2}\, dx$

37. $\displaystyle\int \frac{x^4}{x^4-1}\, dx$

38. $\displaystyle\int \frac{x^3}{x^2+1}\, dx$

39. $\displaystyle\int \frac{x^3}{x^2-1}\, dx$

40. $\displaystyle\int \frac{3x^2-1}{(x-1)(x+2)}\, dx$

41. Use the substitution $u = \sqrt{x+1}$ to find $\displaystyle\int \frac{dx}{x\sqrt{x+1}}$.

42. Use the substitution $u = \sqrt{x-1}$ to find $\displaystyle\int \frac{dx}{x\sqrt{x-1}}$.

43. (a) Show that $\displaystyle\int \frac{dx}{x+c} = \ln|x+c|$.

(b) Let n be an integer such that $n > 1$. Show that
$$\int \frac{dx}{(x+c)^n} = \frac{1}{(1-n)(x+c)^{n-1}}.$$

(c) Use parts (a) and (b) to show that, if $a \neq 0$,
$$\int \frac{dx}{(ax+b)^n} = \begin{cases} a^{-1}\ln|ax+b| & \text{if } n=1 \\ a^{-1}(1-n)^{-1}(ax+b)^{1-n} & \text{if } n>1. \end{cases}$$

44. Show that $\displaystyle\int \frac{dx}{x^2+d^2} = \frac{1}{|d|}\arctan\left(\frac{x}{|d|}\right)$.

45. Show that $\displaystyle\int \frac{x}{x^2+d^2}\, dx = \frac{1}{2}\ln(x^2+d^2)$.

46. Let n be an integer such that $n \geq 1$.

(a) Use integration by parts to show that
$$\int \frac{dx}{(x^2+d^2)^n} = \frac{x}{(x^2+d^2)^n} + 2n\int \frac{x^2}{(x^2+d^2)^{n+1}}\, dx.$$

(b) Use the result in part (a) and the algebraic identity $x^2 = (x^2+d^2) - d^2$ to show that
$$\int \frac{dx}{(x^2+d^2)^{n+1}} = \frac{1}{2nd^2}\frac{x}{(x^2+d^2)^n} + \frac{2n-1}{2nd^2}\int \frac{dx}{(x^2+d^2)^n}.$$

47. Let $n \geq 2$ be an integer. Show that $\displaystyle\int \frac{x}{(x^2+d^2)^n}\, dx = \frac{1}{2(1-n)\,(x^2+d^2)^{n-1}}$.

8.3 TRIGONOMETRIC ANTIDERIVATIVES

The symbolic antidifferentiation methods we've developed so far—u-substitution, integration by parts, and partial fractions—handle many, but not all, classes of functions. In this section we consider functions that involve two special ingredients: powers of trigonometric functions and roots of quadratic expressions. Here are some examples:

$$\int \sin^2 x \cos^3 x \, dx = \frac{\sin^3 x}{3} - \frac{\sin^5 x}{5} + C;$$

$$\int \sqrt{-x^2 - 2x} \, dx = \frac{1}{2}\left((x+1)\sqrt{-x^2 - 2x} + \arcsin(x+1)\right) + C.$$

It's easy enough to *check* these answers by differentiation, but it's less obvious how they were found.

This section contains no completely new or different ideas or methods. We explore, mainly through examples, how methods of earlier sections can be combined with properties of trigonometric and algebraic functions to handle these new classes of antiderivatives—and reveal some unexpected connections between them.

Powers of trigonometric functions

Consider the following antiderivatives, all of which involve positive integer powers of the six trigonometric functions:

$$\int \cos^3 x \, dx \qquad\qquad \int \cos^3 x \sin^4 x \, dx$$

$$\int \cos^2 x \sin^3 x \tan^2 x \, dx \qquad \int \sec^3 x \tan^3 x \, dx$$

Antiderivatives of all these functions can be found in elementary form by combining substitution, integration by parts, and (sometimes) well-chosen trigonometric identities.

Useful properties of trigonometric functions   Several properties of, and relations among, the trigonometric functions will be useful as we hunt for antiderivatives.

- **Two main types:**   All six trigonometric functions are defined in terms of sines and cosines. It follows that *every* product of integer powers of trigonometric functions can be written in the form $\sin^n x \cos^m x$, where n and m are (not necessarily positive) integer powers. For instance,

$$\cos^4 x \sin^3 x \tan^2 x = \cos^4 x \sin^3 x \frac{\sin^2 x}{\cos^2 x} = \cos^2 x \sin^5 x.$$

In mathspeak, "in principle" means "it's possible—but probably a lot of work."

In principle, every function of the form $\sin^n x \cos^m x$ can be antidifferentiated, though the process might be tedious and messy. ◄

Finding trigonometric antiderivatives is less messy when only *nonnegative* powers are involved. If negative powers occur, it may help to rewrite a function in terms of positive powers of secants and tangents. For instance,

$$\cos^{-5} x \sin^2 x = \cos^{-3} x \frac{\sin^2 x}{\cos^2 x} = \sec^3 x \tan^2 x.$$

In this section, therefore, we mainly consider integrals of two types:

$$\int \sin^n x \cos^m x \, dx \qquad \text{and} \qquad \int \sec^n x \tan^m x \, dx,$$

where m and n are nonnegative integers.

- **Reduction formulas:** If either $n = 0$ or $m = 0$ in the preceding forms, then the integral can be handled—with care—using one of the following reduction formulas:

$$\int \sin^n x \, dx = -\frac{\sin^{n-1} x \cos x}{n} + \frac{n-1}{n} \int \sin^{n-2} x \, dx \qquad (n \neq 0)$$

$$\int \cos^n x \, dx = \frac{\cos^{n-1} x \sin x}{n} + \frac{n-1}{n} \int \cos^{n-2} x \, dx \qquad (n \neq 0)$$

$$\int \tan^n x \, dx = \frac{\tan^{n-1} x}{n-1} - \int \tan^{n-2} x \, dx \qquad (n \neq 1)$$

$$\int \sec^n x \, dx = \frac{\sec^{n-2} x \tan x}{n-1} + \frac{n-2}{n-1} \int \sec^{n-2} x \, dx. \qquad (n \neq 1)$$

Large powers may require *repeated* use of the reduction formulas.

- **Using trigonometric identities:** Many integrands can be simplified to one of the preceding forms by using either the **Pythagorean identities**

$$\sin^2 x + \cos^2 x = 1; \qquad \tan^2 x + 1 = \sec^2 x,$$

or the **double-angle formulas**

$$\sin^2 x = \frac{1}{2} - \frac{\cos(2x)}{2}; \qquad \cos^2 x = \frac{1}{2} + \frac{\cos(2x)}{2}.$$

(These could also be called *half-angle formulas*, depending on the direction in which you read the equations.) The first two identities permit us, in effect, to trade sines for cosines, secants for tangents, and vice versa. The double-angle formulas can be used to convert even *powers* of sines and cosines to cosines of even *multiples* of x.

- **Disguised results:** Trigonometric functions are masters of disguise. For example, $f(x) = 2 \sin x \cos x$ and $g(x) = \sin(2x)$ look different but are actually the same function. This ability to change appearance is often helpful, but it can also be misleading. Beware . . . .

Some concrete examples will illustrate the use of these facts in finding antiderivatives.

EXAMPLE 1   Find $\int \cos^3 x \, dx$.

Solution   With the substitution $u = \sin x$ in mind, we reserve one power of the cosine for du and convert the other two powers to sines:

$$\int \cos^2 x \cos x \, dx = \int (1 - \sin^2 x) \cos x \, dx$$

$$= \int (1 - u^2) \, du = \sin x - \frac{\sin^3 x}{3} + C.$$

We could, instead, have used the reduction formula for powers of cosines:

$$\int \cos^3 x \, dx = \frac{\cos^2 x \sin x}{3} + \frac{2}{3} \int \cos x \, dx = \frac{\cos^2 x \sin x + 2 \sin x}{3} + C.$$

Despite first appearances, the two answers are equal. ■

EXAMPLE 2 Find $\int \sin^4 x \, dx$.

Solution We use the double-angle formulas to rewrite the integrand in terms of $\cos(2x)$ and $\cos(4x)$, as follows:

$$\sin^4 x = (\sin^2 x)^2 = \left(\frac{1}{2} - \frac{\cos(2x)}{2} \right)^2 = \frac{1}{4} - \frac{\cos(2x)}{2} + \frac{\cos^2(2x)}{4}.$$

The first two summands are easy to integrate. To handle the last summand we can apply the double-angle formula *again* to get

$$\frac{\cos^2(2x)}{4} = \frac{1}{8} + \frac{\cos(4x)}{8},$$

which is another manageable integrand. Here's the result:

$$\int \sin^4 x \, dx = \frac{x}{4} - \frac{\sin(2x)}{4} + \frac{x}{8} + \frac{\sin(4x)}{32} + C.$$

Other approaches to the same problem may produce equivalent—but different-looking—answers. ◄   ■

Try a TI–89 or other technology, for instance.

EXAMPLE 3 Find $\int \cos^3 x \sin^4 x \, dx$.

Solution One strategy is to convert all the sines to cosines and then attack the result with the reduction formula for powers of cosines. Instead, we substitute $u = \sin x$, $du = \cos x \, dx$. We chip off one power of $\cos x$ to use in the du factor and convert the remaining (two) powers of cosine to sines:

$$\int \cos^3 x \sin^4 x \, dx = \int (1 - \sin^2 x)(\sin^4 x) \cos x \, dx = \int (1 - u^2) u^4 \, du.$$

The last integral is simple; we'll leave it alone.   ■

EXAMPLE 4 Find $\int \sec^3 x \tan^3 x \, dx$.

Solution We substitute $u = \sec x$ and $du = \sec x \tan x \, dx$ and then convert the remaining tangents to secants:

$$\int \sec^3 x \tan^3 x \, dx = \int \sec^2 x \tan^2 x (\sec x \tan x) \, dx$$

$$= \int \sec^2 x (\sec^2 x - 1)(\sec x \tan x) \, dx$$

$$= \int u^2 (u^2 - 1) \, du,$$

and the rest is easy.   ■

Trigonometric substitutions

Trigonometric substitution is a special case of the general idea of u-substitution, but is tailored especially to handle integrands that involve roots of quadratic expressions. The simplest such expressions are

$$\sqrt{a^2 - x^2}, \quad \sqrt{x^2 - a^2}, \quad \text{and} \quad \sqrt{a^2 + x^2},$$

where a is a positive constant. That *trigonometric* functions come up at all in such integrals may be surprising—so far there's nothing obviously trigonometric in sight. Yet it does turn out that carefully chosen substitutions, with trigonometric ingredients, can reduce integrals of the present type to powers of trigonometric functions. A simple example (well, relatively simple) illustrates the method and some of the subtleties involved.

EXAMPLE 5 According to our integral table,

$$\int \sqrt{4 - x^2}\, dx = \frac{x\sqrt{4 - x^2}}{2} + 2\arcsin\left(\frac{x}{2}\right).$$

We could check this formula by differentiation, but how was it found?

Solution If we substitute $x = 2\sin t$, then $dx = 2\cos t\, dt$, and a little algebra produce a simpler-looking integral:

$$\int \sqrt{4 - x^2}\, dx = 2\int \sqrt{4 - 4\sin^2 t}\,\cos t\, dt = 4\int \cos^2 t\, dt.$$

The last integral, a power of cosines, has the form discussed in earlier examples. Using methods just developed (or an integral table) gives

$$4\int \cos^2 t\, dt = 2t + \sin(2t) + C. \tag{1}$$

So far so good, but the final answer (like the original problem) should involve x, not t. To achieve this end we'll need several facts, all of which follow from our having set $x = 2\sin t$ and from the double-angle formula for sines:

$$t = \arcsin\left(\frac{x}{2}\right); \quad 2\cos t = \sqrt{4 - x^2}; \quad \sin(2t) = 2\sin t\cos t = \frac{x\sqrt{4 - x^2}}{2}.$$

Substituting these results into Equation (1) gives

$$\int \sqrt{4 - x^2}\, dx = 2t + \sin(2t) + C = 2\arcsin\left(\frac{x}{2}\right) + \frac{x\sqrt{4 - x^2}}{2} + C,$$

which is our desired destination. ■

Comments on the example Example 5 needs a closer look:

- **Inverse substitution:** Substituting $x = 2\sin t$ and $dx = 2\cos t\, dt$ represents an "inverse" variant of the more usual u-substitution technique. Such substitutions, although legal, require special care.

- **Which values of t?** The equation $x = 2\sin t$ makes sense for *all* real t. In order to write $t = \arcsin(x/2)$, however, we must (and hereafter will) assume that t lies in the interval $[-\pi/2, \pi/2]$. ➡ *That is, t must lie in the range of the arcsine function.*

- **Absolute values:** In the preceding calculation we used without comment the fact that

$$\sqrt{4 - x^2} = \sqrt{4 - 4\sin^2 t} = 2\,|\cos t| = 2\cos t.$$

  The last equation assumes, in effect, that $\cos t$ is nonnegative. Fortunately, that *is* the case—restricting t to the interval $[-\pi/2, \pi/2]$ guarantees this.

- **The bottom line—it works:** Whatever the subtleties of the trigonometric substitution, the bottom line remains: The result *is* a suitable antiderivative, as differentiation can verify.

Trigonometric substitutions: Three types Example 5 illustrates one of three types of trigonometric substitutions. The following table gives the key properties of all three types:

Trigonometric substitutions			
Radical form	**Substitution**	**t-domain**	**Result**
$\sqrt{a^2 - x^2}$	$x = a \sin t$	$[-\pi/2, \pi/2]$	$\sqrt{a^2 - x^2} = a \cos t$
$\sqrt{a^2 + x^2}$	$x = a \tan t$	$[-\pi/2, \pi/2]$	$\sqrt{a^2 + x^2} = a \sec t$
$\sqrt{x^2 - a^2}$	$x = a \sec t$	$[0, \pi],\ x \neq \pi/2$	$\sqrt{x^2 - a^2} = \pm a \tan t$

These substitutions (sometimes combined with completing the square) produce antiderivatives of many functions that involve square roots of quadratic expressions.

From t back to x: Helpful pictures   A successful trigonometric substitution trades a troublesome integral in x for a simpler integral in t, but the problem of translating back to an expression in x always remains. Doing so can be slightly tricky, as Example 5 illustrated. The pictures in Figure 1 may help—both in making a useful substitution and in unraveling it at the end.

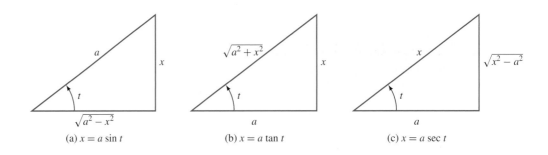

(a) $x = a \sin t$ (b) $x = a \tan t$ (c) $x = a \sec t$

FIGURE 1
Trigonometric substitutions: pictorial aids

From Figure 1(b), for instance, one can read such implications as

$$x = a \tan t \implies \sec t = \frac{\sqrt{a^2 + x^2}}{a}.$$

We use such facts in the following examples.

EXAMPLE 6   Find $\displaystyle \int \frac{dx}{\sqrt{x^2 + 4}}$.

In the assumed t-domain (see the preceding table), $\sec t$ is nonnegative.

Solution   Let $x = 2 \tan t$; then $dx = 2 \sec^2 t\, dt$ and $\sqrt{x^2 + 4} = \sqrt{4 \tan^2 t + 4} = 2 \sec t$. ◄
Therefore,

$$\int \frac{dx}{\sqrt{x^2 + 4}} = \int \frac{2 \sec^2 t}{2 \sec t}\, dt = \int \sec t\, dt = \ln |\sec t + \tan t| + C.$$

To convert the result back to an expression in x, we use the conclusion drawn just before this example:

$$\int \sec t\, dt = \ln \left| \frac{\sqrt{4 + x^2}}{2} + \frac{x}{2} \right| + C. \qquad \blacksquare$$

EXAMPLE 7 Find $\displaystyle\int_0^2 \frac{dx}{\sqrt{x^2+4}}$.

Solution We just calculated the antiderivative, and so the rest is algebra:

$$\int_0^2 \frac{dx}{\sqrt{x^2+4}} = \ln\left|\frac{\sqrt{4+x^2}}{2}+\frac{x}{2}\right|\Bigg]_0^2 = \ln(\sqrt{2}+1).$$

Example 6 also suggests another approach to the same goal: substitution in the *definite integral*. Notice that if $x = 2\tan t$, then $x = 0$ when $t = 0$ and $x = 2$ when $t = \pi/4$. Substituting all this into the original integral gives

$$\int_0^2 \frac{dx}{\sqrt{x^2+4}} = \int_0^{\pi/4} \sec t\, dt = \ln|\sec t+\tan t|\Big]_0^{\pi/4} = \ln(\sqrt{2}+1)$$

just as before. ■

EXAMPLE 8 Find $\displaystyle\int \frac{dx}{\sqrt{x^2+2x+5}}$.

Solution Completing the square in the denominator gives

$$\int \frac{dx}{\sqrt{x^2+2x+5}} = \int \frac{dx}{\sqrt{(x+1)^2+4}}.$$

Next we substitute $u = x+1$ and $du = dx$ to get $\displaystyle\int \frac{du}{\sqrt{u^2+4}}$. The result is familiar—work already done in Examples 6 and 7 completes the problem. ■

BASIC EXERCISES

1. Find each of the following basic trigonometric antiderivatives.

(a) $\displaystyle\int \cos x\, dx$ (b) $\displaystyle\int \sin x\, dx$

(c) $\displaystyle\int \sec^2 x\, dx$ (d) $\displaystyle\int \sec x \tan x\, dx$

2. Find each of the following trigonometric antiderivatives.

(a) $\displaystyle\int \sin x \cos x\, dx$ (b) $\displaystyle\int \tan x\, dx$

(c) $\displaystyle\int \sec^2 x \tan x\, dx$

3. We solved Example 1 in two different ways and, apparently, got two different antiderivatives. Are these antiderivatives really different? Justify your answer.

4. (a) Find $\displaystyle\int \sin^4 x\, dx$ using the reduction formula for powers of sines.

(b) Is the antiderivative you obtained in part (a) the same as the antiderivative found in Example 2? Justify your answer.

5. Let $I = \displaystyle\int \cos^3 x \sin^4 x\, dx$.

(a) Find I by finishing Example 3.

(b) Find I by rewriting the integrand in terms of cosines and then using the reduction formula for powers of cosines.

(c) Are the antiderivatives in parts (a) and (b) the same function? Justify your answer.

6. Find $\displaystyle\int \sec^3 x \tan^3 x\, dx$ by finishing the computations in Example 4.

7. Draw the right triangle that corresponds to the trigonometric substitution $x = a\sin t$ and use it to find an expression in a and x for

(a) $\cos t$ (b) $\tan t$ (c) $\sin(2t)$

8. Draw the right triangle that corresponds to the trigonometric substitution $x = a\tan t$ and use it to find an expression in a and x for

(a) $\cos t$ (b) $\sin t$ (c) $\sec(2t)$

9. Find $\displaystyle\int \frac{dx}{\sqrt{x^2+2x+5}}$ by finishing Example 8.

10. Evaluate $\displaystyle\int_{-1}^1 \frac{dx}{\sqrt{x^2+2x+5}}$.

In Exercises 11–32 find the antiderivative.

11. $\displaystyle\int \sin^2(3x)\, dx$ **12.** $\displaystyle\int \cos^2(x/3)\, dx$

13. $\displaystyle\int \sin^2 x \cos x\, dx$ **14.** $\displaystyle\int \sin^3 x \cos^3 x\, dx$

15. $\int \cos^2 x \sin^3 x \, dx$

16. $\int \cos^3(2x) \sin^2(2x) \, dx$

17. $\int \sin^2 x \cos^2 x \, dx$

18. $\int \cos^4 x \sin^2 x \, dx$

19. $\int \tan^4 x \, dx$

20. $\int \sec^4(3x) \, dx$

21. $\int \sec^2 x \tan^2 x \, dx$

22. $\int \sec^2 x \tan^4 x \, dx$

23. $\int \sec x \tan^2 x \, dx$

24. $\int \sec^3 x \tan^2 x \, dx$

25. $\int \sec x \tan^3 x \, dx$

26. $\int \dfrac{dx}{x^2\sqrt{x^2+1}}$

27. $\int \dfrac{dx}{x^2\sqrt{4-x^2}}$

28. $\int \dfrac{dx}{\sqrt{1+x^2}}$

29. $\int \sqrt{1-x^2} \, dx$

30. $\int \dfrac{dx}{(x^2+4)^2}$

31. $\int x^2\sqrt{1-x^2} \, dx$

32. $\int \dfrac{x^2}{\sqrt{9-x^2}} \, dx$

33. Draw the right triangle that corresponds to the trigonometric substitution $x = a \sec t$ and use it to find an expression in a and x for

   (a) $\cos t$ **(b)** $\sin t$ **(c)** $\tan(2t)$

34. When the trigonometric substitution $x = 2 \sin t$ is used, the expression $\sqrt{4-x^2}$ becomes $2 \cos t$. When the trigonometric substitution $x = 2 \sec t$ is used, the expression $\sqrt{x^2-4}$ becomes $|2 \tan t|$. Why is the absolute value required in one case but not the other?

In Exercises 35–38, evaluate the integral.

35. $\displaystyle\int_3^4 \dfrac{dx}{x^2\sqrt{x^2-4}}$

36. $\displaystyle\int_{-4}^{-3} \dfrac{dx}{x^2\sqrt{x^2-4}}$

37. $\displaystyle\int_1^2 \dfrac{\sqrt{x^2-1}}{x} \, dx$

38. $\displaystyle\int_{-2}^{-1} \dfrac{\sqrt{x^2-1}}{x} \, dx$

FURTHER EXERCISES

In Exercises 39–46, find the antiderivative.

39. $\int \sec x \sin^3 x \, dx$

40. $\int \sin(2x) \cos^2 x \, dx$

41. $\int \sqrt{\cos x} \sin^5 x \, dx$

42. $\int \sqrt{1+\sin x} \, dx$

43. $\int \sqrt{1+x^2} \, dx$

44. $\int \dfrac{\sqrt{4-x^2}}{x^2} \, dx$

45. $\int \dfrac{x+2}{x(x^2+1)} \, dx$

46. $\int x \arcsin x \, dx$

47. Show that $\displaystyle\int \dfrac{\arctan x}{(1+x^2)^{3/2}} \, dx = \dfrac{1+x \arctan x}{\sqrt{1+x^2}}$.

   [HINT: Start by making the substitution $w = \arctan x$.]

48. Evaluate $\displaystyle\int_{-1/2}^{1/2} \dfrac{\arcsin x}{(1-x^2)^{3/2}} \, dx$.

49. (a) Use the addition formula $\cos(u+v) = \cos u \cos v - \sin u \sin v$ to show that $2 \cos u \cos v = \cos(u+v) + \cos(u-v)$.

   (b) Use part (a) to find $\int \cos(ax) \cos(bx) \, dx$.

50. (a) Use the addition formula $\cos(u+v) = \cos u \cos v - \sin u \sin v$ to show that $2 \sin u \sin v = \cos(u-v) - \cos(u+v)$.

   (b) Use part (a) to find $\int \sin(ax) \sin(bx) \, dx$.

51. (a) Use the addition formula $\sin(u+v) = \sin u \cos v + \cos u \sin v$ to show that $2 \sin u \cos v = \sin(u+v) + \sin(u-v)$. [HINT: $\cos(-x) = \cos x$ and $\sin(-x) = -\sin x$.]

   (b) Use part (a) to find $\int \sin(ax) \cos(bx) \, dx$.

52. (a) Use the identity $\sec x = \dfrac{1}{\cos x} = \dfrac{\cos x}{\cos^2 x} = \dfrac{\cos x}{1-\sin^2 x}$ to show that $\displaystyle\int \sec x \, dx = \dfrac{1}{2} \ln \left| \dfrac{1+\sin x}{1-\sin x} \right|$.

   (b) Show that $\displaystyle\int \sec x \, dx = \ln \left| \dfrac{\cos x}{1-\sin x} \right|$.

   [HINT: Multiply the numerator and denominator in the part (a) antiderivative by $1 - \sin x$.]

   (c) Show that $\displaystyle\int \sec x \, dx = \ln |\sec x + \tan x|$.

53. (a) Assume that n is an integer. Use integration by parts with $u = \sin^{n-1} x$ and $dv = \sin x \, dx$ to show that

$$\int \sin^n x \, dx = -\sin^{n-1} x \cos x + (n-1) \int \sin^{n-2} x \cos^2 x \, dx.$$

   (b) Use part (a) and the trigonometric identity $\sin^2 x + \cos^2 x = 1$ to show that if $n \neq 0$, then

$$\int \sin^n x \, dx = -\dfrac{\sin^{n-1} x \cos x}{n} + \dfrac{n-1}{n} \int \sin^{n-2} x \, dx.$$

54. Use the reduction formula in part (b) of Exercise 53 to show that if $n > 0$ is an odd integer, then

$$\int_0^{\pi/2} \sin^n x \, dx = \dfrac{n-1}{n} \cdot \dfrac{n-3}{n-2} \cdots \dfrac{4}{5} \cdot \dfrac{2}{3}.$$

55. Find an expression for $\displaystyle\int_0^{\pi/2} \sin^n x \, dx$ if $n > 0$ is an even integer.

56. Assume that n is a nonzero integer. Derive the reduction formula $\displaystyle\int \cos^n x \, dx = \dfrac{\cos^{n-1} x \sin x}{n} + \dfrac{n-1}{n} \int \cos^{n-2} x \, dx$.

57. Assume that $n \neq 1$ is an integer. Derive the reduction formula $\int \tan^n x \, dx = \dfrac{\tan^{n-1} x}{n-1} - \int \tan^{n-2} x \, dx$.

58. Assume that $n \neq 1$ is an integer. Derive the reduction formula $\int \sec^n x \, dx = \dfrac{\sec^{n-2} x \tan x}{n-1} + \dfrac{n-2}{n-1} \int \sec^{n-2} x \, dx$.

Exercises 59–64 explore the substitution $x = 2 \arctan t$.

59. Let $x = 2 \arctan t$ or, equivalently, $t = \tan(x/2)$.

    (a) Show that $\sin(x/2) = t/\sqrt{1+t^2}$.

    (b) Show that $\cos(x/2) = 1/\sqrt{1+t^2}$.

    (c) Show that $\sin x$ can be written as a rational function of t.

    (d) Show that $\cos x$ can be written as a rational function of t.

60. Use Exercise 59 to show that $\displaystyle \int \frac{dx}{1 + \sin x + \cos x} = \ln \left| 1 + \tan(x/2) \right| + C$.

61. Find $\displaystyle \int \frac{dx}{1 + \cos x}$.

62. Evaluate $\displaystyle \int_0^{\pi/2} \frac{dx}{2 \cos x + 3 \sin x}$

63. Evaluate $\displaystyle \int_{\pi/4}^{\pi} \frac{\sin x}{2 + \cos x} \, dx$

64. Use Exercise 59 to explain why the substitution $t = \tan(x/2)$ can be used to transform any rational expression involving only constants and powers of trigonometric functions into a rational expression in t.

65. Let $I_n = \displaystyle \int \frac{dx}{(1+x^2)^n}$, where $n \geq 1$ is an integer.

    (a) Use integration by parts to find a reduction formula for I_n.

    (b) Use a trigonometric substitution to find a reduction formula for I_n.

8.4   MISCELLANEOUS ANTIDERIVATIVES

Tools for antidifferentiation This chapter has presented several symbolic techniques for finding symbolic antiderivatives. In the following paragraphs we list our short kit of tools and some hints for which tool to use when.

We emphasize, however, that symbolic antidifferentiation is a far less "routine" problem than differentiation. Finding symbolic *derivatives* is mainly a matter of choosing and following one or more straightforward recipes. Finding *antiderivatives* successfully, by contrast, may require skill, experience, educated guesswork, judgment—and sometimes plain good luck.

There is good news, too, and a simple moral: *Checking* by differentiation that "candidate" antiderivatives really work is usually easy. So guess freely—but check carefully.

- **Substitution** Based on the chain rule, substitution is the simplest and most powerful antidifferentiation technique. When it works, substitution transforms "hard" antiderivatives to simpler forms, which may be solvable either at a glance or by reference to an integral table. Try u-substitution *first* before resorting to more laborious methods. The method works best and most simply in cases like

$$\int 2x \cos(x^2) \, dx \quad \text{and} \quad \int e^{\sin x} \cos x \, dx,$$

  where both a substitution $u = u(x)$ and its differential $du = u'(x) \, dx$ are conveniently present.

- **Integration by parts** Based on the product rule for *derivatives*, integration by parts often succeeds with integrals like

$$\int x \sin x \, dx \quad \text{and} \quad \int x^2 e^x \, dx,$$

  which involve products. (The method may succeed with other integrands, too.) As the formula $\int u \, dv = uv - \int v \, du$ suggests, the method effectively trades one integral (on the left) for another (on the right). With any luck (and with judicious choices of u and dv) the right-hand integral may be simpler or more familiar than the left-hand integral.

- **Partial fractions**   The partial fractions method applies only to *rational functions*: quotients of polynomials. The idea is first to use algebra (not calculus) to rewrite such a "total fraction" as a sum of simpler *partial fractions*, as in

$$\frac{3x+4}{(x+1)(x+2)} = \frac{1}{1+x} + \frac{2}{2+x}.$$

  The work is worth doing because the summands on the right are easy to antidifferentiate.

- **Trigonometric methods**   Section 8.3 described a variety of methods, based mainly on trigonometric identities, for antidifferentiating functions that involve powers of trigonometric functions as in

$$\int \cos^2 x \sin^3 x\, dx \quad \text{and} \quad \int \sec^3 x \tan x\, dx.$$

  Perhaps surprisingly, antiderivatives that involve squares and square roots, such as

$$\int \frac{x^2}{\sqrt{1-x^2}}\, dx \quad \text{and} \quad \int \frac{x^2}{\sqrt{1+x^2}}\, dx,$$

  can often be transformed, by means of **trigonometric substitutions**, to antiderivatives that involve powers of trigonometric functions.

Other tools   An integral table, such as the one in this book, makes finding many antiderivatives a routine matter. But not all antiderivatives yield instantly—substitutions or algebraic transformations may be needed.

Modern technology, on the other hand, makes short work of finding almost *any* symbolic antiderivative found in a calculus text. Even when such tools are available, finding *some* antiderivatives "by hand" is worthwhile as an aid to understanding and intuition.

BASIC EXERCISES

In Exercises 1–72, find the antiderivative without using the table of integrals at the back of the book (or its electronic equivalent).

1. $\int \dfrac{\sin x}{(3+\cos x)^2}\, dx$

2. $\int \dfrac{x^2}{x+1}\, dx$

3. $\int x\left(3+4x^2\right)^5 dx$

4. $\int \dfrac{dx}{\sqrt{1-x^2}}$

5. $\int \dfrac{x}{\sqrt[3]{x^2+4}}\, dx$

6. $\int \dfrac{dx}{x(3x-2)}$

7. $\int \dfrac{(\ln x)^2}{x}\, dx$

8. $\int e^x \sin x\, dx$

9. $\int \dfrac{\ln x}{x}\, dx$

10. $\int x\sqrt{x+2}\, dx$

11. $\int \dfrac{x}{3x+2}\, dx$

12. $\int x \cos x\, dx$

13. $\int \sin^2(3x)\cos(3x)\, dx$

14. $\int xe^{3x}\, dx$

15. $\int xe^{3x^2}\, dx$

16. $\int \dfrac{dx}{1+4x^2}$

17. $\int (2-3x)^{10}\, dx$

18. $\int \arctan x\, dx$

19. $\int \dfrac{\sec^2 x}{3+\tan x}\, dx$

20. $\int x \sin x\, dx$

21. $\int \dfrac{dx}{(x-1)(x+2)}$

22. $\int x^2 \ln x\, dx$

23. $\int \dfrac{2x+3}{4x+5}\, dx$

24. $\int \dfrac{x+1}{x^2+1}\, dx$

25. $\int \dfrac{e^x}{\sqrt{1-e^{2x}}}\, dx$

26. $\int \dfrac{\sin x}{2+\cos x}\, dx$

27. $\int \ln x\, dx$

28. $\int x\cos(3x^2)\, dx$

29. $\int \arcsin x\, dx$

30. $\int \dfrac{\sqrt{x}}{1+x}\, dx$

31. $\int \dfrac{dx}{x^2+2x+3}$

32. $\int \dfrac{x}{\sqrt{x-2}}\, dx$

33. $\int \dfrac{dx}{\sqrt{1-4x^2}}$

34. $\int \dfrac{x^3}{1+x^2}\, dx$

35. $\int \tan x\, dx$

36. $\int \cos(2x)\, dx$

37. $\int e^{2x}\sqrt{1+e^x}\, dx$

38. $\int \dfrac{dx}{1+x^2}$

39. $\displaystyle\int \frac{dx}{\sqrt{3-2x-x^2}}$

40. $\displaystyle\int x^2 \arcsin x \, dx$

57. $\displaystyle\int x \arcsin x \, dx$

58. $\displaystyle\int \frac{dx}{9-x^2}$

41. $\displaystyle\int \frac{dx}{x(\ln x)^2}$

42. $\displaystyle\int x \arctan x \, dx$

59. $\displaystyle\int \frac{dx}{\sqrt{2x+3}}$

60. $\displaystyle\int \frac{dx}{(4-x^2)^{3/2}}$

43. $\displaystyle\int \frac{dx}{9x^2-4}$

44. $\displaystyle\int \frac{x+5}{x^2+3x-4} \, dx$

61. $\displaystyle\int \frac{x}{(2x+3)^4} \, dx$

62. $\displaystyle\int x\sqrt{2x+1} \, dx$

45. $\displaystyle\int \frac{x^3}{\sqrt{4-x^2}} \, dx$

46. $\displaystyle\int \frac{dx}{\sqrt[3]{x-1}}$

63. $\displaystyle\int \frac{\tan x}{\sec^2 x} \, dx$

64. $\displaystyle\int \frac{x}{16+9x^2} \, dx$

47. $\displaystyle\int \frac{x}{(x-1)(x+1)} \, dx$

48. $\displaystyle\int x^3 e^{x^2} \, dx$

65. $\displaystyle\int \frac{dx}{e^x-1}$

66. $\displaystyle\int \frac{dx}{\sqrt{2x-x^2}}$

49. $\displaystyle\int \frac{dx}{\sqrt{9+x^2}}$

50. $\displaystyle\int \frac{dx}{2x-x^2}$

67. $\displaystyle\int \frac{dx}{1+\sqrt{x}}$

68. $\displaystyle\int \frac{x^3}{(x^2+1)^2} \, dx$

51. $\displaystyle\int \frac{x^2}{1-3x} \, dx$

52. $\displaystyle\int \frac{x}{(x^2-1)^3} \, dx$

69. $\displaystyle\int x^2 \ln(3x) \, dx$

70. $\displaystyle\int \frac{x}{9+4x^4} \, dx$

53. $\displaystyle\int e^x e^{2x} \, dx$

54. $\displaystyle\int \sqrt{4x-3} \, dx$

71. $\displaystyle\int \sqrt{x} \ln x \, dx$

72. $\displaystyle\int x \sec^2 x \, dx$

55. $\displaystyle\int \ln(1+x^2) \, dx$

56. $\displaystyle\int \sin(\sqrt{x}) \, dx$

FURTHER EXERCISES

In Exercises 73–88, find the antiderivative using your ingenuity and the table of integrals at the back of the book.

73. $\displaystyle\int \frac{7-x}{(x+3)(x^2+1)} \, dx$

74. $\displaystyle\int \frac{x+6}{(x+1)(x^2+4)} \, dx$

75. $\displaystyle\int x \sin^2 x \cos x \, dx$

76. $\displaystyle\int \sin^3 x \cos^3 x \, dx$

77. $\displaystyle\int \frac{dx}{x^3+x}$

78. $\displaystyle\int \tan^4 x \, dx$

79. $\displaystyle\int (x^2+2x+3)^{3/2} \, dx$

80. $\displaystyle\int \sin(3x)\cos(5x) \, dx$

81. $\displaystyle\int \sqrt{1+e^x} \, dx$

82. $\displaystyle\int \frac{dx}{x\left(x+\sqrt[3]{x}\right)}$

83. $\displaystyle\int \frac{dx}{x^3+1}$

84. $\displaystyle\int \frac{dx}{(e^x-e^{-x})^2}$

85. $\displaystyle\int x \tan^2 x \, dx$

86. $\displaystyle\int \cos^3 x \, dx$

87. $\displaystyle\int \sin x \sin(2x) \, dx$

88. $\displaystyle\int \sin^5 x \cos^2 x \, dx$

Beyond Elementary Functions

Integrability in closed form

Not every elementary function has an elementary antiderivative (i.e., an antiderivative that can be written as a combination of the usual elementary functions of calculus). Functions that *do* have elementary antiderivatives are sometimes called **integrable in closed form**; we've seen in this chapter that polynomials, rational functions, the basic trigonometric functions, and many other functions are integrable in closed form, and we've discussed several methods for finding antiderivative formulas.

We'll use the exp *form for the exponential function in this interlude; doing so avoids towers of exponents.*

Many other functions, however, are *not* integrable in closed form. None of the following antiderivatives, for instance, has a closed-form solution: ←

$$\int \sin(x^2)\,dx; \quad \int \frac{\sin x}{\sqrt{x}}\,dx; \quad \int \exp(-x^2)\,dx; \quad \int \frac{dx}{\ln x}.$$

The problem is not that these integrands *lack* antiderivatives—the FTC guarantees that *every* continuous function has an antiderivative. The problem with these functions is that their antiderivatives are not elementary functions.

Recognizing nonelementary antiderivatives Proving from scratch that a function has no elementary antiderivative is a hard problem. Trying and failing (by hand or with technology) to find an antiderivative proves nothing—maybe we (or the software) just weren't smart enough. Given the antiderivative problem $\int \exp(-x^2)\,dx$, for instance, the TI–89 makes no progress at all; *Maple* and *Mathematica* return the strange-looking result

$$\int \exp(-x^2)\,dx = \frac{\sqrt{\pi}}{2}\operatorname{erf}(x).$$

We'll say more about the mysterious "erf" in a moment. The point here is that the right side is *not* elementary in the sense under discussion.

If we know (or take on faith) that a given function is or isn't integrable in closed form, then we can often use antiderivative methods to conclude the same thing for other functions.

> **EXAMPLE 1** Assume that $\exp(-x^2)$ has no elementary antiderivative. For a given nonnegative integer n, does $x^n \exp(-x^2)$ have an elementary antiderivative?
>
> **Solution** For $n = 0$, no—that's our assumption. For $n = 1$, the answer is yes; substituting $u = -x^2$ and $du = -2x\,dx$ (or asking a machine) gives
>
> $$\int x \exp(-x^2)\,dx = -\frac{\exp(-x^2)}{2} + C.$$
>
> This case (and similar experiments with technology) might lead us to guess that the answer is yes for odd n and no for even n.
>
> This guess is correct; integration by parts shows why. Given the integral $\int x^n \exp(x^2)\,dx$ for any $n \geq 2$, integration by parts (with $u = x^{n-1}$ and $dv = x \exp(x^2)\,dx$) gives the reduction formula
>
> $$\int x^n \exp(x^2)\,dx = x^{n-1}\frac{\exp(x^2)}{2} - \frac{n-1}{2}\int x^{n-2}\exp(x^2)\,dx.$$
>
> Observe:
>
> • The left-hand integrand has an elementary antiderivative *if and only if* the right-hand integrand does.

- Applying the formula *repeatedly* reduces the power of x by 2 on each repetition. Thus, beginning with any power x^n, we eventually reach either $n = 0$ or $n = 1$, depending on whether the initial n is even or odd.

These facts imply that our guess is correct. ∎

PROBLEM 1 Make the u-substitution $u = x^2$ to relate the two integrals

$$I_1 = \int \sin(x^2)\,dx; \quad I_2 = \int \frac{\sin x}{\sqrt{x}}\,dx.$$

Conclude that if I_1 has no closed-form solution, then neither does I_2, and vice versa.

PROBLEM 2 Assume that the function $\sin(x^2)$ has no elementary antiderivative.

(a) The function $x \sin(x^2)$ has an elementary antiderivative. Find it.

(b) The function $x^3 \sin(x^2)$ has an elementary antiderivative. Find it.
   [HINT: Try integration by parts with $u = x^2$, $dv = x \sin(x^2)$.]

(c) Show that the function $x^2 \sin(x^2)$ has no elementary antiderivative.
   [HINT: Set $u = x$, $dv = x \sin(x^2)$; then imitate the argument in Example 1.]

Functions defined by integration

Lacking an elementary antiderivative is not a fatal flaw. Many useful mathematical functions are (or can be) defined not by elementary formulas but by integrals.

Logarithms as integrals In earlier work we defined the natural logarithm function as the inverse of the natural exponential function. → Then, using the chain rule and the fact that the exponential function is its own derivative, we found by *calculation* that $(\ln x)' = 1/x$; that is, that $\ln x$ is an antiderivative for $1/x$.

Many, but not all, calculus books take this approach.

   A different approach to the logarithm is to define $\ln x$ (for $x > 0$) as the integral

$$\ln x = \int_1^x \frac{dt}{t}. \tag{1}$$

This definition avoids mentioning exponential functions; it also guarantees (by the FTC rather than by calculation) that $\ln x$ is an antiderivative of $1/x$. The following problems explore some familiar properties of $\ln x$—but from the viewpoint of the integral definition. Solve them using *only* the integral definition, forgetting anything else you may have known about logs. →

Forget only temporarily.

PROBLEM 3 Consider $\ln x$ to be defined for $x > 0$ by Equation (1).

(a) Explain why $\ln x$ is positive if $x > 1$, negative if $0 < x < 1$, and zero if $x = 1$.

(b) Why is $\ln 0$ undefined? Why is $\ln(-1)$ undefined?

(c) Draw (freehand) rough graphs of $1/x$ and $\ln x$ for $x > 0$.

(d) Use the trapezoid-rule sum T_3 (with three subdivisions) to estimate $\ln 2$. (Do this by hand to see the pattern.) Then use technology to find the trapezoid sum T_{30} for the same integral. Are your estimates too big or too small?

(e) Draw a picture to convince yourself geometrically that $\ln(1/3) = -\ln 3$. (Use the graph of $y = 1/x$ and compare appropriate areas.)

PROBLEM 4 Assume that $a > 0$ and $x > 0$. Carry out and justify each of the following steps to derive the identity $\ln(ax) = \ln a + \ln x$.

(a) Show that $\left(\ln(ax) - \ln(x)\right)' = \left(\int_1^{ax} \frac{dt}{t} - \int_1^x \frac{dt}{t}\right)' = 0.$ [HINT: Use the chain rule.]

(b) Explain why the result in part (a) implies that $\ln(ax) = \ln x + C$.

(c) Use an appropriate value of x to show that in part (b), $C = \ln a$.

The error function The *error function* (erf) is defined for all real numbers x by

$$\text{erf}(x) = \frac{2}{\sqrt{\pi}} \int_0^x \exp(-t^2)\, dt. \tag{2}$$

(The name "error function" comes from probability, where the function arises in modeling "errors"—differences between expected and actual values of an observed quantity.) The function erf is well known to *Maple*, *Mathematica*, and their ilk. Figure 1 shows graphs of both erf and its derivative (the integrand). Note that one of the graphs is the proverbial "bell curve":

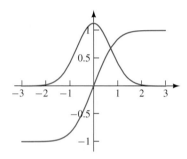

FIGURE 1
$y = \text{erf}(x)$ **and its derivative**

PROBLEM 5

(a) Which graph in Figure 1 shows erf(x)? What is the *other* graph's formula?

(b) Use a trapezoid-rule sum T_{30} (with 30 subdivisions) to estimate erf(3). Does your answer "agree" with Figure 1? How?

(c) Use a trapezoid-rule sum T_{30} (with 30 subdivisions) to estimate erf(10) − erf(3). Does your answer "agree" with Figure 1? How?

The log integral The **log integral function**, denoted Li, is defined for $x > 2$ by

$$\text{Li}(x) = \int_2^x \frac{dt}{\ln t}.$$

This function is useful in number theory; for large x, Li(x) closely approximates the number of primes less than x. (Slightly different definitions of Li exist in other sources.)

PROBLEM 6 In each part following, use the integral definition above.

(a) Explain why $\frac{d}{dx}\text{Li}'(x) = 1/\ln(x)$ for all $x > 2$.

(b) Use technology to plot $\text{Li}'(x) = 1/\ln(x)$ for $2 \leq x \leq 20$. Use your graph to draw (by hand) a rough graph of Li(x) on the same interval.

(c) Use the midpoint rule with 50 subdivisions to estimate Li(100). (The error committed is surprisingly small, even though the interval is quite large.) How many primes are there less than 100?

(d) There are 168 prime numbers less than 1000. Compare this to any convenient estimate of Li(1000).

First-Order Linear Differential Equations

Separable DEs Explicit symbolic solutions to first-order DEs and IVPs are useful, but they can be difficult to find. No single method works every time. We saw in Section 7.4, for example, how to use integration to solve DEs that happen to be **separable**—that is, DEs for which the variables (y and t, say) can be "separated" on opposite sides of an equation. The following DEs are all separable:

$$\frac{dy}{dt} = 3y; \quad \frac{dy}{dt} = 2yt; \quad \frac{dy}{dt} = \cos^2(y)\cos(t); \quad \frac{dy}{dt} = \cos(t^2).$$

Even for separable DEs, finding a symbolic solution can be hard or impossible, since the work may involve one or more difficult integrals (as in the last DE above).

PROBLEM 1 Use separation of variables (as in Section 7.4) to find the solutions

$$y = Ce^{3t}; \quad y = Ce^{t^2}; \quad y = \arctan(\sin(t) + C)$$

for the first three DEs above. (In each case, C is an arbitrary constant.) What makes the fourth DE above harder than the others?

Inseparable DEs Other DEs, including all of the following samples, are **inseparable**:

$$\frac{dy}{dt} = \frac{y}{t} + 1; \quad \frac{dy}{dt} = 2y + \sin t; \quad \frac{dy}{dt} = \cos y + \cos t; \quad \frac{dy}{dt} = \cos(y + t).$$

The variables y and t in these DEs are too "entangled" to separate, and so we need other methods to solve such DEs symbolically. (We *could* use numerical or graphical methods, such as Euler's method and slope fields, but at some cost in accuracy and generality.)

First-order linear DEs A **first-order linear DE** is defined as a DE that can be written in the form

$$f(t)y' + g(t)y = h(t),$$

where f, g, and h are all functions of t. Notice especially that such a DE involves only the first power of y; expressions like y^2 and $\cos(y)$ are not allowed. Notice also that dividing by $f(t)$ and rearranging the DE slightly gives the equivalent form

$$y' = a(t)y + b(t), \tag{1}$$

where $a(t)$ and $b(t)$ are functions of t alone. Throughout this Interlude we'll use the Formula (1) and the notations $a(t)$ and $b(t)$ defined thereby.

EXAMPLE 1 Which of the four inseparable DEs above are linear? Why?

Solution The DE $y' = y/t + 1$ fits template (1), with $a(t) = 1/t$ and $b(t) = 1$, and so it's linear. The DE $y' = 2y + \sin t$ also fits, with $a(t) = 2$ and $b(t) = \sin t$. The other two inseparable DEs above don't fit the pattern and thus are not linear. ■

PROBLEM 2 Decide whether each of the following DEs is (i) linear; (ii) separable. For each linear DE, find $a(t)$ and $b(t)$. (In all parts, k is a constant.)

(a) $y' = y^2 \sin t + \cos t$

(b) $y' = ky + 1$

(c) $y' = kty + 1$

(d) $y' = \dfrac{1}{1+t} y + k$

Like separable DEs, linear DEs may or may not have "nice" solutions in terms of standard elementary functions.

Solving linear DEs: a "formula" The following general solution "formula" for linear DEs is sometimes attributed to work of Euler in the early 1800s. As the quotation marks suggest, the "formula" requires some work and interpretation to produce fully explicit solutions—in particular, it involves finding antiderivatives. ◂ We'll first state the formula and then consider how to use it and how it might have been found.

> **FACT** Consider the first-order linear DE $y' = a(t)y + b(t)$. Let $A(t)$ be any antiderivative of $a(t)$. Solutions to the DE have the form
>
> $$y = e^{A(t)} \int e^{-A(t)} b(t) \, dt. \qquad (2)$$
>
> (The indefinite integral produces an arbitrary constant of integration.)

EXAMPLE 2 Use the formula to find solutions of the DE $y' = y + t$. Which solutions satisfy the initial conditions (1) $y(0) = 1$; (2) $y(0) = -1$?

Solution Here $a(t) = 1$ and $b(t) = t$, and so we can use $A(t) = t$ in Formula (2):

$$y = e^t \int e^{-t} t \, dt = e^t \left(-te^{-t} - e^{-t} + C \right) = -t - 1 + Ce^t.$$

Now $y(0) = C - 1$; thus, setting $C = 2$ gives the solution function $y = 2e^t - t - 1$, which satisfies $y(0) = 1$. Setting $C = 0$ gives the solution $y = -t - 1$, which satisfies the initial condition $y(0) = -1$. (Note that in this one case the solution function is linear.)

Here are two things to notice about the method:

1. The general result involves one arbitrary constant C; it came from the antiderivative $\int e^{-A(t)} b(t) \, dt$.

2. The method requires antidifferentiating two functions: $a(t)$ and $e^{A(t)}b(t)$. Either or both of these antiderivatives may be difficult to find symbolically. ∎

PROBLEM 3 Check by differentiation that $y = -t - 1 + Ce^t$ is indeed a solution to $y' = y + t$. Then use technology to plot solution curves for integer values of C from -3 to 3.

PROBLEM 4 Imitate the calculations in Example 2 to find solutions for the each of following DEs. ◂ In each part, check by differentiation that your claimed solution really works, and use technology to plot solutions for several values of the arbitrary constant C.

Technology or integral tables might help with the integrals.

(a) $y' = -y + t^2$

(b) $y' = -y + \cos t$

(c) $y' = -\dfrac{y}{t} + 1$ (assume $t > 0$)

(d) $y' = -\dfrac{y}{t} + \sin t$ (assume $t > 0$)

PROBLEM 5 Show by differentiation that $y = e^{A(t)} \int e^{-A(t)} b(t) \, dt$ is indeed a solution to $y' = a(t)y + b(t)$, as claimed. [HINTS: The claimed solution is a product, so the product rule applies. The first factor is a composite function, so the chain rule applies. The second factor (which involves an integral sign) is an antiderivative, and so differentiating it is especially easy.]

How Euler really found such a formula is unknowable.

Why it works The preceding example and problem suggest that the method works. The following problems suggest *why* it works and how such a formula might be guessed. ◂

PROBLEM 6 (A simple case) Consider the linear DE $y' = a(t)y$ (for which $b(t) = 0$). Let $A(t)$ be any antiderivative for $a(t)$. Then the DE has solutions

$$y = Ce^{A(t)},$$

where C is an arbitrary constant. Give details for the following supporting arguments.

(a) Show by differentiation that $y = Ce^{A(t)}$ really *does* solve the DE in question.

(b) Explain why $y = Ce^{A(t)}$ is the special case $b = 0$ of Formula (2). [HINT: Antiderivatives of the zero function are constants.]

(c) Note that the DE $y' = a(t)y$ is separable. Apply the separation of variables method to find the solution claimed above. [HINT: Separating variables and integrating both sides gives $\ln|y| = A(t) + k$, where k is a constant. Solve this equation for y.]

PROBLEM 7 (The general case) The preceding problem shows that $y = Ce^{A(t)}$, where C is a *constant*, solves the linear DE $y' = a(t)y + b(t)$ in the special case that $b(t) = 0$. To handle the (more likely) case that $b(t) \neq 0$, we look for a solution of the form $y = C(t)e^{A(t)}$, where $C(t)$ is a *function* of t. Work through the following steps to find a "formula" for $C(t)$.

(a) Show by differentiation that $y = C(t)e^{A(t)}$ is a solution of the DE $y' = a(t)y + b(t)$ if and only if $C'(t) = e^{-A(t)}b(t)$. (This is another way of saying that $C(t) = \int e^{-A(t)}b(t)\,dt$.)

(b) Use the preceding part to derive Formula (2).

PROBLEM 8 Consider the DE $y' = ay + b$, where a and b are *constant* functions. Solve the DE in two ways: (i) by separating variables; (ii) using Formula (2). Show that the two answers are the same.

Population modeling with linear DEs

The first-order linear DE

$$y'(t) = a(t)y(t) + b(t)$$

can be used to model population growth under various conditions. If $y(t)$ describes a population at time t, then $y'(t)$ represents the rate of population growth, and the DE describes how growth occurs. The first summand, $a(t)y(t)$, represents growth with a *time-variable* proportionality factor $a(t)$. (This might be the population's net "birth rate.") The second summand, $b(t)$, can be thought of as a rate of *immigration* (or *emigration* if $b(t) < 0$); the latter rate depends on time but not on the population. For instance, the DE

$$y'(t) = \frac{0.1}{1+t}y(t) + 1 \tag{3}$$

could describe a population with a constant immigration rate (1 population unit per unit of time) and a birth rate ($0.1/(1+t)$) that decreases over time.

PROBLEM 9 Suppose that a population satisfies the DE (3) and the initial condition $y(0) = 50$. Find and plot a formula for $y(t)$. (Assume that $t > 0$ to avoid absolute value signs in the calculation.)

PROBLEM 10 Consider the DE

$$y'(t) = \frac{1}{1+t}y(t) + k,$$

where k is a constant, and the initial condition $y(0) = 10$.

(a) Solve the IVP given by the DE and the initial condition above. (The result involves the constant k.)

(b) Plot on one set of axes solutions to the preceding part for $k = -3, -2, -1, 0, 1, 2, 3$. (Use the time interval $0 \leq t \leq 20$.) Interpret the different curve *shapes* in terms of immigration or emigration.

FUNCTION APPROXIMATION

9.1 TAYLOR POLYNOMIALS

Introduction: the idea of approximation

"Function approximation"—this chapter's title phrase—suggests two possible meanings: approximation *by* functions and approximation *of* functions. In fact, we'll take *both* points of view in this chapter. A recurring problem throughout the chapter is to find (or build) "nice" functions that approximate less well behaved or well understood functions.

Why would we bother to approximate one type of function with another? What makes some functions "nicer" than others? How can we find nice functions that do what we want?

Polynomials: a nice family Polynomials form one especially nice family of functions. Calculus with polynomials is convenient; differentiation and integration could hardly be easier. Just as important, polynomial functions are easy to *evaluate*, even by hand or with crude technology. Finding values of transcendental functions, such as e^{-1}, may take much more work. ➡ The following example shows how a polynomial can approximate another function and it suggests why doing so is useful. (We'll explain later in this section how the polynomial was found.)

Could you find or estimate e^{-1} by hand? How would you try?

EXAMPLE 1 Consider the functions

$$f(x) = e^x \quad \text{and} \quad P_5(x) = 1 + x + \frac{x^2}{2} + \frac{x^3}{6} + \frac{x^4}{24} + \frac{x^5}{120}.$$

(The subscript on the polynomial P corresponds to its degree.) How closely does P_5 approximate f near $x = 0$? Use P_5 to approximate $f(-1) = e^{-1}$ and $f(2) = e^2$.

Solution The approximation is excellent near $x = 0$, as Figure 1 (page 494) shows. The graphs of f and P_5 are hard to distinguish anywhere near $x = 0$. At $x = -1$, for instance, we get ➡

Numbers are rounded to five decimal places.

$$P_5(-1) = 1 - 1 + \frac{1}{2} - \frac{1}{6} + \frac{1}{24} - \frac{1}{120} \approx 0.36667, \quad \text{while} \quad f(-1) = e^{-1} \approx 0.36788.$$

At $x = 2$ the match is less impressive but still respectable: $P_5(2) \approx 7.26667$ and

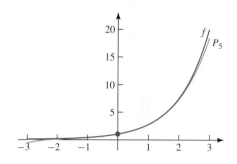

FIGURE 1

A polynomial approximation to $f(x) = e^x$

How do calculators find values of transcendental functions? They use (simpler) approximating functions.

$f(2) \approx 7.38906$. Notice, especially, that finding $P_5(-1)$ and $P_5(2)$ requires only arithmetic, whereas finding $f(-1)$ and $f(2)$ is best left to the mysterious inner workings of a calculator. ◄

We'll explain in a few pages exactly how we chose P_5 to "match" f so closely. To allay suspense, we'll mention now (and you can check easily) that f and P_5 "agree" at $x = 0$, as do their first five derivatives:

$$f(0) = P_5(0); \quad f'(0) = P_5'(0); \quad f''(0) = P_5''(0); \quad \ldots \quad f^{(5)}(0) = P_5^{(5)}(0). \quad \blacksquare$$

Trigonometric polynomials: Another nice family Ordinary polynomials, simple as they are, are not *always* the best choice for approximating a given "target" function f. It may be known, for instance, that f is a **periodic function**—one that repeats itself on domain intervals of a fixed length. (Such functions arise often in modeling physical phenomena, in which repeating or rhythmic behaviors are common.) In such cases, sines and cosines—the simplest and best-understood periodic functions—are better suited for use as basic modeling elements than are ordinary polynomials, which are not periodic.

In Section 9.3 we'll meet another nice function family, the **trigonometric polynomials**, members of which are used to approximate periodic functions. Trigonometric polynomials are functions like

$$1 + 2\cos x - 4\sin(5x) \quad \text{and} \quad 2 - 3\sin x + 4\cos(2x) - 5\sin(7x);$$

that is, sums and constant multiples of the trigonometric functions

$$1, \ \cos x, \ \sin x, \ \cos(2x), \ \sin(2x), \ \cos(3x), \ \sin(3x), \ldots.$$

(In the same way, ordinary polynomials are sums and constant multiples of the power functions $1, x, x^2, x^3, \ldots$.) As a look ahead, the following example illustrates graphically how such an approximation works.

EXAMPLE 2 A "square wave" function f is shown in Figure 2(a); the function might represent a simple alternating electric pulse. The "formula" for f is somewhat inconvenient:

$$f(x) = \begin{cases} \vdots \\ -1 & \text{if } -3\pi \leq x < -2\pi; \\ 1 & \text{if } -2\pi \leq x < -\pi; \\ -1 & \text{if } -\pi \leq x < 0; \\ 1 & \text{if } 0 \leq x < \pi; \\ \vdots \end{cases}$$

Notice that f repeats itself on intervals of length 2π.

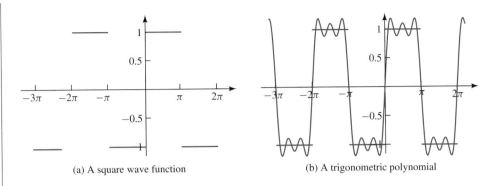

(a) A square wave function

(b) A trigonometric polynomial

FIGURE 2
Approximating a square wave

Figure 2(b) shows the square wave function again, this time together with the trigonometric polynomial

$$q(x) = \frac{4}{\pi} \left(\sin(x) + \frac{\sin(3x)}{3} + \frac{\sin(5x)}{5} \right).$$

We'll show in Section 9.3 how to find trigonometric polynomials like q, which approximate given periodic functions. We can see now, however, that q has advantages over f, including a relatively simple formula built from familiar ingredients and such "nice" properties as continuity and differentiability. ∎

Polynomials, derivatives, and coefficients

Consider the nth degree polynomial function $p(x) = a_0 + a_1(x - x_0) + a_2(x - x_0)^2 + a_3(x - x_0)^3 + \cdots + a_n(x - x_0)^n$, where x_0 and the a_i are all fixed numbers. A polynomial written this way, using powers of $(x - x_0)$, is said to be **expanded about the base point** x_0. Polynomials in this form may look strange at first compared to (say) $p(x) = 1 + 2x + 3x^2 + 4x^3$, which is expanded about $x = 0$. But the first form turns out to be more convenient for some purposes—it pays off nicely in the next (important) example.

EXAMPLE 3 Consider the polynomial

$$p(x) = a_0 + a_1(x - x_0) + a_2(x - x_0)^2 + a_3(x - x_0)^3 + a_4(x - x_0)^4.$$

Find $p(x_0)$, $p'(x_0)$, $p''(x_0)$, $p^{(3)}(x_0)$, and $p^{(4)}(x_0)$ (the value and first four derivatives of p at $x = x_0$).

Solution Differentiating p is easy—but watch the coefficients:

$$p(x) = a_0 + a_1(x - x_0) + a_2(x - x_0)^2 + a_3(x - x_0)^3 + a_4(x - x_0)^4;$$

$$p'(x) = a_1 + 2a_2(x - x_0) + 3a_3(x - x_0)^2 + 4a_4(x - x_0)^3;$$

$$p''(x) = 2a_2 + 2 \cdot 3a_3(x - x_0) + 3 \cdot 4a_4(x - x_0)^2;$$

$$p^{(3)}(x) = 2 \cdot 3a_3 + 2 \cdot 3 \cdot 4a_4(x - x_0);$$

$$p^{(4)}(x) = 2 \cdot 3 \cdot 4a_4.$$

Now the special form of p pays off: substituting $x = x_0$ gives

$$p(x_0) = a_0 = 1a_0; \qquad p'(x_0) = a_1 = 1a_1; \qquad p''(x_0) = 2a_2;$$
$$p^{(3)}(x_0) = 2 \cdot 3a_3 = 6a_3; \qquad p^{(4)}(x_0) = 2 \cdot 3 \cdot 4a_4 = 24a_4.$$

Notice the pleasant pattern: For each i, the ith derivative $p^{(i)}(x_0)$ is a *constant* multiple of a_i. ◄ ∎

We can think of $p(x_0)$ as the zeroth derivative at x_0.

Factorials   The constant multiples $(1, 1, 2, 6, 24)$ in the preceding computation are **factorials**. The **factorial** of i, denoted by $i!$, is defined as the repeated product $i \cdot (i-1) \cdot (i-2) \cdots 2 \cdot 1$. (By convention, $0! = 1$.) ◄ Factorials crop up often in mathematics (often enough to rate their own key on many calculators), especially around repeated operations like those in Example 3. We use factorials in the following theorem to describe the pattern just seen.

This convention looks strange at first glance, but it simplifies calculations.

> **THEOREM 1**   Let $p(x) = a_0 + a_1(x - x_0) + a_2(x - x_0)^2 + \cdots + a_n(x - x_0)^n$.
> Then
> $$p(x_0) = a_0; \qquad p'(x_0) = a_1; \qquad p''(x_0) = 2a_2;$$
> $$p^{(3)}(x_0) = 6a_3; \qquad \cdots \qquad p^{(n)}(x_0) = n!\, a_n.$$
> Equivalently, $a_i = \dfrac{p^{(i)}(x_0)}{i!}$ for all $i \geq 0$.

Theorem 1 restates, a little more generally, what we noticed for derivatives through degree 4 in Example 3. A formal proof, which we omit, uses the technique of mathematical induction.

Read one way, the theorem is a rule for calculating certain derivatives. More important for us, the theorem offers a nice recipe for cooking up polynomials with prescribed derivatives at a point.

EXAMPLE 4   Find a quadratic function $q(x) = a_0 + a_1 x + a_2 x^2$ for which $q(0) = 1$, $q'(0) = 2$, and $q''(0) = 3$. Are there *linear* functions with all three properties? Are there cubic functions?

Solution   Theorem 1 (with $x_0 = 0$) says exactly what to do: Set $a_0 = 1$, $a_1 = 2$, and $a_2 = q''(0)/2 = 3/2$. Thus, the quadratic polynomial we want is $q(x) = 1 + 2x + 3x^2/2$.

No *linear* polynomial $\ell(x) = a_0 + a_1 x$ can work because $\ell''(0) = 0$ for every linear polynomial. But infinitely many *cubic* polynomials satisfy all three requirements: Every function of the form $c(x) = 1 + 2x + 3x^2/2 + a_3 x^3$ satisfies all three conditions, regardless of the value of a_3. Figure 3 shows five functions of this form; the darker curve is the quadratic q (for which $a_3 = 0$):

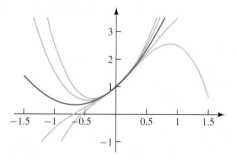

Each curve's equation is $y = 1 + 2x + 3x^2/2 + a_3 x^3$, for $a_3 = -2, -1, 0, 1, 2$. The darker curve has $a_3 = 0$, and so it's quadratic.

FIGURE 3
Several functions that "match" at $x = 0$

Notice the similarity among the curves: Because all five functions shown have the same value and first two derivatives at $x = 0$, they behave similarly near $x = 0$ but spread apart elsewhere. ∎

Polynomials that approximate other functions

An important use of Theorem 1 is for finding polynomial functions (linear, quadratic, or higher-order) that *approximate* a given (non-polynomial) function f near a given domain point $x = x_0$. If f happens to be complicated, inconvenient, or poorly understood, then having a close polynomial approximation p can reduce clutter and simplify calculations. One way to achieve this approximation is to construct polynomials p that "match" the value and one or more derivatives of the "target" function f at $x = x_0$.

Linear approximation: An old story   The linear approximation (or tangent-line approximation) to a function f at a point $x = x_0$ is nothing new. We observed as early as Chapter 1 that a differentiable function is "close to" its tangent line near $x = x_0$ (the point of tangency) and that this line has point–slope equation $y = f(x_0) + f'(x_0)(x - x_0)$. Theorem 1, in its simplest form, says that this linear function has the same value and the same (first) derivative as f at $x = x_0$. One important use of linear approximation is to produce quick estimates for values of functions we don't know exactly.

EXAMPLE 5   Your calculator's **TAN** key is stuck, and you really need $\tan 31°$. You know that $\tan 30° = 1/\sqrt{3}$. What to do?

Solution   We'll use a linear approximation to estimate $\tan 31°$. To this end, let $f(x) = \tan x$ (in radians!); note that $30° = \pi/6$ radians and that $31° = \pi/6 + \pi/180$ radians. Now $f'(x) = \sec^2 x$, and so we have

$$f\left(\frac{\pi}{6}\right) = \frac{1}{\sqrt{3}}; \quad f'\left(\frac{\pi}{6}\right) = \sec^2\left(\frac{\pi}{6}\right) = \frac{4}{3}.$$

Thus,

$$\ell(x) = \frac{1}{\sqrt{3}} + \frac{4}{3}\left(x - \frac{\pi}{6}\right); \quad \ell\left(\frac{\pi}{6} + \frac{\pi}{180}\right) = \frac{1}{\sqrt{3}} + \frac{4}{3}\frac{\pi}{180} \approx 0.6006.$$

This compares well with the true value $\tan 31° \approx 0.6008\ldots$ (as we can check if the **TAN** key comes unstuck). ∎

Quadratic approximation: A second step   Quadratic approximation takes linear approximation a step farther: We look for a quadratic polynomial that matches the value and first *two* derivatives of a given function f at a given point x_0.

The geometric result of quadratic approximation is a *parabola* (not a line, unless the quadratic term happens to be zero) that "fits" the f-graph at and near $x = x_0$. The (little) extra work done in quadratic approximation pays off nicely: The curved shape of a parabola permits a closer fit to a curved graph than any straight line allows. In particular, a quadratic approximation reflects the *concavity* of the target curve: The "tangent parabola" opens upward or downward depending on whether f is concave up or concave down at $x = x_0$. Figure 4 shows the sine function with two tangent parabolas opening in different directions.

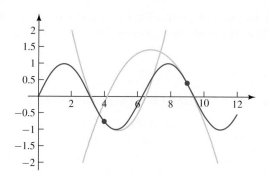

FIGURE 4
The sine curve with two "tangent parabolas"

The following example shows the advantage of quadratic over linear approximation.

EXAMPLE 6   Let $f(x) = 10^x$. Find ℓ and q, the linear and quadratic approximations to f, based at $x = 3$. Use each to estimate $f(3.1)$.

Solution   We need the value and two derivatives of f at $x = 3$. From $f(x) = 10^x$ we get $f'(x) = 10^x \ln 10$ and $f''(x) = 10^x (\ln 10)^2$, and so

$$f(3) = 10^3 = 1000; \quad f'(3) = 1000 \ln 10 \approx 2303; \quad f''(3) = 1000(\ln 10)^2 \approx 5302.$$

Thus,

$$\ell(x) = f(3) + f'(3) \cdot (x - 3) \approx 1000 + 2303(x - 3);$$
$$q(x) = \ell(x) + \frac{f''(3)}{2} \cdot (x - 3)^2 \approx 1000 + 2303(x - 3) + 2651(x - 3)^2,$$

and plugging in numbers gives the following results:

$$\ell(3.1) \approx 1230.3; \qquad q(3.1) \approx 1256.7; \qquad f(3.1) \approx 1258.9.$$

Figure 5 shows graphically how much better q does than ℓ in fitting f near $x = 3$:

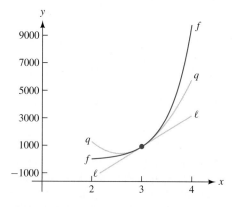

Although q does better than ℓ near $x = 3$, neither does well elsewhere. Note that q opens upward because f is concave up.

FIGURE 5
Linear and quadratic approximations to $f(x) = 10^x$

The picture shows why ℓ fits f so poorly: f bends sharply away from ℓ, even quite near $x = 3$. The second derivative, f'', explains why. Any linear function ℓ has zero second derivative. Our target function f, by contrast, has a huge second derivative:

$f''(3) \approx 5000$. Because f is so drastically nonlinear, it is poorly approximated by *any* straight line—even the best-fitting one.   ∎

Onward and upward: Taylor-made polynomials

There's no need to stop with quadratic polynomials in matching derivatives of a given function f at a point x_0. Higher-order matching is possible with polynomials of degree 3, 4, and higher. The following example illustrates the idea in an unusually simple setting and shows where the function P_5 in Example 1 came from.

EXAMPLE 7   Find the fifth-degree polynomial P_5 that matches $f(x) = e^x$ to order 5 at $x_0 = 0$ — that is,

$$P_5(0) = f(0), \quad P_5'(0) = f'(0), \quad P_5''(0) = f''(0), \dots P_5^{(5)}(0) = f^{(5)}(0).$$

Solution   Here $f'(x) = e^x = f(x)$, and so all derivatives are the same:

$$f(0) = 1 = f'(0) = f''(0) = f^{(3)}(0) = f^{(4)}(0) = f^{(5)}(0).$$

Theorem 1 tells how to find a polynomial

$$P_5(x) = a_0 + a_1 x + a_2 x^2 + a_3 x^3 + a_4 x^4 + a_5 x^5$$

with the same derivatives; we just set $a_i = f^{(i)}(0)/i! = 1/i!$ for all $i \geq 0$. That is,

$$P_5(x) = \frac{1}{0!} + \frac{x}{1!} + \frac{x^2}{2!} + \frac{x^3}{3!} + \frac{x^4}{4!} + \frac{x^5}{5!} = 1 + x + \frac{x^2}{2} + \frac{x^3}{6} + \frac{x^4}{24} + \frac{x^5}{120}.$$

Figure 1 (page 494) shows graphs of P_5 and f.   ∎

The formal definition   The polynomial P_5 constructed in Example 7 is called the **fifth-order Taylor polynomial** for $f(x) = e^x$ based at $x = 0$. It illustrates a more general definition:

> **DEFINITION (Taylor polynomials)**   Let f be any function whose first n derivatives exist at $x = x_0$. The **Taylor polynomial** of order n, based at x_0, is defined by
>
> $$P_n(x) = f(x_0) + f'(x_0)(x - x_0) + \frac{f''(x_0)}{2!}(x - x_0)^2$$
>
> $$+ \frac{f^{(3)}(x_0)}{3!}(x - x_0)^3 + \cdots + \frac{f^{(n)}(x_0)}{n!}(x - x_0)^n.$$

Observe:

- **Linear and quadratic cases**   If $n = 1$, the definition gives the familiar linear approximation; if $n = 2$, we get the quadratic approximation discussed above.

- **Maclaurin polynomials**   Taylor polynomials that happen to be based at $x = 0$ are called **Maclaurin polynomials.** ➡ They look a little cleaner on the page:

$$P_n(x) = f(0) + f'(0)x + \frac{f''(0)}{2}x^2 + \cdots + \frac{f^{(n)}(0)}{n!}x^n.$$

(In Example 7 we found the fifth-order Maclaurin polynomial for $f(x) = e^x$.)

- **Order and degree**   A Taylor polynomial P_n of **order** n is chosen so that its value and first n derivatives at $x = x_0$ agree with those of f. The definition shows that P_n also has **degree** n because it involves powers of $(x - x_0)$ up through the nth power. (If, by chance, $a_n = 0$, then the degree of P_n is less than n.)

Colin Maclaurin (1698–1746) was a Scottish contemporary of Isaac Newton. He studied the mathematics of tides, eclipses, honeycombs, and molasses barrels—along with the theory of calculus.

- **Many symbols, just one variable**   The definition involves many symbols: x, x_0, f, and n. While we are in this thicket, it is helpful to remember that x *is the only variable*. In specific cases all the other symbols take *numerical* values.

Taylor's idea?   The idea underlying Taylor polynomials probably predates Brook Taylor (1685–1731), for whom they are named. According to some authors the idea appears in work of John Bernoulli (1667–1748) published in 1694. The same Bernoulli (one of several famous Bernoullis) is also credited with having discovered—behind the scenes—l'Hôpital's rule.

How well do Taylor polynomials fit?   Taylor polynomials can be used to approximate more complicated functions. Gauging *how closely* a given Taylor polynomial fits its target function is the subject of the next section. Here we simply remark that Taylor polynomials usually fit their target functions better and better as their order increases. We illustrate this principle graphically and numerically in the following example.

EXAMPLE 8   Find P_1, P_3, P_5, and P_7, the Taylor (or Maclaurin) polynomials for $f(x) = \sin x$ based at $x = 0$. Plot everything on one set of axes. How closely does each polynomial approximate $\sin 1$?

Solution   To find P_7 we need the value and the first seven derivatives of $f(x) = \sin x$ at $x = 0$. Here they are in order: $0, 1, 0, -1, 0, 1, 0, -1$. Using Theorem 1 to match derivatives gives

$$P_7(x) = 0 + 1x + 0x^2 - \frac{1}{3!}x^3 + 0x^4 + \frac{1}{5!}x^5 + 0x^6 - \frac{1}{7!}x^7.$$

The formula lets us read off the lower-order Maclaurin polynomials:

$$P_1(x) = x = P_2(x); \qquad P_3(x) = x - \frac{x^3}{6} = P_4(x);$$

$$P_5(x) = x - \frac{x^3}{6} + \frac{x^5}{120} = P_6(x); \quad P_7(x) = x - \frac{x^3}{6} + \frac{x^5}{120} - \frac{x^7}{5040}.$$

Figure 6 shows all five graphs together:

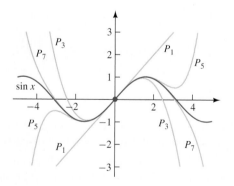

FIGURE 6
Several Taylor approximations to $f(x) = \sin x$

Observe:

- **An odd function**   All the nonzero powers of x in the Maclaurin polynomials are *odd*; this occurs because $\sin x$ itself is an *odd* function. (For similar reasons, all nonzero terms in the Maclaurin polynomials for $\cos x$ are *even*.)

- **Snug fits**   All four Taylor polynomials fit the sine curve well near the base point $x = 0$. Differences among the functions appear only as one moves away (to left or right) from $x = 0$.

- **Better and better**   Higher-order Taylor polynomials approximate the sine function better than lower-order ones. For instance, P_7 "sticks with" the sine curve much farther than P_1 or P_3.

Intuition suggests that P_1, P_3, P_5, and P_7 should give successively closer estimates for $\sin 1 \approx 0.84147$. So they do:

$$P_1(1) = 1; \quad P_3(1) \approx 0.83333; \quad P_5(1) \approx 0.84167; \quad P_7(1) \approx 0.84147. \qquad \blacksquare$$

BASIC EXERCISES

1. Let $f(x) = x^4 - 12x^3 + 44x^2 + 2x + 1$.

   (a) Find the Maclaurin series representation of f.

   (b) Find the Taylor series representation of f expanded about $x = 3$.

2. Find numbers $a_0, a_1, a_2, a_3, a_4,$ and a_5 such that

$$x^4 - 4x^3 + 5x = a_0 + a_1(x-1) + a_2(x-1)^2$$
$$+ a_3(x-1)^3 + a_4(x-1)^4 + a_5(x-1)^5.$$

3. Let $p(x) = 9 - \dfrac{(x-1)}{8} + \dfrac{(x-1)^2}{7} + \dfrac{(x-1)^3}{6} + \dfrac{(x-1)^4}{5}$.
Evaluate

   (a) $p(1)$ **(b)** $p^{(4)}(1)$.

4. Let $q(x) = 6 + \displaystyle\sum_{k=1}^{10} \dfrac{(x-2)^k}{k^2}$. Evaluate

   (a) $q''(2)$ **(b)** $q^{(9)}(2)$.

5. Let $p(x) = \displaystyle\sum_{m=0}^{19} \dfrac{(x+3)^{2m}}{(2m+1)^4}$. Evaluate

   (a) $p^{(11)}(-3)$ **(b)** $p^{(12)}(-3)$.

6. Let $q(x) = \displaystyle\sum_{n=0}^{100} \dfrac{(x-3)^{2n+1}}{n!}$. Evaluate

   (a) $q^{(99)}(3)$ **(b)** $q^{(100)}(3)$.

In Exercises 7–12, first find the sixth-order Maclaurin polynomial of the given function; then find an interval centered at $x = 0$ *in which the approximation error* $|f(x) - P_6(x)|$ *is less than* 0.01.

7. $f(x) = \cos x$ **8.** $f(x) = e^x$

9. $f(x) = \ln(1+x)$ **10.** $f(x) = \sqrt{1+x}$

11. $f(x) = \sinh x$ **12.** $f(x) = \cosh x$

In Exercises 13–24, first find the fifth-order Taylor polynomial for $f(x)$ *based at* x_0; *then find an interval centered at* $x = x_0$ *in which the approximation error* $|f(x) - P_5(x)|$ *is less than* 0.01.

13. $f(x) = \sin x, x_0 = \pi/4$

14. $f(x) = \cos x, x_0 = \pi/4$

15. $f(x) = \sin x, x_0 = \pi/3$

16. $f(x) = \cos x, x_0 = \pi/3$

17. $f(x) = 1/x, x_0 = 1$

18. $f(x) = 1/x, x_0 = 2$

19. $f(x) = \ln x, x_0 = 1$

20. $f(x) = 1/x^2, x_0 = 1$

21. $f(x) = \sqrt{x}, x_0 = 1$

22. $f(x) = \sqrt{x}, x_0 = 4$

23. $f(x) = 1/\sqrt{x}, x_0 = 1$

24. $f(x) = 1/\sqrt{x}, x_0 = 4$

FURTHER EXERCISES

25. Suppose that $P_n(x)$ is the nth-order Taylor polynomial for a function f. Can the degree of P_n be greater than n? Justify your answer.

26. Suppose that $P_n(x)$ is the nth-order Taylor polynomial for a function f. Can the degree of P_n be less than n? Justify your answer.

27. Suppose that $P_5(x) = 3 - x/4 - x^2/5 + 7x^3 - x^4 + 11x^5/9$ is the Taylor polynomial of order 5 for a function f. Find the Taylor polynomial of order 2 for the function f.

28. (a) Show that if f is an even function, then f' is an odd function.

   (b) Suppose that f is an even function. Use part (a) to show that every Maclaurin polynomial for $f(x)$ involves only *even* powers of x.

29. Suppose that f is an odd function. Show that every Maclaurin polynomial for $f(x)$ involves only *odd* powers of x.

30. Suppose that $M_n(x) = \sum_{k=0}^{n} a_k x^n$ is the nth-order Maclaurin polynomial for the function $f(x)$ and that $g(x) = f(x-b)$, where b is a constant. Show that $P_n(x) = \sum_{k=0}^{n} a_k (x-b)^n$ is the nth-order Taylor polynomial for $g(x)$ based at $x_0 = b$.

31. The fourth-order Maclaurin polynomial for $f(x) = \ln(1+x)$ is $P_4(x) = x - x^2/2 + x^3/3 - x^4/4$. Use this fact and Exercise 30 to find the fourth-order Taylor polynomial for $\ln x$ based at $x = 1$.

32. The fourth-order Taylor polynomial for $f(x) = 1/x$ based at $x_0 = 1$ is $P_4(x) = 1 - (x-1) + (x-1)^2 - (x-1)^3 + (x-1)^4$. Use this fact and Exercise 30 to find the fourth-order Maclaurin polynomial for $g(x) = 1/(1+x)$.

33. The fourth-order Taylor polynomial for $f(x) = 1/\sqrt{x}$ based at $x_0 = 1$ is $P_4(x) = 1 - (x-1)/2 + 3(x-1)^2/8 - 5(x-1)^3/16 + 35(x-1)^4/128$. Use this fact and Exercise 30 to find the fourth-order Maclaurin polynomial for $g(x) = 1/\sqrt{1+x}$.

34. Suppose that $M_n(x)$ is the nth-order Maclaurin polynomial for $f(x)$. Show that if k is a constant, then $M_n(kx)$ is the nth-order Maclaurin polynomial for $f(kx)$.

Suppose that $M_n(x)$ is the nth-order Maclaurin polynomial for $f(x)$. Use the fact stated in Exercise 34 to find the fifth-degree Maclaurin polynomial for the functions in Exercises 35–38.

35. $f(x) = \sin(x/2)$ 36. $f(x) = \cos(\pi x)$

37. $f(x) = e^{-x}$ 38. $f(x) = \ln(1 + 2x)$

Suppose that $M_n(x)$ is the nth-order Maclaurin polynomial for $f(x)$ and that m is a positive integer. It can be shown that $M_{mn}(x^m)$ is the mnth-order Maclaurin polynomial for $f(x^m)$. Use this fact to find the sixth-degree Maclaurin polynomial for the functions in Exercises 39–42.

39. $f(x) = e^{x^2}$ 40. $f(x) = \sin(x^2)$

41. $f(x) = (1 + x^2)^{-1}$ 42. $f(x) = \sqrt{1 + x^3}$

43. **(a)** Let $g(x) = xf(x)$. Show that $g^{(n)}(x) = nf^{(n-1)}(x) + xf^{(n)}(x)$.
    (b) Suppose that $M_n(x)$ is the nth-order Maclaurin polynomial for the function $f(x)$ and that $g(x) = xf(x)$. Use part (a) to show that $xM_n(x)$ is the Maclaurin polynomial of degree $n+1$ for $g(x)$.

44. Use part (b) of Exercise 43 to find the sixth-degree Maclaurin polynomial for $f(x) = x \sin x$.

45. Suppose that p and q are the nth-order Taylor polynomials based at x_0 for the functions f and g, respectively. Explain why $r(x) = p(x) + q(x)$ is the nth-order Taylor polynomial based at x_0 for the function $h(x) = f(x) + g(x)$.

46. Use the trigonometric identity $\cos(u + v) = \cos u \cos v - \sin u \sin v$ and the fifth-order Maclaurin polynomials for the sine and cosine functions to find the fifth-order Taylor polynomial for $\cos x$ based at $x = \pi/3$. [HINT: See Exercises 30 and 45.]

47. Suppose that $F'(x) = f(x)$ and that $P_n(x)$ is the nth-order Taylor polynomial for $F(x)$ based at x_0. Show that $\dfrac{d}{dx} P_n(x)$ is the Taylor polynomial of degree $n-1$ for $f(x)$ based at x_0.

48. Suppose that $F'(x) = f(x)$ and that $P_n(x)$ is the nth-order Taylor polynomial for $f(x)$ based at x_0. Show that $F(x_0) + \displaystyle\int_{x_0}^{x} P_n(t)\,dt$ is the Taylor polynomial of degree $n+1$ for $F(x)$ based at x_0.

Use the ideas explored in Exercises 30–48 to find the sixth-degree Maclaurin polynomial of the functions in Exercises 49–54.

49. $f(x) = \cos^2 x$
    [HINT: $\cos^2 x = (1 + \cos(2x))/2$.]

50. $f(x) = \sin x \cos x$
    [HINT: $\sin x \cos x = \sin(2x)/2$.]

51. $f(x) = \ln \left| \dfrac{1+x}{1-x} \right|$
    [HINT: $\ln |u/v| = \ln |u| - \ln |v|$.]

52. $f(x) = \displaystyle\int_{0}^{x} e^{-t^2}\,dt$

53. $f(x) = \arctan x$
    [HINT: $\arctan x = \displaystyle\int_{0}^{x} dt/(1 + t^2)$.]

54. $f(x) = \arcsin x$
    [HINT: $\arcsin x = \displaystyle\int_{0}^{x} dt/\sqrt{1 - t^2}$.]

9.2 TAYLOR'S THEOREM: ACCURACY GUARANTEES FOR TAYLOR POLYNOMIALS

In the preceding section we saw that a Taylor polynomial $P_n(x)$ approximates its target function $f(x)$ closely if x is near the base point $x = x_0$ and that the "goodness" of approximation usually improves as the degree n increases. In this section we make these useful general principles more precise: we show how to *guarantee* that Taylor polynomials approximate their target functions with specified accuracy.

Taylor's theorem lets us give such ironclad assurances. It is in the spirit of—and even resembles—the error-bound formulas for numerical integrals we studied in Chapter 6. (See, e.g., Theorem 3, page 387.) In all of these cases, the errors committed by various approximations are linked to the size of *derivatives* (first-, second-, or higher-order) of the functions in question.

Taylor errors and where they come from

A Taylor polynomial P_n is chosen to match its target function f at the base point x_0 in the sense that

$$f(x_0) = P_n(x_0); \quad f'(x_0) = P_n{}'(x_0); \quad f''(x_0) = P_n{}''(x_0); \quad \ldots \quad f^{(n)}(x_0) = P_n{}^{(n)}(x_0).$$

(In math jargon, f and P_n "agree to order n at x_0.") The real question about approximation, therefore, is how closely f and P_n agree *near*, rather than *at*, $x = x_0$ and how this accuracy depends on n. We focus on these questions in the following example.

EXAMPLE 1 Let $f(x) = \sqrt{x}$. Find the constant, linear, and quadratic Taylor polynomials (P_0, P_1, and P_2) for f at the base point $x_0 = 64$. ➡ How much error does each Taylor polynomial commit in approximating f near $x = 64$?

This base point is convenient because we know the exact value $f(64) = 8$.

Solution Routine calculations give the following (exact) values:

$$f(64) = 8; \quad f'(64) = \frac{1}{16}; \quad f''(64) = -\frac{1}{2048}.$$

Plugging these ingredients into the recipes of Section 9.1 gives the Taylor polynomials

$$P_0(x) = 8; \quad P_1(x) = 8 + \frac{1}{16}(x - 64); \quad P_2(x) = 8 + \frac{1}{16}(x - 64) - \frac{1}{4096}(x - 64)^2.$$

Figure 1 shows P_0, P_1, P_2, and f plotted together.

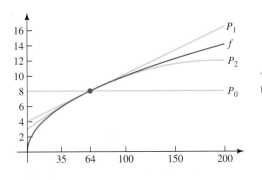

The dark curve is the "target" function $f(x) = \sqrt{x}$.

FIGURE 1
Constant, linear, and quadratic approximations to $f(x) = \sqrt{x}$

The graphs suggest, and numerical values confirm, how well or poorly P_0, P_1, and P_2 approximate f near to and far from $x = 64$:

Three approximations to $f(x) = \sqrt{x}$							
x	**35**	**63**	**63.9**	**64**	**64.1**	**65**	**100**
$P_0(x)$	8.00000	8.00000	8.00000	8	8.00000	8.00000	8.00000
$P_1(x)$	6.18750	7.93750	7.99375	8	8.00625	8.06250	9.00000
$P_2(x)$	5.98218	7.93726	7.99375	8	8.00625	8.06226	8.93750
$f(x)$	5.91608	7.93725	7.99375	8	8.00625	8.06226	8.94427

The graphs and the numbers give the same message: P_2 approximates f best, especially near $x = 64$. Indeed, the graphs of P_2 and f "share ink" in the range $35 \le x \le 100$. ■

Sources of error   Example 1 suggests several lessons about how Taylor approximation errors arise:

- **Error committed by P_0: Look at f'**   The zero-order Taylor approximation is the *constant* function P_0. In effect, P_0 "pretends" that f itself is *constant* near x_0, or, equivalently, that $f'(x) = 0$ for x near x_0. In general, $f'(x)$ measures how much f *differs* from being constant. We expect P_0 to approximate f well if f' is near zero and poorly if f' has large positive or large negative values.

- **Error committed by P_1: Look at f''**   The Taylor polynomial P_1 is the familiar *linear* approximation to f at x_0; P_1 "pretends" that f is *linear* near x_0, or, equivalently, that $f''(x) = 0$ for x near x_0. ◂ Thus, $f''(x)$ measures how much f *differs* from being linear; we expect P_1 to approximate f well if f'' is near zero, and poorly if f'' takes large positive or large negative values near $x = x_0$. (In Example 1, $f''(64) = -1/2048$, and f'' takes similar small negative values near $x = 64$; this explains why P_1 commits relatively little error near $x = 64$.)

A function f is linear if and only if $f''(x) = 0$ for all x.

- **Error committed by P_2: Look at f'''**   The emerging pattern continues. This time, the quadratic Taylor polynomial P_2 "pretends" that f is *quadratic* near x_0, or, equivalently, that $f'''(x) = 0$ for x near x_0. ◂ We expect P_2 to approximate f well if $|f'''(x)|$ is near zero, and poorly if $|f'''(x)|$ takes large positive values near $x = x_0$. (In Example 1, $f'''(64) = 3/262144$, and f''' takes similar small positive values near $x = 64$; this explains why P_2 approximates f so accurately near $x = 64$.)

A function f is quadratic if and only if $f'''(x) = 0$ for all x.

Taylor's theorem

The preceding paragraphs suggest a moral: The error committed by P_n in approximating f is related to the magnitude of the $(n+1)^{\text{th}}$ derivative, $f^{(n+1)}(x)$. The following important theorem makes this message precise. The hypotheses ensure that all the derivatives in question exist.

> **THEOREM 2 (Taylor's theorem)**   Suppose that f is repeatedly differentiable on an interval I containing x_0 and that
> $$P_n(x) = a_0 + a_1(x - x_0) + a_2(x - x_0)^2 + \cdots + a_n(x - x_0)^n$$
> is the nth-order Taylor polynomial based at x_0. Suppose that for all x in I,
> $$\left| f^{(n+1)}(x) \right| \le K_{n+1}.$$
> Then
> $$\left| f(x) - P_n(x) \right| \le \frac{K_{n+1}}{(n+1)!} |x - x_0|^{n+1}. \tag{1}$$

Here are some notes on the theorem:

- **Error on the left**   The left-hand quantity in Inequality (1) is the *error* committed by $P_n(x)$ in approximating $f(x)$. Taylor's theorem guarantees that this error is no larger than the quantity on the right side.

- **Factorials**   The factorial in the denominator on the right side of Inequality (1) grows quickly with n; this helps explain why the error committed by P_n normally decreases as n increases.

- **Powers of the distance**   The positive quantity $|x - x_0|^{n+1}$ on the right side of Inequality (1) is a power of the *distance* from x to x_0. The larger this distance, the larger the possible error.

- **Other versions**   Other versions of Taylor's theorem exist. This version is relatively simple, partly because it involves an *inequality*. In other versions, Inequality (1) is replaced with an *equality* that is slightly more complicated.

- **It's OK to overestimate K_{n+1}**   *Any* upper bound for $|f^{(n+1)}|$ on I can serve as K_{n+1} in Inequality (1). Choosing K_{n+1} as small as possible improves the error bound, but the inequality holds even for larger values. ➙   *But don't underestimate K_{n+1}.*

- **Worst-case scenarios**   The right side of Inequality (1) represents the *worst* case that the hypotheses allow; in many cases the actual error committed is less than the theorem permits.

We'll discuss the theorem's proof after exploring its meaning and some of its uses.

EXAMPLE 2   In Example 1 we found P_1 and P_2, the linear and quadratic approximations to $f(x) = \sqrt{x}$ at $x = 64$. What does Theorem 2 say about the errors these approximations commit on the interval $[50, 80]$? On the interval $[63, 65]$?

Solution   First we need suitable values for K_2 and K_3. Calculation gives

$$f''(x) = -\frac{1}{4x^{3/2}} \quad \text{and} \quad f'''(x) = \frac{3}{8x^{5/2}},$$

and so

$$\left|f''(x)\right| = \frac{1}{4|x|^{3/2}} \quad \text{and} \quad \left|f'''(x)\right| = \frac{3}{8|x|^{5/2}}.$$

A moment's thought (or a look at graphs) shows that $|f''(x)|$ and $|f'''(x)|$ *decrease* as x increases. For the interval $[50, 80]$, therefore, we may use

$$K_2 = |f''(50)| \leq 0.00071 \quad \text{and} \quad K_3 = |f'''(50)| \leq 0.000022.$$

On $[63, 65]$, we may use

$$K_2 = |f''(63)| \leq 0.00050 \quad \text{and} \quad K_3 = |f'''(63)| \leq 0.000012.$$

All ingredients are now ready. Here's what the theorem says:

- For x in $[50, 80]$, the error bounds are

$$\left|\sqrt{x} - P_1(x)\right| \leq \frac{|f''(50)|}{2}(x - 64)^2 \leq \frac{0.00071}{2}(80 - 64)^2 \approx 0.091;$$

$$\left|\sqrt{x} - P_2(x)\right| \leq \frac{|f'''(50)|}{6}(x - 64)^3 \leq \frac{0.000022}{6}(80 - 64)^3 \approx 0.015.$$

(We used $x = 80$ to make $x - 64$ as large as possible.) Thus, P_1 approximates f on $[50, 80]$ to about one-decimal-place accuracy, while P_2 does considerably better.

- For x in $[63, 65]$, the largest possible errors are even less:

$$|\sqrt{x} - P_1(x)| \le \frac{|f''(63)|}{2}(x - 64)^2 \le \frac{0.00050}{2}(65 - 64)^2 = 0.00025;$$

$$|\sqrt{x} - P_2(x)| \le \frac{|f'''(63)|}{6}(x - 64)^3 \le \frac{0.000012}{6}(65 - 64)^3 = 0.000002.$$

Thus, P_1 approximates f on $[63, 65]$ to about three-decimal-place accuracy; again, P_2 does much better.

The table on page 504 shows that these error estimates are actually conservative: P_1 and P_2 behave even better than the theorem predicts. ∎

EXAMPLE 3   In Example 1 of Section 9.1 we studied the 5th-order Maclaurin polynomial

$$P_5(x) = 1 + x + \frac{x^2}{2} + \frac{x^3}{6} + \frac{x^4}{24} + \frac{x^5}{120}$$

for $f(x) = e^x$. What does Theorem 2 say about how well P_5 approximates f on the interval $[-1, 1]$? How close must $P_5(1) = 163/60$ be to the *exact* value of e?

Solution   Finding derivatives of $f(x) = e^x$ is unusually easy:

$$f(x) = f'(x) = f''(x) = \cdots = f^{(6)}(x) = e^x.$$

Recall: It's OK to overestimate K_6.

Since e^x is positive and increases with x, we have $|f^{(6)}(x)| = e^x \le e^1$ for all x in $[-1, 1]$. Hence, we can take $K_6 = e$, or, for easier calculation, we can use $K_6 = 3$. ← Now Theorem 1 gives

$$\left| e^x - P_5(x) \right| \le \frac{3}{6!}|x|^6 = \frac{1}{240}|x|^6$$

for all x in $[-1, 1]$. With $x = 1$ we have

$$\left| e^1 - P_5(1) \right| \le \frac{1}{240} \approx 0.00412,$$

and so $P_5(1) = 163/60 \approx 2.71667$ is guaranteed to differ from e by less than about 4 in the third decimal place. (In fact, $e \approx 2.71828$—closer to $P_5(1)$ than Theorem 2 requires.) ∎

Taylor's theorem, the speed limit law, and the mean value theorem   Taylor's theorem is simplest if $n = 0$. In this case, $P_0(x) = f(x_0)$, and so Taylor's theorem says that

$$\left| f(x) - f(x_0) \right| \le K_1 |x - x_0|,$$

where K_1 is an upper bound for $|f'(x)|$ on I. If $x > x_0$ we can drop the absolute value signs and get

$$f(x) - f(x_0) \le K_1(x - x_0);$$

this is what we called the **speed limit law** in Chapter 1. (The name makes sense because the inequality $|f'(x)| \le K_1$ represents a "speed limit" on the growth of f.)

We studied the mean value theorem in Chapter 4.

This case of Taylor's theorem is also closely related to the **mean value theorem**. ← The mean value theorem says that, for some input t between x_0 and x,

$$f(x) - f(x_0) = f'(t)(x - x_0);$$

this can be seen as an "equality version" of the $n = 0$ case of Taylor's theorem.

Proving Taylor's theorem

We'll prove the theorem assuming (just for convenience) that $x_0 = 0$, and only for the specific case $n = 2$. We won't prove the general case, but the proof's main idea—integrating an inequality repeatedly—applies with only minor changes to handle *any* value of n. We'll use two familiar properties of the definite integral:

- **Bigger integrands, bigger integrals**   If g and h are functions and $g(t) \le h(t)$ for $a \le t \le b$, then $\displaystyle\int_a^b g(t)\,dt \le \int_a^b h(t)\,dt$.

- **Integrating a derivative**   By the FTC,

$$\int_a^b g'(t)\,dt = g(b) - g(a)$$

  for any function g with a continuous derivative g'.

Proof of Taylor's theorem (for $n = 2$ and $x_0 = 0$)   We assume that $\left| f'''(x) \right| \le K_3$ for all x in I (an interval containing zero) and need to show that

$$\left| f(x) - P_2(x) \right| \le \frac{K_3}{6} |x|^3 \tag{2}$$

for all x in I. We'll show this for positive x; a similar argument works for negative x. Because $x > 0$, we can drop the absolute value on the right of Inequality (2); this produces the equivalent double inequality

$$-\frac{K_3}{6} x^3 \le f(x) - P_2(x) \le \frac{K_3}{6} x^3; \tag{3}$$

we'll show that this holds for all positive x in I.

Our assumption that $\left| f'''(t) \right| \le K_3$ for t in I is equivalent to the double inequality

$$-K_3 \le f'''(t) \le K_3. \tag{4}$$

(Calling the input variable t rather than x is harmless—and it's convenient for reasons we'll see in a moment.)

The proof's first step (and main idea) is to integrate Inequality (4) over the interval $[0, x]$, using the integral properties mentioned above. Doing so gives

$$-\int_0^x K_3\,dt \le \int_0^x f'''(t)\,dt \le \int_0^x K_3\,dt;$$

working out the integrals gives

$$-K_3 x \le f''(x) - f''(0) \le K_3 x. \tag{5}$$

The second step is to integrate Inequality (5) over the interval $[0, x]$: ➡   *As before we first change the variable name to t to avoid confusion with x.*

$$-\int_0^x K_3 t\,dt \le \int_0^x f''(t)\,dt - \int_0^x f''(0)\,dt \le \int_0^x K_3 t\,dt,$$

which works out to

$$-K_3 \frac{x^2}{2} \le f'(x) - f'(0) - f''(0)x \le K_3 \frac{x^2}{2}. \tag{6}$$

The final step is no surprise: We Integrate Inequality (6): ➡   *We leave details to the reader.*

$$-\int_0^x K_3 \frac{t^2}{2}\,dt \le \int_0^x \left(f'(t) - f'(0) - f''(0)t \right) dt \le \int_0^x K_3 \frac{t^2}{2}\,dt.$$

Working out these integrals gives

$$-K_3\frac{x^3}{6} \le f(x) - f(0) - f'(0)x - f''(0)\frac{x^2}{2} \le K_3\frac{x^3}{6},$$

which is just another way of stating Inequality (3). This completes the proof.   □

BASIC EXERCISES

1. Let f and P_1 be as in Example 1.

    (a) Plot $|f(x) - P_1(x)|$ over the interval $[50, 80]$. What does the graph say about the maximum *actual* approximation error that occurs over this interval?

    (b) In Example 2 we used Theorem 2 to show that $|f(x) - P_1(x)| \le 0.091$. Is what you observed in part (a) consistent with this result? Justify your answer.

2. Let f and P_5 be as in Example 3.

    (a) Show that $|f^6(x)| \le 8$ for all x in the interval $[-2, 2]$.

    (b) Use part (a) to show that Theorem 2 implies that $|f(x) - P_5(x)| \le |x|^6/90$ if $-2 \le x \le 2$.

    (c) Explain why part (b) implies that $|f(x) - P_5(x)| \le 0.72$ for all x in the interval $[-2, 2]$.

    (d) Plot $|f(x) - P_5(x)|$ over the interval $[-2, 2]$. What does the graph say about the maximum *actual* approximation error that occurs over this interval?

    (e) Is the result in part (c) consistent with what you observed in part (d)? Justify your answer.

3. Let $f(x) = \sin x$ and $P_5(x)$ be the fifth-order Maclaurin polynomial for f.

    (a) What does Theorem 2 imply about the maximum approximation error committed by P_5 over the interval $[-2, 2]$?

    (b) What is the maximum *actual* approximation error committed by P_5 over the interval $[-2, 2]$?

4. Let $f(x) = \cos x$ and $P_6(x)$ be the sixth-order Maclaurin polynomial for f.

    (a) What does Theorem 2 imply about the maximum approximation error committed by P_6 over the interval $[-2, 2]$?

    (b) What is the maximum *actual* approximation error committed by P_6 over the interval $[-2, 2]$?

5. Let $f(x) = 1/\sqrt{x}$ and $P_4(x)$ be the fourth-order Taylor polynomial for f based at $x_0 = 1$.

    (a) What does Theorem 2 imply about the maximum approximation error committed by P_4 over the interval $[1/2, 3/2]$?

    (b) What is the maximum *actual* approximation error committed by P_4 over the interval $[1/2, 3/2]$?

6. Let $f(x) = \ln x$ and $P_4(x)$ be the fourth-order Taylor polynomial for f based at $x_0 = 1$.

    (a) What does Theorem 2 imply about the maximum approximation error committed by P_4 over the interval $[1/2, 3/2]$?

    (b) What is the maximum *actual* approximation error committed by P_4 over the interval $[1/2, 3/2]$?

FURTHER EXERCISES

7. Let $f(x) = \sin x$ and let $M_n(x)$ be the nth-degree Maclaurin polynomial for f. Explain why, for any x, n can be chosen so that $|f(x) - P_n(x)| \le 10^{-6}$.

8. Let $f(x) = e^x$ and let $M_n(x)$ be the nth-degree Maclaurin polynomial for f. Explain why, for any x, n can be chosen so that $|f(x) - P_n(x)| \le 10^{-6}$.

9. Let $f(x) = \sin(x^2)$ and $P_2(x)$ be the quadratic Maclaurin polynomial for f.

    (a) Show that $P_2(x) = x^2$.

    (b) Use Theorem 2 to show that if $0 \le t \le 1/2$, then $|\sin(t^2) - t^2| \le 2.5t^3/3!$.

    (c) It follows from part (a) that $\int_0^{1/2} f(x)\,dx \approx \frac{1}{24}$. Show that part (b) implies that $\left|\int_0^{1/2} f(x)\,dx - \frac{1}{24}\right| \le \frac{5}{768}$.

    (d) Using Theorem 2, it can be shown that if $0 \le t \le 1/2$, then $|\sin(t^2) - t^2| \le 120t^6/6!$. Use this fact to find

another bound on the approximation error committed when $\int_0^{1/2} f(x)\,dx$ is approximated $\int_0^{1/2} P_2(x)\,dx$.

10. Let $f(x) = e^{-x^2}$ and $P_2(x)$ be the quadratic Maclaurin polynomial for f. Find a bound on the approximation error committed if $\int_0^{2/3} f(x)\,dx$ is approximated by $\int_0^{2/3} P_2(x)\,dx$.

In Exercises 11–14, use Theorem 2 to evaluate the limit.

11. $\displaystyle\lim_{x \to 0} \frac{x - \sin x}{x^2}$

12. $\displaystyle\lim_{x \to 0} \frac{x - \sin x}{x^3}$

13. $\displaystyle\lim_{x \to 0} \frac{\cos x - 1 + x^2/2}{x^3}$

14. $\displaystyle\lim_{x \to 0} \frac{e^x - 1 - x}{x^2}$

15. (a) Compute the quadratic Maclaurin polynomial for $f(x) = \sqrt{1+x}$.

    (b) The electrical potential V at a distance r along the axis perpendicular to the center of a uniformly charged disk of radius a and charge density σ is

$$V = 2\pi\sigma\left(\sqrt{r^2 + a^2} - r\right).$$

If r is large in comparison to a, then $a/r \approx 0$. Use part (a) to find an approximation for V.

16. According to the theory of relativity, the mass m of a particle moving with velocity v is $m = m_0/\sqrt{1 - (v/c)^2}$, where m_0 is the rest mass of the particle and c is the velocity of light. Show that $m \approx m_0 + m_0 v^2/2c^2$.

17. Suppose that f is a function such that $f(1) = 1$, $f'(1) = 2$, and $f''(x) = (1 + x^3)^{-1}$ for $x > -1$.

  (a) Estimate $f(1.5)$ using a quadratic Taylor polynomial.

  (b) Find an upper bound on the approximation error made in part (a).

18. During an encounter with a friendly extraterrestrial, it is revealed to you that the answer to life, the universe, and everything else is $f(1)$ for some function f. It is also revealed that $f(0) = 26$, $f'(0) = 22$, $f''(0) = -16$, $f'''(0) = 12$, and $\left|f^{(4)}(x)\right| \le 7x^4$ if $|x| \le 3$.

  (a) Find an upper bound on the value of $f(1)$.

  (b) Find a lower bound on the value of $f(1)$.

19. Let f be a function that has continuous derivatives on an interval containing a and x.

  (a) Explain why $f(x) = f(a) + \int_a^x f'(t)\,dt$.

  (b) Use part (a) and integration by parts to show that

$$f(x) = f(a) + f'(a)(x - a) - \int_a^x (t - x)f''(t)\,dt$$

$$= f(a) + f'(a)(x - a) + \int_a^x (x - t)f''(t)\,dt.$$

  (c) Show that repeated integration by parts leads to this alternative form of Taylor's theorem:

$$f(x) = f(a) + f'(a)(x - a) + \frac{1}{2}f''(a)(x - a)^2$$

$$+ \frac{1}{3!}f'''(a)(x - a)^3 + \cdots + \frac{1}{n!}f^n(a)(x - a)^n$$

$$+ \frac{1}{n!}\int_a^x (x - t)^n f^{(n+1)}(t)\,dt.$$

20. Let $R_n(x) = \dfrac{1}{n!}\int_a^x (x - t)^n f^{(n+1)}(t)\,dt$. (This is the integral form of the remainder in Taylor's theorem derived in Exercise 19.) Show that if $a = 0$ and $\left|f^{(n+1)}(t)\right| \le K_{n+1}$ for all t, then

$$\left|f(x) - P_n(x)\right| = \left|R_n(x)\right| \le \frac{K_{n+1}}{(n+1)!}\,|x|^{n+1}.$$

9.3 FOURIER POLYNOMIALS: APPROXIMATING PERIODIC FUNCTIONS

In earlier sections we've seen how Taylor polynomials are constructed to approximate a target function near a chosen base point x_0. Taylor polynomial approximation is "local" in the sense that it focuses attention on a single base point and its immediate vicinity.

In this section we take another approach to the problem of approximating functions, this time using **Fourier polynomials**. Our approach here is "global": We'll choose Fourier polynomials that approximate a target function as closely as possible over a full domain interval rather than at a single point. To achieve this global approximation we'll use *integrals* over the full interval rather than *derivatives* at a single point.

Trigonometric polynomials

The main tools for Taylor approximation are ordinary polynomials. The corresponding tools for Fourier approximations are **trigonometric polynomials**—functions of the form

$$q_n(x) = a_0 + a_1 \cos x + a_2 \cos(2x) + \cdots + a_n \cos(nx)$$
$$+ b_1 \sin x + b_2 \sin(2x) + \cdots + b_n \sin(nx),$$

where n is a nonnegative integer and the coefficients a_k and b_k are real numbers. Notice that trigonometric polynomials are "linear combinations" (i.e., sums and constant multiples) of the basic trigonometric functions

$$1, \cos x, \sin x, \cos(2x), \sin(2x), \cos(3x), \sin(3x), \ldots$$

in exactly the same way that ordinary polynomials are built from the basic power functions $1, x, x^2, x^3, \ldots$.

Trigonometric polynomials (like ordinary polynomials) may have very different shapes and properties depending on the choices of coefficients a_k and b_k. Figure 1 shows three samples, all plotted over $[-\pi, \pi]$:

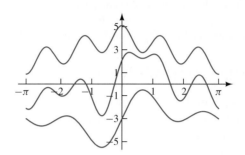

FIGURE 1
Three trigonometric polynomials

The particular formulas of the curves in Figure 1 don't matter for the moment, but observe the variety of shapes that trigonometric polynomials can assume.

Choosing the coefficients   The main idea of Fourier approximation (as for Taylor approximation) is to choose the coefficients a_k and b_k so that the resulting trigonometric polynomial closely approximates a given target function. For example, we saw in Example 2 of Section 9.1 (see especially Figure 2, page 495) the trigonometric polynomial

$$q(x) = \frac{4}{\pi} \left(\sin(x) + \frac{\sin(3x)}{3} + \frac{\sin(5x)}{5} \right)$$

approximates a certain "square wave" function. Note that in this case all the cosine coefficients a_k are zero, while the sine coefficients obey a simple rule:

$$b_k = \frac{4}{\pi} \cdot \frac{1}{k} \quad \text{(for odd } k); \qquad b_k = 0 \quad \text{(for even } k).$$

The main question, of course, is *how* to choose the a_k and the b_k in order to achieve a good fit with a target function. Providing and explaining an answer are the main goals of this section.

Properties of trigonometric polynomials   Understanding how to choose coefficients appropriately starts with basic properties of *all* trigonometric polynomials. We state several such properties in the following paragraphs. We'll be brief and leave some details to the reader.

- **Periodicity**   Periodicity is the most important property of trigonometric functions—and of trigonometric polynomials. Every trigonometric polynomial is 2π-periodic: it repeats itself on every interval of length 2π. This periodic behavior has a simple explanation: Each summand $a_k \cos(kx)$ and $b_k \sin(kx)$ repeats itself (k times, actually) on intervals of length 2π, and any sum of 2π-periodic functions is itself 2π-periodic. The graphs in Figure 1, for example, are shown only for inputs in the x-interval $[-\pi, \pi]$; the same shapes simply repeat themselves on intervals such as $[\pi, 3\pi]$.

  This periodicity property has two practical results: First, we may as well work just on the basic interval $[-\pi, \pi]$ because whatever happens there repeats itself elsewhere.

Second, we should expect good approximation from trigonometric polynomials only if the target function is *also* 2π-periodic.

- **Cosines are even, sines are odd**   Two simple but very useful facts follow from properties of ordinary sines and cosines:

  For each k, $\sin(kx)$ *is an* odd *function;* $\cos(kx)$ *is an* even *function.*

  These properties have consequences that will save us a lot of labor in calculating integrals:

  If g is an odd function, then $\int_{-\pi}^{\pi} g(x)\,dx = 0$. *If h is an even function, then*
  $\int_{-\pi}^{\pi} h(x)\,dx = 2\int_{0}^{\pi} h(x)\,dx$.

- **Zero integrals**   Lots of integrals arise in working with trigonometric polynomials. Luckily, many of these integrals turn out to be zero. In particular,

  $$\int_{-\pi}^{\pi} \sin(kx)\,dx = 0 \quad \text{and} \quad \int_{-\pi}^{\pi} \cos(kx)\,dx = 0$$

  for *every* positive integer k. Both claims are easy to check; either apply u-substitution or note that the first integrand is odd. These results make it easy to integrate *any* trigonometric polynomial. If

  $$q_n(x) = a_0 + a_1 \cos x + b_1 \sin x + \cdots + a_n \cos(nx) + b_n \sin(nx),$$

  then

  $$\int_{-\pi}^{\pi} q_n(x)\,dx = \int_{-\pi}^{\pi} a_0\,dx + \cdots + \int_{-\pi}^{\pi} b_n \sin(nx)\,dx = 2\pi a_0.$$

  (Every integral but the very first is zero, as we've just seen.)

- **More zero (and some nonzero) integrals**   Many other trigonometric integrals turn out to vanish. For instance, if m and n are any integers, then

  $$\int_{-\pi}^{\pi} \cos(mx) \sin(nx)\,dx = 0$$

  because the integrand is odd. Similarly, if m and n are *distinct* integers (i.e., $m \neq n$), then

  $$\int_{-\pi}^{\pi} \sin(mx) \sin(nx)\,dx = 0; \quad \int_{-\pi}^{\pi} \cos(mx) \cos(nx)\,dx = 0.$$

  Both of these claims are readily proved using trigonometric identities. If, on the other hand, $m = n$ in the preceding integrals, we find that

  $$\int_{-\pi}^{\pi} \cos^2(mx)\,dx = \pi \quad \text{and} \quad \int_{-\pi}^{\pi} \sin^2(mx)\,dx = \pi.$$

  Again, both integrals are straightforward calculations.

We'll use these facts to show how to choose trigonometric polynomials to approximate almost any target function f and to suggest why the process works. First comes the main definition:

DEFINITION (Fourier polynomials)   Let $f(x)$ be a function defined on $[-\pi, \pi]$. The **Fourier polynomial** of degree n for f is the trigonometric polynomial

$$q_n(x) = a_0 + a_1 \cos x + a_2 \cos(2x) + \cdots + a_n \cos(nx)$$
$$+ b_1 \sin x + b_2 \sin(2x) + \cdots + b_n \sin(nx),$$

with coefficients given by

$$a_0 = \frac{1}{2\pi} \int_{-\pi}^{\pi} f(x)\,dx;$$

$$a_k = \frac{1}{\pi} \int_{-\pi}^{\pi} f(x) \cos(kx)\,dx \qquad \text{if } k > 0;$$

$$b_k = \frac{1}{\pi} \int_{-\pi}^{\pi} f(x) \sin(kx)\,dx \qquad \text{if } k > 0.$$

Fourier and his polynomials   Fourier polynomials are named for the French mathematician Joseph Fourier (1768–1830), who used them to study the mathematics of various physical problems, including heat diffusion. Over the last 200 years, a large and still growing mathematical theory, **harmonic analysis**, has grown up around the subject.

For a given function f, the definition tells how to find the coefficients of a trigonometric polynomial that approximates f on the interval $[-\pi, \pi]$. After some examples to illustrate *that* Fourier polynomials have this approximation property, we'll suggest *why* they do. The first example illustrates, among other things, that the target function f may even have a discontinuity in the domain interval $[-\pi, \pi]$.

EXAMPLE 1   Figure 2 shows a "square wave" function f and the approximating trigonometric polynomial

$$q_9(x) = \frac{4}{\pi} \sin(x) + \frac{4}{\pi} \frac{\sin(3x)}{3} + \frac{4}{\pi} \frac{\sin(5x)}{5} + \frac{4}{\pi} \frac{\sin(7x)}{7} + \frac{4}{\pi} \frac{\sin(9x)}{9}. \qquad (1)$$

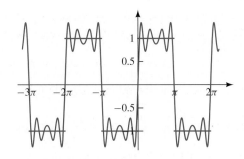

FIGURE 2

A square wave and a Fourier approximation

Figure 2(b), page 495, is similar but shows q_5 rather than q_9.

How was q_9 found? ◄

Solution The target function's "formula" is

$$f(x) = \begin{cases} \ \vdots \\ -1 & \text{if } -3\pi \le x < -2\pi; \\ 1 & \text{if } -2\pi \le x < -\pi; \\ -1 & \text{if } -\pi \le x < 0; \\ 1 & \text{if } 0 \le x < \pi. \\ \ \vdots \end{cases}$$

To find the coefficients we use the integral-based recipes

$$a_0 = \frac{1}{2\pi} \int_{-\pi}^{\pi} f(x)\,dx; \quad a_k = \frac{1}{\pi} \int_{-\pi}^{\pi} f(x) \cos(kx)\,dx; \quad b_k = \frac{1}{\pi} \int_{-\pi}^{\pi} f(x) \sin(kx)\,dx$$

from the definition.

Our first labor-saving observation is that f is an *odd* function. This implies, at one stroke, that *all* of the integrals for a_0 and a_k are zero. As for the b_k, the fact that both $f(x)$ and $\sin(kx)$ are odd functions implies that their product is even, and so

$$b_k = \frac{1}{\pi} \int_{-\pi}^{\pi} f(x) \sin(kx)\,dx = \frac{2}{\pi} \int_{0}^{\pi} \sin(kx)\,dx.$$

The last integral can be found by substituting $u = kx$ and $du = k\,dx$; the result is

$$\frac{2}{\pi} \cdot \frac{-\cos(kx)}{k} \Big]_{0}^{\pi} = \frac{2}{k\pi} \left(-\cos(k\pi) + \cos 0 \right).$$

The value of the last expression depends on whether k is even or odd:

$$b_k = \frac{4}{k\pi} \quad \text{if } k \text{ is odd}; \qquad b_k = 0 \quad \text{if } k \text{ is even}.$$

Thus, the b_k have the pattern

$$b_1, \ b_2, \ b_3, \ b_4, \ b_5, \ \ldots = \frac{4}{\pi}, \ 0, \ \frac{4}{3\pi}, \ 0, \ \frac{4}{5\pi}, \ \ldots,$$

which gives Formula (1). ∎

EXAMPLE 2 Find several Fourier polynomials q_n for $f(x) = x$. Plot them together with f. Do the q_n appear to approximate f? Where?

Solution We'll use the definition to find the coefficients; again, there's a simple pattern. Again the target function $f(x) = x$ is odd, and it follows as above that $a_k = 0$ for all k.

The b_k can be found using integration by parts, a table of integrals, or technology. Here's the result:

$$b_k = \frac{1}{\pi} \int_{-\pi}^{\pi} x \, \sin(kx)\,dx = \frac{-2\cos(k\pi)}{k}.$$

Again the numerical value depends on whether k is odd or even:

$$b_k = -\frac{2}{k} \quad \text{if } k \text{ is even}; \qquad b_k = \frac{2}{k} \quad \text{if } k \text{ is odd}.$$

Thus, the nth Fourier polynomial is

$$q_n(x) = \frac{2}{1} \sin x - \frac{2}{2} \sin(2x) + \frac{2}{3} \sin(3x) - \frac{2}{4} \sin(4x) + \cdots \pm \frac{2}{n} \sin(nx).$$

Figures 3(a) and (b) show graphs of f, q_3, and q_7 on two different domain intervals:

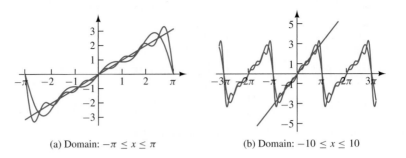

(a) Domain: $-\pi \le x \le \pi$ (b) Domain: $-10 \le x \le 10$

FIGURE 3

A function f and two Fourier polynomials

A close look at Figure 3(a) shows that q_7 (the wigglier curve) does a better job than q_3 of approximating f on $[-\pi, \pi]$. On the other hand, as Figure 3(b) shows, neither q_3 nor q_7 approximates f at all well outside the interval $[-\pi, \pi]$ because q_3 and q_7 are 2π-periodic and $f(x) = x$ is not. The figures illustrate that Fourier polynomials are effective *only* for approximating functions that, like Fourier polynomials themselves, repeat on intervals of length 2π. In the present case, we can think of q_3 and q_7 as approximations not to $f(x) = x$ itself but to the "sawtooth" function that's formed by repeating the basic shape of $y = x$ on each interval of length 2π. ∎

Why Fourier approximation works

Why do Fourier polynomials approximate a given function f on $[-\pi, \pi]$? Why are the coefficient formulas given in the definition "correct"?

Rigorous answers are well beyond our scope, but the basic idea is straightforward. The deep fact is that *every* continuous function f (and even many discontinuous functions) on $[-\pi, \pi]$ can be approximated as closely as we like by *some* trigonometric polynomial. In other words,

$$f(x) \approx q(x) = a_0 + a_1 \cos x + b_1 \sin x + \cdots + a_n \cos(nx) + b_n \sin(nx)$$

for *some* coefficients a_k and b_k. Given this, the remaining question is how to find the coefficients a_k and b_k.

Similar reasoning handles the a_k. (The case $k = 0$ is slightly different.)

Here's the answer for the b_k: ◄ Because $f(x) \approx q(x)$, we'd expect that

$$\int_{-\pi}^{\pi} f(x) \sin(kx)\, dx \approx \int_{-\pi}^{\pi} q(x) \sin(kx)\, dx$$

for all k. We'll show that the *second* integral gives πb_k. It follows (with more work) that the *first* integral has the same value.

$$\int_{-\pi}^{\pi} q(x) \sin(kx)\, dx = \int_{-\pi}^{\pi} \left(a_0 + a_1 \cos x + \cdots + b_n \sin(nx) \right) \sin(kx)\, dx$$

$$= \int_{-\pi}^{\pi} a_0 \sin(kx)\, dx + \int_{-\pi}^{\pi} a_1 \cos x \sin(kx)\, dx + \cdots + \int_{-\pi}^{\pi} b_n \sin(nx) \sin(kx)\, dx.$$

Now comes the big payoff from the integral identities discussed above: They imply that *only one* of the integrals in the last expression is nonzero. The single exception is

$$\int_{-\pi}^{\pi} b_k \sin(kx) \sin(kx)\, dx = b_k \int_{-\pi}^{\pi} \sin^2(kx)\, dx = b_k \pi.$$

This explains the formula for b_k given in the definition.

BASIC EXERCISES

1. Let $f(x) = \cos(kx)$, and let $g(x) = \sin(kx)$, where k is a non-negative integer.

   (a) Explain why, for all x, $f(x + 2\pi) = f(x)$ and $g(x + 2\pi) = g(x)$, regardless of the value of k.

   (b) Use part (a) to explain why every trigonometric polynomial is 2π-periodic.

   (c) What is the smallest period of the trigonometric polynomial $p(x) = \cos(4x) + \sin(8x)$? Show that p actually has this period. [HINT: Plot $p(x)$ first to guess the period.]

2. Let k be a positive integer.

   (a) We claimed that $\int_{-\pi}^{\pi} \cos(kx)\,dx = 0$. Use the FTC to verify this.

   (b) We also claimed that $\int_{-\pi}^{\pi} \sin(kx)\,dx = 0$. Use the FTC to verify this.

3. Suppose that m and n are distinct (i.e., $m \neq n$) positive integers. Use the FTC to show that

   (a) $\int_{-\pi}^{\pi} \cos(mx)\sin(mx)\,dx = 0$

   (b) $\int_{-\pi}^{\pi} \cos(mx)\sin(nx)\,dx = 0$

   (c) $\int_{-\pi}^{\pi} \cos(mx)\cos(nx)\,dx = 0$
   [HINT: $2\cos u \cos v = \cos(u+v) + \cos(u-v)$.]

   (d) $\int_{-\pi}^{\pi} \cos^2(mx)\,dx = \pi$

   (e) $\int_{-\pi}^{\pi} \sin(mx)\sin(nx)\,dx = 0$
   [HINT: $2\sin u \sin v = \cos(u-v) - \cos(u+v)$.]

   (f) $\int_{-\pi}^{\pi} \sin^2(mx)\,dx = \pi$

4. Let $q(x) = a_0 + a_1\cos x + b_1\sin x + \cdots + a_n\cos(nx) + b_n\sin(nx)$. Show that $\int_{-\pi}^{\pi} q(x)\,dx = 2\pi a_0$.

5. Let k be a positive integer and let $q(x) = a_0 + a_1\cos x + b_1\sin x + \cdots + a_n\cos(nx) + b_n\sin(nx)$. Show that $\int_{-\pi}^{\pi} q(x)\cos(kx)\,dx = \pi a_k$.

6. (a) Suppose that f is an odd function and that g is an even function. Show that $h = f \cdot g$ is an odd function.

   (b) Suppose that f and g are even functions. Show that $h = f \cdot g$ is an even function.

   (c) Suppose that f and g are odd functions. Show that $h = f \cdot g$ is an even function.

7. Suppose that f is an even function. Show that, for all $k > 0$, the Fourier polynomial coefficient $b_k = 0$.

8. Suppose that f is an odd function. Show that, for all $k \geq 0$, the Fourier polynomial coefficient $a_k = 0$.

9. Suppose that the Fourier polynomial of degree 5 for a function f is $q_5(x) = 2/\pi + 4\sin(3x) - \cos(5x)$.

   (a) What is the Fourier polynomial of degree 2 for f? Justify your answer.

   (b) What is the Fourier polynomial of degree 3 for f? Justify your answer.

   (c) What is the Fourier polynomial of degree 4 for f? Justify your answer.

10. Let $f(x) = 2\cos(5x) + 3\sin(4x)$ and let q_n be the Fourier polynomial of degree n for f.

    (a) Show that if $n = 3$, $q_n(x) = 0$.

    (b) Find q_4.

    (c) Explain why $q_n(x) = f(x)$ if $n \geq 5$.

FURTHER EXERCISES

11. Suppose that the Fourier polynomial of degree 5 for a function f is $q_5(x) = 3 - 7\cos x + 4\sin x - 6\cos(5x) - 2\sin(5x)$. What is the average value of f over the interval $[-\pi, \pi]$? Justify your answer.

12. Suppose that the Fourier polynomial of degree n for a function f is q_n. If $g(x) = f(x) + 1$, how is the Fourier polynomial of degree n for g related to q_n?

13. Suppose that the Fourier polynomial of degree n for a function f is q_n and that the Fourier polynomial of degree n for a function g is Q_n. Show that the nth-degree Fourier polynomial for $f + g$ is $q_n + Q_n$.

14. Let k be a positive integer. Use integration by parts to show that $\dfrac{1}{\pi}\int_{-\pi}^{\pi} x\sin(kx)\,dx = \dfrac{-2\cos(k\pi)}{k}$, as claimed in Example 2.

15. Consider the function $f(x)$ defined by

$$f(x) = \begin{cases} 1 & \text{if } 2m\pi < x \leq (2m+1)\pi \\ 0 & \text{if } (2m+1)\pi < x \leq (2m+2)\pi \end{cases}$$

    for all integers m. This function is sometimes called a **pulse train** or a **box wave**. (Notice that, although f is discontinuous, it repeats on intervals of length 2π.)

    (a) Use the definition of Fourier polynomials to find formulas for a_k and b_k. [HINT: For any function g, $\int_{-\pi}^{\pi} f(x)g(x)\,dx = \int_{0}^{\pi} g(x)\,dx$.]

    (b) Use the results from part (a) to find the Fourier polynomials q_1, q_3, q_5, and q_7 for f.

    (c) Plot f and the Fourier polynomials from part (b) on the same axes.

(d) Explain the relationship between the Fourier polynomial in this exercise and the one found in Example 2.

16. Consider the function $f(x)$ defined by

$$f(x) = \begin{cases} \pi + x & \text{if } -\pi \leq x < 0 \\ \pi - x & \text{if } 0 \leq x \leq \pi \end{cases}$$

and repeated on intervals of length 2π. This function is sometimes called a **sawtooth wave**.

(a) Use the definition of Fourier polynomials to find formulas for a_k and b_k.

(b) Use your answers from part (a) to find the Fourier polynomials $q_1, q_3, q_5,$ and q_7 for f.

(c) Plot f and the Fourier polynomials from part (b) on the same axes.

In Exercises 17–19, find the Fourier polynomial of degree 7 for the function f.

17. $f(x) = |x|$ **18.** $f(x) = x^2$

19. $f(x) = 4 + 3x + x^2$

20. Suppose that f is a function that has period T and is defined on the interval $[-T/2, T/2]$.

(a) Let $g(x) = f(Tx/2\pi)$. Show that g has period 2π and is defined on the interval $[-\pi, \pi]$.

(b) Let g be as in part (a) and let q_n be the Fourier polynomial of degree n for g. Explain why

$$f(x) \approx q_n\left(\frac{2\pi x}{T}\right) = a_0 + a_1 \cos\left(\frac{2\pi x}{T}\right)$$

$$+ a_2 \cos\left(\frac{4\pi x}{T}\right) + \cdots + a_n \cos\left(\frac{2n\pi x}{T}\right)$$

$$+ b_1 \sin\left(\frac{2\pi x}{T}\right) + \cdots + b_n \sin\left(\frac{2n\pi x}{T}\right).$$

(c) Use part (b) to show that the coefficients in the Fourier polynomial for f are

$$a_0 = \frac{1}{T} \int_{-T/2}^{T/2} f(x) \cos\left(\frac{2\pi x}{T}\right) dx;$$

$$a_k = \frac{2}{T} \int_{-T/2}^{T/2} f(x) \cos\left(\frac{2\pi k x}{T}\right) dx, \quad \text{if } k \geq 1;$$

$$b_k = \frac{2}{T} \int_{-T/2}^{T/2} f(x) \sin\left(\frac{2\pi k x}{T}\right) dx, \quad \text{if } k \geq 1.$$

Splines—Connecting the Dots

How does one draw a "curve" (perhaps with corners) through several given points in a plane? Figure 1 shows three possible curves, called **splines,** through five points:

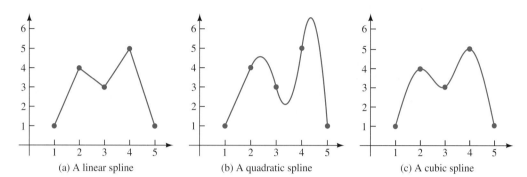

(a) A linear spline   (b) A quadratic spline   (c) A cubic spline

FIGURE 1

Three curves through given points

A human can choose among the many possible curves on aesthetic or other grounds. But computers need precise instructions in their own language—mathematics. Calculus provides the needed tools.

Basic ideas and key words   The idea behind splines is to "tie together" small curve segments, called **spline elements**, to form a single larger curve called a **spline**. The points at which elements are tied together (shown as dots in Figure 1) are called **knots**. Figure 1(a) shows a **linear spline**, a spline formed from line segments. Figure 1(b) shows a **quadratic spline**, a spline formed from parabolic pieces. Figure 1(c) shows a **cubic spline**, a spline formed by knotting cubic arcs. ➡ In all three cases the *individual* elements are smooth and unbroken, and so any spline built from such pieces will also be smooth and un-broken *except* perhaps at the knots. By matching derivatives at the knots we'll iron out the kinks.

A cubic arc is defined by a cubic function $y = a + bx + cx^2 + dx^3$.

PROBLEM 1   Let $S(x)$ be the linear spline function $S(x)$ in Figure 1(a). The first line segment in Figure 1(a) has equation $\ell_1(x) = 3x - 2$. Find linear formulas for ℓ_2, ℓ_3, and ℓ_4 to produce the piecewise-linear formula

$$S(x) = \begin{cases} 3x - 2 & \text{if } 1 \leq x \leq 2; \\ \ell_2(x) & \text{if } 2 < x \leq 3; \\ \ell_3(x) & \text{if } 3 < x \leq 4; \\ \ell_4(x) & \text{if } 4 < x \leq 5. \end{cases}$$

PROBLEM 2   Give a formula for a piecewise-linear function whose graph passes through the points $(0, 0)$, $(1, 1)$, and $(2, 4)$.

Quadratic splines: smoother   Linear splines, although simple, have an ugly kink at each knot (unless, by pure chance, successive slopes match up). One way to avoid such kinks is to use a quadratic spline formed by linking parabolic pieces. ➡

Each piece is defined by a quadratic equation.

517

What makes a curve degenerate? The graph of a quadratic polynomial $q(x) = a + bx + cx^2$, where a, b, and c are constants, is *usually* a parabola. There's one exception: If $c = 0$, then $q(x) = a + bx$ is actually a *linear* polynomial; its graph is a *line*. Any polynomial that happens to have less than its "advertised" power is sometimes called **degenerate**; a line, for instance, can be seen as a "degenerate parabola." Mathematicians are comfortable with degeneracy.

One "parabola" is actually a straight line.

Linear splines are crooked because only *one* line segment can join one knot to the next. Quadratic segments are—literally—more flexible: Given any two successive knots, we can find a parabola that joins them with *any* desired slope at the left knot. Figure 2 shows several parabolas joining $(0, 0)$ and $(1, 1)$, each labeled with its slope at $(0, 0)$: ◄

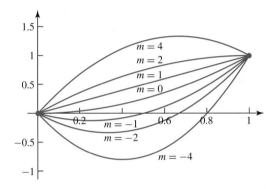

FIGURE 2
Parabolas joining two knots: m is the slope at $(0, 0)$

PROBLEM 3 Find a quadratic equation for each curve in Figure 2.

PROBLEM 4 In each part, find and plot a quadratic polynomial that satisfies the three given conditions.

(a) $Q(0) = 0$; $Q'(0) = 3$; $Q(1) = 5$.

(b) $Q(0) = 0$; $Q'(0) = 3$; $Q(10) = 5$.

(c) $Q(1) = 0$; $Q'(1) = 3$; $Q(3) = 0$.

By choosing slopes judiciously at each knot we can smooth out the resulting curve, as the next example shows.

EXAMPLE 1 Discuss how to find a piecewise-quadratic formula for the spline in Figure 1(b).

Solution To prevent kinks we'll arrange that slopes of "incoming" and "outgoing" elements match at each knot. Smoothness is no problem at the first knot, and so *any* quadratic curve joining $(1, 1)$ to $(2, 4)$ will do. Figure 1(b) starts with a straight line; a close look shows that the formula $S_1(x) = 3x - 2$ works for $1 \le x \le 2$. In particular, $S_1'(2) = 3$. For the second spline element, from $x = 2$ to $x = 3$, we need a quadratic polynomial $S_2(x)$ that joins the second and third knots and also *agrees in slope* with S_1 at $x = 2$. Thus, S_2 must satisfy three conditions:

$$S_2(2) = 4; \qquad S_2'(2) = 3; \qquad S_2(3) = 3. \tag{1}$$

It's easier to find such an S_2 if we first write

$$S_2(x) = a + b(x-2) + c(x-2)^2$$

and then search for a, b, and c. In this case we have $S_2(2) = a$ and $S_2'(2) = b$, and thus conditions (1) mean that $a = 4$ and $b = 3$. This gives $S_2(x) = 4 + 3(x-2) + c(x-2)^2$ with only c still to be found. The remaining condition, $S_2(3) = 3$, gives our answer. Because $S_2(3) = 7 + c$, we have

$$S_2(3) = 3 \iff 7 + c = 3 \iff c = -4.$$

All pieces are now in place: $S_2(x) = 4 + 3(x-2) - 4(x-2)^2$. ■

PROBLEM 5 Argue as in Example 1 to show that the quadratic spline in Figure 1(b) has piecewise-quadratic formula

$$S(x) = \begin{cases} S_1(x) = 3x - 2 & \text{if } 1 \leq x \leq 2; \\ S_2(x) = 4 + 3(x-2) - 4(x-2)^2 & \text{if } 2 < x \leq 3; \\ S_3(x) = 3 - 5(x-3) + 7(x-3)^2 & \text{if } 3 < x \leq 4; \\ S_4(x) = 5 + 9(x-4) - 13(x-4)^2 & \text{if } 4 < x \leq 5. \end{cases}$$

Figure 3 shows the four quadratic functions S_1 through S_4 plotted together (extended beyond the knots, this time); the spline S is the bold curve.

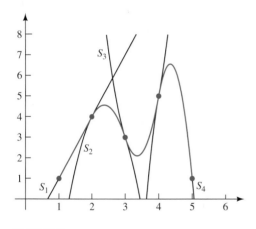

Notice how successive quadratic elements fit together smoothly at each knot.

FIGURE 3
Smooth-fitting quadratic elements

Formal definitions The preceding examples illustrate the first two definitions and suggest the third:

> **DEFINITION** A **linear spline** is a continuous, piecewise-linear function S. A **quadratic spline** is a continuous, piecewise-quadratic function S whose first derivative S' is also continuous. A **cubic spline** S is a continuous, piecewise-quadratic function whose first *two* derivatives, S' and S'' are also continuous.

The definition looks a bit technical, but it means simply that incoming and outgoing spline elements "agree" at each knot, both in their values and in one or more derivatives.

Counting the ways: Degrees of freedom There is just one line through two given points, or through one point with given slope. Either way, a straight line is completely determined by *two* conditions. The algebraic reason is that the generic linear equation $y = a + bx$ involves *two* parameters, a and b. In mathematical language, we have **two degrees of freedom** in choosing a line; specifying two points (or one point and one slope) "uses up" both degrees of freedom.

By contrast, a quadratic polynomial $y = a + bx + cx^2$ involves *three* parameters, and so we have three degrees of freedom in choosing quadratics. The practical result is that a quadratic spline element can be chosen to satisfy *three* conditions. Hitting the left and right knots costs two degrees of freedom, but the third remains available. Our method "spends" the last degree of freedom in specifying a derivative at the left knot.

The pattern continues. Cubic spline elements allow *four* degrees of freedom—hitting the knots costs two, and we spend the others on matching *two* derivatives at the left knot.

Cubic splines: Still smoother Using cubic spline elements, we can arrange that, at each knot, incoming and outgoing elements agree in their values and in their first *two* derivatives. By avoiding abrupt changes in concavity, cubic elements can produce smoother and more pleasing splines than quadratic elements can manage.

EXAMPLE 2 Construct a cubic spline through the knots in Figure 1(c).

Solution Because smoothness is no problem at the first knot, we can start with *any* cubic curve S_1 that joins $(1, 1)$ to $(2, 4)$. Although infinitely many possibilities exist, some choices give better results than others. Moreover, the resulting spline depends very strongly on how the first element is chosen. For instance, Figure 4 shows two cubic splines through the same knots:

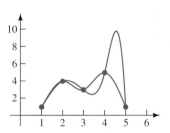

FIGURE 4
Same knots, different cubic splines

Notice that the two splines differ just slightly in the first element but much more thereafter. Given this sensitivity to initial choices, mathematicians have found various ways of choosing the first cubic spline element to produce a pleasing result. The **natural spline method**, for instance, arranges things so that $S''(x_0) = 0 = S''(x_n)$; this means that S has zero concavity at the first and last knots. (The "tamer" curve in Figure 4 is a natural spline.) Full details on natural splining are beyond our scope. ◄ We'll just say, therefore, that the unlikely looking cubic

Numerical analysis textbooks discuss natural and other splining methods.

$$S_1(x) = 1 + \frac{123}{28}(x - 1) - \frac{39}{28}(x - 1)^3$$

can be used to start things off. Given the formula, it's easy to check that

$$S_1(1) = 1, \quad S_1(2) = 4, \quad S_1'(2) = \frac{3}{14}, \quad \text{and} \quad S_1''(2) = -\frac{117}{14}.$$

The first two conditions ensure that S_1 hits the first two knots; the remaining conditions will help us determine S_2.

INTERLUDE

The second spline element is a cubic polynomial that satisfies *four* conditions: ➔

$$S_2(2) = 4; \quad S_2'(2) = \frac{3}{14}; \quad S_2''(2) = -\frac{117}{14}; \quad S_2(3) = 3. \tag{2}$$

The middle two conditions come from matching derivatives of S_1.

Again, writing $S_2(x) = a + b(x-2) + c(x-2)^2 + d(x-2)^3$ simplifies our quest. Because

$$S_2(2) = a, \quad S_2'(2) = b, \quad \text{and} \quad S_2''(2) = 2c,$$

conditions (2) imply that $a = 4$, $b = 3/14$, and $c = -117/28$, and so

$$S_2(x) = 4 + \frac{3}{14}(x-2) - \frac{117}{28}(x-2)^2 + d(x-2)^3$$

with only d left to find. The final condition, $S_2(3) = 3$, does the trick:

$$3 = S_2(3) = 4 + \frac{3}{14} - \frac{117}{28} + d \implies d = \frac{83}{28},$$

and we're done at last:

$$S_2(x) = 4 + \frac{3}{14}(x-2) - \frac{117}{28}(x-2)^2 + \frac{83}{28}(x-2)^3. \qquad ■$$

PROBLEM 6 It turns out that the cubic spline in Figure 1(c) has the piecewise-cubic formula

$$S(x) = \begin{cases} S_1(x) = 1 + \frac{123}{28}(x-1) - \frac{39}{28}(x-1)^3 & \text{if } 1 \le x \le 2; \\ S_2(x) = 4 + \frac{3}{11}(x-2) - \frac{117}{28}(x-2)^2 + \frac{83}{28}(x-2)^3 & \text{if } 2 < x \le 3; \\ S_3(x) = 3 + \frac{3}{4}(x-3) + \frac{33}{7}(x-3)^2 - \frac{97}{28}(x-3)^3 & \text{if } 3 < x \le 4; \\ S_4(x) = 5 - \frac{3}{14}(x-4) - \frac{159}{28}(x-4)^2 + \frac{53}{28}(x-4)^3 & \text{if } 4 < x \le 5. \end{cases}$$

(a) Argue as in Example 2 to verify the formula for S_3.

(b) Write a piecewise-quadratic formula for $S'(x)$; check that $S'(x)$ is indeed continuous at the knots. Plot $S'(x)$ if technology permits.

(c) Write a piecewise-linear formula for $S''(x)$; check that $S''(x)$ is indeed continuous at the knots. Plot $S''(x)$ if technology permits.

Postscript The general idea of linear, quadratic, and cubic splines—our main interest in this section—is simple. Elements of the desired type are linked end to end; as many derivatives as possible are made to agree at knots. The necessary calculations, although routine, can be tedious and repetitive, especially when dozens or hundreds of knots are involved. In real-life practice such calculations are always left to machines, and specialized methods are used to speed up and simplify the machines' work.

FURTHER EXERCISES

1. Let (x_0, y_0) and (x_1, y_1) be any two points with $x_0 \neq x_1$, and let m_0 be any real number. Consider the quadratic function q defined by

$$q(x) = y_0 + m_0(x - x_0) + \frac{y_1 - y_0 - m_0(x_1 - x_0)}{(x_1 - x_0)^2}(x - x_0)^2.$$

(a) Find $q(x)$ in the case $(x_0, y_0) = (2, 4)$, $(x_1, y_1) = (3, 3)$, $m_0 = 3$. Then check that $q(2) = 4$, $q(3) = 3$, and $q'(2) = 3$.

(b) Plot your function q from the previous part. How does the graph show that $q(2) = 4$, $q(3) = 3$, and $q'(2) = 3$?

(c) Show (work with the general formula above) that $q(x_0) = y_0$, $q(x_1) = y_1$, and $q'(x_0) = m_0$. What do these properties mean about the graph of q?

(d) Give an example in which $q(x)$ turns out to be a linear function. Interpret your example in graphical terms.

2. The figure shows two quadratic splines; both pass through the three knots $(0, 1)$, $(1, -2)$, and $(2, 3)$.

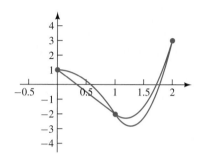

(a) For one of the splines, $S'(0) = -3$. Use this information to find formulas for both elements of this spline.

(b) For the other spline, $S'(0) = 0$. Use this information to find formulas for both elements of this spline.

3. Let (x_0, y_0) and (x_1, y_1) be any two points with $x_0 \neq x_1$, and let d_1 and d_2 be any real numbers. Consider the cubic function c defined by

$$c(x) = y_0 + d_1(x - x_0) + \frac{d_2}{2}(x - x_0)^2 +$$

$$\frac{y_1 - y_0 - d_1(x_1 - x_0) - \frac{d_2}{2}(x_1 - x_0)^2}{(x_1 - x_0)^3}(x - x_0)^3$$

Show by (careful!) calculation that c satisfies the four conditions

$$c(x_0) = y_0; \quad c'(x_0) = d_1; \quad c''(x_0) = d_2; \quad c(x_1) = y_1.$$

These four conditions mean that c is the general cubic spline element that joins (x_0, y_0) to (x_1, y_1) with first two derivatives d_1 and d_2 at (x_0, y_0).

4. In each part, either use the formula in the previous exercise or work from scratch to find a cubic function with the given properties. Plot each result.

(a) $c(0) = 0$; $c'(0) = 0$; $c''(0) = 0$; $c(1) = 0$.

(b) $c(0) = 0$; $c'(0) = 0$; $c''(0) = 0$; $c(1) = 1$.

(c) $c(0) = 1$; $c'(0) = 0$; $c''(0) = -1$; $c(1) = 1/2$.

(d) $c(0) = 1$; $c'(0) = 1$; $c''(0) = 1$; $c(1) = 8/3$.

5. The following figure shows two cubic splines; both pass through the three knots $(0, 1)$, $(1, -2)$, and $(2, 3)$.

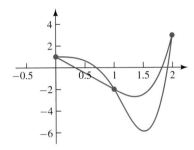

(a) For one of the splines, $S'(0) = -3$ and $S''(0) = 0$. Use this information to find formulas for both elements of this spline.

(b) For the other spline, $S'(0) = 0$ and $S''(0) = 0$. Use this information to find formulas for both elements of this spline.

IMPROPER INTEGRALS

10.1 IMPROPER INTEGRALS: IDEAS AND DEFINITIONS

Each of the following expressions is an **improper integral**:

$$\int_1^\infty \frac{dx}{x^2} \qquad \int_1^\infty \frac{dx}{x} \qquad \int_{-\infty}^\infty \frac{dx}{1+x^2} \qquad \int_0^\infty e^{-x}\,dx \qquad \int_0^1 \frac{dx}{x^2} \qquad \int_0^1 \frac{dx}{\sqrt{x}}$$

Keep these basic examples in mind. They illustrate most of what can go right—and wrong—with improper integrals.

Improper integrals, like improper fractions, commit no breach of morals or manners. The adjective "improper" is a warning sticker attached to integrals that differ somehow from the ordinary $\int_a^b f(x)\,dx$ variety, in which $[a, b]$ is a *finite* interval and $f(x)$ is continuous on all of the interval $[a, b]$. Each of the preceding integrals, examined carefully, should raise some suspicion.

Two improprieties Integrals can commit two types of "impropriety":

- **Infinite intervals** The interval of integration may be infinite, as in the first four sample integrals. This is "improper" because the formal definition of definite integral relies on partitions of a *finite* interval.

- **Infinite integrands** The integrand may be unbounded on the interval of integration, as in the last two samples. ◄ This, too, is "improper" because the integrand is not defined somewhere inside, or at an endpoint of, the interval of integration.

The integrand "blows up," in other words.

Some really offensive integrals, such as $\displaystyle\int_0^\infty \frac{dx}{\sqrt{x}+x^2}$, commit *both* types of impropriety and will need especially strict handling.

Convergence and divergence: Basic ideas and examples

Some improper integrals have a sensible numerical value; they are called **convergent**. For other integrals the impropriety is fatal; these integrals have no sensible finite value and are called **divergent**. We'll give formal definitions after some concrete examples.

EXAMPLE 1 Make sense of $\displaystyle\int_1^\infty \frac{dx}{x^2}$.

Solution   What could such an integral mean? Interpreted geometrically, the integral represents the shaded area in Figure 1:

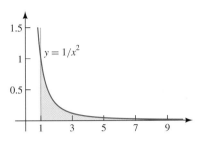

FIGURE 1
The improper integral $\displaystyle\int_1^\infty \dfrac{dx}{x^2}$ as area

The shaded region extends infinitely far to the right, but might its *area* be finite?

The answer, perhaps surprisingly, is yes; the reason involves a limit. For any number $t > 1$, consider the (proper) integral

$$I(t) = \int_1^t \frac{dx}{x^2},$$

which represents the shaded area from $x = 1$ to $x = t$. We can calculate $I(t)$ exactly:

$$I(t) = \int_1^t \frac{dx}{x^2} = \frac{-1}{x}\Bigg]_1^t = 1 - \frac{1}{t}.$$

Clearly, $I(t) \to 1$ as $t \to \infty$. Thus, the total shaded region, although infinitely long, has *finite* area. In summary,

$$\lim_{t \to \infty} \int_1^t \frac{dx}{x^2} = \lim_{t \to \infty}\left(1 - \frac{1}{t}\right) = 1;$$

we say that the integral **converges** to 1. ∎

EXAMPLE 2   Does $\displaystyle\int_1^\infty \dfrac{dx}{x}$ converge or diverge? Why?

Solution   This integral itself resembles the one in Example 1, but a picture (Figure 2) suggests a difference:

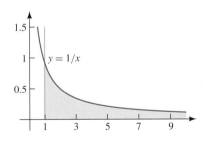

FIGURE 2
The improper integral $\displaystyle\int_1^\infty \dfrac{dx}{x}$ as area

The question, again, is whether the shaded region (unbounded on the right) has finite or infinite area. The picture alone, being finite, doesn't say. To find an answer, we'll calculate much as before. For $t > 1$, the area from $x = 1$ to $x = t$ is

$$\int_1^t \frac{dx}{x} = \ln x\Bigg]_1^t = \ln t.$$

(The shaded area shown in Figure 2 is $\ln 10 \approx 2.3026$.) In this case, $I(t) = \ln t \to \infty$ as $t \to \infty$, and so we conclude that the improper integral **diverges** to infinity. In symbolic shorthand:

$$\lim_{t \to \infty} \int_1^t \frac{dx}{x} = \lim_{t \to \infty} \ln t = \infty. \qquad \blacksquare$$

Looks can deceive   Examples 1 and 2 illustrate a common subtlety of improper integrals: Although the graphs of $y = 1/x^2$ and $y = 1/x$ appear similar, the first graph bounds only *one* unit of area, while the second bounds *infinite* area. In short, graphs tell only a small part of the story for improper integrals.

EXAMPLE 3   Does $\displaystyle\int_{-\infty}^{\infty} \frac{dx}{1 + x^2}$ converge? If so, to what?

Solution   This integral is improper at *both* ends. Breaking the integral into two pieces lets us handle one impropriety at a time:

$$\int_{-\infty}^{\infty} \frac{dx}{1 + x^2} = \int_{-\infty}^{0} \frac{dx}{1 + x^2} + \int_{0}^{\infty} \frac{dx}{1 + x^2}.$$

Let's consider the second summand first. A calculation shows that it converges to $\pi/2$:

$$\lim_{t \to \infty} \int_0^t \frac{dx}{1 + x^2} = \lim_{t \to \infty} \arctan x \Big]_0^t = \lim_{t \to \infty} \arctan t = \frac{\pi}{2}.$$

A similar calculation handles the first summand, but we can avoid needless work if we first notice that the integrand is an *even* function, and thus conclude that the first and second summands have the *same* value. Putting the pieces together shows that the original integral converges to π:

$$\int_{-\infty}^{\infty} \frac{dx}{1 + x^2} = \int_{-\infty}^{0} \frac{dx}{1 + x^2} + \int_{0}^{\infty} \frac{dx}{1 + x^2} = \pi. \qquad \blacksquare$$

Another impropriety: Infinite integrands

The improprieties in Examples 1–3 all involved infinite *intervals* of integration. Much the same strategy applies to integrals with finite *intervals* of integration, but whose *integrands* blow up to infinity within, or at an endpoint of, the interval.

EXAMPLE 4   Does $\displaystyle I_1 = \int_0^1 \frac{dx}{x^2}$ converge? If so, to what? What about $\displaystyle I_2 = \int_{-1}^1 \frac{dx}{x^2}$?

Solution   For I_1 we want to know whether the *vertically* unbounded shaded region in Figure 3 has finite or infinite area.

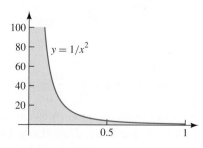

FIGURE 3

The improper integral $\displaystyle\int_0^1 \frac{dx}{x^2}$ as area

To settle the question we find *another* limit of areas, but this time we let t tend to zero from the right. For any $t > 0$, the area from $x = t$ to $x = 1$ is finite, and it's easy to find explicitly:

$$I(t) = \int_t^1 \frac{dx}{x^2} = -\frac{1}{x}\bigg]_t^1 = \frac{1}{t} - 1.$$

Now as $t \to 0^+$, we have $I(t) \to \infty$, too. Here is our conclusion in symbols:

$$\lim_{t \to 0^+} \int_1^t \frac{dx}{x^2} = \lim_{t \to 0^+} \left(\frac{1}{t} - 1\right) = \infty.$$

In words: The integral $\int_0^1 \frac{dx}{x^2}$ **diverges** to infinity.

The new feature of I_2 is that the impropriety (at $x = 0$) is *inside* the interval of integration. We approach the problem in the "usual" way: We break I_2 into two integrals, each with just one impropriety, at an endpoint:

$$I_2 = \int_{-1}^1 \frac{dx}{x^2} = \int_{-1}^0 \frac{dx}{x^2} + \int_0^1 \frac{dx}{x^2}.$$

We've just shown that the second integral diverges; so, therefore, does I_2. ∎

Convergence and divergence: Formal definitions

We applied the same basic idea in each of the preceding examples. First we located an impropriety at ∞, at $-\infty$, or at a finite endpoint of the interval of integration. Then we considered the limit as a variable endpoint tends to the troublesome value from either above or below. In the case of greatest interest, the fine print reads like this:

> **DEFINITION** Consider the integral $I = \int_a^\infty f(x)\,dx$, where f is continuous for $x \geq a$. If the limit
>
> $$L = \lim_{t \to \infty} \int_a^t f(x)\,dx$$
>
> exists and is finite, then I **converges** to L. Otherwise, I **diverges**.

A similar but more general definition of convergence applies to *any* improper integral:

> **DEFINITION** Let the integral $I = \int_a^b f(x)\,dx$ be improper either at a or at b. (The cases $a = -\infty$ and $b = \infty$ are allowed.) If either
>
> $$\lim_{t \to a^+} \int_t^b f(x)\,dx \qquad \text{or} \qquad \lim_{t \to b^-} \int_a^t f(x)\,dx$$
>
> exists and has finite value L, then I **converges** to L. Otherwise, I **diverges**.

Notice, in particular, that every convergent *improper* integral is a limit of *proper* integrals. For example,

$$\int_1^\infty f(x)\,dx = \lim_{t \to \infty} \int_1^t f(x)\,dx$$

if the limit exists.

Improprieties at both ends If an integral is improper at *both* ends, it can be broken into two summands in any convenient manner; the whole integral converges only if *both* summands converge. (We took this approach in Example 3.) The next example illustrates another "double impropriety."

EXAMPLE 5 Does $I = \int_0^\infty \dfrac{dx}{x^2}$ converge? If so, to what?

Solution The integral has improprieties at both ends. To separate them, we write

$$I = \int_0^\infty \frac{dx}{x^2} = \int_0^1 \frac{dx}{x^2} + \int_1^\infty \frac{dx}{x^2} = I_1 + I_2.$$

As we saw in earlier examples, I_2 *converges* to 1, but I_1 *diverges* to ∞. Hence, I itself diverges. (An integral converges only if all of its pieces converge.) ■

Integrands that change sign: Another way to diverge In each example so far the integrand has been a *positive* function. In this case integrals are readily interpreted as areas: The integral converges if the area in question remains finite; otherwise, the integral diverges to infinity.

  If the integrand *changes* sign, the area interpretation is less useful. In this case an integral may diverge without "blowing up."

EXAMPLE 6 Does $I = \int_0^\infty \cos x \, dx$ converge or diverge?

Solution The definitions give a quick answer. Because

$$\int_0^t \cos x \, dx = \sin x \Big]_0^t = \sin t,$$

we have

$$\lim_{t \to \infty} \int_0^t \cos x \, dx = \lim_{t \to \infty} \sin t.$$

The last limit doesn't exist: As $t \to \infty$, $\sin t$ oscillates endlessly between -1 and 1. Hence, I diverges. ■

Substitution in improper integrals Simple u-substitutions sometimes allow us to trade one improper integral for another—or even for a *proper* integral.

EXAMPLE 7 Consider $I_1 = \int_0^1 \dfrac{dx}{x^2}$ and $I_2 = \int_0^1 \dfrac{dx}{\sqrt{x}}$. How does the substitution $u = 1/x$ transform each integral?

As $x \to 0^+$, $u = 1/x \to \infty$.

Solution If $u = 1/x$, then $x = 1/u$ and $dx = -1/u^2 \, du$. Also, $x = 1$ and $x = 0$ correspond, respectively, to $u = 1$ and $u = \infty$. ◄ Making these substitutions in I_1 and I_2 gives two *new* improper integrals:

$$J_1 = \int_1^\infty 1 \, du; \quad J_2 = \int_1^\infty \frac{\sqrt{u}}{u^2} \, du = \int_1^\infty \frac{du}{u^{3/2}}.$$

It's easy to check by direct calculation that I_1 and J_1 diverge to infinity, whereas $I_2 = J_2 = 2$. ■

BASIC EXERCISES

In Exercises 1–4, explain why the integral is improper.

1. $\int_0^\infty x^2 e^{-x^2}\, dx$

2. $\int_0^1 \dfrac{dx}{x^2 - 3x + 2}$

3. $\int_1^4 \dfrac{dx}{x^2 \ln x}$

4. $\int_0^{2\pi} \dfrac{\cos x}{\sqrt{1 + \cos x}}\, dx$

5. Show that $\int_0^1 \dfrac{dx}{\sqrt{x}} = 2$ as claimed in Example 7.

6. Show that $\int_1^\infty \dfrac{du}{u^{3/2}} = 2$ as claimed in Example 7.

In Exercises 7–18, use an antiderivative to evaluate the improper integral.

7. $\int_1^\infty \dfrac{dx}{x^3}$

8. $\int_0^4 \dfrac{dx}{\sqrt{x}}$

9. $\int_1^\infty \dfrac{\ln x}{x^2}\, dx$

10. $\int_e^\infty \dfrac{dx}{x\,(\ln x)^2}$

11. $\int_0^\infty e^{-x}\, dx$

12. $\int_0^\infty x e^{-x}\, dx$

13. $\int_{-\infty}^1 e^x\, dx$

14. $\int_{-\infty}^0 x e^x\, dx$

15. $\int_0^{16} \dfrac{dx}{\sqrt[4]{x^3}}$

16. $\int_0^1 \dfrac{x}{\sqrt{1 - x^2}}\, dx$

17. $\int_3^\infty \dfrac{x}{(x^2 - 4)^3}\, dx$

18. $\int_0^\infty \dfrac{\arctan x}{1 + x^2}\, dx$

19. Suppose that $\int_0^\infty f(x)\, dx = 17$. Does $\int_{100}^\infty f(x)\, dx$ converge? Justify your answer.

20. Suppose that f is continuous everywhere and that $\int_2^\infty f(x)\, dx$ converges. Explain why $\int_5^\infty f(x)\, dx$ also converges.

FURTHER EXERCISES

21. Suppose that $0 \le g(x) \le x^{-2}$ for all $x \ge 1$. Give a graphical argument that explains why $\int_1^\infty g(x)\, dx$ converges. [HINT: See Example 1.]

22. Suppose that $h(x) \ge x^{-1}$ for all $x \ge 1$. Give a graphical argument that explains why $\int_1^\infty h(x)\, dx$ diverges. [HINT: See Example 2.]

23. Suppose that f is an odd function and that $\int_0^\infty f(x)\, dx = 17$. Evaluate $\int_{-\infty}^\infty f(x)\, dx$.

24. (a) Show that $\lim\limits_{a \to \infty} \int_{-a}^a x\, dx = 0$.

   (b) Explain why $\int_{-\infty}^\infty x\, dx$ diverges.

25. Let $I = \int_{-1}^1 \dfrac{dx}{x^3}\, dx$

   (a) Explain why I is an improper integral.

   (b) Does I converge? Justify your answer.

26. Does the integral $\int_0^5 \dfrac{dx}{x - 2}$ converge? Justify your answer.

27. (a) Show that $\int_1^\infty \dfrac{\cos(\sqrt{x})}{\sqrt{x}}\, dx$ diverges.

   (b) Does $\int_0^1 \dfrac{\cos(\sqrt{x})}{\sqrt{x}}\, dx$ also diverge? Justify your answer.

28. Let R be the region between the graphs $y = 1/x$ and $y = 1/(x + 1)$ to the right of $x = 1$. Does R have finite area? Justify your answer.

29. Let S be the region between the graphs $y = 1/x$ and $y = 1/\sqrt{x}$ between $x = 0$ and $x = 1$. Does S have finite area? Justify your answer.

30. The equation $y = \sqrt{R^2 - x^2}$ describes a semicircle of radius R. Use the arc length formula to find the circumference of a circle of radius R.

31. Let $f(x) = 1/x$ and let R be the region under the graph of f and above the x-axis for $x \ge 1$ (i.e., the shaded region in Figure 2). Find the volume of the solid of revolution formed when R is rotated about the x-axis.

32. Let $f(x) = 1/x$ and let R be the region under the graph of f and above the x-axis for $x \ge 1$ (i.e., the shaded region in Figure 2). Find the volume of the solid of revolution formed when R is rotated about the y-axis.

In Exercises 33–48, use an antiderivative to determine whether the improper integral converges. If the integral converges, evaluate it.

33. $\int_0^\infty \dfrac{x}{\sqrt{1 + x^2}}\, dx$

34. $\int_1^\infty \dfrac{dx}{x(1 + x)}$

35. $\int_{-2}^2 \dfrac{2x + 1}{\sqrt[3]{x^2 + x - 6}}\, dx$

36. $\int_\pi^\infty e^{-x} \sin x\, dx$

37. $\int_2^4 \dfrac{x}{\sqrt{|x^2 - 9|}}\, dx$

38. $\int_1^3 \dfrac{dx}{\sqrt[3]{x - 2}}$

39. $\int_2^3 \dfrac{x}{\sqrt{3 - x}}\, dx$

40. $\int_0^2 \dfrac{dx}{\sqrt{4 - x^2}}$

41. $\int_1^\infty \dfrac{dx}{x\,(\ln x)^2}$

42. $\int_0^\infty \dfrac{dx}{(x - 1)^2}$

43. $\int_0^\infty \dfrac{dx}{e^x - 1}$

44. $\int_{-\infty}^\infty e^{-x}\, dx$

45. $\int_{-\infty}^\infty \dfrac{dx}{e^x + e^{-x}}$

46. $\int_0^1 \dfrac{e^{-\sqrt{x}}}{\sqrt{x}}\, dx$

47. $\int_0^{\pi/2} \dfrac{\cos x}{\sqrt{\sin x}}\, dx$

48. $\int_0^{\pi/2} \sec^2 x\, dx$

In Exercises 49–54, find all values of the parameter p for which the integral converges.

49. $\int_1^\infty \dfrac{dx}{x^p}$ [HINT: Consider the cases $p > 1$ and $p \le 1$.]

50. $\int_0^1 \dfrac{dx}{x^p}$

51. $\int_1^e \dfrac{dx}{x(\ln x)^p}$

52. $\int_e^\infty \dfrac{dx}{x(\ln x)^p}$

53. $\int_1^\infty x^p e^x \, dx$

54. $\int_1^\infty x^p e^{-x} \, dx$

55. (a) Explain why $\int_0^1 \dfrac{\cos x}{x} \, dx$ is an improper integral.

(b) Explain why $\int_0^1 \dfrac{\sin x}{x} \, dx$ is *not* an improper integral. [HINT: Use l'Hôpital's rule.]

56. Let $f(x) = \begin{cases} x \ln x & \text{if } x > 0 \\ 0 & \text{if } x = 0. \end{cases}$ Is $\int_0^1 f(x) \, dx$ an improper integral? Justify your answer. [HINT: Use l'Hôpital's rule.]

57. Let $f(x) = x / \ln x$. Is $\int_0^{1/e} f(x) \, dx$ an improper integral? Justify your answer.

58. Suppose that f is a differentiable function, that f' is continuous on $[1, \infty)$, and that $|f(x)| \le e^{-x} \ln x$ if $x \ge 1$. Show that $\int_1^\infty f'(x) \, dx$ converges.

In Exercises 59 and 60, find the value(s) of the parameter C for which the improper integral converges.

59. $\int_0^\infty \left(\dfrac{2x}{x^2+1} - \dfrac{C}{2x+1} \right) dx$

60. $\int_1^\infty \left(\dfrac{Cx}{x^2+1} - \dfrac{1}{2x} \right) dx$

In Exercises 61–66, use a change of variables to transform the improper integral into a proper integral with the same value. (You do not need to evaluate these integrals.)

61. $\int_1^\infty \dfrac{x}{x^3+1} \, dx$ [HINT: Let $u = x^{-1}$.]

62. $\int_0^{\pi/2} \dfrac{\cos x}{\sqrt{\pi - 2x}} \, dx$ [HINT: Let $u = \sqrt{\pi - 2x}$.]

63. $\int_1^\infty \dfrac{dx}{1+x^4}$

64. $\int_0^\infty x^3 e^{-x} \, dx$ [HINT: Let $u = e^{-x}$.]

65. $\int_0^\infty \dfrac{x \ln x}{1+x^4} \, dx$

66. $\int_{\pi/4}^{\pi/2} \sqrt{1 + \tan x} \, dx$
[HINT: Let $u = \tan x = 1/x^2$.]

67. Show that $\int_0^1 \dfrac{dx}{\sqrt{x + x^3}} = \int_1^\infty \dfrac{dx}{\sqrt{x + x^3}}$.

10.2 DETECTING CONVERGENCE, ESTIMATING LIMITS

The preceding section was about the *definitions* of convergence and divergence for improper integrals. An integral of the form $\int_1^\infty f(x) \, dx$, for example, converges to a number L if

$$\lim_{t \to \infty} \int_1^t f(x) \, dx = L;$$

the integral diverges if the limit does not exist.

Determining whether such limits exist, and finding those that do, wasn't hard for the relatively simple integrals in Section 10.1. In more challenging cases, things may go less well. Finding a symbolic expression for $\int_1^t f(x) \, dx$ may be difficult; even if such an expression *is* found, deciding whether the limit exists—let along finding it—may be hard.

All is not lost. As we'll see, we can often determine convergence or divergence even for integrals we can't evaluate exactly. We'll see, too, how to use numerical methods to help us *estimate* values of such integrals.

Detecting convergence or divergence by comparison

Our main tool for deciding whether a given improper integral converges or diverges is called **comparison**. The underlying idea—that larger integrands give larger integrals—could hardly be simpler. It's quite clear, for instance, that if $0 \le f(x) \le g(x)$ for all x in a *finite* interval $[a, b]$, then

$$0 \le \int_a^b f(x) \, dx \le \int_a^b g(x) \, dx.$$

The following theorem says much the same thing for *improper* integrals:

THEOREM 1 (Comparison for nonnegative improper integrals) Let f and g be continuous functions. Suppose that for all $x \geq a$,

$$0 \leq f(x) \leq g(x).$$

- If $\displaystyle\int_a^\infty g(x)\,dx$ *converges*, then so does $\displaystyle\int_a^\infty f(x)\,dx$, and

$$\int_a^\infty f(x)\,dx \leq \int_a^\infty g(x)\,dx.$$

- If $\displaystyle\int_a^\infty f(x)\,dx$ *diverges*, then so does $\displaystyle\int_a^\infty g(x)\,dx.$

(A similar result holds for improper integrals with unbounded integrands.) Figure 1 illustrates both claims.

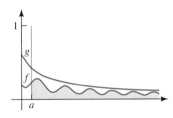

FIGURE 1
The comparison test, graphically

The graphs make both of the theorem's claims easy to believe:

- If the area under the g-graph is *finite*, then so is the (shaded) area under the "lower" f-graph, and the shaded area is less.
- If the area under the f-graph is *infinite*, then so is the area under the "higher" g-graph.

A formal proof (which we omit) makes these ideas precise.

Using the theorem   Using Theorem 1 successfully means finding appropriate "known" integrals with which to compare "unknown" integrals. Finding the "right" comparison may be the main problem.

EXAMPLE 1   Does $I = \displaystyle\int_1^\infty \frac{dx}{x^5+1}$ converge? If so, can anything be said about its limit?

Solution   By definition,

$$I = \lim_{t \to \infty} \left(\int_1^t \frac{dx}{x^5+1} \right),$$

if the limit exists. Unfortunately, the given integrand has a messy antiderivative. Even if we wrote it out, finding a limit (if one exists) would be hard.

Theorem 1 is just what we need; we'll compare I to something simpler:

$$\int_1^\infty \frac{dx}{x^5+1} \quad \text{vs.} \quad \int_1^\infty \frac{dx}{x^5}.$$

We can handle the second integral directly:

$$\lim_{t \to \infty} \int_1^t \frac{dx}{x^5} = \lim_{t \to \infty} \left(\frac{1}{4} - \frac{1}{4t^4} \right) = \frac{1}{4}.$$

Figure 2 shows both integrands.

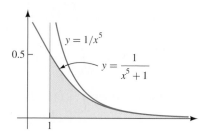

FIGURE 2

Comparing two improper integrals

The picture (or basic algebra) shows that

$$\frac{1}{x^5 + 1} < \frac{1}{x^5}$$

These are the values of x we care about.

for all $x \geq 1$. ◂ Now Theorem 1 applies:

$$I = \int_1^\infty \frac{dx}{x^5 + 1} < \int_1^\infty \frac{dx}{x^5} = \frac{1}{4},$$

and so we conclude that I converges to *some* limit between 0 and 1/4. (We'll estimate the limit more precisely soon.) ∎

Compared to what? Integrals for reference   Using Theorem 1 successfully (as we just did in Example 1) requires some "benchmark" integrals against which to compare. A few important examples (and their close relatives, such as constant multiples) go a long way:

$$\int_0^\infty e^{-x}\, dx = 1; \qquad\qquad \int_1^\infty \frac{dx}{x} \quad \text{diverges}$$

$$\int_1^\infty \frac{dx}{x^p} = \frac{1}{p - 1} \quad \text{(if } p > 1\text{)}; \qquad \int_1^\infty \frac{dx}{x^p} \quad \text{diverges if } p \leq 1$$

We've done some already in examples and exercises.

All four claims are easily checked using the definition of convergence. ◂ The last line is sometimes summarized as the **p-test** for integrals:

$$\int_1^\infty \frac{dx}{x^p} \quad \text{converges if } p > 1 \text{ and diverges otherwise.}$$

Manipulating improper integrals   Convergent improper integrals behave just like ordinary integrals with respect to sums and constant multiples. For instance, if both $\int_1^\infty f(x)\, dx$ and $\int_1^\infty g(x)\, dx$ are convergent integrals, then we can write equations like these:

$$\int_1^\infty \big(3f(x) - 2g(x) \big)\, dx = 3 \int_1^\infty f(x)\, dx - 2 \int_1^\infty g(x)\, dx;$$

$$\int_1^\infty f(x)\, dx = \int_1^{10} f(x)\, dx + \int_{10}^\infty f(x)\, dx.$$

In particular, all of the integrals just mentioned converge.

EXAMPLE 2 Use comparison to decide whether each of the following integrals converges or diverges:

$$\int_0^\infty \frac{dx}{e^x + x}; \qquad \int_{100}^\infty \frac{dx}{\sqrt{x}}; \qquad \int_{100}^\infty \frac{dx}{10\sqrt{x} - 7}.$$

If an integral converges, find an upper bound for its value.

Solution The first integral resembles $\int_0^\infty e^{-x}\,dx$; indeed, the comparison works well:

$$\frac{1}{e^x + x} \le \frac{1}{e^x} \quad \text{if } x \ge 0, \text{ so} \quad \int_0^\infty \frac{1}{e^x + x} \le \int_0^\infty e^{-x} = 1.$$

The second integral resembles another of our benchmarks; thus, we might guess—correctly—that it diverges. Because

$$\int_{100}^\infty \frac{dx}{\sqrt{x}} = \int_1^\infty \frac{dx}{\sqrt{x}} - \int_1^{100} \frac{dx}{\sqrt{x}}$$

and the right side diverges, the left side must also diverge. ➡

The right side is the sum of a divergent integral and a finite number.

The third integral diverges too, and a little algebra with inequalities shows why. Notice first that

$$\frac{1}{10\sqrt{x} - 7} > \frac{1}{10\sqrt{x}}$$

for $x \ge 100$. Since

$$\int_{100}^\infty \frac{dx}{10\sqrt{x}} = \frac{1}{10} \int_{100}^\infty \frac{dx}{\sqrt{x}}$$

diverges, Theorem 1 implies that the third integral must diverge too. ■

Integrands that change sign Does the integral

$$I = \int_1^\infty \frac{\sin x}{x^2}\,dx$$

converge or diverge? Theorem 1 doesn't say—it applies only to *nonnegative* integrands. On the other hand, because

$$\left| \frac{\sin x}{x^2} \right| \le \frac{1}{x^2}$$

for all $x \ge 1$, it follows that

$$\int_1^\infty \left| \frac{\sin x}{x^2} \right| dx \le \int_1^\infty \frac{dx}{x^2} = 1.$$

We would therefore expect that I itself converges, and that

$$|I| = \left| \int_1^\infty \frac{\sin x}{x^2}\,dx \right| \le \int_1^\infty \left| \frac{\sin x}{x^2} \right| dx \le 1.$$

The following theorem justifies these speculations. ➡

We'll see a similar theorem in Chapter 11, where we study infinite series.

> **THEOREM 2 (Absolute comparison for improper integrals)**   Suppose that $\int_a^\infty |f(x)|\,dx$ converges. Then $\int_a^\infty f(x)\,dx$ also converges, and
>
> $$\left| \int_a^\infty f(x)\,dx \right| \leq \int_a^\infty |f(x)|\,dx.$$

Absolute convergence   An improper integral $\int_a^\infty f(x)\,dx$ is called **absolutely convergent** if $\int_a^\infty |f(x)|\,dx$ converges, as in the hypothesis of Theorem 2. The theorem guarantees, in effect, that if an improper integral converges *with* absolute value signs, then it also converges *without* absolute value signs. ◄

There are integrals—seldom encountered in calculus courses—that diverge with absolute values but converge without them.

Tails and estimation

Any improper integral of the form $I = \int_1^\infty f(x)\,dx$ can be rewritten as a sum like the following: ◄

We could have broken the integral elsewhere than at $x = 1000$.

$$\int_1^\infty f(x)\,dx = \int_1^{1000} f(x)\,dx + \int_{1000}^\infty f(x)\,dx.$$

The first summand on the right is an *ordinary* definite integral; we can apply either antidifferentiation or numerical methods to find or estimate its value—without any worries about convergence or divergence. The second integral on the right is called, picturesquely, an **upper tail** (or just a **tail**) of I.

Convergent integrals have skinny tails   If $I = \int_1^\infty f(x)\,dx$ converges to a number L, then we must have

$$\int_1^a f(x)\,dx \approx \int_1^\infty f(x)\,dx = L$$

for large positive a. This means, in turn, that the upper tail $\int_a^\infty f(x)\,dx$ must be *near zero* for sufficiently large a.

EXAMPLE 3   Both $I = \int_0^\infty e^{-x}\,dx$ and $J = \int_1^\infty \dfrac{\sin x}{x^2}\,dx$ converge. For each integral, find an upper tail with absolute value less than 0.001. What do the answers mean about the values of I and J?

Solution   For each integral we need a number a such that

$$\left| \int_a^\infty f(x)\,dx \right| < 0.001.$$

Handling I is relatively easy. For any real number a we have

$$\int_a^\infty e^{-x}\,dx = \lim_{t \to \infty} -e^{-x} \Big]_a^t = e^{-a} = \frac{1}{e^a}.$$

Thus,

$$\int_a^\infty e^{-x}\,dx = \frac{1}{e^a} < 0.001 \iff e^a > 1000 \iff a > \ln 1000 \approx 6.908.$$

Handling J is a little trickier. Because

$$\left| \frac{\sin x}{x^2} \right| \leq \frac{1}{x^2}$$

for all x, Theorems 1 and 2 give

$$\left| \int_a^\infty \frac{\sin x}{x^2}\, dx \right| \le \int_a^\infty \frac{1}{x^2}\, dx = \frac{1}{a}.$$

Finally, $1/a < 0.001$ if $a > 1000$.

The results mean that the ordinary (proper) integrals

$$I_1 = \int_0^7 e^{-x}\, dx \quad \text{and} \quad J_1 = \int_1^{1000} \frac{\sin x}{x^2}\, dx$$

differ from I and J, respectively, by less than 0.001.   ∎

Approximating improper integrals numerically   The result of Example 1—that a certain integral converges to *some* limit less than 1/4—is a bit unsatisfying. Can we say more? The next example illustrates that we can indeed, by combining skinny tails with numerical estimates.

EXAMPLE 4   Estimate the value of $I = \int_1^\infty \dfrac{dx}{x^5+1}$. How good is the estimate?

Solution   Symbolic methods aren't promising (as we said in Example 1), and so we'll take a numerical approach. Because numerical integration methods apply to *proper* integrals, we first break I into a finite piece I_1 and a tail I_2 (which we hope is small): ➤

We could have amputated the tail elsewhere.

$$I = \int_1^\infty \frac{dx}{x^5+1} = \int_1^5 \frac{dx}{x^5+1} + \int_5^\infty \frac{dx}{x^5+1} = I_1 + I_2.$$

To estimate I_1 we can use any numerical method, such as the midpoint rule with 50 subdivisions:

$$I_1 = \int_1^5 \frac{dx}{x^5+1} \approx M_{50} \approx 0.1799.$$

Assuming (for the moment) that the tail integral I_2 is near zero, we have our desired estimate:

$$I = I_1 + I_2 \approx 0.1799 + 0 = 0.1799.$$

The result is (as expected) a positive number less than 1/4 (the bound on I we found in Example 1), but how *good* is this estimate? Note that error can arise from two different sources:

 (i) the error M_{50} commits in estimating I_1;

 (ii) the error due to ignoring the tail I_2.

We discussed the first type of error in Chapter 6; using methods discussed there we can show that this error cannot exceed 0.003. ➤

We omit the details; the idea is the point.

To gauge the second type of error (the process is called **bounding the tail**) we can use Theorem 1—which applies as readily to upper tail integrals as to any others. Because the inequality

$$\frac{1}{x^5+1} < \frac{1}{x^5}$$

holds for $x \ge 5$, Theorem 1 and an easy calculation give

$$I_2 = \int_5^\infty \frac{dx}{x^5+1} < \int_5^\infty \frac{dx}{x^5} = \frac{1}{2500} = 0.0004.$$

Having bounded *both* types of possible error, we conclude that the estimate $I \approx 0.1799$ commits *total* error less than $0.003 + 0.0004 = 0.0034$.   ∎

We assemble several pieces in our last example.

EXAMPLE 5 Show that $I = \int_1^\infty e^{-x^2}\,dx$ converges and estimate its limit. How large could the error be?

Solution Because the integrand doesn't have an elementary antiderivative, we'll compare and estimate. First, I converges by comparison to one of our benchmark integrals. Since

$$e^{-x^2} \le e^{-x} \quad \text{for } x \ge 1,$$

we have

$$\int_1^\infty e^{-x^2}\,dx \le \int_1^\infty e^{-x}\,dx \le \int_0^\infty e^{-x}\,dx = 1.$$

Figure 3 shows both integrands.

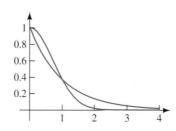

FIGURE 3
Comparing $y = e^{-x}$ and $y = e^{-x^2}$

To estimate I we'll break it into a finite piece and a tail:

$$I = \int_1^5 e^{-x^2}\,dx + \int_5^\infty e^{-x^2}\,dx = I_1 + I_2.$$

To estimate I_1 we can use the midpoint rule; $M_{100} \approx 0.1394$. The midpoint-rule error bound formula shows that M_{100} commits error less than 0.00054 in estimating I_1.

We can use the same comparison as before (this time with $a = 5$) to bound the tail integral I_2:

$$I_2 = \int_5^\infty e^{-x^2}\,dx < \int_5^\infty e^{-x}\,dx = e^{-5} \approx 0.0067.$$

Thus, the total error from *both* sources (numerical integration and ignoring the tail) in the estimate $I \approx 0.1394$ is well under 0.01. ■

BASIC EXERCISES

1. Suppose that $1 < f(x) < g(x)$ for all $x \ge 0$.
   (a) Rank the three values $1, 1/f(x)$, and $1/g(x)$ in increasing order.
   (b) Suppose that $r \ge 1$. Rank the three values $1, 1/\big(f(x)\big)^r$, and $1/\big(g(x)\big)^r$.
   (c) Suppose that $0 < r < 1$. Rank the three values 1, $1/\big(f(x)\big)^r$, and $1/\big(g(x)\big)^r$.

2. Suppose that $0 < f(x) < g(x) < 1$ for all $x \ge 0$.
   (a) Rank the three values $1, 1/f(x)$, and $1/g(x)$ in increasing order.

   (b) Suppose that $r \ge 1$. Rank the three values $1, 1/\big(f(x)\big)^r$, and $1/\big(g(x)\big)^r$.
   (c) Suppose that $0 < r < 1$. Rank the three values 1, $1/\big(f(x)\big)^r$, and $1/\big(g(x)\big)^r$.

3. Let $f(x) = \dfrac{x^2 + 2}{x^2 + 1}$.
   (a) Show that $f(x) > 1$ for all x.
   (b) Use the inequality in part (a) to show that $\int_{-\infty}^\infty f(x)\,dx$ diverges.

4. Let $f(x) = \dfrac{e^x}{1+e^x}$.

   (a) Show that $f(x) > 1/2$ for all $x \geq 0$.
[HINT: $e^x \geq 1$ for all $x \geq 0$.]

   (b) Use the inequality in part (a) to show that $\displaystyle\int_{-\infty}^{\infty} f(x)\,dx$ diverges.

5. Let $f(x) = x + \sin x$.

   (a) Show that $x - 1 \leq f(x)$ for all $x \geq 0$.

   (b) Use the inequality in part (a) to show that $\displaystyle\int_{2}^{\infty} \dfrac{dx}{f(x)}$ diverges.

6. Let $f(x) = x^2 + \sin x$.

   (a) Show that $x^2 - 1 \leq f(x)$ for all $x \geq 0$.

   (b) Does $\displaystyle\int_{2}^{\infty} \dfrac{dx}{f(x)}$ converge? Justify your answer.

7. Let $f(x) = x^2 + \sqrt{x}$.

   (a) Show that $x^2 \leq f(x)$ for all $x \geq 0$.

   (b) Use the inequality in part (a) to show that $\displaystyle\int_{1}^{\infty} \dfrac{dx}{f(x)}$ converges.

8. Let $f(x) = x^2 - \sqrt{x}$.

   (a) Show that $x^2/2 \leq f(x)$ for all $x \geq 2$.

   (b) Does $\displaystyle\int_{3}^{\infty} \dfrac{dx}{f(x)}$ converge? Justify your answer.

9. Does $\displaystyle\int_{2}^{\infty} \dfrac{dx}{x - \sqrt{x}}$ converge? Justify your answer.

10. Does $\displaystyle\int_{1}^{\infty} \dfrac{dx}{x + \sqrt{x}}$ converge? Justify your answer.

In Exercises 11–14, find a value for the parameter a that makes the value of the improper integral less than 10^{-5}.

11. $\displaystyle\int_{a}^{\infty} e^{-x}\,dx$ **12.** $\displaystyle\int_{a}^{\infty} xe^{-x^2}\,dx$

13. $\displaystyle\int_{a}^{\infty} \dfrac{dx}{x^2+1}$ **14.** $\displaystyle\int_{a}^{\infty} \dfrac{dx}{x(\ln x)^3}$

In Exercises 15–18, find a proper integral whose value approximates that of the given (convergent) improper integral within 10^{-5}.

15. $\displaystyle\int_{0}^{\infty} \dfrac{dx}{x^2+e^x}$ **16.** $\displaystyle\int_{1}^{\infty} \dfrac{dx}{x^4\sqrt{2x^3+1}}$

17. $\displaystyle\int_{0}^{\infty} \dfrac{\arctan x}{(1+x^2)^3}\,dx$ **18.** $\displaystyle\int_{0}^{\infty} \dfrac{e^{-x}}{2+\cos x}\,dx$

19. Suppose that $f(x) \geq 0$ for all $x \geq 1$ and that $\displaystyle\int_{1}^{\infty} f(x)\,dx$ converges. Explain why there is a number a such that $\displaystyle\int_{a}^{\infty} f(x)\,dx \leq 10^{-10}$.

20. Suppose that $g(x) \geq 0$ for all $x \geq 1$ and that $\displaystyle\int_{1}^{\infty} g(x)\,dx$ diverges. Explain why there is a number b such that $\displaystyle\int_{1}^{b} g(x)\,dx \geq 10^{10}$.

21. Suppose that $I = \displaystyle\int_{0}^{\infty} f(x)\,dx$ converges and that $\left|\displaystyle\int_{a}^{\infty} f(x)\,dx\right| \leq 0.0001$. Show that $\left|I - \displaystyle\int_{0}^{a} f(x)\,dx\right| \leq 0.0001$.

22. Show that $0.8 < \displaystyle\int_{0}^{\infty} \sin\left(e^{-x}\right)\,dx < 1$.
[HINT: Use the substitution $u = e^{-x}$.]

FURTHER EXERCISES

23. Show that $\displaystyle\int_{0}^{\infty} e^{-x^2}\,dx$ converges. [HINT: See Example 5 but be careful; $e^{-x^2} > e^{-x}$ for some $x \geq 0$.]

24. Consider the integral $I = \displaystyle\int_{2}^{\infty} \dfrac{dx}{(\ln x)^2}$.

   (a) Show that $\ln x \leq \sqrt{x}$ for all $x \geq 2$.

   (b) Does I converge? Justify your answer.

25. Suppose that $p \geq 1$ is a constant and that $I = \displaystyle\int_{2}^{\infty} \dfrac{dx}{(\ln x)^p}$. For which values of p, if any, does I converge? Justify your answer. [HINT: $\displaystyle\lim_{x \to \infty} \dfrac{(\ln x)^p}{x} = 0$.]

26. (a) Show that $0 \leq x/2 \leq \sin x$ if $0 \leq x \leq 1$.
[HINT: $1/2 \leq \cos x$ if $0 \leq x \leq 1$.]

   (b) Use the inequality in part (a) to show that $\displaystyle\int_{0}^{1} \dfrac{dx}{\sqrt{\sin x}}$ converges.

In Exercises 27–30, use the comparison test to determine whether the improper integral converges or diverges.

27. $\displaystyle\int_{0}^{\infty} \dfrac{dx}{x^4+x}$ **28.** $\displaystyle\int_{1}^{\infty} \dfrac{dx}{\sqrt{x-1}}$

29. $\displaystyle\int_{1}^{\infty} \dfrac{dx}{\sqrt[3]{x^6+x}}$ **30.** $\displaystyle\int_{0}^{\infty} \dfrac{dx}{\sqrt{x}(1+x)}$

31. Does $\displaystyle\int_{0}^{\infty} \dfrac{e^{-x}}{\sqrt{x}}\,dx$ converge? Justify your answer.

32. Show that the integral $\displaystyle\int_{0}^{1} \dfrac{\cos x}{x}\,dx$ diverges.

33. (a) Show that $\displaystyle\int_{1}^{\infty} \dfrac{\cos x}{x^2}\,dx$ converges.

   (b) Use integration by parts and part (a) to show that $\displaystyle\int_{1}^{\infty} \dfrac{\sin x}{x}\,dx$ converges.

34. Let $I = \displaystyle\int_{0}^{\infty} \dfrac{dx}{\sqrt{x+x^4}}$. Show that $0 \leq I \leq 3$.

35. Evaluate $\displaystyle\lim_{x \to \infty} \dfrac{\displaystyle\int_{1}^{x} \sqrt{1+e^{-3t}}\,dt}{x}$.

36. Evaluate $\displaystyle\lim_{x \to \infty} e^{x^2} \displaystyle\int_{0}^{x} e^{-t^2}\,dt$.

10.3 IMPROPER INTEGRALS AND PROBABILITY

Many real-world phenomena exhibit forms of random behavior. The number of heads in 10 throws of a coin, the height of a male college student chosen at random, the actual weight of a nominally 16-oz cereal package, and the working life of a 60-watt light bulb are all examples.

Statisticians call such quantities **random variables**. A random variable that can take only *finitely* many values (such as the number of heads in 10 coin tosses) is called **discrete**. A random variable that ranges over an *interval* of real numbers is called **continuous**. Human adult height is a good example: Any number in some interval—(36, 85), say—may conceivably be someone's height in inches. We consider only continuous random variables, even though calculus methods can also be applied in the discrete case.

Random phenomena are, by nature, not perfectly predictable. But one might reasonably want to know the probability that a male college student is between 68 and 75 inches tall or the likelihood of getting less than 15 ounces of Cheerios in a package. Doing so turns out to involve integrals—many of them improper.

Basics of continuous probability

By convention, probabilities are measured on a scale from 0 to 1. An event that is certain to occur has probability 1. (It's certain, for example, that a male college student's height in inches is *some* positive number.) Because a continuous random variable X has infinitely many possible values, it's not interesting to ask for the probability that X is *precisely* equal to some number—that probability is zero. A better question concerns the probability that X lies in some *interval*, such as (68, 75). If this probability is, say, 0.78, then 78% of male college students are between 68 and 75 inches tall.

Associated with every continuous random variable X is a function f called the **probability density function**, or **pdf**, of X. The connection between f and X is given by an integral:

$$\int_a^b f(x)\,dx = \text{probability that } X \text{ lies in } (a, b).$$

The shaded region in Figure 1 represents this probability geometrically.

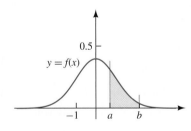

FIGURE 1
Probability that $a < X < b$

Here are some commonsense facts about f and X; some are reflected in Figure 1:

- **Domain** The domain of f is the set of possible values of X. (Often f has an infinite domain—hence the need for improper integrals.)

- **Near, not exact** For a given x, $f(x)$ measures the comparative likelihood that the value of X falls *near* x. The probability that X takes any *single* value x is zero.

- **Certainty**   Because X must take *some* real value, $\int_{-\infty}^{\infty} f(x)\,dx = 1$. (The graph in Figure 1 bounds total area 1.)
- **A positive function**   For all inputs x, $f(x) \geq 0$ (as in Figure 1).

The normal density function

Many real-life random phenomena can be modeled by a **normal** (also called **Gaussian normal**) density function—one of the following form: ➡

Recall: $\exp(\text{stuff}) = e^{\text{stuff}}$.

$$f(x) = \frac{1}{\sqrt{2\pi}\, s} \exp\left(-\frac{(x-m)^2}{2s^2}\right).$$

A random variable whose probability density function has this form is said to be **normally distributed**.

The normal density function looks messy, but in practice it's not too bad. The letters s and m stand for constant parameters called, respectively, the **standard deviation** and the **mean** of the random variable. (Statisticians usually write σ and μ, but we'll stick with English.) Their numerical values depend on the situation being modeled. The mean m, as its name suggests, is a kind of average value of X. The standard deviation s is a measure of how "spread out" X's values are. The graphs in Figure 2 illustrate the geometric effects of varying m and s; all four graphs—versions of the famous "bell curve"—bound the same total (unit) area.

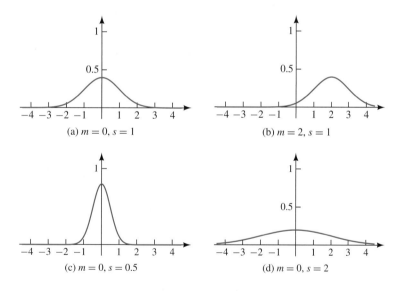

(a) $m = 0$, $s = 1$

(b) $m = 2$, $s = 1$

(c) $m = 0$, $s = 0.5$

(d) $m = 0$, $s = 2$

FIGURE 2
Four normal density functions

Integrating the normal density—properly and improperly

Using the normal density function

$$f(x) = \frac{1}{\sqrt{2\pi}\, s} \exp\left(-\frac{(x-m)^2}{2s^2}\right)$$

requires integrating it over various intervals, either finite or infinite. Because the normal density function has no elementary antiderivative, such integrals *must* be estimated numerically. The following properties of f help simplify the process.

- **Convergence** For any values of m and s, the improper integral $\int_{-\infty}^{\infty} f(x)\,dx$ converges, and its limit is 1. Showing that the integral converges is not too difficult; we showed something similar in Example 5 of Section 10.2 (page 536). Estimating a limit is easy, but showing rigorously that it is *precisely* 1 is considerably harder; we omit the proof.

- **Symmetry** For any values of m and s, the graph of f is "centered" on the line $x = m$; the graph bounds total area 0.5 to each side of this line. This fact can simplify area computations and improve the accuracy of numerical estimates.

EXAMPLE 1 Suppose that the birth weight (in pounds) of a certain population of infants is normally distributed with mean 7.5 lb and standard deviation 1 lb. What percentage of infants can be expected to weigh between 6.5 and 8.5 lb? Over 10 lb?

Solution Both answers are areas bounded by the normal density function with parameter values $m = 7.5$ and $s = 1$. Both areas are shown shaded in Figure 3.

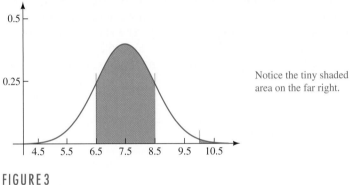

Notice the tiny shaded area on the far right.

FIGURE 3
Normal density, $m = 7.5$, $s = 1$

The answers, written as integrals, are

$$I_1 = \int_{6.5}^{8.5} f(x)\,dx \quad \text{and} \quad I_2 = \int_{10}^{\infty} f(x)\,dx.$$

Applying the midpoint rule to I_1 gives $M_{20} \approx 0.6829$. Thus, given our assumptions, about 68% of babies should weigh within one pound of the mean.

Figure 3 suggests that I_2 is small. To estimate it numerically, we use the fact that the total area to the right of $x = 7.5$ is (exactly!) 0.5. Therefore,

$$I_2 = \int_{10}^{\infty} f(x)\,dx = 0.5 - \int_{7.5}^{10} f(x)\,dx.$$

The last integral can be estimated numerically. The midpoint rule gives $M_{20} \approx 0.4938$, and so $I_2 \approx 0.0062$. Therefore, only around 0.6% of babies can be expected to weigh over 10 lb. ■

Is life really normal? Are real-life phenomena—baby weights, SAT scores, heights of male adults, and so on—*really* normally distributed?

In one sense, certainly not. One obvious problem is that every normal distribution is infinitely "spread out": The variable can take *any* real value—large or small, positive or negative. Few real-life phenomena behave this way. No baby has negative weight, for instance; yet, every normal distribution assigns some tiny but positive probability to this peculiar event.

In another, more practical sense, normal distributions do usefully model many phenomena. "Model" is the key word—in suitable settings the normal density function provides an effective, if imperfect, *approximation* to an underlying reality. Imperfection is no surprise; it comes with the territory of real life. Effectiveness is the real surprise.

EXAMPLE 2 Possible scores on the math and verbal parts of the Scholastic Aptitude Test (SAT) range from 200 to 800. Assume that test scores are normally distributed with mean 500 and standard deviation 100. ➤ What percentage of students score between 400 and 600? Over 750?

Historically, this assumption is not far off.

Solution The situation—and the picture—closely resemble those of Example 1. Setting $m = 500$ and $s = 100$ gives Figure 4; areas of interest are shaded.

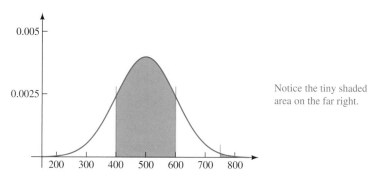

Notice the tiny shaded area on the far right.

FIGURE 4
Normal density, $m = 500$, $s = 100$

The shaded areas can be estimated numerically. We omit details this time because, for reasons we'll soon see, the shaded regions here have exactly the same areas as those in Example 1. We therefore conclude that about 68% of SAT scores fall between 400 and 600, that is, within one standard deviation of the mean; only 0.6% of scores are above 750. ■

Keeping it simple: The standard normal density

Calculations are simplest if $m = 0$ and $s = 1$. With these parameter values, the **standard normal density** function takes a particularly pleasant form

$$n(x) = \frac{1}{\sqrt{2\pi}} \exp\left(\frac{-x^2}{2}\right).$$

In fact, any normal distribution can easily be "standardized"—that is, reinterpreted as a standard normal distribution. In the SAT case, for instance, we could consider how much scores differ from 500 rather than "raw" test scores themselves. Doing so moves the mean to 0; the standard deviation remains at 100. Dividing *these* results by 100 leaves the mean unchanged at 0 but cuts the standard deviation to 1. The results therefore have a *standard normal distribution*.

Statisticians call such altered data **Z-scores**. For example, raw SAT scores of 620 and 450 correspond to Z-scores of 1.20 and −0.5, that is, to scores 1.2 standard deviations above and 0.5 standard deviation below the mean, respectively. Similarly, a 10-lb baby from the population of Example 1 weighs in at 2.5 standard deviations above the mean.

To find the probability of a Z-score between -1 and 1, we integrate the standard normal density:

$$\int_{-1}^{1} \frac{1}{\sqrt{2\pi}} \exp\left(\frac{-x^2}{2}\right) dx.$$

No symbolic antiderivative is available, but numerical integration methods work fine. The midpoint rule with 20 subdivisions gives $M_{20} \approx 0.6829$. This by-now-familiar result offers evidence that Z-scores make mathematical sense.

Why Z-scores work: Substitution in definite integrals   What statisticians call Z-scores *we* call Z-substitution in a definite integral. ◄ We illustrate with an example.

Also known as u-substitution.

EXAMPLE 3   Write integrals, with and without Z-scores, for the probability of an SAT score between 500 and 700. Why do both integrals give the same result?

Solution   With raw scores we use $m = 500$ and $s = 100$; the desired probability is

$$I_1 = \frac{1}{\sqrt{2\pi} \cdot 100} \int_{500}^{700} \exp\left(-\frac{(x - 500)^2}{2 \cdot 100^2}\right) dx.$$

Raw scores of 500 and 700 give Z-scores of 0 and 2.

With Z-scores we want the probability of a result between 0 and 2. ◄ This is the integral

$$I_2 = \frac{1}{\sqrt{2\pi}} \int_0^2 \exp\left(-\frac{x^2}{2}\right) dx.$$

Numerical evidence that $I_1 = I_2$ is easily found: For both integrals, $M_{20} \approx 0.47729$. A symbolic calculation confirms the result. Substituting

$$Z = \frac{x - 500}{100} \quad \text{and} \quad dZ = \frac{dx}{100}$$

in I_1 yields I_2.   ∎

Z-score tables   In olden days (before computers) it was difficult to estimate the definite integrals that arise in the study of normal distributions. The only recourse was to numerical tables; part of one such table follows. For a given input x, the table entry is $A(x) = \int_{-\infty}^{x} n(t)\, dt$, the area shown graphically in Figure 5.

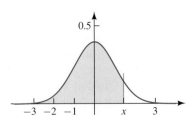

FIGURE 5
Standard normal density; $A(x) = \int_{-\infty}^{x} n(t)\, dt$

Values of $A(x) = \int_{-\infty}^{x} n(t)\,dt$									
x	0.0	0.1	0.2	0.3	0.4	0.5	0.6	0.7	0.8	0.9
$A(x)$	0.5000	0.5398	0.5793	0.6179	0.6554	0.6915	0.7257	0.7580	0.7881	0.8159
x	1.0	1.1	1.2	1.3	1.4	1.5	1.6	1.7	1.8	1.9
$A(x)$	0.8413	0.8643	0.8849	0.9032	0.9192	0.9333	0.9452	0.9554	0.9641	0.9713
x	2.0	2.1	2.2	2.3	2.4	2.5	2.6	2.7	2.8	2.9
$A(x)$	0.9773	0.9822	0.9861	0.9893	0.9918	0.9938	0.9953	0.9965	0.9974	0.9981

The numbers compare well with the graph. As we'd expect, for instance, $A(0) = 0.5$ and $A(2.9) \approx 1.00$.

BASIC EXERCISES

1. Plot the normal probability density function with the given parameter values. On each graph, shade the area corresponding to the probability that the value of a random variable having this distribution will be within one standard deviation of the mean.
   (a) $m = 1, s = 1$
   (b) $m = 2, s = 1/2$
   (c) $m = 0, s = 1/4$

2. (a) Let X be a normally distributed random variable with $m = 0$ and $s = 1$. Explain why the proportion of the possible values of X that lie within three standard deviations of the mean is $\dfrac{1}{\sqrt{2\pi}} \int_{-3}^{3} e^{-x^2/2}\,dx$.

   (b) Let Y be a normally distributed random variable with $m = 0$ and $s = 2$. Write a definite integral whose value is the proportion of the possible values of Y that lie within three standard deviations of the mean.

   (c) Use the substitution $u = x/2$ to show that the integrals in parts (a) and (b) are equal.

   (d) Use numerical integration to estimate the integral in part (a).

3. Let I_1 and I_2 be the integrals discussed in Example 3. Show that I_2 is the result of making the substitution $Z = (x - 500)/100$ in I_1.

4. Use the substitution $z = (x - m)/s$ to show that
$$\frac{1}{\sqrt{2\pi}\,s} \int_{x_1}^{x_2} \exp\left(-\frac{(x-m)^2}{2s^2}\right) dx = \frac{1}{\sqrt{2\pi}} \int_{z_1}^{z_2} e^{-z^2/2}\,dz,$$
where $z_1 = (x_1 - m)/s$ and $z_2 = (x_2 - m)/s$.

5. Let $f(x) = \dfrac{1}{\sqrt{2\pi}\,s} \exp\left(-\dfrac{(x-m)^2}{2s^2}\right)$.

   (a) We showed in Example 5 that $\int_1^\infty e^{-x^2}\,dx$ converges. In fact, the integral $\int_0^\infty e^{-x^2}\,dx$ also converges; its limit can

be shown to be $\sqrt{\pi}/2$. Use these facts to explain why $\int_{-\infty}^{\infty} e^{-x^2}\,dx = \sqrt{\pi}$.

   (b) Use the substitution $u = x/\sqrt{2}$ to show that if $m = 0$ and $s = 1$, then $\int_{-\infty}^{\infty} f(x)\,dx = 1$.

   (c) Use the result in part (b) to show that $\int_{-\infty}^{\infty} f(x)\,dx = 1$ for any real number m and any positive number s. [HINT: Use the substitution $u = (x - m)/s$.]

6. Suppose that an observation of a normally distributed random variable has a Z-score of -1.3.
   (a) Is this observation larger than the mean? Justify your answer.
   (b) How many standard deviations from the mean is this observation?

7. (a) What Z-score corresponds to an SAT score of 600?
   (b) What Z-score corresponds to an SAT score of 450?
   (c) Use part (b) to write an integral whose value is the probability of getting an SAT score greater than 450.
   (d) Use part (a) to write an integral whose value is the probability of getting an SAT score less than 600.
   (e) Use parts (c) and (d) to write an integral whose value is the probability of getting an SAT score between 450 and 600.

8. Let $A(x) = \int_{-\infty}^{x} n(t)\,dt$, where $n(t)$ is the standard normal density function. Explain why $\int_{z_1}^{z_2} n(t)\,dt = A(z_2) - A(z_1)$.

9. Let $A(x) = \int_{-\infty}^{x} n(t)\,dt$, where $n(t)$ is the standard normal density function.
   (a) Explain why $A(-z) = 1 - A(z)$.
   (b) Use part (a) and the table on page 543 to estimate $A(-1.2)$.

In Exercises 10 and 11, use the table on page 543 to compute the probability of getting an SAT score

10. greater than 600.

11. between 350 and 500.

In Exercises 12 and 13, use the table on page 543 to estimate

12. $\dfrac{1}{5\sqrt{2\pi}} \displaystyle\int_{-\infty}^{14} \exp\left(-\dfrac{(x-10)^2}{50}\right)$

13. $\dfrac{1}{5\sqrt{2\pi}} \displaystyle\int_{-\infty}^{4} \exp\left(-\dfrac{(x-10)^2}{50}\right)$

Part of the graph of the probability density function for a random variable X with values in the interval $[0, \infty)$ is shown below. In

Exercises 14–17 use the graph to estimate the probability that a random value of X lies in the interval.

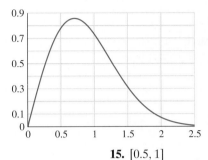

14. $[0, 0.5]$

15. $[0.5, 1]$

16. $[0, 1.5]$

17. $[1.5, \infty)$

FURTHER EXERCISES

18. How high must a student score on the SAT to have a score in the top 10%? In the top 5%? [HINT: Use the tabulated values of $A(x)$ on page 543.]

19. Suppose that the average rainfall in May is 5 inches with a standard deviation of 0.6 inch. What is the probability that the rainfall next May will differ from the average by more than 1 inch? (Assume that May rainfall amounts are normally distributed.)

20. During the grand finale of a circus, a performer is shot from a special cannon. The distance the performer travels varies but is normally distributed with a mean of 150 feet and a standard deviation of 10 feet. The landing net is 30 feet long.

    (a) How far away from the cannon should the nearest edge of the net be placed?

    (b) Given the net position in part (a), what is the probability that the performer will miss the landing net?

21. A stamping machine produces can tops whose diameters are normally distributed with a standard deviation of 0.01 inch. At what "nominal" (i.e., mean) diameter should the machine be set so that no more than 10% of the can tops produced have a diameter greater than 3 inches.

22. Let $A(x) = \displaystyle\int_{-\infty}^{x} n(t)\,dt$, where $n(t)$ is the standard normal probability density function. (See the table on page 543.)

    (a) Since 0.75 lies midway between 0.7 and 0.8, it is reasonable to estimate that $A(0.75) \approx (A(0.7) + A(0.8))/2 = (0.7580 + 0.7881)/2 = 0.77305$. Does this approximation overestimate the exact value? Justify your answer. (Assume that the values of $A(0.7)$ and $A(0.8)$ given in the table are exact.)

    (b) Use a tangent line approximation to estimate $A(0.75)$. [HINT: $A'(x) = n(x)$.]

    (c) Does the estimate you computed in part (b) overestimate the exact value? Justify your answer.

23. The probability density function for the **log-normal** distribution is

$$g(x) = \begin{cases} \dfrac{1}{\sqrt{2\pi}\,\beta x}\, e^{-(\ln x - \alpha)^2/2\beta^2} & \text{if } x > 0, \\ 0 & \text{if } x \le 0. \end{cases}$$

    (a) Explain why $\beta > 0$ must be true for g to be a probability density function.

    (b) Use the substitution $u = (\ln x - \alpha)/\beta$ to show that the probability that $a \le x \le b$ is

$$\int_{a}^{b} g(x)\,dx = A\left(\frac{\ln b - \alpha}{\beta}\right) - A\left(\frac{\ln a - \alpha}{\beta}\right),$$

    where $A(x) = \displaystyle\int_{-\infty}^{x} n(x)\,dx$.

24. Let T be the number of minutes that it takes a randomly selected person to solve a certain puzzle. The probability density function of T is

$$f(x) = \begin{cases} 3/x^4 & \text{if } x \ge 1, \\ 0 & \text{if } x < 1. \end{cases}$$

    (a) Show that $\displaystyle\int_{-\infty}^{\infty} f(x)\,dx = 1$.

    (b) What is the probability that $T > 2$?

    (c) What is the probability that $T < 1.5$?

    (d) Evaluate $\displaystyle\int_{-\infty}^{\infty} xf(x)\,dx$. (This is the **mean** of the random variable T.)

    (e) Find m such that 50% of the people solve the puzzle in m minutes or less. (This is the **median** of the random variable T.)

25. Find k so that $f(x) = k/(1 + x^2)$ is a probability density function. (This is the **Cauchy** distribution.)

26. Find k so that

$$f(x) = \begin{cases} k(1-x^2) & \text{if } 0 \le x \le 1, \\ 0 & \text{otherwise}, \end{cases}$$

is a probability density function.

27. Let $\lambda > 0$ be a constant, and let f be the function

$$f(x) = \begin{cases} \lambda e^{-\lambda x} & \text{if } x \ge 0, \\ 0 & \text{if } x < 0. \end{cases}$$

(a) Show that f is a probability density function.
[NOTE: f is known as the **exponential** density function.]

(b) Evaluate $\displaystyle\int_{-\infty}^{\infty} x f(x)\,dx$. [NOTE: If X is a random variable that has f as its probability density function, then the value of this integral is the **mean** of X.]

28. The exponential density function (see Exercise 27) is often used to model the lifespan of electrical devices. Suppose that 10% of a certain type of device fail after 6 months.

(a) What value of λ corresponds to the observed lifespan?

(b) What is the probability that a randomly selected device will still be functioning after 18 months?

(c) What is the mean lifespan of this device?

*Exercises 29–33 explore some properties of the **gamma** function:*
$$\Gamma(x) = \int_0^\infty t^{x-1} e^{-t}\,dt.$$

29. (a) Evaluate $\Gamma(1)$.

(b) Use integration by parts to show that if $x > 1$, $\Gamma(x) = (x-1)\Gamma(x-1)$. [HINT: For any real number z, $\displaystyle\lim_{t \to \infty} t^z e^{-t} = 0$.]

30. Show that $0 < \Gamma\left(\frac{2}{3}\right) < 2$. [HINT: Be careful; the integral has *two* improprieties.]

31. Use the fact that $\displaystyle\int_0^\infty e^{-x^2}\,dx = \frac{\sqrt{\pi}}{2}$ to evaluate $\Gamma\left(\frac{1}{2}\right)$.
[HINT: Let $u = \sqrt{t}$.]

32. Show that $\Gamma(z+1) = \displaystyle\int_0^\infty t^z e^{-t}\,dt = \int_0^1 \left(\ln(1/x)\right)^z\,dx$.
[HINT: Use the substitution $t = \ln(1/x)$.]

33. It is a fact that $\Gamma(n+1) = \displaystyle\int_0^\infty x^n e^{-x}\,dx = n!$ if $n \ge 0$ is an integer. Use this fact to show that if m and n are positive integers, then $\displaystyle\int_0^1 x^m (\ln x)^n\,dx = \frac{(-1)^n\, n!}{(m+1)^{n+1}}$.
[HINT: Make the substitution $x = e^{-u}$ in the second integral.]

34. Suppose that x and y are uniformly distributed random variables with values in the interval $[0, 2]$. (Uniformly distributed means that every value in the interval is equally probable.)

(a) Consider (x, y) as a point in the xy-plane. Explain why the **sample space** is the square $[0, 2] \times [0, 2]$.

(b) Explain why the probability that $x \le 1$ *and* $y \le 1$ is $1/4$. [HINT: What proportion of the sample space is occupied by the region $[0, 1] \times [0, 1]$.]

35. Suppose that x and y are uniformly distributed random variables with values in the interval $[-2, 2]$. What is the probability that $x^2 + y^2 \le 1$? Justify your answer.

36. Suppose that x and y are uniformly distributed random variables with values in the interval $[0, 1]$. What is the probability that $x + y \le 1/2$? Justify your answer.

37. Suppose that x and y are uniformly distributed random variables with values in the interval $[0, 1]$. What is the probability that $y/x \ge 2$? Justify your answer.

38. Suppose that x and y are uniformly distributed random variables with values in the interval $[0, 2]$. What is the probability that $xy \le 2$? Justify your answer.

39. Suppose that x and y are uniformly distributed random variables with values in the interval $[0, 3]$. What is the probability that $x + y \ge xy$? Justify your answer.

11

INFINITE SERIES

11.1 SEQUENCES AND THEIR LIMITS

This section, on infinite *sequences*, prepares the ground for the next topic—infinite *series*. Convergent series are defined in terms of the simpler, more basic idea of convergent sequences. We start with a brief introduction to sequences—what they are, what it means for them to converge or diverge, and how to find their limits.

Terminology and basic examples

A **sequence** is an infinite list of numbers, of the general form

$$a_1, \ a_2, \ a_3, \ a_4, \ldots, a_k, \ a_{k+1}, \ldots.$$

Read "a sub three" and "a sub k."

Individual entries are called the **terms** of the sequence; a_3 and a_k, ◂ for instance, are the third term and the kth term, respectively. The full sequence is, technically speaking, an **ordered set**; the standard notation

$$\{a_k\}_{k=1}^{\infty}$$

(or simply $\{a_k\}$) uses set (wiggly) brackets to emphasize this view.

Our main interest in sequences is in their **limits**. For the simplest sequences, limits (or the lack thereof) are evident at a glance. The next three examples are of this type.

EXAMPLE 1 Discuss the sequence $\{a_k\}_{k=1}^{\infty}$ defined by the formula $a_k = 1/k$. Does this sequence have a limit?

Solution Sampling some terms—

$$\frac{1}{1}, \frac{1}{2}, \frac{1}{3}, \ldots, \frac{1}{10}, \frac{1}{11}, \ldots, \frac{1}{100}, \frac{1}{101}, \ldots$$

shows (to nobody's surprise) that the sequence **converges** to zero: As k increases, the terms a_k approach zero arbitrarily closely. In symbols,

$$\lim_{k \to \infty} a_k = \lim_{k \to \infty} \frac{1}{k} = 0. \qquad \blacksquare$$

E X A M P L E 2   Suppose that $\{b_j\}_{j=1}^{\infty}$ is defined by

$$b_j = \frac{(-1)^j}{j}, \quad j = 1, 2, 3, \ldots.$$

(We used j, not k, as our **index variable**—the choice is up to us.) What's $\lim_{j \to \infty} b_j$?

Solution   Writing out terms shows a pattern similar to that in Example 1:

$$-\frac{1}{1}, \frac{1}{2}, -\frac{1}{3}, \ldots, \frac{1}{10}, -\frac{1}{11}, \ldots, \frac{1}{100}, -\frac{1}{101}, \ldots.$$

Although the terms oscillate in *sign*, they approach zero more and more closely as j increases. Eventually, all the terms—positive or negative—remain within any specified distance from zero. (All terms past b_{1000}, for instance, are within 0.001 of zero.) Hence, the sequence $\{b_j\}$ **converges** to zero:

$$\lim_{j \to \infty} b_j = \lim_{j \to \infty} \frac{(-1)^j}{j} = 0. \qquad \blacksquare$$

E X A M P L E 3   Does the sequence $\{c_k\}_{k=0}^{\infty}$ with general term $c_k = (-1)^k$ converge?

Solution   No; it **diverges**. Successive terms have the pattern

$$1, -1, 1, -1, 1, \ldots,$$

never settling on a single limit. $\qquad \blacksquare$

E X A M P L E 4   Discuss the **Fibonacci sequence**, defined by the rules

$$F_1 = 1; \quad F_2 = 1; \quad F_{n+2} = F_n + F_{n+1};$$

each term is the sum of its two predecessors. ➥ Such definitions are called **recursive**: Each term is defined by means of earlier terms.

The sequence is named for the Italian mathematician Leonardo Fibonacci (1170–1250), who related it to a rabbit population explosion under certain conditions.

Solution   The first few terms of this sequence are

$$1, 1, 2, 3, 5, 8, 13, 21, 34, 55, \ldots.$$

As the pattern suggests, the sequence **diverges** to infinity, and we write $\lim_{n \to \infty} F_n = \infty$. $\quad \blacksquare$

Lessons from the examples

Sequences have their own notational quirks and conventions. Here are several to watch for:

- **Where to start?**   The sequence in Example 3 began with c_0, not c_1. Other starting points, such as a_2 or even b_{-3}, occasionally arise. In practice, such differences are unimportant. What matters for sequences is their long-run behavior, not the presence or absence of a few initial terms.

- **Index variable names don't matter**   We can define the squaring function by writing either $f(t) = t^2$ or $f(x) = x^2$; the *variable name* makes no difference. In the same way, a sequence's *index name* is arbitrary: $\{a_k\}_{k=1}^{\infty}$ and $\{a_j\}_{j=1}^{\infty}$ mean exactly the same thing.

- **Reindexing** The sequence

$$\frac{1}{1}, \frac{1}{2}, \frac{1}{3}, \ldots, \frac{1}{10}, \frac{1}{11}, \ldots$$

of Example 1 can be described symbolically in various different-looking but still equivalent ways. Here are two:

$$a_k = \frac{1}{k}, \quad k = 1, 2, \ldots \qquad \text{or} \qquad a_j = \frac{1}{j+1}, \quad j = 0, 1, 2, \ldots.$$

Depending on the situation, one description or the other may be preferable.

Sequences as functions Sequences are closely related to functions, as expressions such as

$$a_k = \frac{\sin k}{k} \qquad \text{and} \qquad f(x) = \frac{\sin x}{x}$$

illustrate. The formal definition makes this connection precise.

DEFINITION An **infinite sequence** is a real-valued function that is defined for *positive integer inputs.*

The definition and the preceding examples give us several useful ways to think of a sequence:

- **As a list** As an infinite list of numbers: $a_1, a_2, a_3, \ldots.$
- **As a function** As a function $a(n)$, where n takes only positive integer values. (The rule for a may or may not make sense for other inputs.) Thus, $a(1) = a_1$, $a(2) = a_2$, $a(3) = a_3$, and so on.
- **As a discrete sample** As a "discrete sample" of values of an ordinary calculus-style function $f(x)$ defined for real $x \geq 1$. The function $f(x) = 1/x$, for example, produces the sequence $a_k = 1/k$ in Example 1.

Graphs of sequences, graphs of functions The graph of any function f consists of the points $\left(x, f(x)\right)$ for x in the domain of f. The graph of a *sequence* $\{a_k\}$ is therefore the set of points (k, a_k) as k runs through positive integer values.

EXAMPLE 5 Let a function f and a sequence $\{a_k\}$ be defined by

$$f(x) = \frac{\sin x}{x}; \qquad a_k = \frac{\sin k}{k}.$$

Plot graphs of both. What do the graphs say about limits?

Solution   Figure 1 shows the graphs.

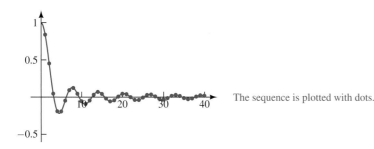

The sequence is plotted with dots.

FIGURE 1
Graphs of $f(x) = \dfrac{\sin x}{x}$ and $a_k = \dfrac{\sin k}{k}$

The graphs illustrate how the sequence $\{a_k\}$ is a "discrete sample" of the continuous function f. They show, too, that as x (or k) tends to infinity, this sequence and function, although oscillating in sign, tend to zero.   ■

Not every sequence comes naturally from a familiar calculus function. Nevertheless, graphs or tables often suggest limits.

EXAMPLE 6   A sequence $\{b_j\}$ has general term

$$b_j = 1 \cdot 2 \cdot 3 \cdot 4 \cdots (j-1) \cdot j = j!$$

(In words: b_j is j **factorial**.) Tabulate $\{b_j\}$; find its limit, if any.

Solution   As the table suggests, $\{b_j\}$ diverges—quickly—to infinity.

As $k \to \infty$, $k! \to \infty$: explosive numerical evidence						
k	1	2	4	8	16	32	64
$k!$	1	2	24	40,320	2.092×10^{13}	2.631×10^{35}	1.269×10^{89}

Can any doubt remain? ➤

■ *Graphs work poorly for this sequence—most windows are too small.*

Sequence limits defined

Sequences are special sorts of functions; limits of sequences, therefore, are mild variants on limits at infinity. An informal definition will suffice. ➤

A formal definition describes more precisely what "approaches" means.

> **DEFINITION (Limit of a sequence)**   Let $\{a_k\}$ be a sequence and L a real number. If a_k approaches L to within any desired tolerance as k increases without bound, then the sequence **converges** to L. In symbols,
>
> $$\lim_{k \to \infty} a_k = L.$$
>
> Otherwise, the sequence **diverges**.

Notice:

- **Divergence to infinity**   If either $a_k \to \infty$ or $a_k \to -\infty$ as k increases without bound, then the sequence diverges to (positive or negative) infinity. We write, for instance,

$$\lim_{k \to \infty} k! = \infty.$$

- **Other divergence behavior**   Example 3 shows that a divergent sequence need not "blow up"; other patterns of "wandering" behavior (or no pattern at all) are possible.

- **Asymptotes**   A sequence, like a function, converges to a finite limit L if and only if $y = L$ is a horizontal asymptote of its graph.

Sequences, functions, and limits   For sequences that are "discrete samples" of familiar functions, we can often use what we know of the underlying function to find limits.

> **FACT**   Let f be a function defined for $x \geq 1$. If $\lim_{x \to \infty} f(x) = L$ and $a_k = f(k)$ for all $k \geq 1$, then $\lim_{k \to \infty} a_k = L$.

Example 5 illustrates this Fact:

$$\lim_{k \to \infty} \frac{\sin k}{k} = \lim_{x \to \infty} \frac{\sin x}{x} = 0.$$

EXAMPLE 7   Does the sequence with general term $a_k = \sin k$ converge or diverge?

Solution   Because $\lim_{x \to \infty} \sin x$ does not exist, the preceding Fact doesn't help. Instead, let's look at the graph (Figure 2).

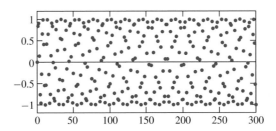

FIGURE 2
The sequence $a_k = \sin k$

Although full of interesting shapes, the graph never settles on a single limit, and so the sequence diverges.   ∎

Finding limits of sequences

Many of the tools we've developed for finding limits of functions will help us find limits of sequences as well.

We studied l'Hôpital's rule in Chapter 4.

Using l'Hôpital's rule   If a sequence has a nice symbolic formula, l'Hôpital's rule can sometimes be applied to find the limit. ◄

EXAMPLE 8 Numerical evidence suggests that if $a_k = 2^k/k^2$, then $\{a_k\}$ diverges to infinity. Use l'Hôpital's rule to show this result symbolically.

Solution l'Hôpital's rule applied to $h(x) = 2^x/x^2$ gives

$$\lim_{x \to \infty} \frac{2^x}{x^2} = \lim_{x \to \infty} \frac{2^x \ln 2}{2x} \quad \text{(differentiate top and bottom separately)}$$

$$= \lim_{x \to \infty} \frac{2^x \ln 2 \cdot \ln 2}{2} \quad \text{(apply l'Hôpital's rule again)}$$

$$= \frac{(\ln 2)^2}{2} \lim_{x \to \infty} 2^x = \infty.$$

Because $h(x) \to \infty$, $a_k \to \infty$ too; the sequence diverges. ∎

EXAMPLE 9 Show that $\lim_{n \to \infty} n^{1/n} = 1$.

Solution l'Hôpital's rule doesn't apply directly, but applying the natural logarithm produces a fraction:

$$a_n = n^{1/n} \implies \ln(a_n) = \frac{\ln n}{n}.$$

l'Hôpital's rule *does* apply to the last quantity:

$$\lim_{n \to \infty} \frac{\ln n}{n} = \lim_{n \to \infty} \frac{1/n}{1} = 0.$$

We've shown that $\ln(a_n) \to 0$ as $n \to \infty$; it follows that $a_n \to 1$, as desired. ∎

New sequence limits from old Limits of sequences, like limits of functions, can be combined in various symbolic ways to give new limits. We'll need some *known* limits, of course, to get started. Here are several useful and important ones:

$$\lim_{n \to \infty} n^{1/n} = 1$$

$$\lim_{n \to \infty} x^{1/n} = 1 \quad \text{(for all } x > 0\text{)}$$

$$\lim_{n \to \infty} \frac{1}{n^k} = 0 \quad \text{(for all } k > 0\text{)}$$

$$\lim_{n \to \infty} r^n = 0 \quad \text{(if } -1 < r < 1\text{)}$$

(We derived the first of these limits in Example 9, using l'Hôpital's rule. The other limits are easier.)

Calculating with limits Plausible-seeming calculations such as

$$\lim_{k \to \infty} \left(\frac{1}{k} + \frac{3k}{k+1} \right) = \lim_{k \to \infty} \frac{1}{k} + 3 \lim_{k \to \infty} \frac{k}{k+1} = 0 + 3 = 3$$

rely implicitly on the following theorem. We've already seen it—in almost identical form—for functions.

THEOREM 1 (Algebra with limits)   Suppose that

$$a_k \to L \quad \text{and} \quad b_k \to M$$

as $k \to \infty$, where L and M are finite numbers. Let c be any real constant. Then

$$ca_k \to cL, \quad a_k \pm b_k \to L \pm M, \quad \text{and} \quad a_k b_k \to LM.$$

If $M \neq 0$, then $\dfrac{a_k}{b_k} \to \dfrac{L}{M}$.

Squeezing limits   For sequences, as for functions, unknown limits can sometimes be found by "squeezing" them between known limits.

THEOREM 2 (The squeeze principle)   Suppose that

$$a_k \leq b_k \leq c_k \quad \text{for all } k \geq 1$$

and that

$$\lim_{k \to \infty} a_k = \lim_{k \to \infty} c_k = L.$$

Then $\displaystyle\lim_{k \to \infty} b_k = L$.

EXAMPLE 10   We've seen graphically that $\displaystyle\lim_{k \to \infty} \frac{\sin k}{k} = 0$. Use squeezing to show this legalistically.

Solution   The squeeze inequality

$$-\frac{1}{k} \leq \frac{\sin k}{k} \leq \frac{1}{k}$$

holds for all integers $k > 0$. Because both the left and the right sides tend to 0 as k tends to infinity, the middle expression must do so too.   ■

Guaranteeing convergence: An existence theorem

The best way to show that a sequence converges is to find a limit. But doing so is sometimes hard. The following example illustrates the difficulty—and also suggests a remedy.

EXAMPLE 11   Does the sequence $a_n = \left(1 + \dfrac{1}{n}\right)^n$ converge?

Solution   The answer isn't obvious from the formula, so let's tabulate some numerical values:

Values of $(1 + 1/n)^n$						
n	16	32	64	128	256	512	1024
$(1 + 1/n)^n$	2.63793	2.67699	2.69734	2.70774	2.71299	2.71563	2.71696

As $n \to \infty$, the terms a_n seem to increase but not to blow up. Thus, apparently, the sequence converges to *some* limit. (The number $e \approx 2.71828$ is a tempting guess.)   ■

A convergence theorem The sequence in Example 11 appears from the table of numbers to be **nondecreasing**:

$$a_1 \le a_2 \le \cdots \le a_k \le a_{k+1} \le \cdots .$$

(For this sequence, *strict* inequalities happen to hold, and so we could use the stronger adjective **increasing**.) Common sense suggests that any nondecreasing sequence should either (i) converge to a limit; or (ii) diverge to ∞. Because the sequence in Example 11 appears to be bounded above (by 3, for example), it's reasonable to guess that it converges. Theorem 3 makes these commonsense impressions precise. It applies to any **monotone** sequence (i.e., any sequence that is either nondecreasing on nonincreasing).

THEOREM 3 Suppose that the sequence $\{a_k\}$ is *nondecreasing* and *bounded above* by a number A. That is,

$$a_1 \le a_2 \le a_3 \le \cdots a_k \le a_{k+1} \le \cdots \le A.$$

Then $\{a_k\}$ converges to some finite limit a, with $a \le A$.

  Similarly, if $\{b_k\}$ is *nonincreasing* and *bounded below* by a number B, then $\{b_k\}$ converges to a finite limit $b \ge B$.

EXAMPLE 12 Consider the increasing sequence $\{a_k\}$ defined recursively by

$$a_1 = 0; \quad a_{k+1} = \sqrt{6 + a_k} \quad \text{if } k \ge 1.$$

Show that the sequence $\{a_k\}$ converges and that $\lim\limits_{k \to \infty} a_k \le 3$.

Solution The first few terms suggest that the sequence is increasing: ➡

We'll take this for granted, though it's not hard to show.

$$a_1 = 0; \quad a_2 = \sqrt{6} \approx 2.45; \quad a_3 \approx 2.91; \quad a_4 \approx 2.98; \quad a_5 \approx 2.997.$$

The sequence also seems to be bounded above by 3. This is easy to show. If $a_k < 3$, then, by definition,

$$a_{k+1} = \sqrt{6 + a_k} < \sqrt{6 + 3} = 3.$$

Thus, *no* term can exceed 3. It follows from the theorem, therefore, that the sequence converges to a limit no greater than 3. ➡ ■

The limit is 3, but that requires further proof.

In Exercises 1–17, find the limit of the sequence or explain why the limit does not exist.

1. $a_k = (-3/2)^k$

2. $a_k = (-0.8)^k$

3. $a_k = (1.1)^k$

4. $a_k = \left(\sqrt{26}/17\right)^k$

5. $a_k = (1/k)^k$

6. $a_k = \sin k$

7. $a_k = \arctan k$

8. $a_k = \cos(1/k)$

9. $a_k = \sin(k\pi)$

10. $a_k = \cos(k\pi)$

11. $a_k = \dfrac{k^2}{k^2 + 3}$

12. $a_k = \sqrt{\dfrac{2k}{k+3}}$

13. $a_m = m^2 e^{-m}$

14. $a_j = \dfrac{\ln j}{\sqrt[3]{j}}$

15. $a_k = \dfrac{k!}{(k+1)!}$

16. $a_k = \ln\left(\dfrac{k}{k+1}\right)$

17. $a_k = 3^{1/k}$

18. (a) For which values of x does $\lim\limits_{n \to \infty} x^n$ exist?

  (b) Find all values of x for which $\lim\limits_{n \to \infty} x^n = 0$.

  (c) Are there any values of x for which $\lim\limits_{n \to \infty} x^n = L \ne 0$? For each such x, find the limit L.

In Exercises 19–24, give an example of a sequence that is

19. convergent but not monotone.

20. bounded but not monotone.

21. monotone but not convergent.

22. nonincreasing and unbounded.

23. nonincreasing and convergent.

24. unbounded but not monotone.

25. Use Theorem 3 to show that the sequence 0.7, 0.77, 0.777, 0.7777, 0.77777, 0.777777, 0.7777777, ... has a limit.

26. Suppose that $\lim_{k\to\infty} a_k = L$, where L is a finite number, and that the terms of the sequence $\{b_k\}$ are defined by $b_k = L - a_k$. Explain why $\lim_{k\to\infty} b_k = 0$.

FURTHER EXERCISES

In Exercises 27–32, find the limit of the sequence or explain why the limit does not exist. [HINT: l'Hôpital's rule may be useful.]

27. $a_n = n\sin(1/n)$

28. $a_k = (2^k + 3^k)^{1/k}$

29. $a_k = \int_0^k e^{-x}\,dx$

30. $a_k = \int_k^\infty \dfrac{dx}{1+x^2}$

31. $a_k = \sqrt{k^2+1} - k$

32. $a_k = \sqrt{k^2+k} - k$

In Exercises 33–36, determine the values of x for which the sequence converges as $k\to\infty$. Evaluate $\lim_{k\to\infty} a_k$ for these values of x.

33. $a_k = e^{kx}$

34. $a_k = (\ln x)^k$

35. $a_k = (\arcsin x)^k$

36. $a_k = 2^{-k}(\arctan x)^k$

37. Show that $\lim_{n\to\infty} x^{1/n} = 1$ for all $x > 0$.

38. Let $a_k = \left(1 + \dfrac{x}{k}\right)^k$, where x is a real number.

    (a) Show that $\lim_{k\to\infty} \ln(a_k) = x$.

    (b) Use part (a) to evaluate $\lim_{k\to\infty} a_k$.

39. Evaluate $\lim_{n\to\infty} \left(1 - \dfrac{1}{2n}\right)^n$.

40. Let $a_n = \sum_{k=1}^{n} \dfrac{k}{n^2}$.

    (a) Evaluate a_{10}.

    (b) Explain why $\lim_{n\to\infty} a_n = \int_0^1 x\,dx = 1/2$.

    [HINT: $k/n^2 = (k/n) \cdot (1/n)$.]

41. Let $a_n = \sum_{k=1}^{n} \dfrac{1}{n+k}$.

    (a) Show that this sequence is increasing (i.e., $a_{n+1} > a_n$).

    (b) Show that $a_n \leq \dfrac{n}{n+1} < 1$.

    (c) What do parts (a) and (b) imply about $\lim_{n\to\infty} a_n$?

    (d) Explain why $\lim_{n\to\infty} a_n > 1/2$. [HINT: $a_1 = 1/2$.]

    (e) Show that $\lim_{n\to\infty} a_n = \ln 2$.

    [HINT: $\dfrac{1}{n+k} = \dfrac{1}{1+k/n} \cdot \dfrac{1}{n}$, and a_n is a Riemann-sum approximation to an integral.]

42. Let $a_n = \sum_{k=1}^{n} k^2/n^3$. Evaluate $\lim_{n\to\infty} a_n$. [HINT: Think of a_n as a Riemann-sum approximation to an integral.]

43. Let $a_n = \dfrac{\sqrt[n]{n!}}{n}$.

    (a) Show that $\ln a_n = \dfrac{1}{n}\sum_{k=1}^{n} \ln k - \dfrac{1}{n}\sum_{k=1}^{n} \ln n$.

    (b) Use part (a) to show that $\ln a_n$ is a right-sum approximation to $\int_0^1 \ln x\,dx$.

    (c) Use part (b) to show that $\lim_{n\to\infty} a_n = e^{-1}$.

44. Let $a_n = 4^n/n!$.

    (a) Find a number N such that $a_{n+1} \leq a_n$ for all $n \geq N$.

    (b) Use part (a) to explain why $\lim_{n\to\infty} a_n$ exists.

    (c) Evaluate $\lim_{n\to\infty} a_n$.

45. Suppose that $\{a_n\}$ is a sequence with the property $|a_{n+1}/a_n| \leq (n+3)/(2n+1)$ for all $n \geq 1$. Show that $\lim_{n\to\infty} a_n = 0$.
[HINT: Start by showing that $|a_{n+1}/a_3| \leq (6/7)^{n-2}$ for all $n \geq 3$.]

In Exercises 46–49, let $a_n = \cos 1 \cdot \cos 2 \cdot \cos 3 \cdot \cos 4 \cdots \cos n$.

46. Is the sequence $\{a_n\}$ bounded? Justify your answer.

47. Is the sequence $\{a_n\}$ monotone? Justify your answer.

48. Is the sequence $\{|a_n|\}$ bounded? Justify your answer.

49. Is the sequence $\{|a_n|\}$ monotone? Justify your answer.

50. Let $a_n = \dfrac{1 \cdot 3 \cdot 5 \cdots (2n-1)}{2 \cdot 4 \cdot 6 \cdots (2n)}$. Use Theorem 3 to show that $\lim_{n\to\infty} a_n$ exists.

51. Show that $\lim_{n\to\infty} \sin\left(\dfrac{\pi}{2^2}\right) \cdot \sin\left(\dfrac{\pi}{3^2}\right) \cdots \sin\left(\dfrac{\pi}{n^2}\right) = 0$.

    [HINT: $0 < \sin x < x$ when $0 < x < 1$.]

52. Does the sequence defined by $a_1 = 1$, $a_{n+1} = 1 - a_n$ converge? Justify your answer.

53. Does the sequence defined by $a_1 = 1$, $a_{n+1} = a_n/2$ converge? Justify your answer.

54. Consider the sequence defined by $a_1 = 1$, $a_{n+1} = \left(\dfrac{n}{n+1}\right) a_n$.

    (a) Show that the sequence converges.

    (b) Find the limit. [HINT: Write out the first few terms and look for a pattern.]

55. For which values of $x \geq 0$ does the sequence defined by $a_1 = x$, $a_{n+1} = \sqrt{a_n}$ converge? Justify your answer.

11.2 INFINITE SERIES, CONVERGENCE, AND DIVERGENCE

An **infinite series** (just **series** for short) is an expression of the form

$$\sum_{k=1}^{\infty} a_k = a_1 + a_2 + a_3 + a_4 + \cdots + a_k + a_{k+1} + \cdots.$$

(Notice the **sigma notation** on the left; we used it earlier for approximating sums for integrals.) A series results from *adding* the terms of a sequence $a_1, a_2, a_3, \ldots$. If, say, $a_k = 1/k^2$, then

$$\sum_{k=1}^{\infty} a_k = \sum_{k=1}^{\infty} \frac{1}{k^2} = \frac{1}{1} + \frac{1}{4} + \frac{1}{9} + \frac{1}{16} + \cdots.$$

If $a_k = k$, then

$$\sum_{k=1}^{\infty} a_k = \sum_{k=1}^{\infty} k = 1 + 2 + 3 + \cdots.$$

Natural questions The notion of an infinite sum raises natural questions:

- What does it *mean* to add infinitely many numbers?
- Which series add up to a finite number? Which series blow up?
- If a series has a finite sum, how can we find (or estimate) it?
- What good are infinite series?

The rest of this chapter addresses these questions.

Improper sums vs. improper integrals Standard examples of infinite series include

$$\text{(i)} \quad \sum_{k=1}^{\infty} \frac{1}{k^2}; \quad \text{(ii)} \quad \sum_{k=1}^{\infty} \frac{1}{k}; \quad \text{(iii)} \quad \sum_{k=1}^{\infty} \frac{1}{\sqrt{k}}; \quad \text{(iv)} \quad \sum_{k=1}^{\infty} \frac{1}{2^k}.$$

Notice the close typographical resemblance to improper integrals:

$$\text{(i)} \quad \int_1^{\infty} \frac{dx}{x^2}; \quad \text{(ii)} \quad \int_1^{\infty} \frac{dx}{x}; \quad \text{(iii)} \quad \int_1^{\infty} \frac{dx}{\sqrt{x}}; \quad \text{(iv)} \quad \int_1^{\infty} \frac{dx}{2^x}.$$

This is no accident; infinite series and improper integrals are closely analogous; we'll often exploit the connection. We'll see, in fact, that items (i) and (iv) of *both* preceding lists converge, whereas (ii) and (iii) diverge.

Why series matter: A look ahead

Understanding series takes some work. To preview why the work is worthwhile, consider the fact (we'll see later why it's true) that for any real number x,

$$\cos x = 1 - \frac{x^2}{2!} + \frac{x^4}{4!} - \frac{x^6}{6!} + \frac{x^8}{8!} - \cdots. \tag{1}$$

If, say, $x = 1$, then

$$\cos 1 = 1 - \frac{1}{2!} + \frac{1}{4!} - \frac{1}{6!} + \frac{1}{8!} - \cdots. \tag{2}$$

So what? Why would we write something familiar—the cosine function—in terms of something exotic—a series?

Finding $\cos x$ is easy for a few special values of x, such as $x = \pi/4$.

One good answer is that the cosine function has no *algebraic* formula, and so finding accurate numerical values of $\cos x$ for arbitrary inputs x is a genuine problem. ◄ Infinite series help solve this problem. Although not quite a formula in the ordinary sense (ordinary formulas don't include dots ...), Equation (1) gives a concrete, computable recipe for *approximating* $\cos x$: Given an input x, calculate, as far out as practically possible, the "infinite polynomial"

$$1 - \frac{x^2}{2!} + \frac{x^4}{4!} - \frac{x^6}{6!} + \frac{x^8}{8!} - \cdots.$$

With any luck, the result should closely approximate the "true" value of $\cos x$.

Equation (2) shows how to approximate $\cos 1$. A calculator readily gives, to seven decimals,

$$1 - \frac{1}{2!} = 0.5000000;$$

$$1 - \frac{1}{2!} + \frac{1}{4!} = 0.5416667;$$

$$1 - \frac{1}{2!} + \frac{1}{4!} - \frac{1}{6!} = 0.5402778;$$

$$1 - \frac{1}{2!} + \frac{1}{4!} - \frac{1}{6!} + \frac{1}{8!} = 0.5403026.$$

The results converge with gratifying speed to the "right" answer—the true value of $\cos 1$ (≈ 0.5403023).

Good questions   These calculations raise good questions:

> *Where did Equation (1) come from? What do all the "dots" really mean? Are similar equations available for* other *functions—sine, arctangent, logarithmic, and so on? How many terms are needed to guarantee accuracy to, say, five decimals?*

We'll answer all of these questions in this chapter.

Definitions and terminology

Working successfully with series requires some up-front investment in definitions and technical language. After stating terms and definitions, we show by example why they're reasonable.

Series language   Let

$$\sum_{k=1}^{\infty} a_k = a_1 + a_2 + a_3 + \cdots + a_k + a_{k+1} + \cdots$$

be an infinite series. (In most cases the index variable starts at 1, but sometimes it's convenient to start k at 0 or elsewhere.) ➤ The summand a_k is called the kth **term** of the series, and the nth **partial sum**, usually denoted by S_n, is the (finite) sum of all terms *through index n*:

Sometimes we'll use other names, such as j, for the index variable.

$$S_n = a_1 + a_2 + a_3 + \cdots + a_{n-1} + a_n = \sum_{k=1}^{n} a_k. \tag{3}$$

The nth **tail**, denoted by R_n, is the (infinite) sum of all terms *beyond index n*:

$$R_n = a_{n+1} + a_{n+2} + a_{n+3} + \cdots = \sum_{k=n+1}^{\infty} a_k.$$

As the notation R_n suggests, the nth tail is a *remainder*—what's left after adding terms through index n. In symbols,

$$\sum_{k=1}^{\infty} a_k = \sum_{k=1}^{n} a_k + \sum_{k=n+1}^{\infty} a_k = S_n + R_n.$$

(We did the same thing in Chapter 10 with improper integrals:

$$\int_{1}^{\infty} a(x)\,dx = \int_{1}^{n} a(x)\,dx + \int_{n}^{\infty} a(x)\,dx,$$

where the last integral is another type of upper tail.)

The crucial definition of convergence involves the partial sums S_n:

DEFINITION If $\lim_{n \to \infty} S_n = S$, for some finite number S, then the series $\sum_{k=1}^{\infty} a_k$ **converges** to the limit S. (S is also called the **sum** of the series.) Otherwise, the series **diverges**.

Notice the following aspects of the definition:

- **Divergent series** A divergent series is one for which the sequence of partial sums does *not* converge to a finite limit S. One possibility is that the partial sums S_n blow up to infinity. Another possibility is that the partial sums remain bounded but never settle on a specific limit.

- **Improper integrals, improper sums** The definition says, in symbols, that

$$\sum_{k=1}^{\infty} a_k = \lim_{n \to \infty} \sum_{k=1}^{n} a_k$$

if the limit exists. Convergence for improper integrals means much the same thing:

$$\int_{x=1}^{\infty} f(x)\,dx = \lim_{n \to \infty} \int_{x=1}^{n} f(x)\,dx$$

if *this* limit exists. ➤ An infinite series is an improper sum in exactly the sense that an integral may be improper.

In each case we take the limit of something proper.

- **Convergence, partial sums, and tails**   To say that $\sum a_k$ converges to the sum S means that $S_n \to S$ as $n \to \infty$. Since for all n, $S = S_n + R_n$, it follows that $R_n \to 0$ as $n \to \infty$.

- **Two sequences—keep them straight**   Every series $\sum a_k$ involves *two* sequences: the sequence $\{a_k\}$ of *terms* and the sequence $\{S_n\}$ of *partial sums*. Keeping these related but different sequences separate is essential. We'll take special care to do so in the following examples.

EXAMPLE 1   Does the series $\displaystyle\sum_{k=0}^{\infty} \frac{1}{2^k} = 1 + \frac{1}{2} + \frac{1}{4} + \frac{1}{8} + \cdots$ converge? If so, to what limit?

Solution   Because the index k begins at 0, not 1, so does the sequence of partial sums. Direct calculation yields

$$S_0 = 1 = 2 - 1;$$

$$S_1 = 1 + \frac{1}{2} = \frac{3}{2} = 2 - \frac{1}{2};$$

$$S_2 = 1 + \frac{1}{2} + \frac{1}{4} = \frac{7}{4} = 2 - \frac{1}{4};$$

$$S_3 = 1 + \frac{1}{2} + \frac{1}{4} + \frac{1}{8} = \frac{15}{8} = 2 - \frac{1}{8};$$

$$S_4 = 1 + \frac{1}{2} + \frac{1}{4} + \frac{1}{8} + \frac{1}{16} = \frac{31}{16} = 2 - \frac{1}{16}.$$

The pattern is easy to see: For any $n \geq 1$,

$$S_n = 2 - \frac{1}{2^n}.$$

This explicit formula for S_n lets us answer the question posed above: Because $S_n \to 2$ as $n \to \infty$, the series converges to 2.

$R_n = 2 - S_n.$

A numerical table of partial sums and tails supports this conclusion: ◄

\multicolumn{12}{c}{**Partial sums and tails of** $1 + \dfrac{1}{2} + \dfrac{1}{4} + \cdots$}
n	0	1	2	3	4	5	6	7	8	9	10
S_n	1	1.5	1.75	1.875	1.938	1.969	1.984	1.992	1.996	1.998	1.999
R_n	1	0.5	0.25	0.125	0.063	0.031	0.016	0.008	0.004	0.002	0.001

As the numbers show, the partial sums S_n converge to 2 while the tails R_n tend to 0.   ∎

EXAMPLE 2   Does $\displaystyle\sum_{k=1}^{\infty} (-1)^k = -1 + 1 - 1 + 1 - \cdots$ converge?

Solution   No. Successive partial sums are $-1,\ 0,\ -1,\ 0,\ \ldots$. Thus, the sequence $\{S_n\}$ diverges and so does the series.   ∎

EXAMPLE 3   Does the **harmonic series** $\sum_{k=1}^{\infty} \frac{1}{k}$ converge? ➡

The harmonic series will be one of our most important examples.

Solution   The answer depends on the partial sums S_n. By definition,

$$S_n = 1 + \frac{1}{2} + \frac{1}{3} + \frac{1}{4} + \frac{1}{5} + \frac{1}{6} + \cdots + \frac{1}{n}.$$

Because no simple formula for S_n comes to mind (in fact, there *is* no such simple formula), we'll investigate the partial sums numerically. Here are some results, computed to three decimal places:

$$S_1 = 1; \quad S_{20} = 3.598; \quad S_{50} = 4.499; \quad S_{100} = 5.187; \quad S_{1000} = 7.485.$$

In fact, the series diverges. We'll show this fact soon by comparing the series to an integral.

The numerical evidence is ambiguous; the S_n's seem to keep growing, although slowly. Whether the sequence of partial sums $\{S_n\}$ converges or diverges is not yet clear. ➡ ■

When does a series (or an integral) converge?

Whether a given infinite series converges or diverges is a delicate question. ➡ Such series as

So was the analogous question for improper integrals.

$$\sum_{k=1}^{\infty} \frac{1}{k} = 1 + \frac{1}{2} + \frac{1}{3} + \cdots \quad \text{and} \quad \sum_{k=1}^{\infty} \frac{1}{k^2} = 1 + \frac{1}{4} + \frac{1}{9} + \cdots$$

pose the same puzzle: Although successive *terms* of both series tend to zero, the *number* of terms is infinite. Convergence or divergence hinges on which of these conflicting tendencies "wins" in the long run.

The same dilemma arose for the improper *integrals*

$$\int_1^{\infty} \frac{1}{x}\, dx \quad \text{and} \quad \int_1^{\infty} \frac{1}{x^2}\, dx,$$

and it's worth recalling exactly what happened. In each case, the "partial integral" $I(t)$ is easily found. For the first integral, we found

$$I(t) = \int_1^t \frac{dx}{x} = \ln t \to \infty.$$

For the second,

$$I(t) = \int_1^t \frac{dx}{x^2} = 1 - \frac{1}{t} \to 1.$$

Thus, the two similar-looking integrals led to opposite results: The first integral diverges to infinity and the second converges to 1. Deciding whether the corresponding *series* converge or diverge is a bit harder because no convenient formulas for S_n are available.

Convergence and divergence: Graphical views   For a series $\sum a_k$, plotting both the terms $\{a_k\}$ and the partial sums $\{S_n\}$ on the same axes illustrates the connection between the two— and sometimes suggests whether the series converges or diverges. The two series $\sum 1/k$ and $\sum 1/k^2$ generate Figures 1(a) and 1(b), respectively.

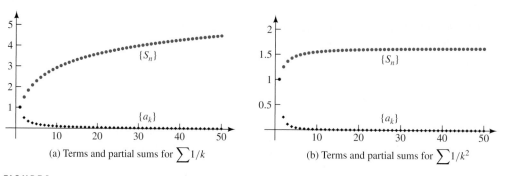

(a) Terms and partial sums for $\sum 1/k$ (b) Terms and partial sums for $\sum 1/k^2$

FIGURE 1
Terms and partial sums for two series

The two pictures give different impressions. The pictured behavior of partial sums suggests (but doesn't prove) that the first series diverges; the second series appears to converge because the partial sums seem to level off, perhaps approaching a limit near 1.7.

Geometric series: The nicest kind

The most important family of infinite series are the **geometric series**. They have the form

$$\sum_{k=0}^{\infty} ar^k = a + ar + ar^2 + ar^3 + ar^4 + \cdots;$$

here a is called the **leading term** and r is the **ratio**, because each term is r times the preceding term. An especially simple example is

$$\sum_{k=0}^{\infty} \left(\frac{1}{2}\right)^k = 1 + \frac{1}{2} + \frac{1}{4} + \frac{1}{8} + \cdots,$$

with $a = 1$ and $r = 1/2$. We saw in Example 1, page 558, that this series converges to 2.

Partial sums of geometric series   Geometric series have a great advantage: It's easy to tell whether they converge and, if so, to find their sums. This is because geometric series—unlike many others—have a simple, explicit formula for the partial sums S_n. The formula depends on a beautiful algebraic fact. If $r \neq 1$ and $n \geq 0$, then

$$1 + r + r^2 + r^3 + \cdots + r^n = \frac{1 - r^{n+1}}{1 - r}. \tag{4}$$

Multiplying Equation (4) by a gives a formula for S_n:

$$S_n = a + ar + ar^2 + ar^3 + \cdots + ar^n = a\frac{1 - r^{n+1}}{1 - r}. \tag{5}$$

From Formula (5) follows the whole story of convergence and divergence for geometric series. As always, the question is how the sequence $\{S_n\}$ of partial sums behaves. In this case, the issue has to do entirely with the power r^{n+1}. We'll need the following facts:

$$\lim_{m \to \infty} r^m = \begin{cases} 0 & \text{if } |r| < 1 \\ 1 & \text{if } r = 1 \\ \text{does not exist} & \text{if } |r| > 1 \text{ or } r = -1 \end{cases}$$

The conclusion for geometric series follows, and it's worth emphasizing:

> **THEOREM 4 (Convergence and divergence of geometric series)** If $|r|<1$, the geometric series
>
> $$\sum_{k=0}^{\infty} ar^k = a + ar + ar^2 + ar^3 + \cdots$$
>
> converges to $\dfrac{a}{1-r}$. If $a \neq 0$ and $|r| \geq 1$, the series diverges.

EXAMPLE 4 The series $\dfrac{1}{3} - \dfrac{1}{6} + \dfrac{1}{12} - \dfrac{1}{24} + \cdots$ converges. To what limit?

Solution The series is geometric, with $a = 1/3$ and $r = -1/2$. By Theorem 4, it converges to

$$\frac{a}{1-r} = \frac{1/3}{1+1/2} = \frac{2}{9} \approx 0.2222222.$$

A numerical look at partial sums and tails supports this computation. ➡

$R_n = \dfrac{2}{9} - S_n.$

Partial sums and tails of $\dfrac{1}{3} - \dfrac{1}{6} + \dfrac{1}{12} - \dfrac{1}{24} + \cdots$								
n	0	1	2	3	4	5	...	10
S_n	0.3333	0.1667	0.2500	0.2083	0.2292	0.2188	...	0.2223
R_n	−0.1111	0.0555	−0.0278	0.0139	−0.0069	0.0035	...	−0.0001

∎

Telescoping series Most series *other* than the geometric variety do *not* admit explicit formulas for partial sums. Telescoping series are among the unusual (and pleasant) exceptions. The next example illustrates how telescoping series work, and explains the name.

EXAMPLE 5 Show that $\displaystyle\sum_{k=1}^{\infty} \frac{1}{k(k+1)}$ converges and find its limit.

Solution A little algebra inside the summation sign lets us rewrite the series in a more convenient form: ➡

$$\sum_{k=1}^{\infty} \frac{1}{k(k+1)} = \sum_{k=1}^{\infty} \left(\frac{1}{k} - \frac{1}{k+1} \right).$$

To convince yourself that the two sides are equal, find a common denominator on the right side.

Writing out some terms shows the "telescoping" pattern: ➡

$$S_n = \sum_{k=1}^{n} \left(\frac{1}{k} - \frac{1}{k+1} \right) = \left(\frac{1}{1} - \frac{1}{2} \right) + \left(\frac{1}{2} - \frac{1}{3} \right) + \left(\frac{1}{3} - \frac{1}{4} \right) + \cdots + \left(\frac{1}{n} - \frac{1}{n+1} \right)$$

$$= 1 - \frac{1}{n+1}.$$

The summands collapse like an old-fashioned spyglass.

Now it's clear that, as $n \to \infty$, $S_n \to 1$; that's the sum. Again the numbers agree: ➡

Watch the tails go to zero.

Partial sums and tails of $\displaystyle\sum_{k=1}^{\infty} \frac{1}{k^2 + k}$							
n	1	2	3	4	5	6	...	11
S_n	0.5000	0.6667	0.7500	0.8000	0.8333	0.8571	...	0.9167
R_n	0.5000	0.3333	0.2500	0.2000	0.1667	0.1429	...	0.0833

∎

Algebra with series

As with functions and sequences, combining series algebraically produces new series. Combining *convergent* series produces new *convergent* series, with limits related in the expected way.

> **THEOREM 5**   Suppose that $\sum a_k$ converges to S and that $\sum b_k$ converges to T. Let c be any constant. Then
>
> $$\sum_{k=1}^{\infty} (a_k + b_k) \quad \text{converges to } S + T$$
>
> and
>
> $$\sum ca_k \quad \text{converges to } cS.$$
>
> In short:
>
> $$\sum (a_k + b_k) = \sum a_k + \sum b_k \quad \text{and} \quad \sum ca_k = c \sum a_k.$$

After all, the limit of a series is defined as the limit of the sequence of partial sums.

These reasonable-looking properties of convergent *series* follow directly from the analogous properties of convergent *sequences*. ◄

A little series algebra, cleverly applied, can immensely simplify finding the limits of certain series.

EXAMPLE 6   Evaluate $\displaystyle\sum_{k=0}^{\infty} \frac{4 + 2^k}{3^k}$.

Solution   The series is the sum of two (convergent) geometric series. Applying Theorems 4 and 5 gives

$$\sum_{k=0}^{\infty} \frac{4 + 2^k}{3^k} = \sum_{k=0}^{\infty} \frac{4}{3^k} + \sum_{k=0}^{\infty} \left(\frac{2}{3}\right)^k = 4 \cdot \frac{3}{2} + 3 = 9.$$ ∎

EXAMPLE 7   Calculate the tail R_{10} for the geometric series $\displaystyle\sum_{k=0}^{\infty} \frac{3}{2^k}$.

Solution   A little algebra is all we need:

$$R_{10} = \sum_{k=11}^{\infty} \frac{3}{2^k} = \frac{3}{2^{11}} + \frac{3}{2^{12}} + \frac{3}{2^{13}} + \cdots$$

$$= \frac{3}{2^{11}} \left(1 + \frac{1}{2} + \frac{1}{2^2} + \frac{1}{2^3} + \cdots\right) \quad \text{(factoring out the constant)}$$

$$= \frac{3}{2^{11}} \cdot 2 = \frac{3}{2^{10}} = \frac{3}{1024}. \quad \text{(summing the geometric series)}$$ ∎

Detecting divergent series

We are mainly interested in *convergent* series and their limits. Theorem 5, for instance, applies only to convergent series, and so it's important to recognize *divergence* when it occurs. One strategy is to use Theorem 5 indirectly.

EXAMPLE 8   Suppose that $\sum a_k$ diverges, and let $c \neq 0$ be a constant. Explain why $\sum c \, a_k$ diverges, too.

Solution   If $\sum c \, a_k$ were *convergent*, then, by Theorem 5, the series

$$\sum \frac{1}{c} c \, a_k = \sum a_k$$

must *also* converge, which contradicts our assumption.   ■

The *n*th term test   The following theorem describes another useful divergence detector:

> **THEOREM 6 (The *n*th term test for divergence)**   If $\lim_{n \to \infty} a_n \neq 0$, then $\sum a_n$ diverges.

The theorem holds because, for a series $\sum a_k$ to converge, the partial sums S_n must converge to a limit. For this to occur the difference $S_n - S_{n-1}$ between *successive* partial sums must tend to zero. ➡ But this difference is just the *n*th term:

Otherwise the partial sum sequence wouldn't "level off."

$$S_n - S_{n-1} = (a_1 + a_2 + \cdots + a_{n-1} + a_n) - (a_1 + a_2 + \cdots + a_{n-1}) = a_n.$$

Thus, the terms of a convergent series must tend to zero, as the theorem says.

What the theorem does not say   It's especially important to notice that the *n*th term test does *not* guarantee that $\sum a_n$ converges whenever $a_n \to 0$. For example, the harmonic series $\sum 1/k$ "passes" the *n*th term test, but it *diverges*. ➡

In practice, the *n*th term test is an effective but rather blunt instrument—it sometimes detects divergence, but it *never* detects convergence. We'll develop tools for that purpose in the next section.

We've said this several times, and we'll show it rigorously in the next section.

EXAMPLE 9   Does $\displaystyle\sum_{k=1}^{\infty} \frac{k}{k+1000}$ converge?

Solution   No. Because $a_k = \dfrac{k}{k+1000} \to 1$ as $k \to \infty$, the series diverges.   ■

EXAMPLE 10   Assuming that $\displaystyle\sum_{k=0}^{\infty} \frac{3^k}{k!}$ converges (as it really does), find $\displaystyle\lim_{k \to \infty} \frac{3^k}{k!}$.

Solution   The limit is zero: By Theorem 6, the terms of *every* convergent series must tend to zero.   ■

BASIC EXERCISES

1. Consider the series $\sum_{k=0}^{\infty} a_k = \sum_{k=0}^{\infty} \frac{1}{5^k}$.

   (a) Evaluate $a_1, a_2, a_5, a_{10}, S_1, S_2, S_5,$ and S_{10}.

   (b) Show that the sequence $\{a_k\}$ is decreasing and bounded below.

   (c) Show that the sequence $\{S_n\}$ is increasing and bounded above. What does this imply about the sequence of partial sums?

   (d) Find the sum of the series (i.e., $\lim_{n \to \infty} S_n$).

   (e) Evaluate $R_1, R_2, R_5,$ and R_{10}.

   (f) Show that $\{R_n\}$ is decreasing and bounded below.

   (g) Evaluate $\lim_{n \to \infty} R_n$.

2. Consider the series $\sum_{k=0}^{\infty} a_k = \sum_{k=0}^{\infty} (-0.8)^k$.

   (a) Evaluate $a_1, a_2, a_5, a_{10}, S_1, S_2, S_5,$ and S_{10}.

   (b) Find the sum of the series (i.e., $\lim_{n \to \infty} S_n$).

   (c) Evaluate $R_1, R_2, R_5,$ and R_{10}.

   (d) Is the sequence $\{a_k\}$ decreasing? Is it bounded?

   (e) Is the sequence $\{S_n\}$ increasing? Justify your answer.

   (f) Show that the sequence $\{R_n\}$ is neither increasing nor decreasing.

   (g) Show that the sequence $\{|R_n|\}$ is decreasing.

   (h) Evaluate $\lim_{n \to \infty} R_n$.

3. Consider the series $\sum_{k=0}^{\infty} a_k = \sum_{k=0}^{\infty} \frac{1}{k + 2^k}$.

   (a) Show that the sequence $\{a_k\}$ is decreasing.

   (b) Explain why $a_k \leq 2^{-k}$ for all $k \geq 0$.

   (c) Show that the sequence $\{S_n\}$ is increasing.

   (d) Use part (c) to show that $S_n \leq 2 - 2^{-n} < 2$. [HINT: $\sum_{k=0}^{n} 2^{-k}$ is a geometric series.]

   (e) Show that $\sum_{k=0}^{\infty} a_k$ converges.

4. Consider the series $\sum_{j=0}^{\infty} a_j = \sum_{j=0}^{\infty} \frac{1}{2 + 3^j}$.

   (a) Evaluate $S_1, S_2, S_5,$ and S_{10}.

   (b) Show that the sequence $\{S_n\}$ is increasing and bounded above.

   (c) Does $\sum_{j=0}^{\infty} a_j$ converge? Justify your answer.

5. Consider the series $\sum_{k=0}^{\infty} \frac{1}{k!}$.

   (a) Explain why $\frac{1}{k!} \leq \frac{1}{2^{k-1}}$ if $k \geq 1$. [HINT: Explain why $1/3! < 1/(2 \cdot 2)$ and $1/4! < 1/(2 \cdot 2 \cdot 2)$.]

   (b) Show that the sequence of partial sums $\{S_n\}$ is increasing.

   (c) Use parts (a) and (b) to show that the series converges.

6. The series $\sum_{k=0}^{\infty} a_k = \sum_{k=0}^{\infty} \frac{1}{k!}$ converges to $e \approx 2.718282$.

   (a) Evaluate $a_1, a_2, a_5, a_{10}, S_1, S_2, S_5,$ and S_{10}.

   (b) Show that $\{a_k\}$ is a decreasing sequence.

   (c) Show that $\{S_n\}$ is an increasing sequence.

   (d) Show that $R_n > 0$ for all $n \geq 0$.

   (e) Show that $\{R_n\}$ is a decreasing sequence.

   (f) Find a value of n for which S_n differs from e by less than 0.001.

   (g) Find a value of n for which S_n differs from e by less than 10^{-5}.

   (h) Use parts (e) and (g) to show that $R_{50} < 10^{-5}$.

In Exercises 7 and 8, use the fact that $\sum_{m=1}^{\infty} \frac{1}{m^4} = \frac{\pi^4}{90}$ to evaluate the series.

7. $\sum_{i=0}^{\infty} \frac{1}{(i+1)^4}$

8. $\sum_{k=3}^{\infty} \frac{1}{k^4}$

9. This exercise is about partial sums of the geometric series $\sum_{k=0}^{\infty} ar^k$.

   (a) Find a formula for S_n when $r = 1$. Explain why the formula for S_n (Equation 5) does not hold in this case.

   (b) Show that $S_n - rS_n = (1 - r)S_n$ for any r.

   (c) Use part (b) to show that Equation 5 holds if $r \neq 1$.

10. Use Equation 5 to evaluate $3 + 6 + 12 + 24 + 48 + 96 + \cdots + 3072$.   [HINT: $3072 = 3 \cdot 2^{10}$.]

In Exercises 11–18, find the limit of the series.

11. $\frac{1}{16} + \frac{1}{32} + \frac{1}{64} + \frac{1}{128} + \cdots + \frac{1}{2^{i+4}} + \cdots$

12. $2 - 5 + 9 + \frac{1}{3} + \frac{1}{9} + \frac{1}{27} + \frac{1}{81} + \cdots + \frac{1}{3^n} + \cdots$

13. $\sum_{n=0}^{\infty} e^{-n}$

14. $\sum_{k=3}^{\infty} \left(\frac{e}{\pi} \right)^k$

15. $\sum_{m=2}^{\infty} (\arctan 1)^m$

16. $\sum_{i=10}^{\infty} \left(\frac{2}{3} \right)^i$

17. $\sum_{j=5}^{\infty} \left(-\frac{1}{2} \right)^j$

18. $\sum_{j=0}^{\infty} \frac{3^j + 4^j}{5^j}$

19. Show that $\sum_{k=1}^{\infty} \frac{1}{2 + \sin k}$ diverges.

20. Does $\sum_{k=0}^{\infty} (-1)^k$ converge? Justify your answer.

In Exercises 21–24, find an expression for the partial sum S_n of the series. Use this expression to determine whether the series converges and, if so, to find its limit.

21. $\sum_{k=0}^{\infty} \left(\arctan(k+1) - \arctan k \right)$

22. $\displaystyle\sum_{j=1}^{\infty} \frac{j}{(j+1)!}$

23. $\displaystyle\sum_{m=1}^{\infty} \left(\frac{1}{\sqrt{m}} - \frac{1}{\sqrt{m+2}}\right)$

24. $\displaystyle\sum_{j=1}^{\infty} \ln\left(1+\frac{1}{j}\right)$ [HINT: $\displaystyle\sum_{j=1}^{\infty}\big(\ln(j+1) - \ln j\big)$]

25. Suppose that the partial sums of the series $\displaystyle\sum_{k=1}^{\infty} a_k$ are
$$S_n = \sum_{k=1}^{n} a_k = 5 - \frac{3}{n}.$$

(a) Evaluate $\displaystyle S_{100} = \sum_{k=1}^{100} a_k$.

(b) Evaluate $\displaystyle\sum_{k=1}^{\infty} a_k$.

(c) Evaluate $\displaystyle\lim_{k\to\infty} a_k$.

(d) Show that $a_k > 0$ for all $k \geq 1$. [HINT: $a_{n+1} = S_{n+1} - S_n$.]

26. Let $\displaystyle H_n = \sum_{k=1}^{n} \frac{1}{k}$ and let $\displaystyle S_n = \sum_{k=0}^{n} \frac{1}{2k+1}$.

(a) Explain why $\displaystyle\lim_{n\to\infty} H_n = \infty$.

(b) Show that $S_n \geq \frac{1}{2} H_n$.

(c) What do the results in parts (a) and (b) imply about $\displaystyle\sum_{k=0}^{\infty} 1/(2k+1)$? Justify your answer.

FURTHER EXERCISES

In Exercises 27–29, use the fact that $\displaystyle\sum_{i=1}^{\infty} \frac{1}{i^2} = 1 + \frac{1}{4} + \frac{1}{9} + \frac{1}{16} + \frac{1}{25} + \cdots = \frac{\pi^2}{6}$ *to find the limit of the series.*

27. $\displaystyle\sum_{j=1}^{\infty} \frac{1}{(2j)^2}$

28. $\displaystyle\sum_{k=0}^{\infty} \frac{1}{(2k+1)^2}$ [HINT: See the previous exercise.]

29. $\displaystyle\sum_{m=1}^{\infty} \frac{(-1)^{m+1}}{m^2} = 1 - \frac{1}{4} + \frac{1}{9} - \frac{1}{16} + \frac{1}{25} - \frac{1}{36} + \cdots$

30. Express $\displaystyle\sum_{m=3}^{\infty} \frac{2^{m+4}}{5^m}$ as a rational number.

For each of the series in Exercises 31–36, find all values of x for which the series converges, then state the limit as a simple expression involving x. (Assume that $x^0 = 1$ for all x.)

31. $\displaystyle\sum_{k=0}^{\infty} x^k$

32. $\displaystyle\sum_{m=2}^{\infty} \left(\frac{x}{5}\right)^m$

33. $\displaystyle\sum_{j=5}^{\infty} x^{2j}$

34. $\displaystyle\sum_{k=1}^{\infty} x^{-k}$

35. $\displaystyle\sum_{n=3}^{\infty} (1+x)^n$

36. $\displaystyle\sum_{j=4}^{\infty} \frac{1}{(1-x)^j}$

37. Find the limit of the sequence defined by $S_1 = 1$, $S_{n+1} = S_n + 1/3^n$. [HINT: Write out the first few terms to see the pattern.]

38. Find the limit of the sequence defined by $a_1 = 4$, $a_{n+1} = a_n - 1/2^n$.

In Exercises 39–52, determine whether the series converges or diverges. If a series converges, find its limit. Justify your answers.

39. $\displaystyle\sum_{n=0}^{\infty} \frac{n+1}{2n+1}$

40. $\displaystyle\sum_{j=0}^{\infty} (\ln 2)^j$

41. $\displaystyle\sum_{n=2}^{\infty} \frac{2}{n^2-1}$

42. $\displaystyle\sum_{n=1}^{\infty} \left(1+\frac{1}{n}\right)^n$

43. $\displaystyle\sum_{j=1}^{\infty} \sqrt[j]{\pi}$

44. $\displaystyle\sum_{k=1}^{\infty} \frac{1}{\ln(10^k)}$

45. $\displaystyle\sum_{j=2}^{\infty} \frac{3^j}{4^{j+1}}$

46. $1 - \dfrac{1}{2} - \dfrac{1}{3} - \dfrac{1}{4} - \dfrac{1}{5} - \cdots$

47. $\dfrac{1}{100} + \dfrac{1}{200} + \dfrac{1}{300} + \cdots$

48. $2 - 2 + 2 - 2 + 2 - 2 + \cdots$

49. $\dfrac{3}{10} - \dfrac{3}{20} + \dfrac{3}{40} - \dfrac{3}{80} + \dfrac{3}{160} - \dfrac{3}{320} + \cdots$

50. $1 - \dfrac{1}{2} + \dfrac{1}{2} - \dfrac{1}{3} + \dfrac{1}{3} - \cdots$

51. $1 - 1 + 2 - 1 - 1 + 3 - 1 - 1 - 1 + 4 - 1 - 1 - 1 - 1 + \cdots$

52. $\dfrac{4}{7^{10}} + \dfrac{4}{7^{12}} + \dfrac{4}{7^{14}} + \dfrac{4}{7^{16}} + \dfrac{4}{7^{18}} + \cdots$

53. A rubber ball rebounds to two-thirds the height from which it falls. If it is dropped from a height of 4 feet and is allowed to continue bouncing indefinitely, what is the total distance it travels?

54. Let $\displaystyle S_n = \sum_{k=1}^{n} \frac{1}{\sqrt{k}}$.

(a) Evaluate $\displaystyle\lim_{k\to\infty} \frac{1}{\sqrt{k}}$.

(b) Show that $S_n \geq \dfrac{n}{\sqrt{n}} = \sqrt{n}$ for all $n \geq 1$. [HINT: If $k \leq n$, then $1/k \geq 1/n$.]

(c) Use part (b) to show that $\displaystyle\sum_{k=1}^{\infty} \frac{1}{\sqrt{k}}$ diverges.

55. Use the previous exercise and the fact that $\ln x \le \sqrt{x}$ for all $x \ge 1$ to show that $\sum_{k=2}^{\infty} \frac{1}{\ln k}$ diverges.

56. Let $\{a_k\}$ be an increasing sequence such that $a_1 > 0$ and $a_k \le 100$ for all $k \ge 1$.

    (a) Does $\lim_{k \to \infty} a_k$ exist? Justify your answer.

    (b) Show that $\sum_{k=1}^{\infty} a_k$ diverges.

57. Let $\{a_k\}$ be a sequence of positive terms such that $\sum_{k=1}^{n} a_k \le 100$ for all $n \ge 1$. Explain why $\lim_{k \to \infty} a_k = 0$ must be true.

58. Suppose that the partial sums of the series $\sum_{j=1}^{\infty} b_j$ are
$$S_n = \sum_{j=1}^{n} b_j = \ln\left(\frac{2n+3}{n+1}\right).$$

    (a) Evaluate $\lim_{n \to \infty} S_n$.

    (b) Does the series converge? Justify your answer.

    (c) Show that $b_j < 0$ for all $j \ge 1$.

59. Suppose that the partial sums of the series $\sum_{k=1}^{\infty} a_k$ satisfy the inequality $\frac{6 \ln n}{\ln(n^2 + 1)} < S_n < 3 + ne^{-n}$ for all $n \ge 100$.

    (a) Does the series converge? If so, to what limit? Justify your answers.

    (b) What, if anything, can be said about $\lim_{k \to \infty} a_k$? Explain.

60. Let $S_n = \sum_{k=1}^{n} a_k$, and suppose that $0 \le S_n \le 100$ for all $n \ge 1$.

    (a) Give an example of a sequence $\{a_k\}$ that satisfies these conditions but $\sum_{k=1}^{\infty} a_k$ diverges.

    (b) Show that if $a_k > 0$ for all $k \ge 1$, then $\sum_{k=1}^{\infty} a_k$ converges.

    (c) Show that if $a_k > 0$ for all $k \ge 10^6$, then $\sum_{k=1}^{\infty} a_k$ converges.

61. Suppose that $\sum_{k=1}^{\infty} a_k$ diverges.

    (a) Explain why $a_k > 0$ for all $k \ge 1$ implies that $\lim_{n \to \infty} S_n = \infty$.

    (b) Give an example of a divergent series for which $\lim_{n \to \infty} S_n$ does not exist.

62. Let $H_n = \sum_{k=1}^{n} \frac{1}{k}$, and let $I_n = \int_{1}^{n+1} \frac{dx}{x}$.

    (a) Let L_n be the left Riemann-sum approximation, with n equal subdivisions, to I_n. Show that $L_n = H_n$.

    (b) Use part (a) to show that the harmonic series diverges. [HINT: Start by comparing L_n and I_n.]

63. Let $H_n = \sum_{k=1}^{n} \frac{1}{k}$, and let $a_m = \sum_{j=1}^{2^{m-1}} \frac{1}{2^{m-1} + j}$. Then $H_{2^n} = 1 + \sum_{m=1}^{n} a_m$. [NOTE: a_m is the sum of a "block" of 2^{m-1} consecutive terms of the harmonic series—those from $n = 2^{m-1} + 1$ through $n = 2^m$.]

    (a) Show that $a_1 = 1/2$, $a_2 = 7/12$, and $a_3 = 533/840$.

    (b) Show that $H_8 = 1 + a_1 + a_2 + a_3 = 761/280$.

    (c) Show that $a_k \ge 1/2$ for all $k \ge 1$.

        [HINT: $\frac{1}{2^{m-1} + j} \ge \frac{1}{2^{m-1} + 2^{m-1}}$ if $1 \le j \le 2^{m-1}$.]

    (d) Use part (c) to show that $\lim_{n \to \infty} H_n = \infty$ (i.e., the harmonic series diverges).

64. Consider the series $\sum_{k=1}^{\infty} \frac{1}{k^p}$ with $p > 1$. This exercise outlines a proof that this series converges.

    (a) Let $S_n = \sum_{k=1}^{n} \frac{1}{k^p}$. Show that the sequence of partial sums $\{S_n\}$ is increasing.

    (b) Show that $S_{2m+1} = 1 + \sum_{k=1}^{m} \frac{1}{(2k)^p} + \sum_{k=1}^{m} \frac{1}{(2k+1)^p}$.

    (c) Explain why $S_{2m+1} < 1 + 2 \sum_{k=1}^{m} \frac{1}{(2k)^p}$.
        [HINT: $1/(x+1) < 1/x$ if $x > 0$.]

    (d) Show that $S_{2m+1} < 1 + 2^{1-p} S_{2m+1}$.
        [HINT: First show that $S_{2m+1} < 1 + 2^{1-p} S_m$.]

    (e) Show that $\{S_n\}$ is bounded above.

11.3   TESTING FOR CONVERGENCE; ESTIMATING LIMITS

In theory, the question of convergence is simple: The series $\sum a_k$ converges to the sum S if the sequence $\{S_n\}$ of partial sums tends to S. The trouble, in practice, is that a simple, explicit formula for S_n is often unavailable. It might seem, then, that testing for convergence—let alone finding a limit—would be difficult or impossible. Surprisingly, that isn't so. All the convergence tests of this section and the next (comparison test, integral test, ratio test, and so on) offer clever, indirect ways of testing whether $\{S_n\}$ converges.

Nonnegative series A series $\sum a_k$ for which $a_k \geq 0$ for all k is called **nonnegative**. For such a series, the sequence $\{S_n\}$ of partial sums is nondecreasing: ➤

Convince yourself; it's easy but important.

$$S_1 \leq S_2 \leq S_3 \leq \cdots \leq S_n \leq S_{n+1} \leq \cdots.$$

Nondecreasing sequences are easier to study than others for one main reason: A nondecreasing sequence either (i) converges or (ii) blows up to infinity. ➤ (An arbitrary sequence can diverge without blowing up.) This means, in practice, that the question of convergence or divergence for a *nonnegative* series $\sum a_k$ boils down to the following simpler question:

Theorem 3, page 553, says so.

Do the partial sums S_n blow up or remain bounded as $n \to \infty$?

The answer may be obvious at a glance if a simple formula for S_n is available. ➤ If not, our best strategy may be to *compare* the given series to something that is better understood (another series, an integral—whatever works). Sometimes an obvious comparison suggests itself.

We have such a formula for geometric series.

EXAMPLE 1 Does $\displaystyle\sum_{k=0}^{\infty} \frac{1}{2^k + 1}$ converge?

Solution We saw in the preceding section that the geometric series $\sum_{k=0}^{\infty} 1/2^k$ converges to 2. For all k, it's clear that

$$\frac{1}{2^k + 1} < \frac{1}{2^k},$$

and so each partial sum of $\sum_{k=0}^{\infty} 1/(2^k + 1)$ is *less than* the corresponding partial sum of $\sum_{k=0}^{\infty} 1/2^k$. ➤ Therefore, the partial sums of the original series cannot blow up; they must tend to some limit less than 2. Numerical evidence agrees. For the original series, calculation gives

Adding smaller summands gives a smaller sum.

$$S_{10} \approx 1.263523536; \quad S_{20} \approx 1.264498827; \quad S_{100} \approx 1.264499780. \qquad \blacksquare$$

The comparison test: One series vs. another

In Example 1 we showed that one series converges by comparing it to another series that is *known* to converge. The following theorem makes this idea precise.

THEOREM 7 (Comparison test for nonnegative series) Consider two series $\sum a_k$ and $\sum b_k$, with $0 \leq a_k \leq b_k$ for all k.

- If $\sum b_k$ *converges*, so does $\sum a_k$, and $\sum a_k \leq \sum b_k$.
- If $\sum a_k$ *diverges*, so does $\sum b_k$.

Observe:

- **What we'd expect** The theorem's assertions should seem reasonable. A formal proof uses the fact that *every* partial sum of $\sum a_k$ is less than the corresponding partial sum of $\sum b_k$.

Sequences are "discrete" versions of functions; series are "discrete" versions of integrals.

• **Comparing integrals, comparing series** An almost identical comparison theorem applies to improper integrals. ◄ Theorem 1, Section 10.2, says that if $0 \le a(x) \le b(x)$ for all $x \ge 1$, then

$$\int_1^\infty a(x)\,dx \le \int_1^\infty b(x)\,dx.$$

We did this explicitly in Example 1.

• **Successful comparisons** In a "successful" comparison, either both series converge or both diverge. To use the comparison test, therefore, it's necessary first to *guess* whether the series in question converges or diverges. ◄

Comparing tails Theorem 7 says that if $0 \le a_k \le b_k$ for all $k \ge 1$ and $\sum b_k$ converges, then $\sum a_k$ converges too, and

$$\sum_{k=1}^\infty a_k \le \sum_{k=1}^\infty b_k.$$

A similar inequality holds for upper tails of convergent series. If N is *any* positive integer, $0 \le a_k \le b_k$ for all $k \ge N$, and $\sum b_k$ converges, then $\sum a_k$ converges too, and

$$\sum_{k=N+1}^\infty a_k \le \sum_{k=N+1}^\infty b_k.$$

We'll "compare tails" in the following example to estimate the numerical value of a limit.

EXAMPLE 2 We've seen (using comparison) that $\sum_{k=0}^\infty \dfrac{1}{2^k+1}$ converges, and a calculator gives $S_{100} \approx 1.264499781$. How closely does S_{100} approximate the *true* sum S?

Solution The error in using S_{100} to estimate the limit comes from ignoring the tail R_{100}. In symbols,

$$S = \sum_{k=0}^{100} a_k + \sum_{k=101}^\infty a_k = S_{100} + R_{100}.$$

Clearly, $\dfrac{1}{2^k+1} < \dfrac{1}{2^k}$ for all k; in particular, the inequality holds for $k \ge 101$. Now we compare tails:

$$R_{100} = \sum_{k=101}^\infty \frac{1}{2^k+1} < \sum_{k=101}^\infty \frac{1}{2^k}.$$

The comparison is worthwhile because the right side is a *geometric* series, which we can calculate exactly:

$$\sum_{k=101}^\infty \frac{1}{2^k} = \frac{1}{2^{101}} + \frac{1}{2^{102}} + \frac{1}{2^{103}} + \cdots = \frac{1}{2^{101}}\left(1 + \frac{1}{2} + \frac{1}{2^2} + \cdots\right) = \frac{1}{2^{101}} \cdot 2 = \frac{1}{2^{100}}.$$

This shows that we commit *very* little error by ignoring the tail. The error, R_{100}, is less than $1/2^{100} \approx 8 \times 10^{-31}$. ∎

The integral test: Series vs. integrals

The *idea* of comparison is easy. Harder, in practice, is deciding what to compare a series *to*. Our only reliable "benchmark" series, so far, are geometric series. The **integral test** enlarges our stock of benchmark series considerably.

Comparing areas   Thinking of the terms of a nonnegative *series* $\sum_{k=1}^{\infty} a_k$ as rectangular areas, each with base 1, helps clarify the close connection with the *integral* $\int_1^{\infty} a(x)\,dx$. Figure 1 shows two ways of doing so.

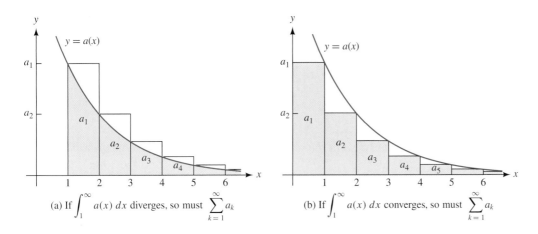

$$(a)\ \text{If } \int_1^{\infty} a(x)\,dx \text{ diverges, so must } \sum_{k=1}^{\infty} a_k \qquad (b)\ \text{If } \int_1^{\infty} a(x)\,dx \text{ converges, so must } \sum_{k=1}^{\infty} a_k$$

FIGURE 1
Relating improper integrals to infinite series

Look closely:

- **Total areas**   The successive rectangles have *heights* $a(1) = a_1$, $a(2) = a_2$, $a(3) = a_3$, and so on; each *base* is 1. The respective *areas*, therefore, are a_1, a_2, a_3, and so on. Thus,

  The series $a_1 + a_2 + a_3 + a_4 + \cdots$ represents the total "left-rule" rectangular area from 1 to ∞.

- **An important inequality**   The shaded area in Figure 1(a) represents the integral $\int_1^{\infty} a(x)\,dx$. Thus, if the integral diverges, so must the series. Here is the message of Figure 1(a) in inequality form:

$$a_1 + a_2 + a_3 + \cdots \geq \int_1^{\infty} a(x)\,dx. \tag{1}$$

  If the right side diverges to infinity, then so must the left.

- **A decreasing integrand**   The reasoning that led to Inequality (1) requires that $a(x)$ be *decreasing* for $x \geq 1$, as shown in the pictures. (We collect such technical hypotheses carefully in the following theorem.)

- **Bounding a series from above** Figure 1(b) shows an integral bounding a series from *above*. This time, comparing areas gives

$$a_2 + a_3 + \cdots \le \int_1^\infty a(x)\,dx,$$

or, equivalently,

$$a_1 + a_2 + a_3 + \cdots \le a_1 + \int_1^\infty a(x)\,dx. \tag{2}$$

If the right side converges, so must the left.

Combining Inequalities (1) and (2) gives *upper* and *lower* bounds for the series:

$$\int_1^\infty a(x)\,dx \le \sum_{k=1}^\infty a_k \le a_1 + \int_1^\infty a(x)\,dx. \tag{3}$$

In particular:

The integral $\int_1^\infty a(x)\,dx$ and the series $\sum_{k=1}^\infty a_k$ either both converge or both diverge.

The final picture, Figure 2, relates the tails of an integral and of a series.

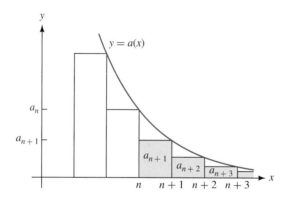

The shaded part is the tail R_n.

FIGURE 2
Comparing tails: Why $\displaystyle\sum_{k=n+1}^\infty a_k \le \int_n^\infty a(x)\,dx$

Comparing areas in Figure 2 shows that, for all n, the tail R_n satisfies

$$R_n = \sum_{k=n+1}^\infty a_k \le \int_n^\infty a(x)\,dx.$$

Lessons from the pictures Figures 1 and 2 show a particular function $a(x)$, but the conclusions we drew hold for *any* function $a(x)$ that is both positive and decreasing. It's time to collect our observations in a theorem.

THEOREM 8 (Integral test for positive series) Suppose that, for all $x \geq 1$, the function $a(x)$ is continuous, positive, and decreasing. Consider the series and the integral

$$\sum_{k=1}^{\infty} a_k \quad \text{and} \quad \int_1^{\infty} a(x)\,dx,$$

where $a_k = a(k)$ for integers $k \geq 1$.

- If either diverges, so does the other.
- If either converges, so does the other. In this case, we have

$$\int_1^{\infty} a(x)\,dx \leq \sum_{k=1}^{\infty} a_k \leq a_1 + \int_1^{\infty} a(x)\,dx.$$

and

$$R_n = \sum_{k=n+1}^{\infty} a_k \leq \int_n^{\infty} a(x)\,dx.$$

Convergent and divergent p-series Series of the form $\sum_{k=1}^{\infty} 1/k^p$ are called **p-series**. They form an important family of examples.

EXAMPLE 3 For which p does the p-series $\displaystyle\sum_{k=1}^{\infty} \frac{1}{k^p}$ converge?

Solution We saw in Chapter 10 that the improper integral $\int_1^{\infty} dx/x^p$ converges if and only if $p > 1$. By the integral test, the same is true for the series. In particular, the harmonic series $\sum 1/k$ diverges. ∎

This useful result deserves special mention:

FACT (The p-test for series) The p-series $\displaystyle\sum_{k=1}^{\infty} \frac{1}{k^p}$ converges if and only if $p > 1$.

Harmonic divergence The fact that the harmonic series

$$1 + \frac{1}{2} + \frac{1}{3} + \frac{1}{4} + \frac{1}{5} + \cdots$$

diverges to infinity—even though the terms themselves tend to zero—has fascinated mathematicians for many centuries. One early proof (unrelated to integrals) is attributed to Nicole Oresme, a 14th-century French bishop, scientist, and mathematician.

EXAMPLE 4 The p-series $\sum_{k=1}^{\infty} 1/k^3$ converges to some limit S. (The integral test says so.) How large must n be to ensure that S_n differs from S by less than 0.0001?

Solution   We need to choose n so that the tail R_n is less than 0.0001. By the last inequality in Theorem 8,

$$R_n \leq \int_n^\infty \frac{dx}{x^3} = \frac{1}{2n^2}.$$

(The last step is an easy calculation.) This quantity (and hence also R_n) are less than 0.0001 if $n \geq 71$. Hence, $S_{71} \approx 1.20196$ differs from the true limit by less than 0.0001. ■

EXAMPLE 5   Does $\displaystyle\sum_{k=1}^\infty \frac{1}{10k+1}$ converge?

Solution   The integral test says no.

$$\int_1^\infty \frac{1}{10x+1}\,dx = \lim_{n\to\infty} \int_1^n \frac{1}{10x+1}\,dx = \lim_{n\to\infty} \frac{\ln(10x+1)}{10}\bigg]_1^n = \infty.$$

The comparison test *also* says no. Since $10k+1 \leq 11k$ for all $k \geq 1$, we have

$$\frac{1}{10k+1} \geq \frac{1}{11k} \implies \sum_{k=1}^\infty \frac{1}{10k+1} \geq \sum_{k=1}^\infty \frac{1}{11k} = \frac{1}{11}\sum_{k=1}^\infty \frac{1}{k}.$$

Because the last series diverges to infinity, so must the first. ■

The ratio test: Comparison with a geometric series

In a geometric series $a + ar + ar^2 + ar^3 + ar^4 + \cdots$, the ratio of successive terms is r; by Theorem 4, page 561, the series converges if and only if $|r| < 1$.

The **ratio test** is also based on ratios of successive terms; it amounts to a lightly disguised form of comparison with a geometric series.

THEOREM 9 (Ratio test for positive series)   Suppose that $a_k > 0$ for all k and that

$$\lim_{k\to\infty} \frac{a_{k+1}}{a_k} = L.$$

- If $L < 1$, then $\sum a_k$ *converges*.
- If $L > 1$, then $\sum a_k$ *diverges*.
- If $L = 1$, the test is inconclusive—either convergence or divergence is possible.

Two points deserve emphasis:

- **Best cases**   The ratio test works best for series like $\sum 1/k!$, $\sum r^k$, and $\sum 1/(2^k+3)$, in which the index k appears in an exponent or a factorial.

The limit is 1 for every p-series.

- **Other cases**   For many series, unfortunately, the ratio a_{k+1}/a_k either tends to 1 or has *no* limit. ◄ In such cases, the ratio test tells us nothing.

EXAMPLE 6 Show that $\sum\limits_{k=0}^{\infty} \dfrac{1}{k!}$ converges. Guess its limit.

Solution The ratio test works nicely. Since

$$\lim_{k \to \infty} \frac{a_{k+1}}{a_k} = \lim_{k \to \infty} \frac{k!}{(k+1)!} = \lim_{k \to \infty} \frac{1}{k+1} = 0,$$

the series converges. In fact, it converges very, very fast. Here are some representative partial sums:

$$S_5 \approx 2.716667; \quad S_{10} \approx 2.718281; \quad S_{30} \approx 2.7182818284590452353602874 7135.$$

Is e involved somehow? It certainly seems so—S_{30} agrees with e in all decimal places shown. In fact, this series can be shown to converge to e. We explore this phenomenon further in later sections. ∎

EXAMPLE 7 Does $\sum\limits_{k=0}^{\infty} \dfrac{100^k}{k!}$ converge?

Solution Yes, by the ratio test:

$$\lim_{k \to \infty} \frac{a_{k+1}}{a_k} = \lim_{k \to \infty} \frac{100^{k+1}}{(k+1)!} \cdot \frac{k!}{100^k} = \lim_{k \to \infty} \frac{100}{k+1} = 0.$$

Notice what the result means: Even though 100^k grows very fast, $k!$ grows even faster. ∎

Why the ratio test works   To illustrate the connection between the ratio test and geometric series, and to give the idea of a proof, let's suppose, for instance, that

$$\lim_{k \to \infty} \frac{a_{k+1}}{a_k} = \frac{1}{2}.$$

Why must $\sum a_k$ converge?

The idea is that $a_{k+1} \approx a_k/2$ for large k, and so the given series should be comparable to a *geometric* series. Suppose, for instance, that $a_{k+1} < 0.6a_k$ for all $k \ge 1000$. ➡ Then

Such an inequality must hold for large k because the ratio limit is 1/2.

$$a_{1001} < (0.6)a_{1000}, \quad a_{1002} < (0.6)a_{1001} < (0.6)^2 a_{1000}, \quad a_{1003} < (0.6)^3 a_{1000}, \ldots.$$

Therefore,

$$a_{1000} + a_{1001} + a_{1002} + \cdots < a_{1000} \left(1 + (0.6) + (0.6)^2 + (0.6)^3 + \cdots \right).$$

The preceding inequality is the point; it shows that an upper tail of the original series $\sum a_k$ converges by comparison with the geometric series in the last line. (The divergence part of the ratio test can be proved using similar ideas.)

BASIC EXERCISES

1. Consider the series $\sum\limits_{k=0}^{\infty} a_k$, where $a_k = \dfrac{1}{k+2^k}$.

   (a) Use the comparison test to show that the series converges.

   (b) Show that $0 \le R_{10} \le 2^{-10}$.

   (c) Compute an estimate of the limit of the series that is guaranteed to be within 0.001 of the exact value.

   (d) Is your estimate in part (c) an overestimate? Justify your answer.

2. Consider the series $\displaystyle\sum_{j=0}^{\infty} \frac{1}{2+3^j}$.

   (a) Show that the series converges.

   (b) Estimate the limit of the series within 0.01.

   (c) Is your estimate in part (b) an overestimate? Justify your answer.

In Exercises 3 and 4, suppose that $a(x)$ is continuous, positive, and decreasing for all $x \geq 1$ and that $a_k = a(k)$ for all integers $k \geq 1$.

3. Rank the values $\displaystyle\int_1^n a(x)\,dx$, $\displaystyle\sum_{k=1}^{n-1} a_k$, and $\displaystyle\sum_{k=2}^{n} a_k$ in increasing order.   [HINT: Draw a picture.]

4. Rank the values $\displaystyle\int_n^\infty a(x)\,dx$, $\displaystyle\sum_{k=n+1}^{\infty} a_k$, and $\displaystyle\int_{n+1}^\infty a(x)\,dx$ in increasing order.

Suppose that $a(x)$ is continuous, positive, and decreasing for all $x \geq 1$ and that $a_k = a(k)$ for all integers $k \geq 1$. In Exercises 5–8, draw a carefully annotated picure that shows that

5. $\displaystyle\int_1^{n+1} a(x)\,dx \leq \sum_{k=1}^{n} a_k$

6. $\displaystyle\sum_{k=2}^{n} a_k \leq \int_1^n a(x)\,dx$

7. $\displaystyle\sum_{k=n+1}^{\infty} a_k \leq a_{n+1} + \int_{n+1}^\infty a(x)\,dx$

8. $\displaystyle\sum_{k=n+1}^{\infty} a_k \leq \int_n^\infty a(x)\,dx$

In Exercises 9–12, use the integral test to find upper and lower bounds on the limit of the series.

9. $\displaystyle\sum_{k=1}^{\infty} \frac{1}{k^3}$

10. $\displaystyle\sum_{k=1}^{\infty} \frac{1}{k\sqrt{k}}$

11. $\displaystyle\sum_{j=1}^{\infty} je^{-j}$

12. $\displaystyle\sum_{k=0}^{\infty} \frac{1}{k^2+1}$

13. Consider the series $\displaystyle\sum_{k=1}^{\infty} \frac{e^{\sin k}}{k^2}$.

   (a) $\displaystyle\int_1^\infty \frac{e^{\sin k}}{k^2}$ is a convergent improper integral. Does it follow from this fact and Theorem 8 that the series converges? Justify your answer.

   (b) Show that this series converges.

14. Use the fact that $\ln x \leq \sqrt{x}$ for all $x \geq 1$ to show that $\displaystyle\sum_{n=2}^{\infty} \frac{1}{(\ln n)^2}$ diverges.

15. Use the fact that $\ln x \leq x$ for all $x \geq 1$ to show that $\displaystyle\sum_{n=2}^{\infty} \frac{1}{(\ln n)^2}$ diverges.   [HINT: $x \ln x \geq (\ln x)^2$ if $x \geq 1$.]

16. Show that $\displaystyle\sum_{k=3}^{\infty} \frac{1}{(\ln k)^k}$ converges.   [HINT: $\ln 3 \approx 1.0986$.]

In Exercises 17–20, use the comparison test to show that the series converges. Then find upper and lower bounds on the limit of the series.

17. $\displaystyle\sum_{n=1}^{\infty} \frac{1}{n^2 + \sqrt{n}}$

18. $\displaystyle\sum_{j=0}^{\infty} \frac{1}{j + e^j}$

19. $\displaystyle\sum_{m=1}^{\infty} \frac{1}{m\sqrt{1+m^2}}$

20. $\displaystyle\sum_{k=1}^{\infty} \frac{k}{(k^2+1)^2}$

In Exercises 21–24, use the ratio test to show that the series converges.

21. $\displaystyle\sum_{j=0}^{\infty} \frac{j^2}{j!}$

22. $\displaystyle\sum_{k=1}^{\infty} \frac{2^k}{k!}$

23. $\displaystyle\sum_{n=1}^{\infty} \frac{n^2}{2^n}$

24. $\displaystyle\sum_{m=1}^{\infty} \frac{m!}{(2m)!}$

25. Consider the series $\displaystyle\sum_{k=1}^{\infty} a_k = \sum_{k=1}^{\infty} \frac{\ln k}{k}$.

   (a) Use the integral test to show that the series diverges. [HINT: Be careful, the integrand is not monotone.]

   (b) Use the comparison test to show that the series diverges. [HINT: $1 - x^{-1} \leq \ln x$ for all $x > 0$.]

   (c) Can the ratio test be used to show that the series diverges? Explain.

26. Consider the series $\displaystyle\sum_{k=1}^{\infty} a_k = \frac{1}{2} + \frac{1}{3} + \frac{1}{2^2} + \frac{1}{3^2} + \frac{1}{2^3} + \frac{1}{3^3} + \cdots$.

   (a) Explain why $\displaystyle\lim_{k\to\infty} a_{k+1}/a_k$ does not exist.

   (b) What, if anything, does the ratio test say about the convergence of $\displaystyle\sum_{k=1}^{\infty} a_k$?

   (c) Show that the series converges, and evaluate its limit. [HINT: Rewrite the series as the sum of two convergent series.]

27. (a) What, if anything, does the ratio test say about the convergence of the series $\dfrac{1}{2} + \dfrac{1}{2} + \dfrac{1}{4} + \dfrac{1}{4} + \dfrac{1}{8} + \dfrac{1}{8} + \cdots$?

   (b) Does the series in part (a) converge or diverge? Justify your answer.

28. Give an example of a divergent series $\displaystyle\sum_{k=1}^{\infty} a_k$ such that $a_k > 0$ and $a_{k+1}/a_k < 1$ for all $k \geq 1$.

29. Use the ratio test to show that $\displaystyle\sum_{n=1}^{\infty} n^{-n}$ converges.

30. Use the ratio test to show that the series $\displaystyle\sum_{n=1}^{\infty} \frac{n^n}{n!}$ diverges.

FURTHER EXERCISES

In Exercises 31–36, determine whether the series converges or diverges. If the series converges, find a number N such that the partial sum S_N approximates the sum of the series within 0.001. If the series diverges, find a number N such that $S_N \geq 1000$.

31. $\displaystyle\sum_{k=0}^{\infty} \frac{1}{k^2+3}$

32. $\displaystyle\sum_{m=1}^{\infty} \frac{\arctan m}{m}$

33. $\displaystyle\sum_{k=0}^{\infty} \frac{1}{2+\cos k}$

34. $\displaystyle\sum_{m=2}^{\infty} \frac{\ln m}{m^3}$

35. $\displaystyle\sum_{k=0}^{\infty} \frac{k}{k^6+17}$

36. $\displaystyle\sum_{k=2}^{\infty} \frac{1}{k(\ln k)^5}$

In Exercises 37 and 38, let $S_n = \displaystyle\sum_{k=1}^{n} a_k$, $T_n = \displaystyle\sum_{k=1}^{n} b_k$, and assume that $0 \leq a_k \leq b_k$ for all $k \geq 1$.

37. (a) Suppose that $\displaystyle\sum_{k=1}^{\infty} b_k$ converges. Explain why there is a number M such that $S_n \leq T_n \leq M$ for all $n \geq 1$.

(b) Explain why $\{S_n\}$ is an increasing sequence.

(c) Explain why parts (a) and (b) together imply that $\displaystyle\sum_{k=1}^{\infty} a_k$ converges.

38. (a) Suppose that $\displaystyle\sum_{k=1}^{\infty} a_k$ diverges. Explain why $\displaystyle\lim_{n\to\infty} S_n = \infty$.

(b) Suppose that $\displaystyle\sum_{k=1}^{\infty} a_k$ diverges. Use part (a) to show that $\displaystyle\sum_{k=1}^{\infty} b_k$ diverges.

39. Suppose that $a(x)$ is continuous, positive, and decreasing for all $x \geq 1$, that $a_k = a(k)$ for all integers $k \geq 1$, and that $\displaystyle\int_1^{\infty} a(x)\,dx$ converges.

(a) Explain why the sequence of partial sums $\{S_n\}$ is an increasing sequence.

(b) Explain why $\displaystyle\int_1^{n} a(x)\,dx \leq \int_1^{\infty} a(x)\,dx$.

(c) Use parts (a) and (b) to show that the sequence of partial sums $\{S_n\}$ converges.

40. (a) Where in the proof of the integral test (Theorem 8) is the assumption that $a(x)$ is a decreasing function used?

(b) Suppose that the requirement that $a(x)$ be decreasing for all $x \geq 1$ is replaced by the "weaker" requirement that $a(x)$ be decreasing for all $x \geq 10$. How does this change in assumptions affect the conclusions of Theorem 8?

41. Does the series $1 + \dfrac{1}{1\cdot3} + \dfrac{1}{1\cdot3\cdot5} + \dfrac{1}{1\cdot3\cdot5\cdot7} + \cdots + \dfrac{1}{1\cdot3\cdot5\cdot7\cdots(2k+1)} + \cdots$ converge? Justify your answer.

42. Does the series $\dfrac{1}{2} + \dfrac{1}{2\cdot4} + \dfrac{1}{2\cdot4\cdot6} + \cdots + \dfrac{1}{2\cdot4\cdot6\cdots(2k)} + \cdots$ converge? Justify your answer.

43. Show that $\displaystyle\sum_{k=3}^{\infty} \frac{1}{(\ln k)^{\ln k}}$ converges.

[HINT: $\ln k > e^2$ if $k > 1619$.]

44. Suppose that $a_n \geq 0$ for all $n \geq 1$ and that $\displaystyle\sum_{n=1}^{\infty} a_n$ converges.

Show that $\displaystyle\sum_{n=1}^{\infty} |\sin a_n|$ converges.

[HINT: $|\sin x| \leq |x|$ for all x.]

In Exercises 45–52, determine whether the series converges. If it converges, find upper and lower bounds on its limit. Justify your answers.

45. $\displaystyle\sum_{n=1}^{\infty} \frac{\arctan n}{1+n^2}$

46. $\displaystyle\sum_{m=1}^{\infty} \frac{m^3}{m^5+3}$

47. $\displaystyle\sum_{j=1}^{\infty} \frac{1}{100+5j}$

48. $\displaystyle\sum_{k=2}^{\infty} \frac{1}{k\ln k}$

49. $\displaystyle\sum_{n=1}^{\infty} \frac{1}{n\,3^n}$

50. $\displaystyle\sum_{n=2}^{\infty} \frac{1}{\sqrt[3]{n^2-1}}$

51. $\displaystyle\sum_{j=0}^{\infty} \frac{j!}{(j+2)!}$

52. $\displaystyle\sum_{n=0}^{\infty} \frac{n!}{(2n)!}$

53. Suppose that $\{a_k\}$ is a sequence of positive terms and that r is a constant such that $a_{k+1}/a_k \leq r < 1$ for all $k \geq 1$.

(a) Show that $a_2 \leq a_1 r$.

(b) Show that $a_{k+1} \leq a_1 r^k$ for every $k \geq 1$.

(c) Use part (b) to show that $\displaystyle\sum_{k=1}^{\infty} a_k$ converges.

[HINT: Use the comparison test and the formula for the sum of geometric series.]

(d) Show that $R_n = \displaystyle\sum_{k=n+1}^{\infty} a_k \leq \frac{a_{n+1}}{1-r}$.

54. Use Exercise 53 to find an integer N such that the partial sum S_N approximates the sum of the series $\displaystyle\sum_{n=1}^{\infty} n^2/2^n$ within 0.0005.

55. Use Exercise 53 to find an integer N such that the partial sum S_N approximates the sum of the series $\displaystyle\sum_{n=1}^{\infty} (n!)^2/(2n)!$ within 0.0005.

11.4   ABSOLUTE CONVERGENCE; ALTERNATING SERIES

Not-necessarily-positive series   The integral, comparison, and ratio tests apply only to *nonnegative* series. Some interesting series, however, have both positive and negative terms.

E X A M P L E 1   Does the **alternating harmonic series**

$$\sum_{k=1}^{\infty} \frac{(-1)^{k+1}}{k} = 1 - \frac{1}{2} + \frac{1}{3} - \frac{1}{4} + \frac{1}{5} - \cdots$$

converge or diverge? To what limit?

Solution   As always, the question is how partial sums behave. Tabulating some of them shows a pattern.

Partial sums of $1 - \frac{1}{2} + \frac{1}{3} - \frac{1}{4} + \frac{1}{5} - \cdots$										
n	1	2	3	4	5	6	7	8	9	10	...
S_n	1	0.500	0.833	0.583	0.783	0.617	0.760	0.635	0.746	0.646	...
n	51	52	53	54	55	56	57	58	59	60	...
S_n	0.703	0.684	0.702	0.684	0.702	0.684	0.702	0.685	0.702	0.685	...

Successive partial sums seem to hop back and forth across some limiting value. Plots of partial sums and terms (Figure 1) exhibit the same pattern:

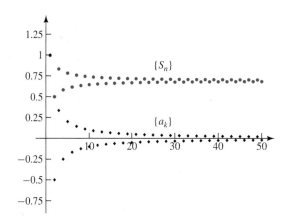

Blue dots show partial sums; black dots show terms.

FIGURE 1

Terms and partial sums for $1 - \frac{1}{2} + \frac{1}{3} - \frac{1}{4} + \frac{1}{5} - \cdots$

Because the terms alternate in sign, the partial sums successively rise and fall, alternately overshooting and undershooting the limiting value, which is apparently around 0.69. (It can be shown—with considerable effort—that the exact limit is $\ln 2 \approx 0.69315$.) ∎

Absolute vs. conditional convergence

The alternating harmonic series illustrates the phenomenon of **conditional convergence**. Although

$$1 - \frac{1}{2} + \frac{1}{3} - \frac{1}{4} + \frac{1}{5} - \cdots$$

converges (as Example 1 suggests), the *ordinary* harmonic series

$$1 + \frac{1}{2} + \frac{1}{3} + \frac{1}{4} + \frac{1}{5} + \cdots$$

diverges, as we saw from the integral test.

EXAMPLE 2   Does $\displaystyle\sum_{k=1}^{\infty} \frac{\sin k}{k^2}$ converge? Does $\displaystyle\sum_{k=1}^{\infty} \frac{|\sin k|}{k^2}$? Estimate limits.

Solution   The first series, like the alternating harmonic series, has both negative and positive terms—although in no regular order this time. Plotting terms and partial sums (Figure 2) suggests (but doesn't prove) that this series converges.

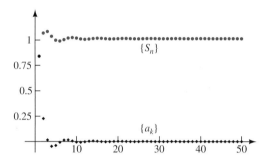

FIGURE 2

Terms and partial sums for $\displaystyle\sum_{k=1}^{\infty} \frac{\sin k}{k^2}$

The partial sums wander up and down just slightly but still appear to approach a horizontal asymptote, perhaps near $y = 1$.

   The second series also seems to converge, judging from Figure 3. This time the limit appears to be near 1.25.

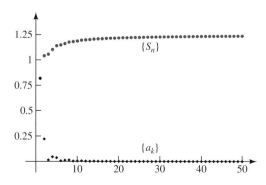

FIGURE 3

Terms and partial sums for $\displaystyle\sum_{k=1}^{\infty} \frac{|\sin k|}{k^2}$

If $\sum a_k$ is the first series, $\sum |a_k|$ is the second.

The second series comes from the first by taking the absolute value of each term. ← The new series is nonnegative, and so the comparison test applies. Because

$$0 \le \frac{|\sin k|}{k^2} \le \frac{1}{k^2}$$

for all $k \ge 1$ and $\sum_{k=1}^{\infty} 1/k^2$ converges, so must $\sum_{k=1}^{\infty} |\sin k|/k^2$. ∎

Examples 1 and 2 illustrate the phenomena of **conditional convergence** and **absolute convergence**, respectively. The formal definitions are as follows:

DEFINITION (Absolute and conditional convergence)   Let $\sum a_k$ be any series.

- If $\sum |a_k|$ converges, then $\sum a_k$ **converges absolutely**.
- If $\sum |a_k|$ diverges but $\sum a_k$ converges, then $\sum a_k$ **converges conditionally**.

The wacky world of conditional convergence   Conditionally convergent series have some surprising properties. Here is one of the oddest:

Let $\sum a_k$ be conditionally convergent, and let L be any real number. Then the terms of $\sum a_k$ can be reordered in such a way that the resulting series converges to L.

(For more details, see your instructor.) Notice how drastically this peculiar property of conditionally convergent series upsets the naive hope that addition is commutative even for infinite sums.

Pluses and minuses of pluses and minuses   Let $\sum_{k=1}^{\infty} a_k$ be any series. If it happens that $a_k \ge 0$ for all k, the advantage is simplicity: The partial sums are nondecreasing. The disadvantage, as the harmonic series shows, is that the partial sums may tend to infinity.

Mixing positive and negative terms may cost something in simplicity, but it's an advantage for convergence. As the alternating harmonic series shows, positive and negative terms can offset each other, thus helping the cause of convergence.

Absolute convergence implies ordinary convergence   We saw in Chapter 10 that if $\int_1^{\infty} |f(x)|\, dx$ converges, then so must $\int_1^{\infty} f(x)\, dx$, and $\left| \int_1^{\infty} f(x)\, dx \right| \le \int_1^{\infty} |f(x)|\, dx$. (See Theorem 2, page 534.) The same principle applies to infinite series:

THEOREM 10   If $\sum_{k=1}^{\infty} |a_k|$ converges, then so does $\sum_{k=1}^{\infty} a_k$, and

$$\left| \sum_{k=1}^{\infty} a_k \right| \le \sum_{k=1}^{\infty} |a_k|.$$

The theorem's inequality should be believable—it is the infinite version of the fact that the absolute value of a sum can't exceed the sum of the absolute values. The idea of a rigorous proof is to write the original series as a sum of two new series, one entirely positive and the other entirely negative. Using the comparison test, one can show that each of these new series converges.

The theorem shows that, as Figure 3 suggested, the series $\sum_{k=1}^{\infty} \sin k / k^2$ of Example 2 does indeed converge, because $\sum_{k=1}^{\infty} |\sin k| / k^2$ does. Our limit estimates from Example 2 are also consistent with the theorem:

$$1 \approx \left| \sum_{k=1}^{\infty} \frac{\sin k}{k^2} \right| \leq \sum_{k=1}^{\infty} \frac{|\sin k|}{k^2} \approx 1.25.$$

EXAMPLE 3 For which values of x does the **power series**

$$\sum_{k=1}^{\infty} k x^k = x + 2x^2 + 3x^3 + 4x^4 + \cdots$$

converge? (A power series is something like an "infinite polynomial"; we discuss power series carefully in the next section.)

Solution First we use the ratio test to check for *absolute* convergence:

$$\lim_{k \to \infty} \left| \frac{(k+1)x^{k+1}}{kx^k} \right| = \lim_{k \to \infty} \left| \frac{k+1}{k} \right| \cdot |x| = 1 \cdot |x| = |x|.$$

If $|x| < 1$, the original series converges absolutely. By Theorem 10, it also converges *without* absolute value signs.

If $|x| \geq 1$, the series fails the nth term test, and so diverges. ∎

Estimating limits

Estimating a limit for any series—nonnegative or not—depends upon keeping the upper tail small. Theorem 10, combined with earlier estimates, can help.

EXAMPLE 4 For the series $\sum_{k=1}^{\infty} (-1)^{k+1} / k^3$, we find (using technology) that $S_{100} \approx 0.901542$. How closely does S_{100} approximate S, the true sum of the series?

Solution Because

$$S = \sum_{k=1}^{100} \frac{(-1)^{k+1}}{k^3} + \sum_{k=101}^{\infty} \frac{(-1)^{k+1}}{k^3} = S_{100} + R_{100},$$

we need only estimate R_{100}, as follows:

$$|R_{100}| = \left| \sum_{k=101}^{\infty} \frac{(-1)^{k+1}}{k^3} \right| \leq \sum_{k=101}^{\infty} \frac{1}{k^3} \qquad \text{(by Theorem 10)}$$

$$< \int_{100}^{\infty} \frac{1}{x^3} \, dx. \qquad \text{(by the integral test)}$$

The last integral is easy to calculate:

$$\int_{100}^{\infty} \frac{1}{x^3} \, dx = \frac{-1}{2x^2} \Big]_{100}^{\infty} = \frac{1}{20{,}000} = 0.00005,$$

and so the estimate $S \approx S_{100} \approx 0.901542$ is good to at least four decimal places. ∎

Alternating series

For series with both positive and negative terms, testing for absolute convergence is usually the best option. In the special (but surprisingly useful) case that the terms alternate in sign, we can sometimes do better.

> **DEFINITION**   An **alternating series** is one whose terms alternate in sign, that is, a series of the form
>
> $$c_1 - c_2 + c_3 - c_4 + c_5 - c_6 + \cdots,$$
>
> where each c_i is positive.

The **alternating harmonic series**

$$1 - \frac{1}{2} + \frac{1}{3} - \frac{1}{4} + \frac{1}{5} - \cdots$$

See Example 1, especially Figure 1.

illustrates the best possibility. ◄ Because successive terms alternate in sign and decrease in size, successive partial sums straddle smaller and smaller intervals. If the terms also tend to zero, then the partial sums narrow down on a limit. The following theorem makes these observations formal; it also gives a convenient error bound.

> **THEOREM 11 (Alternating series test)**   Consider the series
>
> $$\sum_{k=1}^{\infty} (-1)^{k+1} c_k = c_1 - c_2 + c_3 - c_4 + \cdots,$$
>
> where
>
> $$\text{(i)}\quad c_1 \geq c_2 \geq c_3 \geq \cdots \geq 0; \qquad \text{and} \qquad \text{(ii)}\quad \lim_{k \to \infty} c_k = 0.$$
>
> Then the series converges, and its sum S lies *between* any two successive partial sums S_n and S_{n+1}. In particular,
>
> $$|S - S_n| < c_{n+1}$$
>
> for all $n \geq 1$.

A formal proof is slightly tricky, but the underlying idea is simple: Because the limit S lies *between* successive partial sums, adding another term to any partial sum always "overshoots" the limit, which explains the final inequality.

Using the theorems: Miscellaneous examples

Combining Theorems 10 and 11 with results from earlier sections enables us to handle many not-necessarily-positive series, detecting convergence or divergence and estimating limits. The following examples illustrate some useful tricks of this trade.

EXAMPLE 5   (An alternating p-series: another look)   What does Theorem 11 say about $\sum_{k=1}^{\infty} (-1)^{k+1}/k^3$ and its 100th partial sum, $S_{100} \approx 0.9015422$?

Solution In this context, $c_k = 1/k^3$. Now Theorem 11 says not only that the series converges—which we already knew—but also that

$$|S - S_{100}| < c_{101} = \frac{1}{101^3} \approx 0.000001.$$

Thus, $S_{100} \approx 0.9015422$ lies within 0.000001 of the true limit S. Equivalently, S lies between $S_{100} \approx 0.9015422$ and $S_{101} \approx 0.9015432$. ■

EXAMPLE 6 Does $\displaystyle\sum_{j=1}^{\infty}(-1)^j \frac{j}{j+1}$ converge or diverge? Why?

Solution The alternating series test looks tempting at first glance. But it doesn't apply, because

$$\lim_{j \to \infty} \frac{j}{j+1} = \lim_{j \to \infty} \frac{1}{1 + 1/j} = 1,$$

not zero as Theorem 11 requires. The nth term test does apply, however. (Maybe we should call it the "jth term test" here.) Since

$$\frac{j}{j+1} \to 1 \quad \text{as} \quad j \to \infty,$$

it follows that the jth term has no limit as $j \to \infty$. ➧ Therefore, the given series diverges. ■

Successive terms are alternately near 1 and −1, and so the terms diverge.

EXAMPLE 7 Does the series

$$1 + 2 + 3 + 4 + 5 - \frac{1}{6} + \frac{1}{7} - \frac{1}{8} + \frac{1}{9} - \cdots$$

converge? If so, find or estimate the limit.

Solution The alternating series test doesn't apply right out of the box because the first five terms break the desired pattern. The problem isn't fatal, however. Basic series algebra lets us group our terms into two blocks, as follows:

$$(1 + 2 + 3 + 4 + 5) - \left(\frac{1}{6} - \frac{1}{7} + \frac{1}{8} - \frac{1}{9} + \cdots \right).$$

The first block is finite, so convergence isn't an issue; its sum is 15. The second block clearly satisfies all hypotheses of the alternating series test and so converges to some limit L. Any partial sum of the second block, moreover, differs from L by less than the magnitude of the next term (by the last line of Theorem 11).

   The entire series therefore converges to $S = 15 - L$, and any partial sum differs from S by no more than the next term. The partial sum $S_9 = 1 + 2 + \cdots + 1/9 \approx 14.962$, for instance, overshoots the true limit by less than $1/10$. In other words, the exact limit S satisfies $14.862 \le S \le 14.962$. ■

EXAMPLE 8 Does $\displaystyle\sum_{n=1}^{\infty} \frac{\sin n}{n^3 + n^2 + n + 1 + \cos n}$ converge or diverge? Why?

Solution The problem is easier than it looks. Hoping for absolute convergence, we start by taking absolute values:

$$\left| \frac{\sin n}{n^3 + n^2 + n + 1 + \cos n} \right| = \frac{|\sin n|}{n^3 + n^2 + n + 1 + \cos n}.$$

The general appearance of numerator and denominator suggests a comparison. A simple inequality make the job much easier:

$$\frac{|\sin n|}{n^3 + n^2 + n + 1 + \cos n} \le \frac{1}{n^3}.$$

Because the p-series $\sum 1/n^3$ converges, so must the absolute-value version of the given series. Finally, Theorem 10 guarantees that the original series also converges. ∎

BASIC EXERCISES

Exercises 1–4 are about the series in Example 7.

1. Does the series converge conditionally or absolutely? Justify your answer.

2. Does the partial sum S_{15} overestimate the limit of the series? Justify your answer.

3. $S_{60} \approx 14.902$. Use this result to find upper and lower bounds for S.

4. The alternating harmonic series can be shown to converge to $\ln 2$. Use this fact to find the limit of the series exactly.

5. Consider the series in Example 8.
   (a) Compute S_{50}.
   (b) Explain why $|R_{50}| \le \int_{50}^{\infty} \frac{dx}{x^3}$.
   (c) Use parts (a) and (b) to find upper and lower bounds for the limit of the series.

6. Suppose that $\sum_{k=1}^{\infty} a_k$ converges absolutely. Show that $\sum_{k=1}^{\infty} \frac{a_k}{k}$ also converges.

7. Suppose that $\sum_{k=1}^{\infty} \frac{a_k}{k}$ converges. Must $\sum_{k=1}^{\infty} a_k$ also converge? Justify your answer.

8. Show that $\sum_{k=2}^{\infty} (-1)^k \frac{k}{k^2 - 1}$ converges conditionally.

In Exercises 9–12, show that the series converges. Then compute an estimate of the limit that is guaranteed to be in error by no more than 0.005.

9. $\sum_{k=1}^{\infty} \frac{(-1)^k}{k^4}$

10. $\sum_{k=1}^{\infty} \frac{(-1)^k}{k^2 + 2^k}$

11. $\sum_{k=0}^{\infty} \frac{(-3)^k}{(k^2)!}$

12. $\sum_{k=5}^{\infty} (-1)^k \frac{k^{10}}{10^k}$

13. Suppose that $a_k \ge 0$ for all $k \ge 1$. Is it possible that $\sum_{k=1}^{\infty} a_k$ converges conditionally? Justify your answer.

14. Does $\sum_{n=2}^{\infty} (-1)^n \frac{n}{2n - 1}$ converge? Justify your answer.

FURTHER EXERCISES

15. For which values of p, if any, does $\sum_{k=2}^{\infty} \frac{\ln k}{k^p}$ converge? Justify your answer.

16. For which values of p, if any, does $\sum_{k=2}^{\infty} (-1)^k \frac{\ln k}{k^p}$ converge? Justify your answer.

17. For which values of p, if any, does $\sum_{k=2}^{\infty} (-1)^k \frac{\ln k}{k^p}$ converge absolutely? Justify your answer.

18. For which values of p, if any, does $\sum_{k=2}^{\infty} (-1)^k \frac{\ln k}{k^p}$ converge conditionally? Justify your answer.

In Exercises 19–26, determine whether the series converges absolutely, converges conditionally, or diverges. If the series converges, find upper and lower bounds on its limit. Justify your answers.

19. $\sum_{j=1}^{\infty} \frac{(-1)^{j+1}}{j^2}$

20. $\sum_{k=1}^{\infty} \frac{(-1)^k}{\sqrt{k}}$

21. $\sum_{n=1}^{\infty} \frac{(-3)^n}{n^3}$

22. $\sum_{k=4}^{\infty} (-1)^k \frac{\ln k}{k}$

23. $\displaystyle\sum_{k=0}^{\infty}(-1)^k\frac{k}{2k+1}$

24. $\displaystyle\sum_{m=0}^{\infty}(-1)^m\frac{m^3}{2^m}$

25. $\displaystyle\sum_{j=0}^{\infty}(-1)^j\frac{j!}{(j^2)!}$

26. $\displaystyle\sum_{n=1}^{\infty}(-1)^{n+1}\frac{\arctan n}{n}$

27. Consider the series $\displaystyle\sum_{k=1}^{\infty}(-1)^{k+1}a_k$. Suppose that the terms of the sequence $\{a_k\}$ are positive and decreasing for all $k\geq 10^9$ and $\lim_{k\to\infty}a_k=0$ but that $a_{10^9}>a_1$. Explain why the series converges. [HINT: Theorem 11 doesn't apply directly.]

28. Does $\quad 1-\dfrac{1}{2^3}+\dfrac{1}{3^2}-\dfrac{1}{4^3}+\dfrac{1}{5^2}-\dfrac{1}{6^3}+\dfrac{1}{7^2}-\dfrac{1}{8^3}+\cdots\quad$ converge? Justify your answer.

29. Suppose that $\displaystyle\sum_{j=1}^{\infty}b_j$ converges to a number S and that $b_j\geq 0$ for all $j\geq 1$. Show that $\displaystyle\sum_{j=1}^{\infty}(-1)^{j+1}b_j$ also converges.

30. Suppose that $0\leq b_{j+1}\leq b_j$ for all $j\geq 1$ and that the partial sum $\displaystyle\sum_{j=1}^{100}b_j$ approximates $S=\displaystyle\sum_{j=1}^{\infty}b_j$ within 0.005. Explain why $0\leq\displaystyle\sum_{j=1}^{\infty}(-1)^{j+1}b_j-\displaystyle\sum_{j=1}^{100}(-1)^{j+1}b_j\leq 0.005$.

31. Give an example of a convergent series $\displaystyle\sum_{k=1}^{\infty}a_k$ with the property that $\displaystyle\sum_{k=1}^{\infty}(a_k)^2$ diverges.

32. Suppose that $\displaystyle\sum_{k=1}^{\infty}a_k$ diverges. Is it possible that $\displaystyle\sum_{k=1}^{\infty}|a_k|$ converges? Justify your answer.

33. **A proof of the alternating series test.** Let $S_n=\displaystyle\sum_{k=1}^{n}(-1)^{k+1}c_k$ denote the partial sum of the first n terms of a series satisfying the hypotheses of the alternating series test (Theorem 11).

(a) Show that the sequence of even partial sums, S_2, S_4, S_6, S_8, ... is increasing.

(b) Show that the sequence of odd partial sums, S_1, S_3, S_5, S_7, ..., is decreasing.

(c) Show that $S_{2m}\leq S_{2m-1}$ for any integer $m\geq 1$.

(d) Use part (c) to show that the sequence of even partial sums and the sequence of odd partial sums both converge. [NOTE: Although both sequences converge, we must still show that they converge to the same limit.]

(e) Show that $\lim_{m\to\infty}(S_{2m+1}-S_{2m})=0$. From this it follows that there is a real number S such that $\lim_{n\to\infty}S_n=S$.

(f) Explain why $0<S-S_{2m}<c_{2m+1}$ and $0<S_{2m+1}-S<c_{2m+2}$.

11.5 POWER SERIES

A **power series** is a series of the form

$$a_0+a_1(x-x_0)+a_2(x-x_0)^2+a_3(x-x_0)^3+\cdots+a_n(x-x_0)^n+\cdots=\sum_{k=0}^{\infty}a_k(x-x_0)^k.$$

Here x is a variable, the constants a_k are called the **coefficients**, and the constant x_0 is called the **base point**. For convenience we'll often use $x_0=0$; in this case the appearance is slightly simpler:

$$a_0+a_1x+a_2x^2+a_3x^3+\cdots+a_nx^n+\cdots=\sum_{k=0}^{\infty}a_kx^k.$$

A power series may converge for some values of x and diverge for others. We illustrate these words and ideas with examples.

EXAMPLE 1 **(A geometric power series)** One of the simplest and most useful power series is

$$S(x)=1+x+x^2+x^3+x^4+x^5+\cdots=\sum_{k=0}^{\infty}x^k.$$

(In this case, $x_0 = 0$ and $a_k = 1$ for all $k \geq 0$.) For which real numbers x does $S(x)$ converge?

Solution   Setting $x = 1$ gives the series

$$S(1) = 1 + 1 + 1^2 + 1^3 + 1^4 + \cdots = 1 + 1 + 1 + 1 + 1 + \cdots,$$

which clearly diverges. If, instead, $x = 1/2$, then the series converges to 2:

$$S(1/2) = 1 + \frac{1}{2} + \frac{1}{4} + \frac{1}{8} + \cdots + \frac{1}{2^n} + \cdots = 2.$$

Indeed, $S(x)$ is a geometric series (in powers of x) for *any* value of x, and so the series converges if and only if $|x| < 1$. We even know the limit:

$$\text{If } |x| < 1, \text{ then } \quad 1 + x + x^2 + x^3 + x^4 + \cdots \quad \text{converges to} \quad \frac{1}{1-x}.$$

Thus, $S(x)$ converges if $-1 < x < 1$ and diverges otherwise.   ■

EXAMPLE 2   We'll show in the next section that, for any number x,

$$e^x = 1 + x + \frac{x^2}{2!} + \frac{x^3}{3!} + \frac{x^4}{4!} + \cdots .$$

Interpret the right side in the language of power series. For which x does the series converge?

Solution   Writing the series in the form

$$1 + x + \frac{x^2}{2!} + \frac{x^3}{3!} + \cdots = \sum_{k=0}^{\infty} \frac{1}{k!} x^k$$

shows the pattern of coefficients. (Here $x_0 = 0$ and $a_k = 1/k!$.)

We'll use the ratio test to decide where the series converges. (Since the ratio test applies only to *positive* series, we'll check for *absolute* convergence.) For any input x,

$$\lim_{k \to \infty} \frac{|a_{k+1} x^{k+1}|}{|a_k x^k|} = \lim_{k \to \infty} \frac{|x|^{k+1}}{(k+1)!} \cdot \frac{k!}{|x|^k} = \lim_{k \to \infty} \frac{|x|}{k+1} = 0.$$

Recall: If a series converges with absolute value signs, then it converges without.

Because $0 < 1$, the ratio test guarantees that this series converges absolutely—and therefore also in the ordinary sense—for *all* values of x.   ◄   ■

Power series and polynomials   Power series are, roughly speaking, "infinite-degree" polynomials. Notice these points of similarity:

- **The terms are power functions**   For both polynomials and power series, each summand is of the form $a_k x^k$ with k a nonnegative integer.

- **Partial sums are ordinary polynomials**   Every partial sum S_n of the power series $\sum_{k=0}^{\infty} a_k x^k$ has the form

$$S_n = a_0 + a_1 x + a_2 x^2 + a_3 x^3 + \cdots + a_n x^n,$$

  that is, an ordinary polynomial of degree n.

An important proviso!

- **Easy to use**   Polynomials are easy to differentiate and integrate, term by term. So are power series—if due care is taken for convergence. ◄ We'll return soon to this theme and see why it matters.

Choosing base points   The polynomial functions

$$p(x) = 1 + 2x + 3x^2 \quad \text{and} \quad q(x) = 1 + 2(x-5) + 3(x-5)^2$$

are not much different—the q-graph is found by sliding the p-graph 5 units to the right. The differences between p and q have to do with different choices of the **base point**: q is said to be **expanded** about $x = 5$ whereas p is expanded about $x = 0$.

The same notion of base point applies to power series. For example, the two power series

$$\sum_{k=0}^{\infty} 2^k x^k \quad \text{and} \quad \sum_{k=0}^{\infty} 2^k (x-1)^k$$

are written in powers of $(x-0)$ and powers of $(x-1)$, and so their respective base points are $x = 0$ and $x = 1$. As with polynomials, the mathematical difference is small: The second series is the result of shifting the first series one unit to the right.

Power series as functions

Any power series

$$S(x) = \sum_{k=0}^{\infty} a_k (x - x_0)^k = a_0 + a_1(x - x_0) + a_2(x - x_0)^2 + a_3(x - x_0)^3 + \cdots$$

defines, in a natural way, a *function* of x. For a given input x, $S(x)$ is the limit—if one exists—of the power series.

Domains of power series   Any function given by a "formula" in x has a natural domain: the set of x for which the formula makes sense. Power series are no different. The domain of a power series function $S(x)$ is the set of inputs x for which the series converges; this set is also called the **interval of convergence**. We've seen, for instance, that $\sum_{k=0}^{\infty} x^k$ converges for x in $(-1, 1)$, whereas $\sum_{k=0}^{\infty} x^k / k!$ converges for x in $(-\infty, \infty)$.

EXAMPLE 3   A function $S(x)$ is defined by the power series

$$S(x) = \sum_{k=0}^{\infty} 2^k x^k = 1 + 2x + 4x^2 + 8x^3 + \cdots.$$

What's the domain of S? Is a simpler formula available for S?

Solution   We *could* use the ratio test. Instead, let's think of $S(x)$ as a geometric series $1 + r + r^2 + r^3 + \cdots$, with $r = 2x$:

$$S(x) = 1 + 2x + (2x)^2 + (2x)^3 + \cdots.$$

A geometric series converges if $|r| < 1$; *this* one converges, therefore, if $|2x| < 1$, that is, if $|x| < 1/2$. In that case the limit is $1/(1-r) = 1/(1-2x)$. To summarize: The power series $S(x)$ has interval of convergence $(-1/2, 1/2)$; for x in that interval,

$$S(x) = 1 + 2x + (2x)^2 + (2x)^3 + \cdots = \frac{1}{1 - 2x}.$$

Thus, the function S has the simple formula $S(x) = 1/(1 - 2x)$—but the formula holds only if $|x| < 1/2$. ∎

Finding the interval of convergence   Given a power series, the first task is to find its interval of convergence. For many series, the ratio test is all that's needed. We illustrate with several (important) examples.

EXAMPLE 4   Where does $S(x) = \sum_{k=0}^{\infty}(x-5)^k$ converge?

Solution   The series is geometric in powers of $r = (x-5)$:

$$S(x) = 1 + (x-5) + (x-5)^2 + (x-5)^3 + \cdots,$$

and so it converges if (and only if) $|r| = |x-5| < 1$. Thus, the convergence interval is $(4, 6)$—the result of shifting $(-1, 1)$ five units to the right. ∎

EXAMPLE 5   Show that $1 + 2x + 3x^2 + 4x^3 + 5x^4 + \cdots = \sum_{k=1}^{\infty} kx^{k-1}$ converges only for x in $(-1, 1)$. Guess a limit.

Solution   We'll use the ratio test to check for *absolute* convergence. The ratio of successive terms is

$$\frac{|a_{k+1}x^{k+1}|}{|a_k x^k|} = \frac{(k+1)|x|^k}{k|x|^{k-1}} = |x| \cdot \frac{k+1}{k}.$$

This ratio tends to $|x|$ as $k \to \infty$, and so the series converges absolutely if $|x| < 1$.

If $|x| = 1$, the ratio test is inconclusive. But it's easy to see that if $x = \pm 1$, then

$$\left|kx^{k-1}\right| = |k| \to \infty \quad \text{as} \quad k \to \infty.$$

Thus, by the nth term test, the series *diverges* if $x = \pm 1$. (It diverges for the same reason if $|x| > 1$.) The interval of convergence is therefore $(-1, 1)$, as claimed.

To guess a limit, let's recall from Example 1 that the equation

$$S(x) = 1 + x + x^2 + x^3 + x^4 + \cdots = \frac{1}{1-x} \tag{1}$$

holds for $|x| < 1$. Differentiating all three quantities in Equation (1) suggests a natural guess for the limit of the series at hand:

$$S'(x) = 1 + 2x + 3x^2 + 4x^3 + \cdots = \frac{1}{(1-x)^2}.$$

With $x = 1/2$, for example, our guess is that

$$\sum_{k=1}^{\infty} k \left(\frac{1}{2}\right)^{k-1} = 1 + \frac{2}{2} + \frac{3}{4} + \frac{4}{8} + \cdots = \frac{1}{(1-1/2)^2} = 4.$$

Numerical evidence is promising: For the preceding series, $S_{20} \approx 3.999958038$. ∎

Good questions   Our guess in Example 5 is reasonable (and correct), but it raises some good questions:

- Is it legitimate to differentiate a series term by term?
- On what interval does the resulting series converge?

- Is $S'(x)$ the correct derivative of $S(x)$?
- Can we *antidifferentiate* a series function term by term?

We explore these questions informally in the following examples and return to them more formally in the next section.

EXAMPLE 6 **(Antidifferentiating a power series)** Antidifferentiating the geometric series

$$S(x) = 1 + x + x^2 + x^3 + \cdots = \sum_{k=1}^{\infty} x^{k-1}$$

term by term gives the new series

$$T(x) = x + \frac{x^2}{2} + \frac{x^3}{3} + \frac{x^4}{4} + \frac{x^5}{5} + \cdots = \sum_{k=1}^{\infty} \frac{x^k}{k}.$$

Where does $T(x)$ converge? Guess a limit.

Solution Because $S(x)$ converges for $|x| < 1$, we might expect the same of $T(x)$. The ratio test supports this guess:

$$\lim_{k \to \infty} \frac{\left|a_{k+1} x^{k+1}\right|}{\left|a_k x^k\right|} = \lim_{k \to \infty} |x| \cdot \frac{k}{k+1} = |x|,$$

and so $T(x)$ converges (absolutely) on the interval $(-1, 1)$.

What happens at the endpoints? Setting $x = \pm 1$ in $T(x)$ produces two series we've seen before:

$$T(1) = \sum_{k=1}^{\infty} \frac{1}{k}; \qquad T(-1) = \sum_{k=1}^{\infty} \frac{(-1)^k}{k}.$$

As we saw earlier, the first series diverges; the second converges conditionally (by the alternating-series theorem). Thus, T converges for x in $[-1, 1)$.

Because $T(x)$ came from $S(x)$ by antidifferentiation, it's reasonable to guess a similar relationship for limits:

$$1 + x + x^2 + x^3 + \cdots = \frac{1}{1-x} \implies x + \frac{x^2}{2} + \frac{x^3}{3} + \cdots = -\ln(1-x).$$

Numerical evidence suggests that we're right. If $x = 1/2$, the series gives $S_{20} \approx 0.69314714$; that's not far from $-\ln 1/2 \approx 0.69314718$. ∎

EXAMPLE 7 Where does the power series $\sum_{k=0}^{\infty} k! \, x^k$ converge?

Solution Every power series converges at its base point—in this case, $x = 0$. But if $x \neq 0$, the ratio test (applied to absolute values) gives

$$\lim_{k \to \infty} \frac{\left|a_{k+1} x^{k+1}\right|}{\left|a_k x^k\right|} = \lim_{k \to \infty} \frac{(k+1)! \, |x|^{k+1}}{k! \, |x|^k} = \lim_{k \to \infty} (k+1) \cdot |x| = \infty,$$

and so the series diverges for *all* $x \neq 0$. This power series has, in a sense, the smallest possible domain: It converges *only* if $x = 0$. ∎

Lessons from the examples

The preceding examples illustrate several useful properties of power series and their convergence sets.

In Example 7 the power series converged only at the base point $x = 0$. We'll call the set $\{0\}$ an interval of radius 0.

- **Domains are intervals**   In each example we've seen, the convergence set turned out to be an *interval*, centered at the base point. ← This is no accident—*every* power series converges on an interval centered at the base point. The radius of this interval is called the **radius of convergence**; it can be zero, finite, or infinite.

The ratio test shows this.

- **Any radius of convergence is possible**   Every positive number R is a possible radius of convergence for a power series. Indeed, the series $\sum_{k=0}^{\infty} x^k / R^k$ has radius of convergence precisely R. ←

- **At endpoints, anything can happen**   The series in Example 1 converges on the *open* interval $(-1, 1)$; the series in Example 6 converges on the "half-open" interval $[-1, 1)$. In fact, an interval of convergence may include either, both, or neither of its endpoints—any combination is possible. (In practice, what matters most is the *radius* of convergence; what happens at the endpoints can be interesting, but it's usually less important.)

The following theorem summarizes our observations.

THEOREM 12   Let $S(x) = \displaystyle\sum_{k=0}^{\infty} a_k (x - x_0)^k$ be a power series. The set of x for which $S(x)$ converges is an interval centered at x_0; endpoints may or may not be contained in the interval. The radius of convergence may be zero, finite, or infinite.

The idea of proof   The theorem says that if $S(x)$ converges for $x = C$, then $S(x)$ also converges for $|x| < |C|$. Suppose, for instance, that $\sum_{k=0}^{\infty} a_k x^k$ converges for $x = C = 1$. Then $\sum_{k=0}^{\infty} a_k$ converges. By the nth term test, $a_k \to 0$ as $k \to \infty$, and so the a_k's must be bounded in absolute value. This means that there is a number $M > 0$ such that $|a_k| \le M$ for all k, which implies that

$$\left| a_k x^k \right| \le M |x|^k$$

for all k. Now if $|x| < 1$, then the geometric series $\sum_{k=0}^{\infty} M |x|^k$ converges; by comparison, so does $\sum_{k=0}^{\infty} |a_k x^k|$. (A general proof for any value of C isn't much harder.)

Power series convergence, graphically   The nth partial sum of the power series $S(x) = \sum_{k=0}^{\infty} a_k x^k$ is the polynomial

$$p_n(x) = a_0 + a_1 x + a_2 x^2 + a_3 x^3 + \cdots + a_n x^n.$$

To say that the power series converges for x in $(-R, R)$ means that, for any x in that interval, there's a number $S(x)$ such that $p_n(x) \to S(x)$ as $n \to \infty$.

Sorting out exactly what this means and when it happens is a worthy challenge. Figure 1 gives a graphical sense of the situation for the geometric power series $S(x) = \sum_{k=0}^{\infty} x^k$.

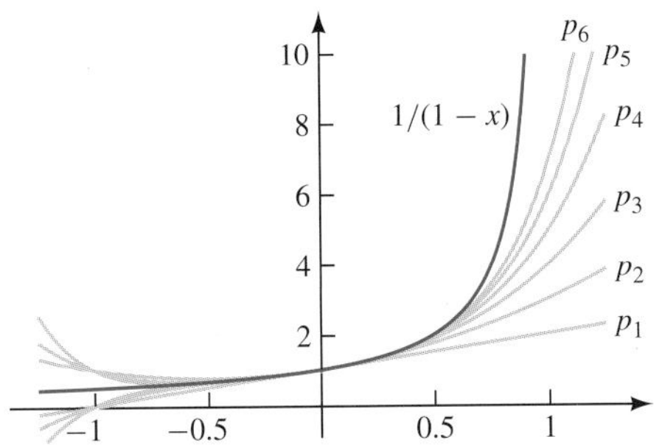

FIGURE 1

On $(-1, 1)$, $\displaystyle\sum_{k=0}^{\infty} x^k$ **converges to** $\dfrac{1}{1-x}$

The graphs labeled p_1 through p_6 represent the first six partial sums. Over the interval $(-1, 1)$, they appear to approach the graph of the limiting function (shown darker) more and more closely. Outside the interval $(-1, 1)$ the polynomial graphs seem to diverge, failing to approach any common limiting function.

BASIC EXERCISES

1. The power series $\displaystyle\sum_{k=1}^{\infty} x^k/k$ has radius of convergence 1. Plot the partial sum polynomials of degree 1, 2, 4, 6, 8, and 10 over the interval $[-2, 2]$. Is the interval of convergence of the series apparent? Explain.

2. The power series $\displaystyle\sum_{k=1}^{\infty} (x-3)^k/k$ has radius of convergence 1. Plot the partial sum polynomials of degree 1, 2, 4, 6, 8, and 10 over the interval $[1, 5]$. Is the interval of convergence of the series apparent? Explain.

In Exercises 3–6, find the radius of convergence of the power series.

3. $\displaystyle\sum_{j=1}^{\infty} \left(\frac{x}{2}\right)^j$

4. $\displaystyle\sum_{k=1}^{\infty} \frac{x^k}{k\,3^k}$

5. $\displaystyle\sum_{k=1}^{\infty} \frac{x^k}{\sqrt{k}}$

6. $\displaystyle\sum_{n=0}^{\infty} \frac{x^n}{n!+n}$

In Exercises 7–10, find the radius and the interval of convergence of the power series.

7. $\displaystyle\sum_{n=0}^{\infty} (x-2)^n$

8. $\displaystyle\sum_{n=2}^{\infty} \frac{(x-3)^{2n}}{n^4}$

9. $\displaystyle\sum_{n=2}^{\infty} \frac{(x+5)^n}{n \ln n}$

10. $\displaystyle\sum_{n=1}^{\infty} \frac{(x+1)^n}{n}$

FURTHER EXERCISES

11. Show that $\displaystyle\sum_{k=0}^{\infty} \frac{x^k}{R^k}$ converges on $(-R, R)$, where $R > 0$ is a constant.

12. Show that $\displaystyle\sum_{k=1}^{\infty} \frac{x^k}{k R^k}$ converges on $[-R, R)$, where $R > 0$ is a constant.

13. Show that $\displaystyle\sum_{k=1}^{\infty} \frac{x^k}{k^2 R^k}$ converges on $[-R, R]$, where $R > 0$ is a constant.

14. Concoct a power series that converges on $(-R, R]$, where $R > 0$ is a constant.

In Exercises 15–20, give an example of a power series that has the given interval as its interval of convergence.
[HINT: See Exercises 11–14.]

15. $[-4, 4)$

16. $(-3, 3]$

17. $[-1, 5]$

18. $(-4, 0)$

19. $(8, 16]$

20. $[-11, -3]$

In Exercises 21–24, suppose that the power series $\sum_{k=0}^{\infty} a_k x^k$ converges only if $-2 < x \le 2$.

21. Explain why the radius of convergence of the power series is 2.

22. Explain why the power series $\sum_{k=0}^{\infty} a_k (x-1)^k$ has radius of convergence 2.

23. Show that the interval of convergence of $\sum_{k=0}^{\infty} a_k (x-3)^k$ is $(1, 5]$.

24. Find the interval of convergence of $\sum_{k=0}^{\infty} a_k (x+1)^k$.

25. Suppose that $\sum_{k=0}^{\infty} a_k (x-b)^k$ converges only if $-11 \le x < 17$.

    (a) What is the radius of convergence of the power series?

    (b) Determine the value of b.

26. Suppose that $\sum_{n=1}^{\infty} a_n$ converges and that $|x| < 1$. Show that $\sum_{n=1}^{\infty} a_n x^n$ converges absolutely.

Suppose that the power series $\sum_{k=0}^{\infty} a_k x^k$ converges if $x = -3$ and diverges if $x = 7$. In Exercises 27–32, indicate whether the statement must be true, cannot be true, or may be true. Justify your answers.

27. The power series converges if $x = -10$.

28. The power series diverges if $x = 3$.

29. The power series converges if $x = 6$.

30. The power series diverges if $x = 2$.

31. The power series diverges if $x = -7$.

32. The power series converges if $x = -4$.

Suppose that the power series $\sum_{k=0}^{\infty} a_k (x+2)^k$ converges if $x = -7$ and diverges if $x = 7$. In Exercises 33–38, indicate whether the statement must be true, cannot be true, or may be true. Justify your answers.

33. The power series converges if $x = -8$.

34. The power series converges if $x = 1$.

35. The power series converges if $x = 3$.

36. The power series diverges if $x = -11$.

37. The power series diverges if $x = 5$.

38. The power series diverges if $x = -5$.

In Exercises 39–42, let $f(x) = \sum_{n=0}^{\infty} \dfrac{2x^n}{3^n + 5}$.

39. Explain why 10 is not in the domain of f.

40. Which of the numbers 0.5, 1.5, 3, and 6 are in the domain of f? Justify your answer.

41. Estimate $f(1)$ within 0.01 of its exact value.

42. Estimate $f(-1.5)$ within 0.01 of its exact value.

In Exercises 43–46, let $h(x) = \sum_{k=0}^{\infty} \dfrac{(x-2)^k}{k! + k^3}$.

43. What is the domain of h? Justify your answer.

44. Estimate $h(0)$ within 0.005 of its exact value.

45. Estimate $h(1)$ within 0.005 of its exact value.

46. Estimate $h(3)$ within 0.005 of its exact value.

In Exercises 47–50, let $g(x) = \sum_{n=1}^{\infty} \dfrac{(x+4)^n}{n^3 5^n}$.

47. What is the domain of g?

48. Estimate $g(0)$ within 0.005 of its exact value.

49. Estimate $g(1)$ within 0.005 of its exact value.

50. Estimate $g(-5)$ within 0.005 of its exact value.

11.6   POWER SERIES AS FUNCTIONS

Any power series

$$S(x) = \sum_{k=0}^{\infty} a_k (x - x_0)^k = a_0 + a_1(x - x_0) + a_2(x - x_0)^2 + a_3(x - x_0)^3 + \cdots$$

can be thought of as a function of x; its domain is the series' interval of convergence—an interval centered at x_0. In this section we explore the remarkable—and useful—properties of functions defined by power series. (For simplicity we'll mainly use $x_0 = 0$.)

Calculus with power series

Given *any* power series $S(x)$, convergent or divergent, it's easy to differentiate or antidifferentiate term by term to produce *new* series we'll call $D(x)$ and $A(x)$:

$$S(x) = \sum_{k=0}^{\infty} a_k x^k = a_0 + a_1 x + a_2 x^2 + a_3 x^3 + \cdots;$$

$$D(x) = \sum_{k=1}^{\infty} k a_k x^{k-1} = a_1 + 2a_2 x + 3a_3 x^2 + \cdots;$$

$$A(x) = \sum_{k=0}^{\infty} a_k \frac{x^{k+1}}{k+1} = a_0 x + a_1 \frac{x^2}{2} + a_2 \frac{x^3}{3} + a_3 \frac{x^4}{4} + \cdots.$$

Important questions arise:

- If the series S has radius of convergence r, is the same true of D and A?
- Even if we assume that S, D, and A all converge on $(-r, r)$, must D be the derivative of S in the ordinary calculus sense? Must A be an antiderivative of S?

The following theorem answers all these questions in the affirmative. Conveniently, everything works just as we'd hope.

> **THEOREM 13 (Derivatives and antiderivatives of power series)** Let $S(x)$ be a power series with radius of convergence $r > 0$. Let $D(x)$ and $A(x)$ be defined as above.
>
> - Both D and A have radius of convergence r.
> - If $|x - x_0| < r$, $D(x) = S'(x)$.
> - If $|x - x_0| < r$, $A'(x) = S(x)$.

The theorem says, among other things, that a function S given by a power series is differentiable and that its derivative is another power series S' with the same radius of convergence as S. The same principle applies to S', to S'', and so on to show that S has *infinitely many* derivatives—all available by repeated term-by-term differentiation.

EXAMPLE 1 For any x, $e^x = 1 + x + \dfrac{x^2}{2!} + \dfrac{x^3}{3!} + \cdots$. Explain why.

Solution Let $S(x)$ represent the preceding series. We saw in the last section that $S(x)$ converges for *all* x. By Theorem 13, S' can be found by differentiating S term by term. In this case, differentiation produces a curious result:

Differentiating S term by term leaves S unchanged.

But *every* differentiable function S for which $S' = S$ has the form $S(x) = Ce^x$. Since $S(0) = 1$, it follows that $C = 1$; thus, $S(x) = e^x$, as claimed.   ∎

Writing known functions as power series

In earlier examples we've written several functions, including $1/(1 - x)$ and e^x, in power series form. Can *other* familiar functions be "represented" as power series? If so, how?

Theorem 13 suggests some answers. But first let's address an even *more* basic question.

Why bother? Examples with the sine function   Why write a function as a power series? What good, for instance, is the equation

$$\sin x = x - \frac{x^3}{3!} + \frac{x^5}{5!} - \frac{x^7}{7!} + \frac{x^9}{9!} - \cdots, \tag{1}$$

We'll explain in the next section why the equation holds.

which holds for all real numbers x? ◄ The next two examples suggest some answers.

Trigonometric functions lack finite algebraic formulas; power series are the next best thing. Using them lets us find approximate—but very accurate—values of trigonometric (and other transcendental) functions.

EXAMPLE 2   **(Transcendental family values)**   Use the series expression (1) to approximate sin 1 accurately.

Solution   Substituting $x = 1$ into Equation (1) gives

$$\sin 1 = 1 - \frac{1}{3!} + \frac{1}{5!} - \frac{1}{7!} + \frac{1}{9!} - \frac{1}{11!} + \cdots,$$

an alternating series with terms decreasing (rapidly!) to zero. By the alternating series theorem, each partial sum S_n of such a series differs from the limit by no more than the size of the next term, a_{n+1}. In particular, the partial sum

$$1 - \frac{1}{3!} + \frac{1}{5!} - \cdots - \frac{1}{11!} \approx 0.8414709846$$

differs from sin 1 by no more than $1/13! \approx 2 \times 10^{-10}$. In other words, our estimate is *guaranteed* accurate to at least nine decimal places—not bad for so little work!   ∎

As we've seen often, many integrals cannot be calculated in "closed form," by elementary antidifferentiation. Numerical methods—such as the midpoint rule—may help. Infinite series, being easy to integrate, offer another way to transcend our troubles.

EXAMPLE 3   **(Hard integrals made easy)**   Find, in series form, an antiderivative for $\sin(x^2)$. Use it to estimate $I = \int_0^1 \sin(x^2)\,dx$. (For comparison, the midpoint rule applied to I gives $M_{50} \approx 0.31025$.)

Solution   The function $\sin(x^2)$ has no *elementary* antiderivative, but it's easy to find an antiderivative in *series* form. To do so, we replace x with x^2 in the sine series (1) to produce the new series

$$\sin(x^2) = x^2 - \frac{x^6}{3!} + \frac{x^{10}}{5!} - \frac{x^{14}}{7!} + \cdots.$$

(Because the original series converges for all x, so does this one.)

It, too, converges for all x.

Antidifferentiating term by term gives yet another power series: ◄

$$\int_0^x \sin(t^2)\,dt = \frac{x^3}{3} - \frac{x^7}{7 \cdot 3!} + \frac{x^{11}}{11 \cdot 5!} - \frac{x^{15}}{15 \cdot 7!} + \cdots.$$

The result is not an elementary function, but it's an honest antiderivative for $\sin(x^2)$, and so we can find our definite integral in the obvious way:

$$\int_0^1 \sin(x^2)\,dx = \frac{x^3}{3} - \frac{x^7}{7 \cdot 3!} + \frac{x^{11}}{11 \cdot 5!} - \frac{x^{15}}{15 \cdot 7!} + \cdots \Bigg]_0^1$$

$$= \frac{1}{3} - \frac{1}{7 \cdot 3!} + \frac{1}{11 \cdot 5!} - \frac{1}{15 \cdot 7!} + \cdots.$$

The alternating series theorem applies to the last series, and so the estimate

$$\int_0^1 \sin(x^2)\,dx \approx \frac{1}{3} - \frac{1}{7 \cdot 3!} + \frac{1}{11 \cdot 5!} - \frac{1}{15 \cdot 7!} \approx 0.3102681578$$

is in error by no more than $1/(19 \cdot 9!) \approx 1.5 \times 10^{-7}$ (the size of the next term). The midpoint rule estimate agrees with the result through four decimal places. ∎

New series from old: Help from algebra and calculus Power series can be found for many familiar functions by applying simple algebra or calculus operations to a few standard known series. Differentiating the sine series, for instance, gives the new series

$$\cos x = 1 - \frac{x^2}{2!} + \frac{x^4}{4!} - \frac{x^6}{6!} + \cdots.$$

which, like the old one, converges for all x.

We can also start with another famous series,

$$\frac{1}{1-x} = 1 + x + x^2 + x^3 + x^4 + x^5 + \cdots, \tag{2}$$

which converges for $|x| < 1$. With a little algebraic ingenuity we can produce many other useful series, all converging on the same set. Replacing x with $-x$ in Equation (2) gives the alternating series

$$\frac{1}{1+x} = 1 - x + x^2 - x^3 + x^4 - x^5 + \cdots. \tag{3}$$

Replacing x with x^2 in Equation (3) gives still another alternating series:

$$\frac{1}{1+x^2} = 1 - x^2 + x^4 - x^6 + x^8 - x^{10} + \cdots.$$

Integrating *this* series term by term gives another striking result:

$$\arctan x = x - \frac{x^3}{3} + \frac{x^5}{5} - \frac{x^7}{7} + \frac{x^9}{9} - \cdots.$$

Setting $x = 1$ in this last series yields a *really* remarkable result:

$$\frac{\pi}{4} = 1 - \frac{1}{3} + \frac{1}{5} - \frac{1}{7} + \frac{1}{9} - \cdots.$$

(Caution is needed—these simple arguments show only that the series for $\arctan x$ is valid if $-1 < x < 1$. Showing carefully that the series converges to $\pi/4$ at the endpoint $x = 1$ requires further argument.)

Multiplying power series Convergent power series can be multiplied together, something like polynomials, to form new convergent series. As always with series, convergence is a question. Here's the answer: The product of two power series converges wherever both factors converge.

EXAMPLE 4 We've seen already that

$$\frac{1}{1-x} = 1 + x + x^2 + x^3 + \cdots \quad \text{and} \quad \frac{1}{1+x} = 1 - x + x^2 - x^3 + \cdots.$$

Multiply these series. Where does the new series converge? What familiar function does the result represent?

Solution   Symbolically, the problem looks like this:

$$(1 + x + x^2 + x^3 + \cdots) \cdot (1 - x + x^2 - x^3 + \cdots) = a_0 + a_1 x + a_2 x^2 + a_3 x^3 + \cdots.$$

We want numerical values for the constants on the right.

Both factors have infinitely many summands, and so ordinary expansion quickly gets out of hand. To avoid this, we collect like powers right from the start. It's clear, for instance, that $a_0 = 1 \cdot 1 = 1$; no other combination of factors yields a constant result. Similarly, tracking the first and second powers of x gives

$$a_1 = 1 \cdot (-1) + 1 \cdot 1 = 0; \qquad a_2 = 1 \cdot 1 + 1 \cdot (-1) + 1 \cdot 1 = 1.$$

Continuing this process reveals a simple pattern:

$$(1 + x + x^2 + x^3 + \cdots) \cdot (1 - x + x^2 - x^3 + \cdots) = 1 + x^2 + x^4 + x^6 + \cdots.$$

The result is a geometric series in powers of x^2; it converges for $|x^2| < 1$, that is, if $-1 < x < 1$.

What familiar function does the product series represent? Because the two factors represent the functions $1/(1-x)$ and $1/(1+x)$, it follows that the product *series* represents the product *function* $1/(1-x^2)$.   ∎

EXAMPLE 5   Find a power series for $\ln(1+x)$; use it to estimate $\ln(1.5)$ with error guaranteed to be less than 0.0001.

Solution   Integrating both sides of Equation (3) gives

$$\ln(1+x) = \int \frac{1}{1+x}\, dx = \int \left(1 - x + x^2 - x^3 + \cdots\right) dx$$

$$= x - \frac{x^2}{2} + \frac{x^3}{3} - \frac{x^4}{4} + \cdots.$$

(We used $C = 0$ as the constant of integration because our "target" function $\ln(1+x)$ has the value 0 when $x = 0$.) The new series, like the old, converges for x in $(-1, 1)$. To estimate $\ln(1.5)$ we plug in $x = 0.5$:

$$\ln(1.5) = 0.5 - \frac{0.5^2}{2} + \frac{0.5^3}{3} - \frac{0.5^4}{4} + \cdots = \sum_{k=1}^{\infty} (-1)^{k+1} \frac{0.5^k}{k}.$$

Now the alternating series theorem applies. To achieve our target accuracy, we can use any partial sum S_n for which

$$c_{n+1} = \frac{0.5^{n+1}}{n+1} < 0.0001.$$

It's easy to see that $n = 10$ works, with room to spare. In fact, $S_{10} \approx 0.405435$; this compares closely with the value $\ln 1.5 \approx 0.405465$ reported by a calculator.   ∎

A brief atlas of power series   For easy reference, we collect a short list of "standard" power series for basic calculus functions. In the next section we'll see how to show rigorously that the series really converge to the stated limits.

A power series sampler		
Function	**Series**	**Convergence interval**
$\sin x$	$x - \dfrac{x^3}{3!} + \dfrac{x^5}{5!} - \dfrac{x^7}{7!} + \dfrac{x^9}{9!} - \cdots$	$(-\infty, \infty)$
$\cos x$	$1 - \dfrac{x^2}{2!} + \dfrac{x^4}{4!} - \dfrac{x^6}{6!} + \dfrac{x^8}{8!} - \cdots$	$(-\infty, \infty)$
e^x	$1 + x + \dfrac{x^2}{2!} + \dfrac{x^3}{3!} + \dfrac{x^4}{4!} + \dfrac{x^5}{5!} + \cdots$	$(-\infty, \infty)$
$\dfrac{1}{1-x}$	$1 + x + x^2 + x^3 + x^4 + x^5 + \cdots$	$(-1, 1)$
$\dfrac{1}{1+x}$	$1 - x + x^2 - x^3 + x^4 - x^5 + \cdots$	$(-1, 1)$
$\dfrac{1}{1+x^2}$	$1 - x^2 + x^4 - x^6 + x^8 - x^{10} + \cdots$	$(-1, 1)$
$\arctan x$	$x - \dfrac{x^3}{3} + \dfrac{x^5}{5} - \dfrac{x^7}{7} + \dfrac{x^9}{9} - \cdots$	$[-1, 1]$

What's next? A power series for any function

As we've seen, knowing a power series expression for one function can lead, via various manipulations, to power series for related functions. A good question remains:

> *Given any function f, how can we find a power series "from scratch," without knowing a related series to begin with?*

We answer this question in the next section.

BASIC EXERCISES

In Exercises 1–3, $f(x) = \displaystyle\sum_{k=0}^{\infty} \left(\dfrac{x}{2}\right)^k$.

1. What is the radius of convergence of the power series for f?

2. According to Theorem 13, $f'(x) = \displaystyle\sum_{k=1}^{\infty} \dfrac{k\, x^{k-1}}{2^k}$. What is the radius of convergence of the series for f'?

3. According to Theorem 13, $F(x) = \displaystyle\sum_{k=0}^{\infty} \dfrac{x^{k+1}}{(k+1)\, 2^k}$ is an antiderivative of f. What is the radius of convergence of the series for F?

4. Let $f(x) = \ln(1+x)$. Show that the power series for f and f' have the same radius of convergence but not the same interval of convergence.

In Exercises 5–8, use the power series representation of $1/(1-x)$ to produce a power series representation of the function f.

5. $f(x) = \dfrac{x^2}{1+x}$

6. $f(x) = \dfrac{1}{1-x^2}$

7. $f(x) = \dfrac{1}{(1+x)^2}$

8. $f(x) = \dfrac{x}{1-x^4}$

In Exercises 9–12, find a power series representation of the function and the radius of convergence of the power series. Then plot the function and the fifth-order polynomial that is a partial sum of the power series on the same axes. [HINT: Write out the first few terms of the series before trying to find the form of the general term.]

9. $f(x) = \arctan(2x)$

10. $f(x) = \cos(x^2)$

11. $f(x) = x^2 \sin x$

12. $f(x) = \ln\left(1 + \sqrt[3]{x}\right)$

13. Use the partial sum of a series to estimate $1/\sqrt{e}$ with an error less than 0.005.

14. Use the partial sum of a series to estimate $\displaystyle\int_0^{0.2} x e^{-x^3}\, dx$ with an error less than 10^{-5}.

15. Use power series to show that $\lim_{x\to 0} \dfrac{(\sin x - x)^3}{x(1 - \cos x)^4} = -\dfrac{2}{27}$.

16. Show that $\lim_{x\to 0^+} \dfrac{x - \sin x}{(x \sin x)^{3/2}} = \dfrac{1}{6}$.

FURTHER EXERCISES

In Exercises 17–22, find a power-series representation of the function and the radius of convergence of the power series.

17. $f(x) = \dfrac{1}{2 + x}$

18. $f(x) = \sin\left(\sqrt{x}\right)$

19. $f(x) = \sin x + \cos x$

20. $f(x) = \ln(1 + x^2)$

21. $f(x) = (x^2 - 1)\sin x$

22. $f(x) = \ln\left(\dfrac{1 + x}{1 - x}\right)$

23. Use power series to show that $y = e^x$ is a solution of the DE $y' = y$.

24. Use power series to show that $y = 2e^x$ is the solution of the IVP $y' = y$, $y(0) = 2$.

25. Use power series to show that $y = e^{3x}$ is the solution of the IVP $y' = 3y$, $y(0) = 1$.

26. Use power series to show that $y = \sin x$ is a solution of the DE $y'' = -y$.

27. Use power series to show that $y = 1/(1 - x)$ is the solution of the IVP $y' = y^2$, $y(0) = 1$.

28. (a) Show that $f(x) = \tan x$ is the solution of the IVP $f'(x) = 1 + (f(x))^2$, $f(0) = 0$.

(b) Use part (a) to find the first four nonzero terms in the power series representation of $\tan x$.

In Exercises 29–34, use power series to evaluate the limit. Check your answer using l'Hôpital's rule.

29. $\lim_{x\to 0} \dfrac{\sin x}{x}$

30. $\lim_{x\to 0} \dfrac{e^x - 1}{x}$

31. $\lim_{x\to 0} \dfrac{1 - \cos x}{x^2}$

32. $\lim_{x\to 0} \dfrac{\arctan x}{x}$

33. $\lim_{x\to 0} \dfrac{\ln(1 + x) - x}{x^2}$

34. $\lim_{x\to 0} \dfrac{1 - \cos^2 x}{x}$ [HINT: $1 - \cos^2 x = (1 - \cos(2x))/2$.]

35. Evaluate $\displaystyle\sum_{n=1}^{\infty} \dfrac{n}{2^n}$ exactly.   [HINT: If $f(x) = \displaystyle\sum_{n=0}^{\infty} x^n$, then $f'(x) = \displaystyle\sum_{n=1}^{\infty} nx^{n-1}$.]

36. Show that $\dfrac{1}{x - 1} = \displaystyle\sum_{k=1}^{\infty} \dfrac{1}{x^k}$ if $|x| > 1$.

[HINT: $1/(x - 1) = x/(x - 1) - 1$.]

37. (a) Use the formula for the sum of a geometric series to show that

$$\sum_{k=1}^{\infty} \dfrac{x^k}{k} = -\ln|1 - x|.$$

(b) What is the interval of convergence of the series in part (a)?

(c) Show that if $N \geq 1$, then

$$0 < \ln 2 - \sum_{k=1}^{N} \dfrac{1}{k\,2^k} \leq \dfrac{1}{(N+1)2^N}.$$

[HINT: $-\ln(1/2) = \ln 2$.]

38. Use a power series to show that $x - x^2/2 < \ln(1 + x) < x$ for all x in the interval $(0, 1)$.

39. Use power series to show that $1 - \cos x < \ln(1 + x) < \sin x$ if $0 < x < 1$.

40. Let $f(x) = \dfrac{1}{1 + x^4}$.

(a) Find a power series representation of f.

(b) What is the interval of convergence of the series in part (a)?

(c) Use the series found in part (a) to evaluate $\displaystyle\int_0^{0.5} f(x)\,dx$ with an error no greater than 0.001.

41. (a) Find the power series representation of an antiderivative of e^{-x^2}.

(b) Use the result from part (a) to estimate $\displaystyle\int_0^1 e^{-x^2}\,dx$ within 0.005 of its exact value.

42. Estimate $\displaystyle\int_0^1 \cos(x^2)\,dx$ with an error no greater than 0.005.

43. Estimate $\displaystyle\int_0^1 \sqrt{x}\sin x\,dx$ with an error no greater than 0.001.

44. Estimate $\displaystyle\int_0^{1000} \dfrac{e^{-10x}\sin x}{x}\,dx$ with an error no greater than 5×10^{-5}.

In Exercises 45–50, find the first four nonzero terms in the power series representation of the function.

45. $f(x) = e^{2x}\ln(1 + x^3)$

46. $f(x) = \arctan x \sin(4x)$

47. $f(x) = \dfrac{e^x}{1 - x}$

48. $f(x) = \dfrac{\cos x}{1 + x^2}$

49. $f(x) = e^{\sin x}$

50. $f(x) = \ln(1 + \sin x)$

In Exercises 51–54, find the elementary function represented by the power series by manipulating a more familiar power series (e.g., the series for $\cos x$, $\sin x$, $(1 - x)^{-1}$).

51. $\displaystyle\sum_{k=1}^{\infty} kx^{k-1}$

52. $\displaystyle\sum_{k=0}^{\infty} \frac{x^k}{(k+1)!}$

53. $\displaystyle\sum_{k=1}^{\infty} (-1)^{k+1} x^k$

54. $\displaystyle\sum_{k=1}^{\infty} \frac{(2x)^k}{k}$

11.7 TAYLOR SERIES

In the preceding section we saw some of the practical advantages of writing a function as a power series. We also saw ways of using a power series for one function to derive power series for other functions. In this section we show, given a suitable function, how to find its power series "from scratch" by matching derivatives. The idea is not new—we did the same thing in Section 9.1, where we found Taylor polynomials. Here we extend these ideas to find **Taylor series**—"infinite Taylor polynomials."

Does every function have a power series? Mathematical life would be simpler if *every* function $f(x)$ could be written as a power series—ideally, a series that converges to $f(x)$ for all x. Alas, it isn't so. At least two things can go wrong:

- **Smaller domains**   A power series may have a smaller domain than the function it represents. For instance, the series equation

$$\frac{1}{1 + x^2} = 1 - x^2 + x^4 - x^6 + x^8 - x^{10} + \cdots$$

holds if—but only if—$|x| < 1$, even though the left side is defined (and well behaved) for *all* real numbers x.

- **No series at all**   A function may have no series at all. As Theorem 13 (page 591) says, every power series can be differentiated repeatedly on its interval of convergence. ➜ Thus, a function that *has* a power series with base point $x = x_0$ must itself be repeatedly differentiable at $x = x_0$. This means, for instance, that $f(x) = |x|$ (which is not differentiable at $x = 0$) has *no* power series with base point $x = 0$.

Each derivative is another power series.

Coefficients and derivatives at the base point

A simple equation relates coefficients and derivatives of a power series. (We saw the same principle at work for Taylor polynomials.) If

$$S(x) = a_0 + a_1(x - x_0) + a_2(x - x_0)^2 + a_3(x - x_0)^3 + a_4(x - x_0)^4 + \cdots,$$

then, clearly, $S(x_0) = a_0$. Differentiating S repeatedly gives

$$S'(x) = a_1 + 2a_2(x - x_0) + 3a_3(x - x_0)^2 + 4a_4(x - x_0)^3 + \cdots;$$

$$S''(x) = 2a_2 + 6a_3(x - x_0) + 12a_4(x - x_0)^2 + \cdots;$$

$$S'''(x) = 6a_3 + 24a_4(x - x_0) + \cdots;$$

$$\vdots$$

These equations show that

$$S(x_0) = a_0; \quad S'(x_0) = a_1; \quad S''(x_0) = 2a_2; \quad S'''(x_0) = 6a_3; \quad \ldots.$$

The Fact resembles Theorem 1, page 496.

The following Fact captures the general pattern: ◄

FACT If $S(x) = \displaystyle\sum_{k=0}^{\infty} a_k(x - x_0)^k$, then $a_k = \dfrac{S^{(k)}(x_0)}{k!}$ for all $k \geq 0$.

This Fact amounts to a recipe for cooking up a power series for a given function expanded about a given base point x_0.

EXAMPLE 1 Assuming that $f(x) = \sin x$ *has* a power series based at $x = 0$, *find* it.

Solution To find the desired series

$$f(x) = a_0 + a_1 x + a_2 x^2 + a_3 x^3 + \cdots,$$

we need "only" find the coefficients $a_0, a_1, a_2, a_3, \ldots$. Finding infinitely many of *anything* sounds difficult, but in this case it's easier than it seems. For the sine function (and for many functions of interest) the coefficients follow a simple pattern. The first few derivatives of $f(x) = \sin x$ reveal the pattern:

$$f(0), \; f'(0), \; f''(0), \; f'''(0), \; f^{(4)}(0), \ldots = 0, 1, 0, -1, 0, \ldots.$$

By the preceding Fact, the series coefficients have a similar pattern:

$$a_0, a_1, a_2, a_3, a_4, a_5, a_6, \ldots = 0, 1, 0, -\frac{1}{3!}, 0, \frac{1}{5!}, 0, \ldots.$$

Thus, the sine series is the one we've seen before:

$$\sin x = x - \frac{x^3}{3!} + \frac{x^5}{5!} - \frac{x^7}{7!} + \cdots.$$

It's not hard (using the ratio test) to show that this series converges for *all x*, as we'd hope. ∎

Taylor series

Using the preceding Fact as a recipe for the coefficients, we can write a power series for *any* function f that has repeated derivatives at $x = x_0$.

DEFINITION (Taylor series) Let f be any function with infinitely many derivatives at $x = x_0$. The **Taylor series** for f is the series $\displaystyle\sum_{k=0}^{\infty} a_k(x - x_0)^k$ with coefficients given by

$$a_k = \frac{f^{(k)}(x_0)}{k!}, \quad k = 0, 1, 2, \ldots.$$

If $x_0 = 0$, then the series looks a little simpler; such series are called **Maclaurin series** after the 17th-century Scottish mathematician Colin Maclaurin.

EXAMPLE 2   The function $f(x) = \ln x$ is not defined at $x = 0$, and so there's no Maclaurin series. Expand f about $x = 1$.

Solution   Derivatives of $f(x) = \ln x$ are easy for either a human or a machine to calculate. Here are several:

$$f'(x) = \frac{1}{x}; \quad f''(x) = -\frac{1}{x^2}; \quad f'''(x) = \frac{2}{x^3}; \quad f^{(4)}(x) = -\frac{6}{x^4}.$$

At $x = 1$, therefore, we have

$$f(1) = 0; \quad f'(1) = 1; \quad f''(1) = -1; \quad f'''(1) = 2; \quad f^{(4)}(1) = -6, \quad \ldots,$$

so the Taylor coefficients are

$$a_0 = 0; \quad a_1 = 1; \quad a_2 = -\frac{1}{2}; \quad a_3 = \frac{1}{3}; \quad a_4 = -\frac{1}{4}; \quad \ldots.$$

The Taylor series is therefore

$$(x-1) - \frac{(x-1)^2}{2} + \frac{(x-1)^3}{3} - \frac{(x-1)^4}{4} + \cdots = \sum_{k=1}^{\infty} \frac{(-1)^{k+1}}{k}(x-1)^k. \qquad \blacksquare$$

Finding Taylor series: Help from technology   Computing the necessary derivatives to find a Taylor series by hand can be tedious and error prone. Technology can help. The following, for instance, is how one computer algebra system finds Taylor polynomials (partial sums of Taylor series) in the cases illustrated in the preceding examples:

```
> taylorpoly( sin(x), x=0, 7 );
            3        5           7
      x - 1/6 x + 1/120 x - 1/5040 x
> taylorpoly( ln(x), x=1, 7 );
                  2          3           4
    x - 1 - 1/2 (x - 1) + 1/3 (x - 1) - 1/4 (x - 1)
               5          6           7
      + 1/5 (x - 1) - 1/6 (x - 1) + 1/7 (x - 1)
```

Converging to the right place: Taylor's theorem

Any function f that's infinitely differentiable at $x = x_0$ has a Taylor series—ideally, one with a large radius of convergence. One possible problem remains: The series might converge at x but perhaps to a limit *other* than $f(x)$.

Taylor's theorem guarantees that this unfortunate event seldom occurs. We met Taylor's theorem in Section 9.2 (see Theorem 2, page 504); we used it there to predict how closely a Taylor polynomial approximates its "target" function f. That question is relevant here because Taylor *polynomials* are partial sums of Taylor *series*.

THEOREM 14 (Taylor's theorem)   Suppose that f is repeatedly differentiable on an interval I containing x_0 and that

$$P_n(x) = a_0 + a_1(x - x_0) + a_2(x - x_0)^2 + \cdots + a_n(x - x_0)^n$$

is the nth-order Taylor polynomial based at x_0. Suppose that for all x in I,

$$\left| f^{(n+1)}(x) \right| \le K_{n+1}.$$

Then $\left| f(x) - P_n(x) \right| \le \dfrac{K_{n+1}}{(n+1)!} |x - x_0|^{n+1}.$

Observe:

- **Estimating difference**   Taylor's theorem estimates the *difference* between $f(x)$ and $P_n(x)$. Unless K_{n+1} grows very quickly with n, this difference tends to 0 as $n \to \infty$, and so the series converges to $f(x)$.

- **Why does it hold?**   We discussed the theorem's proof at length in Section 9.2; an important ingredient is the mean value theorem.

EXAMPLE 3   For $f(x) = \sin x$, show that the Maclaurin series converges to $\sin x$ for *every* value of x.

Solution   For $f(x) = \sin x$, derivatives of *all* orders are sines or cosines or their opposites, and so the inequality

$$\left| f^{(n+1)}(x) \right| \leq 1$$

holds for all x and for all n. Thus, we can use $K_{n+1} = 1$ for all n in Taylor's theorem:

$$\left| \sin x - P_n(x) \right| \leq \frac{1 \cdot |x|^{n+1}}{(n+1)!}.$$

The last quantity tends to 0 as $n \to \infty$, regardless of the value of x. Hence, the series converges, for all x, to $\sin x$. The graphs in Figure 1 suggest what this convergence means geometrically. ◄

The same figure appears, with additional comments, on page 500.

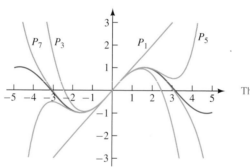

The dark curve is $y = \sin x$.

FIGURE 1
Maclaurin polynomials converging to $f(x) = \sin x$

Notice how the higher-order Maclaurin polynomials approximate the sine function better than lower-order ones. For instance, P_7, "sticks with" the sine curve much farther than P_1 or P_3. ■

BASIC EXERCISES

1. Let $f(x) = \int_3^x \sqrt{t}e^{-t}\, dt$.

  (a) Show that if $x \approx 3$, then

$$f(x) \approx \sqrt{3}e^{-3}(x-3) - \frac{5}{12}\sqrt{3}e^{-3}(x-3)^2 + \frac{23}{216}\sqrt{3}e^{-3}(x-3)^3.$$

  (b) Use Theorem 14 to bound the error made when $f(3.5)$ is estimated using the polynomial in part (a).

2. Let $f(x) = \sqrt{1+x}$.

  (a) Find the first three nonzero terms in the Maclaurin series for f.

  (b) Use Theorem 14 to bound the approximation error made if the Maclaurin polynomial from part (a) is used to estimate $f(1)$.

3. Let $f(x) = e^{2x}$. What is the coefficient of x^{100} in the Maclaurin series representation of f?

4. Let $f(x) = \dfrac{x}{1 - x^3}$.

(a) Find the Maclaurin series for $f(x)$.

(b) What is the interval of convergence of the series in part (a)?

(c) Use part (a) to find a power series for $f''(x)$.

(d) Use part (a) to find the Maclaurin series for $\displaystyle\int_0^x f(t)\, dt$.

5. (a) Find the Maclaurin series representation of
$$f(x) = \frac{1}{2 + x}. \quad [\text{HINT:}\ \frac{1}{2+x} = \frac{1}{2} \cdot \frac{1}{1 + (x/2)}.]$$

(b) Find $f^{(259)}(0)$ exactly.

6. Suppose that f is a function such that $f(1) = 1$, $f'(1) = 2$, and $f''(x) = 1/(1 + x^3)$ for $x > -1$.

(a) Estimate $f(1.5)$ using a quadratic Taylor polynomial.

(b) Find an upper bound on the approximation error made in part (a).

FURTHER EXERCISES

7. Use Theorem 14 to show that the Maclaurin series for e^x converges to e^x for all x.

8. Use Theorem 13 to show that the Maclaurin series for $1/(1 + x)$ converges to $1/(1 + x)$ if $-1/2 < x < 1$. [HINT: Consider the cases $-1/2 < x < 0$ and $0 \le x < 1$ separately.]

9. Suppose that f is a function such that $f^{(n)}$ exists for all $n \ge 1$.

(a) Explain why the Maclaurin series for f converges to f if $\left| f^{(n)}(x) \right| \le n$ for all $n \ge 1$ and all x.

(b) Does Theorem 13 guarantee that the Maclaurin series for f converges to f if $\left| f^{(n)}(x) \right| \le 2^n$ for all $n \ge 1$ and all x?

10. Let $f(x) = \begin{cases} \dfrac{1 - \cos x}{x^2} & \text{if } x \ne 0, \\ \dfrac{1}{2} & \text{if } x = 0. \end{cases}$ Evaluate $f^{(100)}(0)$.

11. Let $f(x) = \begin{cases} x^{-1} \sin x & \text{if } x \ne 0, \\ 1 & \text{if } x = 0. \end{cases}$

(a) Find the Maclaurin series representation of f.

(b) What is the interval of convergence of the power series found in part (a)?

(c) Use the series in part (a) to estimate $f'''(1)$ with an error no greater than 0.005.

12. Let $f(x) = \begin{cases} e^{-1/x^2} & \text{if } x \ne 0, \\ 0 & \text{if } x = 0. \end{cases}$

(a) Use the definition of the derivative (and l'Hôpital's rule) to show that $f'(0) = 0$.

(b) Using methods similar to those in part (a), it can be shown that $f^{(k)}(0) = 0$ for all integers $k \ge 0$. Use this fact to find the Maclaurin series for f.

(c) What is the radius of convergence of the series in part (b)?

(d) For which values of x does the series in part (b) converge to $f(x)$?

Sequences   Infinite series are formed by adding—in a special sense and with due care for convergence—the terms of an infinite **sequence**. To get started, we defined and studied sequences and their limits in their own right.

Series and convergence   An infinite series **converges** if its sequence of partial sums, S_n, formed from only finitely many terms, converge to a limit S. For a given series, the sequence of partial sums may be hard to understand or handle directly. For **geometric series**, however, partial sums (and therefore limits) are easy to calculate. Geometric series are our simplest and most important examples.

Convergence tests   The first question to ask about a series is whether it converges or diverges. As with improper integrals, the question can be subtle. To answer it, we developed several tests for convergence and divergence, including the **comparison test**, the **integral test**, and the **ratio test**. All of these tests apply only to series of **nonnegative** terms.

Absolute convergence   Series that contain both positive and negative terms need special care; they raise the notions of **absolute** versus **conditional** convergence. The **alternating series test** handles certain series with terms that alternate in sign.

Estimating limits Many convergent series are difficult to sum exactly. With help from technology, we can calculate specific partial sums and then estimate the error committed in ignoring the **upper tail**.

Series as functions "Infinite polynomials"—**power series**—are useful and convenient; many standard calculus functions can be written in power series form. Power series, their **intervals of convergence**, and their uses are the subjects of the last two sections of the chapter.

Taylor series; Taylor's theorem Taylor series, constructed by matching derivatives of a target function to those of a power series, give useful "formulas" for many calculus functions, including transcendental ones such as $\sin x$ and e^x. **Taylor's theorem** guarantees that, given appropriate conditions, the Taylor series of a function converges to the "right place"—the value $f(x)$.

REVIEW EXERCISES

In Exercises 1–8, find the limit of the sequence or explain why the limit does not exist.

1. $a_k = \left(\dfrac{\pi}{e}\right)^k$

2. $a_k = e^{-k}$

3. $a_k = (\arcsin 1)^k$

4. $a_k = \dfrac{k}{\sqrt{k}+10}$

5. $a_n = \dfrac{n+2}{n^3+4}$

6. $a_k = \dfrac{\ln(1+k^2)}{\ln(4+3k)}$

7. $a_k = \dfrac{\cos k}{\ln(k+1)}$

8. $a_k = e^{-k}\sin k$

In Exercises 9–34, determine whether the series converges absolutely, converges conditionally, or diverges. If the series converges, find an upper bound on its limit. If the series diverges, explain why.

9. $\displaystyle\sum_{k=0}^{\infty}\left(\dfrac{1}{k!}\right)^2$

10. $\displaystyle\sum_{n=3}^{\infty}\dfrac{1}{n(\ln n)^3}$

11. $\displaystyle\sum_{n=1}^{\infty}\left(\sum_{k=1}^{n}k^{-1}\right)$

12. $1-\dfrac{1}{2}+\dfrac{1}{2}-\dfrac{1}{4}+\dfrac{1}{3}-\dfrac{1}{6}+\dfrac{1}{4}-\dfrac{1}{8}+\dfrac{1}{5}-\dfrac{1}{10}+\cdots$

13. $\displaystyle\sum_{j=1}^{\infty}\dfrac{j}{5^j}$

14. $\displaystyle\sum_{j=1}^{\infty}\dfrac{j}{j^4+j-1}$

15. $\displaystyle\sum_{m=0}^{\infty}e^{-m^2}$

16. $\displaystyle\sum_{m=1}^{\infty}\dfrac{m^3}{m^4-7}$

17. $\displaystyle\sum_{k=1}^{\infty}\dfrac{k!}{(k+1)!-1}$

18. $\displaystyle\sum_{j=2}^{\infty}\dfrac{\ln j}{j^2}$

19. $\displaystyle\sum_{k=1}^{\infty}\dfrac{\sqrt{k}}{k^2+1}$

20. $\displaystyle\sum_{k=0}^{\infty}\dfrac{(-2)^k}{7^k+k}$

21. $\displaystyle\sum_{k=0}^{\infty}\dfrac{(-1)^k}{(k+1)2^k}$

22. $\displaystyle\sum_{m=8}^{\infty}\dfrac{\sin m}{m^3}$

23. $\displaystyle\sum_{n=1}^{\infty}\dfrac{\cos(n\pi)}{n}$

24. $\displaystyle\sum_{k=0}^{\infty}\dfrac{(-3)^k}{k^3+3^k}$

25. $\displaystyle\sum_{k=1}^{\infty}\dfrac{k^\pi}{k^e}$

26. $\displaystyle\sum_{m=2}^{\infty}\dfrac{1}{(\ln 3)^m}$

27. $\displaystyle\sum_{j=0}^{\infty}\left(\dfrac{1}{2^j}+\dfrac{1}{3^j}\right)^2$

28. $\displaystyle\sum_{k=1}^{\infty}\left(\int_k^{k+1}\dfrac{dx}{x^2}\right)$

29. $\displaystyle\sum_{m=0}^{\infty}\left(\int_0^m e^{-x^2}\,dx\right)$

30. $\displaystyle\sum_{n=1}^{\infty}\dfrac{\ln n}{\ln(3+n^2)}$

31. $\displaystyle\sum_{n=1}^{\infty}\sin(1/n)$

32. $\displaystyle\sum_{n=1}^{\infty}\dfrac{\sin(1/n)}{n}$

33. $\displaystyle\sum_{n=1}^{\infty}e^{-1/n}$

34. $\displaystyle\sum_{n=1}^{\infty}\left(1-e^{-1/n}\right)$

35. **(a)** Estimate a lower bound for $n!$ by comparing $\ln(n!)$ and $\displaystyle\int_1^n \ln x\,dx$.

   (b) Let b be a positive number. Use part (a) to find an integer N such that $b^N/N! < 1/2$.

36. Let $a_k = \displaystyle\int_k^{\infty}\dfrac{dx}{2x^2-1}$.

   (a) Evaluate $\displaystyle\lim_{k\to\infty} a_k$.   [HINT: $1/2x^2 \le 1/(2x^2-1) \le 1/x^2$ if $x \ge 1$.]

   (b) Does $\displaystyle\sum_{k=1}^{\infty}(-1)^{k+1}a_k$ converge absolutely? Justify your answer.

   (c) Does $\displaystyle\sum_{k=1}^{\infty}(-1)^{k+1}a_k$ converge? Justify your answer.

In Exercises 37–48, find the interval of convergence (endpoint behavior too!) of the power series.

37. $\displaystyle\sum_{m=1}^{\infty} \frac{x^m}{m^2+1}$

38. $\displaystyle\sum_{n=1}^{\infty} n^n x^n$

39. $\displaystyle\sum_{k=1}^{\infty} (3x)^k$

40. $\displaystyle\sum_{m=0}^{\infty} \frac{(3x)^m}{m!}$

41. $\displaystyle\sum_{n=1}^{\infty} \frac{(3x)^n}{n}$

42. $\displaystyle\sum_{j=1}^{\infty} \frac{(3x)^j}{j^2}$

43. $\displaystyle\sum_{m=0}^{\infty} \left(\frac{x-3}{2}\right)^m$

44. $\displaystyle\sum_{j=0}^{\infty} \frac{(x-2)^j}{j!}$

45. $\displaystyle\sum_{k=1}^{\infty} \frac{(x-1)^k}{k4^k}$

46. $\displaystyle\sum_{n=1}^{\infty} \frac{(x-1)^n}{\sqrt{n}}$

47. $\displaystyle\sum_{i=1}^{\infty} \frac{(x+5)^i}{i(i+1)}$

48. $\displaystyle\sum_{m=1}^{\infty} \frac{2^m(x-1)^m}{m}$

Consider a power series of the form $\displaystyle\sum_{k=0}^{\infty} a_k(x-1)^k$. In Exercises 49–53, indicate whether the statement must *be true,* cannot *be true, or* may *be true. Justify your answers.*

49. The power series converges only if $|x| > 2$.

50. The power series converges for all values of x.

51. If the radius of convergence of the power series is 3, the power series converges if $-2 < x < 4$.

52. The interval of convergence of the power series is $[-5, 5]$.

53. If the interval of convergence of the power series is $(-7, 9)$, the radius of convergence is 7.

54. (a) Evaluate $\displaystyle\lim_{x \to 1^-} \sum_{k=0}^{\infty} (-1)^k x^k$.

(b) Explain why the result in part (a) does *not* mean that $\displaystyle\sum_{k=0}^{\infty} (-1)^k$ converges.

In Exercises 55–58, use power series to evaluate the limit. Check your answers using l'Hôpital's rule.

55. $\displaystyle\lim_{x \to 0} \frac{1 - \cos x}{x}$

56. $\displaystyle\lim_{x \to 0} \frac{e^x - e^{-x}}{x}$

57. $\displaystyle\lim_{x \to 0} \frac{x - \arctan x}{x^3}$

58. $\displaystyle\lim_{x \to 1} \frac{\ln x}{x - 1}$

In Exercises 59–62, find the Maclaurin series for f and determine its radius of convergence.

59. $f(x) = 2^x = e^{x \ln 2}$

60. $f(x) = \cos^2 x = \frac{1}{2}(1 + \cos(2x))$

61. $f(x) = \dfrac{5+x}{x^2+x-2} = \dfrac{2}{x-1} - \dfrac{1}{x+2}$

62. $f(x) = \sin^3(x) = \frac{1}{4}(3 \sin x - \sin(3x))$

63. Determine the coefficients a_k such that $\dfrac{1}{1-x} = \displaystyle\sum_{k=0}^{\infty} a_k(x-2)^k$.

[HINT: $\dfrac{1}{1-x} = -\dfrac{1}{1+(x-2)}$.]

64. Is $1 - x + \dfrac{x^2}{2} - \dfrac{x^4}{8} + \dfrac{x^5}{15} - \dfrac{x^6}{240} + \cdots$ the Maclaurin series representation of the function f shown in the graph below? Justify your answer.

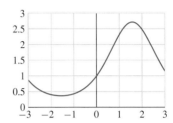

65. Suppose that f is a function that is positive, increasing, and concave down on the interval $[-2, 2]$.

(a) Is the coefficient of x^2 in the Maclaurin series representation of f positive? Justify your answer.

(b) Let $g(x) = 1/\sqrt{1 + f(x)}$. Is the coefficient of x^2 in the Maclaurin series representation of g positive? Justify your answer.

66. Let $H_n = \displaystyle\sum_{k=1}^{n} \frac{1}{k}$ be the nth partial sum of the harmonic series, let $a_n = H_n - \ln n$, and let $f(x) = \ln(x+1) - \ln x - \dfrac{1}{x+1}$.

(a) Show that $\ln(n+1) < H_n < 1 + \ln n$. [HINT: Use the integral test.]

(b) Show that the sequence $\{a_n\}$ is decreasing. [HINT: Explain why $\ln(n+1) - \ln n = \displaystyle\int_{n}^{n+1} x^{-1}\,dx > (n+1)^{-1}$.]

(c) Use part (c) to show that $\displaystyle\lim_{n \to \infty} a_n$ exists. (This limit, denoted by γ, is called **Euler's constant**; $\gamma \approx 0.57722$.)

(d) Show that $f(x) = -\displaystyle\int_{x}^{\infty} f'(t)\,dt$.

(e) Use part (d) to show that $f(x) > \dfrac{1}{2(x+1)^2}$.

[HINT: If $x > 0$, then $-f'(x) = \dfrac{1}{x(x+1)^2} > \dfrac{1}{(x+1)^3}$.]

(f) Show that $a_n - \gamma = \displaystyle\sum_{k=n}^{\infty} (a_k - a_{k+1})$.

(g) Use parts (e) and (f) to show that $\dfrac{1}{2(n+1)} < a_n - \gamma < \dfrac{1}{2n}$. [HINT: $f(k) = a_k - a_{k+1}$.]

*Exercises 67–76 explore some properties of the **binomial series**:*

$$(1+x)^r = 1 + \sum_{n=1}^{\infty} \frac{r(r-1)(r-2)\cdots(r-n+1)}{n!} x^n,$$

where r is a constant.

67. Show that the binomial series converges if $|x| < 1$.

68. Show that $\sqrt{1+x} \approx 1 + \frac{1}{2}x - \frac{1}{8}x^2 + \frac{1}{16}x^3 - \frac{5}{128}x^4 + \frac{7}{256}x^5 \mp \cdots$.

In Exercises 69–72, find the first four nonzero terms of a power series representation of the function.

69. $f(x) = (1 + x^4)^3$

70. $f(x) = \sqrt[3]{1 - x^2}$

71. $f(x) = (1 + x^2)^{-3/2}$

72. $f(x) = \arcsin x$.

73. Estimate $\displaystyle\int_0^{0.4} \sqrt{1 + x^3}\, dx$ with an error less than 5×10^{-4}.

74. Let $f(x) = (1 + x)^r$.

    (a) Show that $(1 + x)f'(x) = rf(x)$.

    (b) Let $g(x) = (1 + x)^{-r} f(x)$. Show that $g'(x) = 0$.

    (c) Use part (b) to show that $f(x) = (1 + x)^r$.

75. **(a)** Show that
$$\arcsin x = \int_0^x \frac{dt}{\sqrt{1 - t^2}} = x + \sum_{n=1}^{\infty} \frac{1 \cdot 3 \cdot 5 \cdots (2n - 1)}{2 \cdot 4 \cdot 6 \cdots (2n)} \frac{x^{2n+1}}{2n + 1}.$$

    (b) If $n \geq 1$ is an integer, then $\displaystyle\int_0^{\pi/2} \sin^{2n+1} x\, dx = \frac{2 \cdot 4 \cdot 6 \cdots (2n)}{3 \cdot 5 \cdot 7 \cdots (2n + 1)}$. Use this fact to evaluate $\displaystyle\int_0^1 \frac{x^{2n+1}}{\sqrt{1 - x^2}}\, dx.$

    (c) Use parts (a) and (b) to show that $\displaystyle\int_0^1 \frac{\arcsin x}{\sqrt{1 - x^2}}\, dx = \sum_{k=0}^{\infty} \frac{1}{(2k + 1)^2}.$

    (d) Use the substitution $u = \arcsin x$ to show that $\displaystyle\int_0^1 \frac{\arcsin x}{\sqrt{1 - x^2}}\, dx = \frac{\pi^2}{8}.$

    (e) Use parts (c) and (d) to show that $\displaystyle\sum_{k=1}^{\infty} \frac{1}{k^2} = \frac{\pi^2}{6}.$

76. Let $f(x) = (1 + x)^r$. In this exercise we prove that the binomial series converges to f.

    (a) Use part (e) of Exercise 19 in Section 9.2 to show that $R_n(x) = f(x) - P_n(x)$ is $R_n(x) = \dfrac{r \cdot (r - 1) \cdot (r - 2) \cdots (r - n)}{n!} \displaystyle\int_0^x \frac{(x - t)^n}{(1 + t)^{n+1-r}}\, dt.$

    (b) Suppose that $0 \leq t \leq x < 1$. Show that $\dfrac{|x - t|}{1 + t} \leq |x|.$

    (c) Use part (b) to show that if $0 \leq x < 1$, then $|R_n(x)| \leq \dfrac{|(r - 1) \cdot (r - 2) \cdots (r - n)|}{n!} x^n \left|(1 + x)^r - 1\right|.$

    (d) Use part (c) to show that the binomial series converges to f if $0 \leq x < 1$.

    (e) Adapt the reasoning in parts (b)–(d) to show that the binomial series converges to f if $-1 < x < 0$.

77. **(A proof that e is irrational.)** Assume that $e = m/n$, where m and n are positive integers.

    (a) Explain why $m! \left| \dfrac{1}{e} - \displaystyle\sum_{k=0}^{m} \frac{(-1)^k}{k!} \right| \leq \dfrac{m!}{(m + 1)!} = \dfrac{1}{m + 1}.$

    (b) Explain why $m!/e$ is an integer.

    (c) Explain why $m! \displaystyle\sum_{k=0}^{m} \frac{(-1)^k}{k!}$ is an integer.

        [HINT: Start by explaining why $m!/k!$ is an integer if k is an integer and $0 \leq k \leq m$.]

    (d) Parts (a)–(c) imply that $N = m! \left| \dfrac{1}{e} - \displaystyle\sum_{k=0}^{m} \frac{(-1)^k}{k!} \right|$ is an integer that is less than or equal to $1/(m + 1)$. Explain why it follows that $N = 0$.

    (e) Explain why the conclusion of part (d) is impossible and therefore e cannot be a rational number.

        [HINT: $\displaystyle\sum_{k=m+1}^{\infty} (-1)^k / k! \neq 0.$]

78. **(Cauchy condensation theorem.)** Let $\sum_{n=1}^{\infty} a_n$ be a series of positive terms such that $a_{n+1} \leq a_n$ for all n.

    (a) Let $m \geq 1$ be an integer. Explain why
$$2^{m-1} a_{2^m} \leq a_{2^{m-1}+1} + a_{2^{m-1}+2} + \cdots + a_{2^m} \leq 2^{m-1} a_{2^{m-1}}.$$

    (b) Use part (a) to show that $\dfrac{1}{2} \displaystyle\sum_{k=1}^{m} 2^k a_{2^k} \leq \sum_{k=2}^{2^m} a_k.$

    (c) Use part (b) to show that if $\dfrac{1}{2} \displaystyle\sum_{k=1}^{\infty} 2^k a_{2^k}$ diverges, then $\displaystyle\sum_{k=1}^{\infty} a_k$ diverges.

    (d) Let m and n be integers such that $n \leq 2^m$. Use part (a) to show that $\displaystyle\sum_{k=2}^{n} a_k \leq \sum_{k=1}^{m} 2^{k-1} a_{2^{k-1}}.$

    (e) Use part (d) to show that if $\displaystyle\sum_{k=0}^{\infty} 2^k a_{2^k}$ converges, then $\displaystyle\sum_{k=2}^{\infty} a_k$ converges.

79. Use Exercise 78 to prove that the harmonic series diverges.

80. Use Exercise 78 and the properties of geometric series to prove that $\displaystyle\sum_{k=1}^{\infty} 1/k^p$ converges if $2^{1-p} < 1$ (i.e., $p > 1$) and diverges otherwise.

81. Use Exercise 78 and the result of the previous exercise to show that $\displaystyle\sum_{k=2}^{\infty} \frac{1}{k(\ln k)^p}$ converges if $p > 1$ and diverges otherwise.

82. Use the fact that $\displaystyle\int_0^{\infty} t^n e^{-t}\, dt = n!$ to show that $\displaystyle\int_0^{\infty} e^{-t} \sin(xt)\, dt = \frac{x}{1 + x^2}$ if $|x| < 1$.

83. **(a)** Use the trigonometric identity $\tan x = \cot x - 2\cot(2x)$ to show that
$$\sum_{k=1}^{n} \frac{1}{2^k} \tan\left(\frac{x}{2^k}\right) = \frac{1}{2^n} \cot\left(\frac{x}{2^n}\right) - \cot x.$$

    (b) Use part (a) to show that
$$\sum_{k=1}^{\infty} \frac{1}{2^k} \tan\left(\frac{x}{2^k}\right) = \frac{1}{x} - \cot x.$$

84. Let $I_n = \int_0^{\pi/4} \tan^n x \, dx$ for any integer $n \geq 0$.

(a) Show that $\{I_n\}$ is a nonincreasing sequence (i.e., $I_{n+1} \leq I_n$).

(b) Show that $I_n + I_{n-2} = \dfrac{1}{n-1}$.

(c) Use parts (a) and (b) to show that $\dfrac{1}{2(n+1)} \leq I_n \leq \dfrac{1}{2(n-1)}$. [HINT: Part (b) implies that $I_{n+2} + I_n = 1/(n+1)$.]

(d) Show that, if $n \geq 2$, $I_n = \dfrac{1}{n-1} - \int_0^{\pi/4} \tan^{n-2} x \, dx$.

(e) Use part (d) to show that $I_{2n} = (-1)^n \left(\dfrac{\pi}{4} + \sum_{k=1}^{n} \dfrac{(-1)^k}{2k-1} \right)$ for all integers $n \geq 1$.

(f) Use parts (c) and (e) to show that $\sum_{k=1}^{\infty} \dfrac{(-1)^{k+1}}{2k-1} = \dfrac{\pi}{4}$.

(g) Use part (d) to show that $I_{2n+1} = (-1)^n \left(\dfrac{1}{2} \ln 2 + \sum_{k=1}^{n} \dfrac{(-1)^k}{2k} \right)$ for all integers $n \geq 1$.

(h) Use parts (c) and (g) to show that $\sum_{k=1}^{\infty} \dfrac{(-1)^k}{k} = -\ln 2$.

Fourier Series

Fourier series are the infinite analogues of **Fourier polynomials**, which we defined and explored briefly in Section 9.3. (In the same sense, Taylor series are the infinite version of Taylor polynomials.) Compare the following definition with that on page 512:

DEFINITION Let $f(x)$ be a function defined on $[-\pi, \pi]$. The **Fourier series** for f is the series

$$a_0 + a_1 \cos x + b_1 \sin x + a_2 \cos(2x) + b_2 \sin(2x) + \cdots$$
$$+ a_n \cos(nx) + b_n \sin(nx) + \cdots$$

with coefficients given by

$$a_0 = \frac{1}{2\pi} \int_{-\pi}^{\pi} f(x)\, dx;$$

$$a_k = \frac{1}{\pi} \int_{-\pi}^{\pi} f(x) \cos(kx)\, dx \qquad \text{if } k > 0;$$

$$b_k = \frac{1}{\pi} \int_{-\pi}^{\pi} f(x) \sin(kx)\, dx \qquad \text{if } k > 0.$$

Naturally enough, Fourier series inherit many of their properties from Fourier polynomials.

PROBLEM 1 Let f and g be functions defined on $[-\pi, \pi]$; assume that f is odd and g is even.

(a) Show that the Fourier series of f has $a_k = 0$ for all $k \geq 0$.

(b) Show that the Fourier series of g has $b_k = 0$ for all $k \geq 1$.

PROBLEM 2 Show that the Fourier series for $f(x) = x$ has the form

$$\frac{2}{1} \sin x - \frac{2}{2} \sin(2x) + \frac{2}{3} \sin(3x) - \frac{2}{4} \sin(4x) + \frac{2}{5} \sin(5x) - \cdots.$$

What's the general "formula" for b_n?

PROBLEM 3 Show that the Fourier series for $f(x) = |x|$ has the form

$$\frac{\pi}{2} - \frac{4}{\pi} \cos x - \frac{4}{9\pi} \cos(3x) - \frac{4}{25\pi} \cos(5x) - \frac{4}{49\pi} \cos(7x) - \cdots.$$

(Note that $b_k = 0$ for all $k \geq 1$.)

Convergence and divergence of Fourier series Whether a Fourier series converges or diverges for a given value of x is, in general, a difficult question—much more so than for Taylor series. Although Taylor series converge for x in a particular *interval*, and diverge elsewhere, the convergence set of a Fourier series may be quite complicated. Under suitably strict conditions, however, a Fourier series *does* converge, and to the "right place"—to the function value $f(x)$ itself. The following theorem illustrates one set of (technical-looking) conditions under which this happens.

THEOREM 15 Let f be continuous on $[-\pi, \pi]$, with $f(-\pi) = f(\pi)$ and only finitely many local maximum and minimum points in the interval $[-\pi, \pi]$. Then the Fourier series of f converges to $f(x)$ for all x in $[-\pi, \pi]$.

PROBLEM 4 This problem is about the function $f(x) = |x|$. Observe that Theorem 15 applies to f.

(a) Apply Theorem 15 with $x = 0$ to show that

$$1 + \frac{1}{9} + \frac{1}{25} + \frac{1}{49} + \cdots = \frac{\pi^2}{8}.$$

(b) Apply Theorem 15 with $x = \pi/4$; what series identity results?

(c) Apply Theorem 15 with $x = \pi/2$; what series identity results?

PROBLEM 5 Experiment with Theorem 15, the function $f(x) = x^2$, and an appropriate value of x to show that

$$\sum_{k=1}^{\infty} \frac{1}{k^2} = 1 + \frac{1}{4} + \frac{1}{9} + \cdots = \frac{\pi^2}{6}.$$

V VECTORS AND POLAR COORDINATES

V.1 VECTORS AND VECTOR-VALUED FUNCTIONS

A **vector** is a quantity that has both magnitude and direction. Vectors can "live" in any dimension—in the real line, in the xy-plane, in xyz-space, or in higher-dimensional spaces. This section introduces vectors and vector-valued functions in the xy-plane.

Vectors in the plane can be thought of either as arrows (a natural metaphor for objects with length and direction) or as 2-tuples:

- **Vectors as arrows** A vector v can be described by an arrow in the xy-plane. Two arrows with the same length and the same direction describe the same vector—regardless of where the arrows are placed in the plane.

    It's common to denote vectors either with bold type, as in v, or with an arrow on top, as in $\overrightarrow{PQ}$. ➡ Ordinary numbers (also called **scalars** when vectors are in the picture) appear in ordinary type.

    The latter is convenient when writing by hand.

- **Vectors as 2-tuples** A vector v can be written as a 2-tuple of real numbers, as in $v = (a, b)$. The numbers a and b are called the **components** (or, alternatively, the **coordinates**) of the vector v.

Figure 1 illustrates these views of vectors and some connections between them.

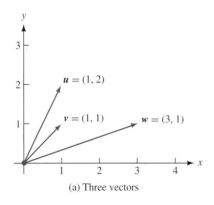
(a) Three vectors

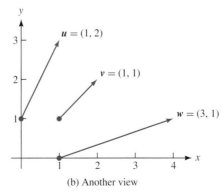
(b) Another view

FIGURE 1
Vectors as arrows and as 2-tuples

Following are some notes on the pictures; watch for some useful vocabulary and notation.

- **Different tail points, same vectors**   The two pictures show exactly the same vectors *u*, *v*, and *w*. The only difference concerns where in the plane the vectors are placed; the vectors' tail points are shown as dark dots in the pictures,

- **Vectors based at the origin**   In Figure 1(a) all three vectors have their tails at the origin so each vector is of the form $\overrightarrow{OP}$, where O is the origin and P is the point at the tip. In this special case, the *components* of each vector are the same as the Cartesian *coordinates* of the tip point P. For this reason a vector of the form $\boldsymbol{v} = \overrightarrow{OP}$ is called the **position vector** of the point P.

- **What the components say**   In Figure 1(b), each vector's components are *not* the Cartesian coordinates of its tip point. Instead, they describe the **displacement** between the vector's head and its tail. The vector $\boldsymbol{w} = (3, 1)$, for instance, involves moving 3 units to the right and 1 unit up.

- **The vector from P to Q**   If $P = (a, b)$ and $Q = (c, d)$ are any points in the plane, then the vector from P to Q, denoted by $\overrightarrow{PQ}$, has components $(c - a, d - b)$. In Figure 1(b), for instance, the vector $\boldsymbol{w}$ starts at $(1, 0)$ and ends at $(4, 1)$, and so its coordinates are indeed $(3, 1)$.

- **Finding lengths**   The **length** (or **magnitude**) of a vector is the distance between its head and its tail. If $\boldsymbol{v} = (a, b)$, then, by the distance formula in the xy-plane,

$$\text{length of } \boldsymbol{v} = |\boldsymbol{v}| = \sqrt{a^2 + b^2}.$$

Both cases involve measuring sizes.

A vector with length 1 is called a **unit vector**. Notice the use of the absolute value notation for the length of a vector. ◂

Algebra with vectors

Vectors permit algebraic operations much like those with ordinary numbers. Here are some simple samples of operations with 2-tuples:

$$(1, 2) + (2, 1) = (3, 3); \qquad -3 (1, 2) = (-3, -6); \qquad 3 (1, 2) - (3, 4) = (0, 2).$$

The first operation illustrates the **sum** of two vectors; the result is another vector. The second operation illustrates **scalar multiplication**; a vector is multiplied by a scalar to produce a new vector. The third operation combines scalar multiplication with the **difference** of two vectors.

Here are the formal rules:

DEFINITION (Operations on vectors)   Let $\boldsymbol{v} = (a, b)$ and $\boldsymbol{w} = (c, d)$ be plane vectors and let r be a scalar. Then

$$\boldsymbol{v} + \boldsymbol{w} = (a, b) + (c, d) = (a + c, b + d)$$

is **sum** of $\boldsymbol{v}$ and $\boldsymbol{w}$. **Scalar multiplication** of $\boldsymbol{v}$ by r is defined by

$$r\boldsymbol{v} = r (a, b) = (ra, rb).$$

More enlightening than these simple formulas are their geometric meanings. Consider addition first.

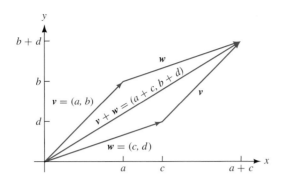

FIGURE 2
Adding vectors

Figure 2 describes addition geometrically. If $v = (a, b)$ and $w = (c, d)$, then the sum $v + w = (a + c, b + d)$ is the vector obtained by putting the tail of one vector at the head of the other. ➡ Figure 2 also illustrates the **parallelogram rule**: The sum $v + w$ is the *diagonal* of the parallelogram with sides v and w.

The order doesn't matter; vector addition is commutative.

Figure 3 describes scalar multiplication:

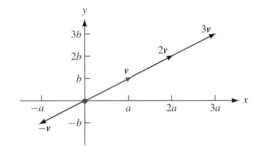

FIGURE 3
Scalar-multiplying vectors

(All vectors shown have tails at the origin.) This picture illustrates the product of a vector v with a scalar a: The result, av, is a new vector with direction either the same as v (if $a > 0$) or opposite to v (if $a \leq 0$); the length of av is $|a|$ times the length of v. ➡

In other words, multiplication by a stretches v by the factor $|a|$.

Figure 4 shows several combined vector operations:

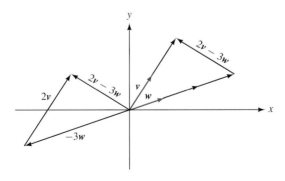

FIGURE 4
Algebra with vectors

Notice, especially, how the **difference** of two vectors appears:

> *If v and w have the same tail point, then the difference $v - w$ points from the tip of w to the tip of v.*

(The picture shows two copies of the difference vector $2v - 3w$.)

Vector algebraic operations enjoy many of the pleasant properties that we've come to expect of operations with ordinary numbers. Theorem 1 collects several such properties.

THEOREM 1 (Properties of vector operations)   Let u, v, and w be vectors and let r and s be scalars. The following algebraic properties hold:

    (a) $u + v = v + u$ (commutativity);

    (b) $u + (v + w) = (u + v) + w$ (associativity);

    (c) $(r + s)v = rv + sv$,   and $r(v + w) = rv + rw$ (distributivity).

All parts of the theorem can be shown to hold using similar properties of ordinary addition and multiplication.

Standard basis vectors   The two-dimensional vectors

$$i = (1, 0) \quad \text{and} \quad j = (0, 1)$$

are especially simple—but especially useful. Both i and j have length 1, and they point in perpendicular directions—along the x- and y-axes, respectively. The vectors i and j are known as the **standard basis vectors** for $\mathbb{R}^2$. They deserve this important-sounding name for a good reason:

> *Every vector $v = (a, b)$ in $\mathbb{R}^2$ can be written as a sum of scalar multiples of i and j.*

If, say, $v = (2, 3)$, then

$$v = (2, 3) = (2, 0) + (0, 3) = 2(1, 0) + 3(0, 1) = 2i + 3j.$$

A sum of the form $rv + sw$, where r and s are scalars, is called a **linear combination** of v and w. As the preceding calculation illustrates, *every* vector (a, b) in $\mathbb{R}^2$ can be written as a linear combination of i and j:

$$(a, b) = ai + bj.$$

Indeed, some authors use the ij-notation for almost all 2-vectors.

Vector-valued functions

A **vector-valued function** is one that produces vectors as outputs. Consider, for example, the function defined by the rule

$$f(t) = (\cos t, \sin t).$$

For this function, the notation $f : \mathbb{R} \to \mathbb{R}^2$ makes sense because f accepts a single number t as input and produces the vector $(\cos t, \sin t)$ as output. Notice too that, because the function is vector-valued, we use the boldface symbol f rather than an ordinary f to denote it. The two functions inside the parentheses are called **component functions**, or **coordinate functions**.

It's often helpful to think of a vector-valued function as a curve traced out by a collection of position vectors $f(t)$, one for each input t. (All tails are pinned at the origin.) Figure 5 shows several of these position vectors for the vector-valued function $f(t) = (t, 3 + \sin t)$.

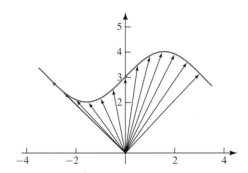

FIGURE 5
Position vectors tracing a curve

Parametric curves and vector-valued functions. Parametric curves are close kin to vector-valued functions. The following example illustrates the relationship.

EXAMPLE 1 A vector-valued function $f : \mathbb{R} \to \mathbb{R}^2$ is defined by $f(t) = (\cos t, \sin t)$. How is f related to a parametric curve? Which curve?

Solution Setting

$$x = \cos t; \qquad y = \sin t; \qquad 0 \le t \le 2\pi$$

gives a parametric form of the unit circle $x^2 + y^2 = 1$, traversed once counterclockwise starting from the "east pole" $(1, 0)$. Thus, for any input t the vector $f(t) = (\cos t, \sin t)$ can be thought of as the *position vector* for a point on the unit circle. ➡ As t ranges through the domain $(-\infty, \infty)$, the tip of the vector $f(t)$ traces the unit circle again and again, counterclockwise. ∎

The tail is pinned at the origin.

Calculus with vector-valued functions

In single-variable calculus, derivatives describe rates and directions of change. ➡ The same useful interpretation proves true for derivatives of vector-valued functions:

Positive derivatives correspond to change in the increasing direction.

> **DEFINITION (Derivative of a vector-valued function)** Consider the vector-valued function $f(t) = \big(f_1(t), f_2(t)\big)$. The derivative of f, denoted f', is the vector-valued function defined by
>
> $$f'(t) = \frac{d}{dt}\big(f_1(t), \ f_2(t) \big) = \big(f_1'(t), \ f_2'(t) \big).$$

Calculating this new derivative is no harder than finding a list of *ordinary* derivatives. For example,

$$f(t) = (\cos t, \sin t) \implies f'(t) = (-\sin t, \cos t);$$

we simply differentiate each component function separately.

What the vector derivative means: Velocity The definition is easy to use symbolically, but what does the derivative *mean*?

The short answer is **velocity**. More precisely, let $f : I \to \mathbb{R}^2$ be a vector-valued function such that $f'(t) \neq (0, 0)$ for all t in an interval I. (This guarantees that f traces out a curve C.) If t_0 is in I, then the derivative vector $f'(t_0)$ has two main meanings:

- **Direction**   The vector $f'(t_0)$ is tangent to the curve C at the point $t = t_0$; it points in the direction of increasing t.
- **Speed**   The scalar $|f'(t_0)|$ (the magnitude of the derivative vector) tells the instantaneous speed (in units of distance per unit of time t) at which $f(t)$ moves along C.

Taken together, these observations mean that if $f(t)$ describes the position of a moving particle (a vector quantity) at time t, then the derivative $f'(t)$ describes the particle's **velocity**—another vector quantity—at the same time. We'll explain in a moment *why* these interpretations hold; first, we illustrate their uses by example.

E X A M P L E 2   Find and interpret the velocity vector to the curve $p(t) = (\cos t, \sin t)$ (i.e., the unit circle) at the points where $t = 0$ and where $t = \pi/4$.

Solution   The symbolic calculations are simple. Since $p'(t) = (-\sin t, \cos t)$, we have

$$p'(0) = (0, 1) \quad \text{and} \quad p'\left(\frac{\pi}{4}\right) = \left(-\frac{\sqrt{2}}{2}, \frac{\sqrt{2}}{2}\right).$$

These vectors are indeed tangent to the curve at the points in question. At $t = 0$, for instance, our function gives $p(0) = (1, 0)$. This is the circle's "east pole"; at this point, the curve is *vertical*, as is the velocity vector $p'(0) = (0, 1)$. (The velocity vector points *upward* because the curve is traversed counterclockwise.) Similarly, the derivative vector $p'(\pi/4)$ is tangent to the circle at the "northeast" point $p(\pi/4) = (\sqrt{2}/2, \sqrt{2}/2)$.

Another calculation shows that both of these vectors have the same length, 1. In fact,

$$|p'(t)| = |(-\sin t, \cos t)| = \sqrt{\sin^2 t + \cos^2 t} = 1$$

for *all* inputs t. This means that the curve is traversed at a *constant* speed of 1 unit of distance per unit of time. ∎

Speed and arc length   Recall that to find the total distance an object travels over a time interval we *integrate* the object's speed function over the interval.

In the present setting, this means that we can find the length of a curve C traced out by a vector-valued position function $p(t)$ by integrating the speed function over the t-interval in question. (Note that the speed at any time t is a scalar, so the speed function is an ordinary scalar-valued function.)

Suppose, then, that a curve C is given by the position function

$$p(t) = \big(x(t), y(t)\big); \qquad a \leq t \leq b.$$

We said above that the speed at which C is traversed at time t is given by

$$|p'(t)| = \sqrt{x'(t)^2 + y'(t)^2}.$$

This formula shows that, for most curves, the speed is a *nonconstant* function of t. In the (rare) exceptional case, we speak of a **constant-speed parametrization**.

> **FACT (Arc length by integration)** Consider the curve C given by the position function
> $$p(t) = (x(t), y(t))$$
> for $a \le t \le b$. Then
> $$\text{arc length of } C = \int_a^b \sqrt{x'(t)^2 + y'(t)^2}\, dt.$$

EXAMPLE 3 Find the length of the unit circle using the parametrization $p(t) = (\cos t, \sin t), 0 \le t \le 2\pi$.

Solution We found in Example 2 that for $p(t) = (\cos t, \sin t)$ the speed function has constant value 1. Thus, we have
$$\text{arc length} = \int_0^{2\pi} 1\, dt = 2\pi,$$
the (familiar) circumference of a unit circle. ∎

Understanding vector derivatives It's useful and enlightening to think of vector derivatives as velocities. But why does this interpretation make sense? Why, for instance, does the derivative vector $f'(t_0)$ point in the direction *tangent* to the curve defined by f at the point $t = t_0$?

A full answer requires more theory than we're ready for now, but the general idea is not difficult. If h is a small positive number, then by the definition of derivative (written in vector form) we have
$$f'(t_0) \approx \frac{f(t_0 + h) - f(t_0)}{h}.$$
The equation deserves a close look:

- The numerator on the right is the difference between two vectors. Such a vector *difference* is a vector that *joins* the two nearby points $f'(t_0 + h)$ and $f(t_0)$ on the curve and so points "in the direction of the curve." (The denominator h is a scalar and so affects only the length, not the direction, of the difference.)

- The preceding approximate equality has a close analogue for lengths:
$$|f'(t_0)| \approx \frac{|f(t_0 + h) - f(t_0)|}{h}.$$

The numerator on the right measures the *distance* between $f'(t_0 + h)$ and $f(t_0)$, whereas the denominator measures the "time elapsed," and so the quotient gives a natural estimate of the curve's speed at time t_0.

Vector antiderivatives and integrals Vector *antiderivatives*, like derivatives, are found component-by-component. In symbols: If $f(t) = (f_1(t), f_2(t))$, then
$$\int f(t)\, dt = \int (f_1(t), f_2(t))\, dt = \left(\int f_1(t)\, dt, \int f_2(t)\, dt \right).$$

EXAMPLE 4 Let $f(t) = (2t, \cos t)$. Show that
$$\int f(t)\, dt = (t^2 + C_1, \sin t + C_2),$$
where C_1 and C_2 are arbitrary constants.

Solution   We antidifferentiate separately in each component:

$$\int (2t, \cos t)\, dt = \left(\int 2t\, dt, \int \cos t\, dt \right) = (t^2 + C_1, \sin t + C_2).$$

The answer (as usual for antiderivatives) is easily checked by differentiation.   ■

Modeling motion   The vector tools and ideas we've developed allow us to model two-dimensional motion—the movement of objects in the xy-plane. The key idea is that vector-valued acceleration, velocity, and position functions are all derivatives and antiderivatives of each other.

Starting with acceleration   In modeling physical motion, it's often most natural to begin with information about **acceleration** (the derivative of velocity) and to work from there to formulas for velocity and position. Acceleration arises naturally in physical settings because of its close connection to force. **Newton's second law** puts the matter like this:

> *A force acting on an object produces an acceleration that is directly proportional to the force and inversely proportional to the object's mass.*

Forces (gravitational force, air drag, sliding friction, a rocket's thrust, etc.) can often be measured directly and, thanks to Newton's law, converted into information about acceleration.

Zero acceleration   In the simplest possible case, an object's acceleration is the zero vector. In this case it's easy to find velocity and position functions by antidifferentiation. For velocity,

$$\boldsymbol{a}(t) = (0, 0) \Longrightarrow \boldsymbol{v}(t) = \int (0, 0)\, dt = \left(\int 0\, dt, \int 0\, dt \right) = (C_1, C_2),$$

where C_1 and C_2 are constants. For position,

$$\boldsymbol{v}(t) = (C_1, C_2) \Longrightarrow \boldsymbol{p}(t) = \int (C_1, C_2)\, dt = \left(\int C_1\, dt, \int C_2\, dt \right)$$
$$= (C_1 t + C_3, C_2 t + C_4),$$

where C_3 and C_4 are arbitrary constants.

In specific cases, the four constants can be evaluated using additional information (such as the velocity and position of the particle at time $t = 0$). Notice, however, that regardless of the numerical values of these constants, the calculation illustrates an interesting physical fact:

> *In the absence of outside forces, an object has* zero *acceleration, constant* velocity *(and speed), and a* linear *position function.*

EXAMPLE 5   At time $t = 0$, a particle in the plane has velocity vector $(1, 2)$ and position vector $(3, 4)$. (These stipulations are called **initial conditions**.) No external force acts, so the acceleration is $\boldsymbol{a}(t) = (0, 0)$ at all times. Describe the particle's movement. Where is the particle at time $t = 100$?

Solution   As we just calculated, the particle has constant velocity function $\boldsymbol{v}(t) = (C_1, C_2)$ and linear position function $\boldsymbol{p}(t) = t(C_1, C_2) + (C_3, C_4)$. The initial conditions give $C_1 = 1$, $C_2 = 2$, $C_3 = 3$, and $C_4 = 4$, and so

$$\boldsymbol{v}(t) = (1, 2) \quad \text{and} \quad \boldsymbol{p}(t) = (3, 4) + t(1, 2).$$

Thus, the particle moves away from $(3, 4)$ with constant speed $\sqrt{5}$, in the direction of the vector $(1, 2)$. At $t = 100$ the particle's position is $\boldsymbol{p}(100) = (103, 204)$.   ■

Constant acceleration In the next simplest case, an object's acceleration is a nonzero *constant* vector. This case is especially important; it arises when a constant force (such as gravity) acts on an object, producing a constant acceleration. Suppose, then, that an object has constant acceleration vector $a(t) = (a_1, a_2)$. Then

$$v(t) = \int (a_1, a_2)\, dt = t(a_1, a_2) + (C_1, C_2),$$

where C_1 and C_2 are arbitrary constants. Because $v(0) = (C_1, C_2)$, the constants represent the object's initial velocity.

To find position we antidifferentiate again:

$$p(t) = \int \left(t(a_1, a_2) + (C_1, C_2) \right) dt = \frac{t^2}{2}(a_1, a_2) + t(C_1, C_2) + (C_3, C_4),$$

where C_3 and C_4 are again arbitrary constants. Since $p(0) = (C_3, C_4)$, these constants represent the object's initial position. In particular:

Constant acceleration begets linear velocity and quadratic position.

EXAMPLE 6 An object starts from rest at the origin and has constant acceleration $a = (1, 2)$. What happens? What are the object's position and velocity at $t = 100$ seconds?

Solution The calculations preceding this example, together with the initial conditions, mean that the object's velocity and position functions are

$$v(t) = t(1, 2) \quad \text{and} \quad p(t) = \frac{t^2}{2}(1, 2).$$

Thus, the particle moves along the curve defined by $p(t)$. At $t = 100$ the particle's velocity is $100(1, 2) = (100, 200)$; its speed is $100\sqrt{5} \approx 223.6$ units per second; and its position is $p(100) = 5000(1, 2) = (5000, 10000)$. ➡   ■   *It's gone a long way.*

Gravity and acceleration It's common knowledge that Earth's gravity pulls everything toward the center of the Earth, producing a "downward" acceleration. It's also true—but less obvious—that the acceleration due to gravity has the same magnitude for all objects. (This magnitude is usually denoted by g; it is called the **acceleration due to gravity**. In the metric system, $g \approx 9.8$ meters per second per second. In the English system, $g \approx 32$ feet per second per second.)

As a result, gravity induces the same acceleration on all objects near the surface of the Earth. In vector language this means that the acceleration due to gravity is a *constant* vector of length g, pointing toward Earth's center.

BASIC EXERCISES

In Exercises 1–8, use the vectors $u = (1, 2)$, $v = (2, 3)$, *and* $w = (-2, 1)$ *to find*

1. $u + v$

2. $u - v$

3. $2u - 3v$

4. $u + v/2 + w/3$

5. $|u|$

6. $|w|$

7. $|3u + 2w|$

8. $|2v - 3u|$

Consider the three vectors $u = (a, b)$, $v = (c, d)$, *and* $w = (e, f)$, *and let* r *be a scalar. By the definition of vector addition,* $u + v = (a, b) + (c, d) = (a + c, b + d) = (c + a, d + b) = (c, d) + (a, b) = v + u$. *(The middle step holds because ordinary addition of numbers is commutative.) This argument shows that*

vector addition is commutative, as asserted in Theorem 1. Use a similar argument to show the claims in Exercises 9 and 10.

9. $u + (v + w) = (u + v) + w$

10. $r(v + w) = rv + rw$

11. In this section we mentioned the parallelogram rule for vector addition: The sum $v + w$ is the diagonal of the parallelogram with adjacent sides v and w.

(a) Draw and label a picture to illustrate this fact for the vectors $v = (2, 1)$ and $w = (1, 3)$. (Base both vectors at the origin.) What are the components of the diagonal vector?

(b) The parallelogram you drew in part (a) actually has *two* diagonals. You drew a northeast-pointing diagonal in part (a). Now draw the northwest-pointing diagonal vector. What are its components? How is it related to *v* and *w*?

(c) Draw the southeast-pointing diagonal vector. What are its components? How is it related to *v* and *w*?

12. Let *v* and *w* be nonparallel vectors. What basic fact from Euclidean geometry is expressed by the relation $|v+w| \leq |v|+|w|$?

*Let **u** and **v** be the vectors pictured below. In Exercises 13–16, draw the vector **w**.*

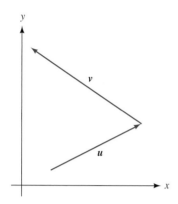

13. $w = u + v$.

14. $w = u - v$.

15. $w = v - u$.

16. $w = u + u$.

*Let **u** and **v** be the vectors pictured below. In Exercises 17–20, draw the vector **w**.*

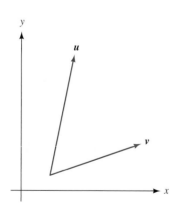

17. $w = u + v$.

18. $w = u - v$.

19. $w = v - u$.

20. $w = v + v$.

21. Let ℓ be the line through the point $P(1, 2)$ in the direction of the vector $v = (2, 3)$.

(a) The line ℓ is the image of the vector-valued function $L(t) = (1, 2) + t(2, 3)$ for all real t. Draw a picture to explain why; label the points on ℓ that correspond to $t = 0$, $t = 1$, $t = 2$, and $t = -1$.

(b) What is the image of the function $L(t) = (1, 2) + t(2, 3)$ if the domain is restricted to $t \geq 0$?

(c) What is the image of the function $L(t) = (1, 2) + t(2, 3)$ if the domain is restricted to $-1 \leq t \leq 1$?

22. Repeat Exercise 21, but let ℓ be the line through (a, b) in the direction of (c, d). (Assume that c and d are not both zero.)

23. Suppose that the motion of a particle is described by the parametric equations $x(t) = 1 - 2t + t^3$, $y(t) = 2t^4$. What is the particle's speed at time $t = 1$?

24. This exercise is about Example 6.

(a) By hand, sketch the "curve" defined by $p(t)$ for $0 \leq t \leq 100$. What very simple shape does the "curve" seem to have? [HINT: The curve is parametrized by $x(t) = p_1(t)$; $y(t) = p_2(t)$; $0 \leq t \leq 100$.]

(b) Eliminate the variable t in the parametric equations for the curve $p(t)$; what's the resulting equation? Does it agree with what you found in part (a)?

(c) Find the arc length of the curve $p(t)$ from $t = 0$ to $t = 100$.

25. Redo Exercise 24 but assume that the acceleration vector is $(1, -1)$.

26. At time $t = 0$ seconds a particle is at the origin and has velocity vector $(4, 4)$. It undergoes constant acceleration $a(t) = (0, -1)$.

(a) Find formulas for the velocity function $v(t)$ and the position function $p(t)$.

(b) Plot the path taken by the particle $p(t)$ for $0 \leq t \leq 10$. What familiar shape does the path seem to have?

(c) Eliminate the variable t in the parametric equations for the curve $p(t)$; what's the resulting equation? Does it agree with what you found in part (c)?

(d) Find the arc length of the curve $p(t)$ from $t = 0$ to $t = 10$. [HINT: Set up the integral by hand; either solve it exactly using a table of integrals or use technology to approximate the integral numerically.]

27. A projectile is at $(0, 0)$ at time 0 seconds. It has initial speed 100 meters per second, initial angle $\pi/3$, and travels under free fall conditions.

(a) Find equations for the velocity and position functions.

(b) When does the projectile touch down?

(c) Plot the projectile's trajectory from takeoff to landing.

(d) At what time is the projectile at maximum height? How high is this? [HINT: Find the time at which the velocity vector is horizontal.]

(e) Find the speed and the velocity at the moment the projectile is highest.

28. A projectile is at $(0, 0)$ at time 0 seconds. It has initial speed s_0 meters per second, initial angle α, and travels under free fall conditions.

(a) Find equations for the velocity and position functions.

(b) At what time is the projectile highest? [HINT: The answer depends on both s_0 and α.]

(c) Find the speed and the velocity at the moment when the projectile is highest.

(d) Find the speed and the velocity at the moment when the projectile lands.

29. Suppose that the velocity of an object at time t seconds is $(5t^2 + 3t - 4, 1 - t)$ meters/second. At time $t = 2$ the object is at the point with coordinates $(1, 0)$.

(a) Find the object's acceleration vector at time $t = 1$.

(b) Find the position of the object at time $t = 0$.

(c) Express the distance traveled by the object between time $t = 0$ and time $t = 2$ as an integral.

30. Suppose that the motion of a particle is described by the parametric equations $x = t^3 - 3t$, $y = t^2 - 2t$.

(a) Does the particle ever come to a stop? If so, when and where?

(b) Is the particle ever moving straight up or down? If so, when and where?

(c) Is the particle ever moving horizontally left or right? If so, when and where?

31. A particle moves in the xy-plane with constant acceleration vector $\boldsymbol{a}(t) = (0, -1)$. At time $t = 0$ the particle is at the point $\boldsymbol{p}(0) = (0, 0)$ and has velocity $\boldsymbol{v}(0) = (1, 0)$.

(a) Give a formula for the particle's velocity function $\boldsymbol{v}(t)$.

(b) Give a formula for the particle's position function $\boldsymbol{p}(t)$.

(c) Write an integral whose value is the distance traveled by the particle between $t = 0$ and $t = 5$.

In Exercises 32–36, plot the curve and estimate its arc length by eye. Set up the arc length integral. If possible, evaluate it exactly by antidifferentiation; otherwise, estimate the answer using a midpoint approximating sum with 20 subdivisions.

32. $\boldsymbol{r}(t) = (3 + t, 2 + 3t), 0 \le t \le 1$.

33. $\boldsymbol{r}(t) = (\cos(2t), \sin(2t)); 0 \le t \le \pi$

34. $\boldsymbol{r}(t) = (\sin(3t), \cos(3t)); 0 \le t \le 2\pi/3$

35. $\boldsymbol{r}(t) = (3 \sin t, \cos t); 0 \le t \le 2\pi$

36. $\boldsymbol{r}(t) = (t \cos t, t \sin t); 0 \le t \le 4\pi$

37–41. Find a formula for the speed at time t of the vector-valued functions given in Exercises 32–36.

42. Consider the vector-valued function $\boldsymbol{p}(t) = (\sin t, t)$.

(a) Draw (by hand) a rough plot of the curve defined by $\boldsymbol{p}(t)$; let $0 \le t \le 4\pi$.

(b) Find the position, velocity, and acceleration vectors at $t = 0$.

(c) Find a vector equation for the tangent line ℓ at $t = \pi$ to the curve defined by $\boldsymbol{p}(t)$.

(d) On one set of axes, plot both the curve and the tangent line from part (c).

(e) Find the arc length from $t = 0$ to $t = 4\pi$.

43. Find the length of the logarithmic spiral $r = e^\theta$ between $\theta = 0$ and $\theta = \pi$.

V.2 POLAR COORDINATES AND POLAR CURVES

Any point P in the xy-plane has a familiar and natural "address": its **rectangular** (or **Cartesian**) **coordinates**. The point P in Figure 1(a), for instance, has rectangular coordinates $(4, 3)$.

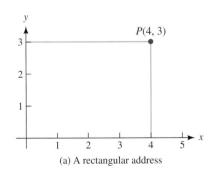

(a) A rectangular address

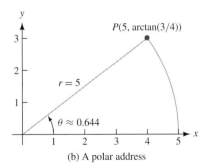

(b) A polar address

FIGURE 1

Polar and rectangular coordinates

The x- and y-coordinates of P measure the distances from P to the two perpendicular coordinate axes. To reach $P(4, 3)$ from the origin, one moves 4 units right and 3 units up.

Polar coordinates offer another way of locating a point in the plane. In the **polar coordinate system**, a point P has coordinates r and θ; they tell, respectively, the *distance* from the origin O to P and the *angle* (in radians!) from the positive x-axis to the ray from O to P. Figure 1(b) shows that the point P with rectangular coordinates $(4, 3)$ has polar coordinates $(5, \arctan(3/4)) \approx (5, 0.644)$.

Polar coordinate systems

A **rectangular coordinate system** in the Euclidean plane starts with an origin O and two perpendicular coordinate axes. Usually the x-axis is horizontal and the y-axis is vertical; x-coordinates increase to the right and y-coordinates increase upward.

A polar coordinate system starts with different ingredients: an origin O, called the **pole**, and a ray (i.e., a half-line) beginning at the origin, called the **polar axis**. The polar axis normally points to the right, along the positive x-axis. With these ingredients and a unit for measuring distance, we can assign polar coordinates (r, θ) to any point P:

r *is the distance from O to P; θ is any angle from the polar axis to the segment $\overline{OP}$.*

Figure 2 shows several points with their polar coordinates, plotted on a polar grid.

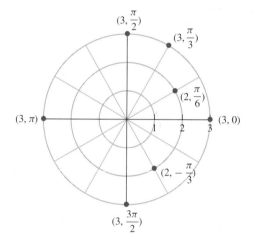

FIGURE 2
Points on a polar grid

Polar vs. rectangular grids   A rectangular coordinate system leads naturally to a rectangular grid, with vertical lines $x = a$ and horizontal lines $y = b$. In a polar system, holding the coordinates r and θ constant produces, respectively, concentric circles and radial lines. The result is a weblike polar grid. Figure 3 shows grids of both types.

On the scales   In a rectangular coordinate system, the two axes often have different scales of measurement. As a result, on a graph, a vertical inch and a horizontal inch may represent different distances. In particular, "circles" may look far from round.

A polar coordinate system, by contrast, has just one axis—the polar axis. As a result, distance does not depend on direction, and circles look round.

Polar coordinates: Not unique   A point in the plane—$P(4, 3)$, for instance—has just one possible pair of *rectangular* coordinates. Different rectangular coordinate pairs (x_1, y_1) and (x_2, y_2) correspond to different points in the plane.

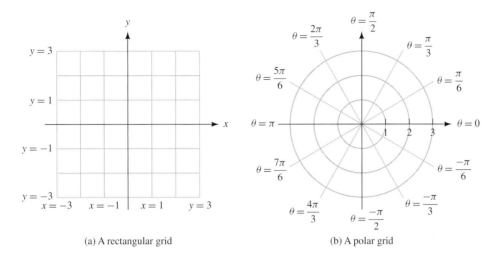

(a) A rectangular grid   (b) A polar grid

FIGURE 3
Two types of grids

Polar coordinates, by contrast, are *not* unique. Every point in the plane has many possible pairs of polar coordinates. For example, all of the polar coordinate pairs

$$\left(2, \frac{\pi}{4}\right), \quad \left(2, \frac{9\pi}{4}\right), \quad \left(2, -\frac{7\pi}{4}\right), \quad \left(2, -\frac{15\pi}{4}\right), \quad \left(-2, \frac{5\pi}{4}\right), \quad \left(-2, -\frac{3\pi}{4}\right),$$

(and many others) represent the same point—the one with rectangular coordinates $(\sqrt{2}, \sqrt{2})$. Notice especially the last two pairs: A negative r-coordinate means that, to locate the point P, one moves r units in the direction *opposite* the θ-direction. The point $(-2, 5\pi/4)$, for instance, lies 2 units from the origin on the ray $\theta = \pi/4$. (That's the ray opposite $\theta = 5\pi/4$.) The origin O allows even more freedom: It's represented by *every* pair of the form $(0, \theta)$, regardless of θ.

This ambiguity of polar coordinates arises for a simple reason. All angles that differ by integer multiples of 2π determine the same direction. In practice, this ambiguity can be annoying but is seldom a serious problem. Two simple rules help:

• **Multiples of 2π**   For any r and θ, the pairs (r, θ) and $(r, \theta + 2\pi)$ describe the same point.

• **Negative r**   For any r and θ, the pairs (r, θ) and $(-r, \theta + \pi)$ describe the same point.

Polar coordinates on Earth.   The polar grid somewhat resembles an overhead view of Earth with the observer looking "down" at the North Pole. In cartographers' language, lines of **longitude** (or **meridians**) converge at the pole; the concentric circles are lines of **latitude** (or **parallels**). For hundreds of years, the **prime meridian** (polar axis, in calculus language), for which $\theta = 0$, has been taken to be the line of longitude that passes through the Greenwich Observatory located just east of London, England. For the same reason, Greenwich Mean Time (the time of day along the prime meridian) is used worldwide as a reference point.

Why is Greenwich "prime" rather than, say, India or Arabia, where navigation and timekeeping flourished even in antiquity? There is no intrinsic reason; Greenwich just happened to be a center of attention when the terms were defined—an early (and quite literal) instance of Eurocentrism.

Polar coordinates in the *plane*, it should be said, aren't perfectly suited to measuring the (almost) spherical Earth. In practice, geographers use a related system called *spherical* coordinates.

Polar graphs

The ordinary graph of an equation in x and y is the set of points (x, y) whose coordinates satisfy the equation. The graph of $x^2 + y^2 = 1$, for instance, is the circle of radius 1 about the origin. The point $(2, 3)$ does not lie on this graph because $2^2 + 3^2 \neq 1$.

   The idea of a **polar graph** is similar but not quite identical. The graph of an equation in r and θ is the set of points whose *polar* coordinates r and θ satisfy the equation. For instance, the polar point $(3, 0)$ lies on the graph of $r = 2 + \cos\theta$ (because $3 = 2 + \cos 0$), but the polar point $(2, \pi)$ does not.

A warning   Because a point in the plane has more than one pair of polar coordinates, polar plotting requires extra care. At first glance, for instance, the point P with polar coordinates $(-3, \pi)$ seems *not* to satisfy the polar equation $r = 2 + \cos\theta$. A closer look, however, shows that P can also be written with polar coordinates $(3, 0)$—which *do* satisfy the given equation. Here's the moral:

> *A point P lies on the graph of a polar equation if P has* any *pair of polar coordinates that satisfy the equation.*

Drawing polar graphs

The simplest polar graphs come from functions—usually of the form $r = f(\theta)$. Given such a function and a specific θ-domain, it's a routine matter to tabulate points and then plot them. We illustrate by example.

E X A M P L E 1   Plot the equation $r = 2 + \cos\theta$ for $0 \le \theta \le 2\pi$.

Solution   Let $f(\theta) = 2 + \cos\theta$. We want the r–θ graph of f. First we tabulate some values.

Values of $r = f(\theta) = 2 + \cos\theta$												
θ	0	$\frac{\pi}{6}$	$\frac{\pi}{3}$	$\frac{\pi}{2}$	$\frac{2\pi}{3}$	$\frac{5\pi}{6}$	π	$\frac{7\pi}{6}$	$\frac{4\pi}{3}$	$\frac{3\pi}{2}$	$\frac{5\pi}{3}$	$\frac{11\pi}{6}$	2π
r	3	2.87	2.5	2	1.5	1.14	1	1.14	1.5	2	2.5	2.87	3

Next we plot the data (polar "graph paper" makes the job easier) and fill in the gaps smoothly. Figure 4 shows the result.

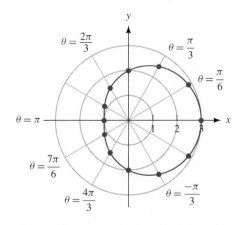

FIGURE 4

A polar graph: $r = 2 + \cos\theta$

Polar graphs: A sampler

Several polar graphs follow. We have already seen the simplest polar graphs of all—those of the equations $r = a$ and $\theta = b$.

Cardioids and limaçons Graphs of the form $r = a \pm b \cos\theta$ and $r = a \pm b \sin\theta$, where a and b are positive numbers, are called **limaçons**; if $a = b$, the term **cardioid** ("heartlike") is used. (Figure 4 shows a limaçon.) The graphs in Figure 5 illustrate the variety of limaçons and show the effects of the constants a and b.

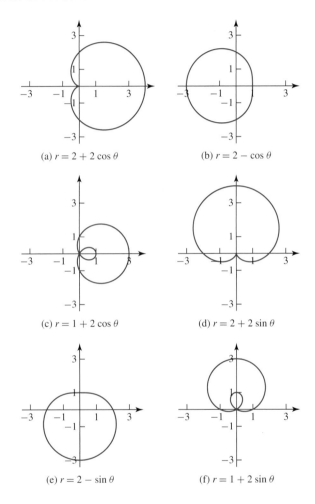

(a) $r = 2 + 2\cos\theta$

(b) $r = 2 - \cos\theta$

(c) $r = 1 + 2\cos\theta$

(d) $r = 2 + 2\sin\theta$

(e) $r = 2 - \sin\theta$

(f) $r = 1 + 2\sin\theta$

FIGURE 5
Six limaçons

Notice some features of these graphs.

- **What θ-range?** The graphs were drawn by letting θ vary through the interval $[0, 2\pi]$. Since all functions involved are 2π-periodic, any other interval of length 2π would produce the same result. (We could have used the interval $-\pi \le t \le \pi$, for example.)

- **Symmetry** Three of the preceding limaçons—those that involve the cosine function— are symmetric about the x-axis, that is, the line $\theta = 0$. (The other three are symmetric about the y-axis.) This symmetry occurs because the cosine function is *even*: For any θ, $\cos\theta = \cos(-\theta)$. The other graphs are symmetric about the y-axis because the sine function is odd.

- **Inner loops**   Each of the limaçons $r = 1 + 2\cos\theta$ and $r = 1 + 2\sin\theta$ has an inner loop. A close look at the graphs and the formulas reveals that these loops correspond to *negative* values of r. For $r = 1 + 2\cos\theta$, for instance, we have $r = 0$ when $\theta = 2\pi/3$ or $\theta = 4\pi/3$, and $r < 0$ for $2\pi/3 < \theta < 4\pi/3$. For these θ-values, therefore, the curve is drawn on the *opposite* side of the origin.

Roses   Equations of the form $r = a\cos(k\theta)$ and $r = a\sin(k\theta)$, where a is a constant and k is a positive integer, produce graphs called **roses**. The graphs in Figure 6 help explain the name.

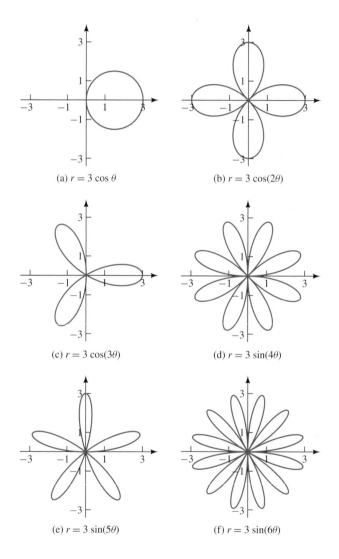

(a) $r = 3\cos\theta$

(b) $r = 3\cos(2\theta)$

(c) $r = 3\cos(3\theta)$

(d) $r = 3\sin(4\theta)$

(e) $r = 3\sin(5\theta)$

(f) $r = 3\sin(6\theta)$

FIGURE 6
A bouquet of roses

Notice the following features of the graphs:

- **Symmetry**   Like limaçons (and for the same reason), all roses are symmetric about an axis—"cosine roses" about the x-axis and "sine roses" about the y-axis.
- **The rose's radius**   The coefficient a in $r = a\cos(k\theta)$ and $r = a\sin(k\theta)$ determines the rose's "radius."

- **How many petals?**   The coefficient k in $r = a \cos(k\theta)$ and $r = a \sin(k\theta)$ determines the number of "petals": k if k is odd and $2k$ if k is even. But here's a subtlety best revealed by plotting some roses by hand:

    If k is odd, then each petal is traversed twice *for* $0 \le \theta \le 2\pi$.

    In other words, for odd k, the rose $r = a \cos(k\theta)$ (or $r = a \sin(k\theta)$) has k *double* petals.

Trading polar and rectangular coordinates

How are the polar coordinates (r, θ) of a point P related to the rectangular coordinates (x, y) of the same point? How can either type of coordinates be found from the other? Figure 7 gives a useful view.

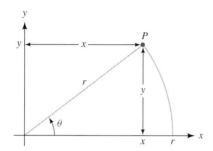

FIGURE 7
Relating polar and rectangular coordinates

The picture illustrates many relations among x, y, r, and θ. Among the simplest are these:

$$x = r\cos\theta; \qquad y = r\sin\theta; \qquad r^2 = x^2 + y^2; \qquad \tan\theta = \frac{y}{x}.$$

(The last equation holds only if $x \ne 0$.)

These relations let us convert from one type of coordinates to the other. Equations in x and y, for instance, are easy to rewrite in terms of r and θ, as the next two examples illustrate.

EXAMPLE 2   Find a polar equation for the straight line $y = mx + b$.

Solution   Substituting $x = r\cos\theta$ and $y = r\sin\theta$ into the equation for the line gives

$$y = mx + b \iff r\sin\theta = m(r\cos\theta) + b \iff r = \frac{b}{\sin\theta - m\cos\theta}.$$

(One moral is that rectangular coordinates are better suited to straight lines than are polar coordinates!)   ■

EXAMPLE 3   The graph of $r = 3\cos\theta$ *looks* like a circle. Is it a circle? Which circle?

Solution   It *is* a circle. Changing to rectangular coordinates shows why:

$$r = 3\cos\theta \implies r^2 = 3r\cos\theta \implies x^2 + y^2 = 3x.$$

As expected, the last equation does define a circle. To decide *which* circle, complete the square:

$$x^2 + y^2 = 3x \iff x^2 - 3x + y^2 = 0 \iff \left(x - \frac{3}{2}\right)^2 + y^2 = \frac{9}{4}.$$

The circle therefore has radius $3/2$ and center at $(3/2, 0)$—just as Figure 6 suggests.   ■

BASIC EXERCISES

In Exercises 1–4, a point is given in rectangular coordinates. Give three different pairs of polar coordinates for the point; r should be negative for at least one pair. [NOTE: Because only familiar angles are involved, give exact answers rather than decimal approximations.]

1. $(\pi, 0)$ **2.** $(0, \pi)$

3. $(1, 1)$ **4.** $(-1, 1)$

In Exercises 5–8, a point is given in rectangular coordinates. Give three different pairs of polar coordinates for the point; r should be negative for at least one pair. [NOTE: Round your answers to four decimal places.]

5. $(1, 2)$ **6.** $(-1, 2)$

7. $(1, 4)$ **8.** $(10, 1)$

In Exercises 9–12, a point is given in polar coordinates. Plot the point and label it with its rectangular coordinates. [NOTE: Give exact answers rather than decimal approximations.]

9. $(2, \pi/4)$ **10.** $(-2, 5\pi/4)$

11. $(1, 13\pi/6)$ **12.** $(42, 0)$

In Exercises 13–16, a point is given in polar coordinates. Plot the point and label it with its rectangular coordinates. [NOTE: Round your answers to four decimal places.]

13. $(1, 1)$ **14.** $(-1, 2)$

15. $(2, \pi/4)$ **16.** $(3, \arctan 2)$

In Exercises 17–20, rewrite the equation in r and θ as an equivalent equation in x and y; then plot the graph.

17. $r = 2 \sec\theta$ **18.** $r = 4$

19. $\theta = \pi/3$ **20.** $r = 2\sin\theta$

In Exercises 21–24, rewrite the equation in x and y as an equivalent equation in r and θ; then plot the graph.

21. $x^2 + y^2 = 9$ **22.** $y = 4$

23. $y = 2x$ **24.** $(x-1)^2 + y^2 = 1$

25. We claimed in this section that for any numbers r and θ the pairs $(r, \theta), (r, \theta + 2\pi)$, and $(-r, \theta + \pi)$ all describe the same point in the plane.

(a) Show that the claim is true if $r = 1$ and $\theta = 0$.

(b) Show that the claim is true if $r = -1$ and $\theta = \pi/4$.

(c) The point with rectangular coordinates $(1, 0)$ can be written in polar coordinates as $(1, 2k\pi)$, where k is any integer, or as $(-1, (2k-1)\pi)$, where k is any integer. In the same sense, describe all the possible polar coordinates of the point with rectangular coordinates $(1, 1)$.

26. Consider the limaçon $r = 2 + \sin\theta, 0 \leq \theta \leq 2\pi$.

(a) Make a table of values like that in Example 1—let θ range from 0 to 2π in steps of $\pi/6$. (Round r-values to three decimals.)

(b) Plot the points calculated in part (a) on a copy of the polar grid shown in Example 1. Join the points with a smooth curve.

(c) What is the axis of symmetry of this limaçon?

(d) The r values in the table in Example 1 are symmetric about $\theta = \pi$. What similar type of symmetry does your table from part (a) show?

27. Consider the cardioid $r = 1 + \cos\theta, 0 \leq \theta \leq 2\pi$.

(a) Make a table of values like that in Example 1—let θ range from 0 to 2π in steps of $\pi/6$. (Round r values to three decimals.)

(b) Plot the points calculated in part (a) on a copy of the polar grid shown in Example 1. Join the points with a smooth curve.

(c) What is the axis of symmetry of this cardioid?

(d) How does the table of values in part (a) reflect the cardioid's symmetry?

28. Consider the limaçon $r = 1 - 2\cos\theta, 0 \leq \theta \leq 2\pi$.

(a) Make a table of values like that in Example 1—let θ range from 0 to 2π in steps of $\pi/6$. (Round r values to three decimals.)

(b) Plot the points calculated in part (a) on a copy of the polar grid shown in Figure 4. Join the points with a smooth curve.

(c) For what values of θ is $r = 0$? How do these values appear on the graph?

(d) On what θ-interval is $r < 0$? How does this interval show up on the graph?

FURTHER EXERCISES

In Exercises 29–34, draw a graph of the given polar equation. (The models of cardioids and limaçons shown in Figure 5 should be helpful. Be sure to label your graphs with appropriate units.)

29. $r = 3 + 3\cos\theta$ **30.** $r = 3 - \cos\theta$

31. $r = 1 + \sqrt{3}\cos\theta$ **32.** $r = 4 + 4\sin\theta$

33. $r = 4 - 2\sin\theta$ **34.** $r = 2 - 4\sin\theta$

In Exercises 35–40, sketch the given polar rose. (The models of roses shown in Figure 6 should be helpful. Be sure to label your graphs with appropriate units.)

35. $r = 2\sin\theta$ **36.** $r = 2\sin(2\theta)$

37. $r = 2\sin(3\theta)$ **38.** $r = 2\cos(4\theta)$

39. $r = 2\cos(5\theta)$

40. $r = 2\cos(1001\theta)$ (rough sketch is fine!)

In Exercises 41–48, plot the graph of the equation. (In each case, the graph is some sort of spiral.)

41. $r = \theta, 0 \leq \theta \leq 2\pi$ (an Archimedean spiral)

42. $r = 2\theta, -2\pi \leq \theta \leq 0$ (another Archimedean spiral)

43. $r = \ln(\theta), 1 \le \theta \le 4\pi$ (a logarithmic spiral)

44. $r = e^{\theta/2}, -\pi \le \theta \le \pi$ (an exponential spiral)

45. $r = \theta^2, 0 \le \theta \le 2\pi$ (a quadratic spiral)

46. $r = 1/\theta, 1/2 \le \theta \le 2$ (a hyperbolic spiral)

47. $r = \sqrt{\theta}, 1/4 \le \theta \le 4$ (a parabolic spiral)

48. $r = 1/\sqrt{\theta}, 1/4 \le \theta \le 4$ (a Lituus spiral)

49. The polar equation $r = a \cos\theta + b \sin\theta$, where a and b are real numbers, describes a circle. Find the radius and the center of this circle.

50. Suppose that $g(\theta) = f(\theta + \alpha)$. Describe how the graph of the polar equation $r = g(\theta)$ is related to that of $r = f(\theta)$ if $\alpha > 0$.

51. (a) Plot the curve $r = \cos^2\theta, 0 \le \theta \le 2\pi$.

(b) Find an equation for this curve in Cartesian coordinates.

52. (a) Plot the curve $r = 1/(1 + \cos\theta), 0 \le \theta < \pi$.

(b) Find an equation for this curve in Cartesian coordinates.

53. Find an equation for the distance between the points (r_1, θ_1) and (r_2, θ_2).

54. This exercise concerns limaçons of the form $r = 1 + a \cos\theta$, where a is any real constant. The following questions are open-ended. Answer them by experimenting with plots for various values of a: positive, negative, large, small, and so forth.

(a) For which positive values of a does the graph have an inner loop?

(b) What happens at $a = 1$?

(c) What happens as $a \to 0$?

(d) How are the graphs for a and $-a$ (e.g., $r = 1 + 0.5 \cos\theta$ and $r = 1 - 0.5 \cos\theta$) related to each other?

(e) What happens as $a \to \infty$?

CALCULUS IN POLAR COORDINATES

In the last section we explored the polar coordinate system, polar equations, and the geometry of polar curves. This section is about calculus with polar curves. The two main problems that stand out for polar curves are the same as those for rectangular curves: finding the slope at a point and finding the area enclosed.

To visualize these problems, consider Figure 1:

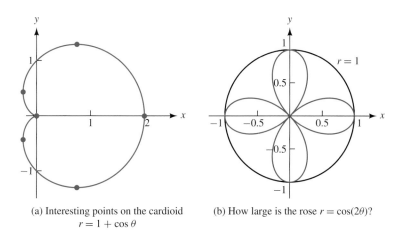

(a) Interesting points on the cardioid $r = 1 + \cos\theta$

(b) How large is the rose $r = \cos(2\theta)$?

FIGURE 1
Two polar curves

Among the questions we'll address are these:

- **Horizontal and vertical tangents** Where—exactly—does the cardioid have horizontal and vertical tangent lines? The points marked with dots on the cardioid seem to be of interest, but what are their coordinates?

- **Slope at any point**   What's the slope of either curve at *any* point?
- **Areas**   How much area does one petal of the rose enclose? What fraction of the circular area does the rose enclose?

Slopes on polar curves

Both curves just shown are of this type.

Let's consider curves of the form $r = f(\theta)$, where f is a differentiable function. ◄ For a given θ_0 we want the slope at the polar point (r_0, θ_0).

An obvious (but wrong) guess   Let's acknowledge, just for the record, that the desired slope is *not* $f'(\theta_0)$. A look at either Figure 1(a) or 1(b) quickly confirms this. For $f(\theta) = 1 + \cos\theta$, for instance, $f'(\theta) = -\sin\theta$, which varies only between -1 and 1. Yet it's clear at a glance that slopes on the cardioid take *all* real values; at three points, moreover, the cardioid has vertical tangent lines. Thus, $f'(\theta)$ is not the slope we seek.

Polar curves and parametric equations   We showed in earlier work how to find the slope at a point on a curve defined by parametric equations. Here is the result, for reference:

> **FACT (Slope of a parametric curve)**   Let a smooth curve C be given by parametric equations $x = f(t)$, $y = g(t)$, with $a \leq t \leq b$. If $f'(t) \neq 0$, then the slope of C at (x, y) is given by
> $$\frac{dy}{dx} = \frac{g'(t)}{f'(t)} = \frac{dy/dt}{dx/dt}.$$

To *use* the result, we need first to write a polar curve $r = f(\theta)$ in parametric form. That is surprisingly easy to do. The key facts are the conversion formulas from polar to rectangular coordinates:

$$x = r\cos\theta; \qquad y = r\sin\theta.$$

For points on our polar curve $r = f(\theta)$, therefore,

$$x = r\cos\theta = f(\theta)\cos\theta; \qquad y = r\sin\theta = f(\theta)\sin\theta.$$

These are the desired parametric equations. We've rewritten the curve, originally given in r–θ form, as a pair of parametric equations with θ as a parameter. All that's left is to apply the preceding slope formula. We need $dx/d\theta$ and $dy/d\theta$; let's compute them now. By the product rule,

$$x = f(\theta)\cos\theta \implies \frac{dx}{d\theta} = f'(\theta)\cos\theta - f(\theta)\sin\theta;$$

$$y = f(\theta)\sin\theta \implies \frac{dy}{d\theta} = f'(\theta)\sin\theta + f(\theta)\cos\theta.$$

Everything is in place. Here's the result, stated with appropriate technical hypotheses:

> **FACT (Slope of a polar curve)** Let a curve C be given in polar coordinates by a function $r = f(\theta)$, $\alpha \le \theta \le \beta$, where f and f' are continuous on (α, β) and not simultaneously zero. Then, for θ in (α, β), the slope of C at $(r, \theta) = \big(f(\theta), \theta\big)$ is given by
>
> $$\frac{dy}{dx} = \frac{dy/d\theta}{dx/d\theta} = \frac{f'(\theta)\sin\theta + f(\theta)\cos\theta}{f'(\theta)\cos\theta - f(\theta)\sin\theta}$$
>
> wherever the denominator is not zero.

Using this Fact, we can answer the questions on tangents raised at the beginning of this section. The next example shows how. It also illustrates some of the caution needed in working with polar coordinates.

EXAMPLE 1 Consider the cardioid $r = 1 + \cos\theta$ shown on page V-19. ➡ Where is the curve horizontal? Where is it vertical? What happens at the origin?   *Take a close look.*

Solution Here $f(\theta) = 1 + \cos\theta$ and $f'(\theta) = -\sin\theta$, thus,

$$\frac{dy/d\theta}{dx/d\theta} = \frac{f'(\theta)\sin\theta + f(\theta)\cos\theta}{f'(\theta)\cos\theta - f(\theta)\sin\theta} \qquad \text{(by the Fact)}$$

$$= \frac{(-\sin\theta)\sin\theta + (1 + \cos\theta)\cos\theta}{-\sin\theta\cos\theta - (1 + \cos\theta)\sin\theta} \qquad \text{(substituting)}$$

$$= \frac{(1 - 2\cos\theta)(1 + \cos\theta)}{(2\cos\theta + 1)\sin\theta}. \qquad \text{(by algebra)}$$

Now the cardioid can have a horizontal tangent line only if the numerator $dy/d\theta$ is zero. The preceding calculation shows when this occurs:

$$(1 - 2\cos\theta)(1 + \cos\theta) = 0 \iff \cos\theta = \frac{1}{2} \quad \text{or} \quad \cos\theta = -1.$$

For θ in $[0, \pi]$, one condition or the other holds only if $\theta = \pi/3$ or $\theta = \pi$. (By symmetry, it's enough to look only on the upper half of the cardioid.)

  The figure shows that, in fact, the upper half of the cardioid is horizontal *only* at $\theta = \pi/3$, that is, at the point with polar coordinates $(3/2, \pi/3)$ and rectangular coordinates $(3/4, 3\sqrt{3}/4) \approx (0.75, 1.30)$. Symmetry dictates that the cardioid is also horizontal at $\theta = 5\pi/3$.

  What happens at $\theta = \pi$, where the cardioid has a "cusp"? At $\theta = \pi$, both $dx/d\theta$ and $dy/d\theta$ are zero, and so the slope expression is undefined. It can be shown using l'Hôpital's rule, however, that the expression tends to zero as θ tends to π. Thus, the cardioid can reasonably be said to have *zero* slope at $\theta = \pi$.

  A vertical tangent line can occur only where the denominator $dx/d\theta$ is zero, that is, if

$$\sin\theta\,(2\cos\theta + 1) = 0 \iff \sin\theta = 0 \quad \text{or} \quad \cos\theta = -\frac{1}{2}.$$

For θ in $[0, \pi]$, one or the other holds if $\theta = 0$ or $\theta = 2\pi/3$. As the figure shows, both $\theta = 0$ and $\theta = 2\pi/3$ correspond to points of vertical tangency on the cardioid. These points have polar coordinates $(2, 0)$ and $(1/2, 2\pi/3)$, respectively.

Superimposing the cardioid on a polar grid supports all the preceding calculations.

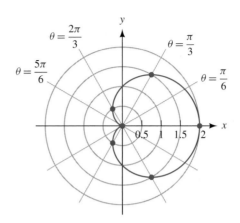

FIGURE 2

Horizontal and vertical points on the cardioid $r = 1 + \cos\theta$

EXAMPLE 2   Let $a > 0$ be any positive constant. What does the slope formula say about the circle $r = a$?

Solution   For the circle $r = a$, we have $f(\theta) = a$ and $f'(\theta) = 0$. Now the Fact says that

$$\frac{dy}{dx} = \frac{f'(\theta)\sin\theta + f(\theta)\cos\theta}{f'(\theta)\cos\theta - f(\theta)\sin\theta}$$

$$= \frac{a\cos\theta}{-a\sin\theta} = -\cot\theta = -\frac{1}{\tan\theta}.$$

Why perpendicular? Because the slopes are negative reciprocals.

Note what the last expression means: The tangent line to the circle at any point $P(a, \theta)$ is perpendicular to the ray from the origin to P. ◄

EXAMPLE 3   Let $r = f(\theta)$ describe a polar curve C; suppose that $f(\theta_0) = 0$. What does the formula

$$\frac{dy}{dx} = \frac{f'(\theta)\sin\theta + f(\theta)\cos\theta}{f'(\theta)\cos\theta - f(\theta)\sin\theta}$$

say about the slope of C at the point $(0, \theta_0)$?

Remember that f and f' aren't simultaneously zero.

Solution   The slope formula becomes much simpler if $f(\theta_0) = 0$. In that case, ◄

$$\frac{dy}{dx} = \frac{f'(\theta_0)\sin\theta_0}{f'(\theta_0)\cos\theta_0} = \tan\theta_0.$$

Note what this means: If a smooth polar curve passes through the origin at $\theta = \theta_0$, its tangent line is simply the ray $\theta = \theta_0$. For the rose $r = \cos(2\theta)$, for example, $r = 0$ when $\theta = \pi/4$, $\theta = 3\pi/4$, $\theta = 5\pi/4$, and $\theta = 7\pi/4$. As Figure 3 shows, the curve passes through the origin in just these directions.

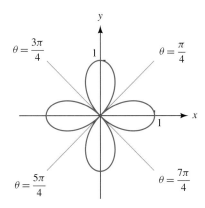

FIGURE 3
**The rose $r = \cos(2\theta)$: Tangent lines at
the origin**

■

Finding area in polar coordinates

The standard area problem in rectangular coordinates concerns the area defined by an ordinary $y = f(x)$-style graph for $a \le x \le b$. In polar coordinates, the standard area problem is a little different: to find the area bounded by a polar curve $r = f(\theta)$ for $\alpha \le \theta \le \beta$. Figure 4 illustrates the "generic" situations (the areas in question are striped).

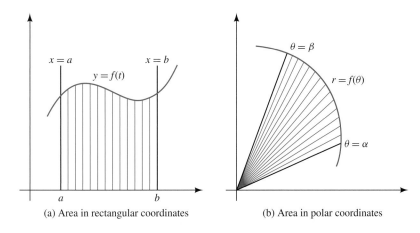

(a) Area in rectangular coordinates   (b) Area in polar coordinates

FIGURE 4
Generic areas

The different styles of shading in Figures 4(a) and (b)—one vertical, the other radial—were chosen intentionally. They reflect two slightly different approaches to finding areas. To emphasize both similarities and differences between the rectangular and polar situations, the next two subsections are written in "parallel."

Area, Cartesian-style   In rectangular coordinates, area is approximated by subdividing the region into thin *vertical strips*, each one based on the x-axis. Each strip corresponds to a small subinterval—say, $[x_i, \ x_i + \Delta x]$—obtained by partitioning the domain interval $a \le x \le b$ into n equal pieces. If Δx is small, then the strip is approximately a rectangle with base Δx and height $f(x_i)$. Therefore,

$$\text{Area of one strip} \approx f(x_i)\,\Delta x.$$

Adding all n strips gives

$$\text{Total area} \approx \sum_{i=1}^{n} f(x_i)\,\Delta x.$$

Therefore, if we take the limit as $\Delta x \to 0$ we get

$$\text{Total area} = \lim_{n \to \infty} \sum_{i=1}^{n} f(x_i)\,\Delta x = \int_a^b f(x)\,dx.$$

Look carefully at Figure 4(b), the polar picture: it shows 20 radial wedges.

Area, polar-style   In polar coordinates, by contrast, area is approximated by subdividing the region into thin *pie-shaped wedges*, each with its vertex at the origin. ← Each wedge corresponds to a small subinterval—say, $[\theta_i,\ \theta_i + \Delta\theta]$—obtained by partitioning the domain interval $\alpha \le \theta \le \beta$ into n equal pieces. If $\Delta\theta$ is small, then each wedge is approximately a sector of a circle with radius $f(\theta_i)$; the wedge makes the angle $\Delta\theta$ at the origin.

Now, a brief detour. Just ahead, we'll need to know the area of a **circular sector** as just described. The answer, although perhaps less familiar than the area of a rectangle, is easy to find. A full circle of radius R (for which θ runs from 0 to 2π) encloses total area πR^2. A wedge making angle $\Delta\theta$ at the origin represents the fraction $\Delta\theta/(2\pi)$ of the total circle, and so

$$\text{Area of wedge} = \pi R^2 \cdot \frac{\Delta\theta}{2\pi} = \frac{R^2}{2}\,\Delta\theta.$$

If, say, $R = f(\theta_i)$, then a circular wedge has area $f(\theta_i)^2\,\Delta\theta/2$.

Back to the main road. Our detour showed that in the "generic" case illustrated earlier,

$$\text{Area of one wedge} \approx \frac{f(\theta_i)^2}{2}\,\Delta\theta.$$

Adding all n wedges gives

$$\text{Total area} \approx \sum_{i=1}^{n} \frac{f(\theta_i)^2}{2}\,\Delta\theta.$$

Taking the limit as $\Delta\theta \to 0$ gives

$$\text{Total area} = \lim_{n \to \infty} \sum_{i=1}^{n} \frac{f(\theta_i)^2}{2}\,\Delta\theta = \int_\alpha^\beta \frac{f(\theta)^2}{2}\,d\theta.$$

The formula is worth remembering:

FACT (Area in polar coordinates)   Let f be a continuous function, and let R be the region in the xy-plane bounded by the polar curve $r = f(\theta)$ and the rays $\theta = \alpha$ and $\theta = \beta$. Then

$$\text{Area of } R = \int_\alpha^\beta \frac{f(\theta)^2}{2}\,d\theta.$$

EXAMPLE 4   What's the area of one leaf of the polar rose $r = f(\theta) = \cos(2\theta)$? What fraction of the circle $r = 1$ does the entire rose cover?

Solution As Figure 3 shows, the eastward-pointing leaf lies between $\theta = -\pi/4$ and $\theta = \pi/4$. To ease calculations, we integrate from $\theta = 0$ to $\theta = \pi/4$ and double the result. Here is the integral calculation:

$$\int_0^{\pi/4} \frac{\cos^2(2\theta)}{2}\, d\theta = \frac{\theta}{4} + \frac{\sin(4\theta)}{16}\bigg]_0^{\pi/4} = \frac{\pi}{16}.$$

Thus, *one* leaf has area $\pi/8$; all four leaves have area $\pi/2$—exactly half the area of the circle $r = 1$. ∎

A rose with any other n ...   A more general (and more surprising) result than that of the last example is true:

> *For any even n, the rose $r = \cos(n\theta)$ encloses total area $\pi/2$—half the area of the enclosing circle.*

EXAMPLE 5 How much area does the cardioid $r = 1 + \cos\theta$ enclose?

Solution With the integral formula, the answer is easy:

$$\begin{aligned}
\text{Area} &= \frac{1}{2}\int_0^{2\pi} f(\theta)^2\, d\theta \\
&= \frac{1}{2}\int_0^{2\pi} \left(1 + 2\cos\theta + \cos^2\theta\right) d\theta \\
&= \frac{1}{2}\left(\theta + 2\sin\theta + \frac{\theta}{2} + \frac{\sin(2\theta)}{4}\right)\bigg]_0^{2\pi} \\
&= \frac{3\pi}{2} \approx 4.71.
\end{aligned}$$
∎

BASIC EXERCISES

1. We claimed in this section that
$$x = f(\theta)\cos\theta \implies \frac{dx}{d\theta} = f'(\theta)\cos\theta - f(\theta)\sin\theta;$$
$$y = f(\theta)\sin\theta \implies \frac{dy}{d\theta} = f'(\theta)\sin\theta + f(\theta)\cos\theta.$$

Use the product rule to show that these claims are valid.

2. We claimed in Example 1 that on the cardioid $r = 1 + \cos\theta$, the slope dy/dx at (r, θ) is
$$\frac{(1 - 2\cos\theta)(1 + \cos\theta)}{(2\cos\theta + 1)\sin\theta}.$$

Use the Fact on page V-21 to verify this.

3. Consider the spiral $r = \theta$, $0 \le \theta \le 4\pi$.
   (a) Draw the spiral.
   (b) At what points does the spiral have horizontal tangent lines?
   (c) At what points does the spiral have vertical tangent lines?

(d) Write an equation in rectangular form for the line tangent to the spiral at the polar point $(1, 1)$. Draw this tangent line on your graph.

4. Consider the limaçon $r = 1 + 2\sin\theta$.
   (a) Draw the limaçon.
   (b) Write an equation in rectangular form for the line tangent to the limaçon at the polar point $(1, 0)$. Draw this tangent line on your graph.
   (c) Find the slope of the line tangent to the limaçon at the polar point $(0, 7\pi/6)$. Draw this tangent line on your graph.
   (d) Find the slope of the line tangent to the limaçon at the polar point $(0, 11\pi/6)$. Draw this tangent line on your graph.
   (e) At what points does the limaçon have horizontal tangent lines?

In Exercises 5–8, draw the region bounded by the given polar curves and find its area.

   5. $r = 1$, $\theta = 0$, $\theta = \pi$

6. $r = 3$, $\quad \theta = 0$, $\quad \theta = \pi/2$

7. $r = a$, $\quad \theta = 0$, $\quad \theta = \pi$

8. $r = 1$, $\quad \theta = 0$, $\quad \theta = \beta$

In Exercises 9–15, draw the given region and find its area.

9. The outer loop of the limaçon $r = 1 + 2\sin\theta$.

10. The inner loop of the limaçon $r = 1 + 2\cos\theta$.

11. The area enclosed by the cardioid $r = 1 - \cos\theta$ that is to the right of the vertical line $\theta = \pi/2$.

12. The area enclosed by the cardioid $r = 1 - \cos\theta$ that is to the left of the vertical line $\theta = \pi/2$ and above the horizontal line $\theta = 0$.

13. The area inside the cardioid $r = 1 + \cos\theta$ and outside the circle $r = \sin\theta$.

14. The region bounded by $r = \sec\theta$, $\theta = 0$, and $\theta = \pi/4$.

15. The region bounded by $r = \sec\theta$, $\theta = 0$, and $\theta = \arctan m$.

16. Consider the $2n$-leafed rose $r = \cos(n\theta)$, where n is even.

    (a) Find the area of one leaf.

    (b) Find the area of all $2n$ leaves. What fraction of the circle $r = 1$ does the entire rose fill up?

17. Consider the n-leafed rose $r = \cos(n\theta)$, where n is odd.

    (a) Find the area of one leaf.

    (b) Find the area of n leaves. What fraction of the circle $r = 1$ does the entire rose fill up?

18. Draw and then calculate the area bounded by one "turn" of the spiral $r = \theta$, that is, from $\theta = 0$ to $\theta = 2\pi$.

19. Draw and then calculate the area bounded by one "turn" of the exponential spiral $r = e^\theta$, that is, from $\theta = 0$ to $\theta = 2\pi$.

20. Draw and then calculate (or estimate numerically) the area bounded by one "turn" of the logarithmic spiral $r = \ln\theta$, that is, from $\theta = 2\pi$ to $\theta = 4\pi$.

21. Use polar coordinates to find the area of the region inside the circle $r = 1$ and to the right of $x = 1/2$.

22. Let $0 < a < 1$. Use polar coordinates to find the area of the region inside the circle $r = 1$ and to the right of $x = a$.

FURTHER EXERCISES

In this section we used the idea that a polar curve can also be thought of as a parametric curve. Specifically, the curve defined by the polar function $r = f(\theta)$, $\alpha \le \theta \le \beta$, can also be defined by the parametric equations $x = f(\theta)\cos\theta$, $y = f(\theta)\sin\theta$, $\alpha \le \theta \le \beta$. Consider, for example, the unit circle $r = 1$, $0 \le \theta \le 2\pi$. Then $f(\theta) = 1$, and so the parametric form is simply $x = f(\theta)\cos\theta = \cos\theta$, $y = f(\theta)\sin\theta = \sin\theta$, $0 \le \theta \le 2\pi$.

In Exercises 23–30, a curve is given in polar form. Rewrite the curve in parametric form and then plot it. Do you see the "expected" results?

23. $r = 2$, $\quad 0 \le \theta \le 2\pi$

24. $r = 2$, $\quad 0 \le \theta \le \pi$

25. $r = \sec\theta$, $\quad -\pi/4 \le \theta \le \pi/4$

26. $r = \csc\theta$, $\quad \pi/4 \le \theta \le 3\pi/4$

27. $r = \theta$, $\quad 0 \le \theta \le 2\pi$

28. $r = \cos\theta$, $\quad 0 \le \theta \le \pi$

29. $r = \cos(2\theta)$, $\quad 0 \le \theta \le 2\pi$

30. $r = 1 + \cos\theta$, $\quad 0 \le \theta \le 2\pi$

31. The formula for the slope of a polar curve requires that f and f' not be simultaneously zero. Show that if this condition holds, then $dy/d\theta$ and $dx/d\theta$ are not simultaneously zero.

[HINT: Recall that $dy/d\theta = f'(\theta)\sin\theta + f(\theta)\cos\theta$ and $dx/d\theta = f'(\theta)\cos\theta - f(\theta)\sin\theta$. Look at $(dy/d\theta)^2 + (dx/d\theta)^2$.]

32. This open-ended problem is about the family of limaçons of the form $r = 1 + a\cos\theta$.

    (a) After plotting limaçons for various values of a, determine which ones have a "dimple" (i.e., a pushed-in section on either the right or the left) and which ones don't. [HINT: Any limaçon with a dimple has three vertical tangent lines.]

    (b) The graph of $r = 1 + a\cos\theta$ has no dimple if $|a| \le 1/2$. Show this using derivatives. [HINT: Which values of a lead to three vertical tangent lines?]

MULTIVARIABLE CALCULUS: A FIRST LOOK

M.1 THREE-DIMENSIONAL SPACE

Single-variable calculus is done mainly in the two-dimensional xy-plane. The Euclidean plane, also known as $\mathbb{R}^2$, is the natural home of such familiar calculus objects as the graph $y = f(x)$, tangent lines to that graph at various points, and the various regions whose areas we might measure by integration.

To do *multivariable* calculus, we need more room. The graph of $z = f(x, y)$, where f is a function of two input variables, lives in *three*-dimensional xyz-space. With three dimensions to work in, we will "see" not only this graph but a variety of multivariable analogues of derivatives and integrals. This section explores three-dimensional Euclidean space, or $\mathbb{R}^3$, where we'll spend most of our time.

In another sense, of course, we spend *all* of our time in $\mathbb{R}^3$. In everyday usage, "space" connotes three physical dimensions. The intuition we gain from living in three spatial dimensions is often useful in mentally picturing and manipulating the objects of multivariable calculus. Familiar as it is, however, three-dimensional space poses special problems for visualization. Two-dimensional pictures (on paper or on a computer screen) of three-dimensional objects are *always* more or less distorted or incomplete. Minimizing such problems is an active science (and an art) in its own right; doing so means carefully controlling viewpoint, perspective, shading, lighting, and other factors.

This chapter's main subject is multivariable calculus, not computer graphics (although we sometimes mention computer graphics) or technical drawing, and so we draw pictures to illustrate ideas as simply as possible, not necessarily to look as lifelike as possible. It's worth remarking, however, that many of the basic tools and methods of computer graphics draw directly on the ideas we'll develop in this chapter.

Cartesian coordinates in three dimensions

The *idea* of Cartesian coordinates is the same in both two and three dimensions, but the pictures look a little different. Compare these (Figure 1):

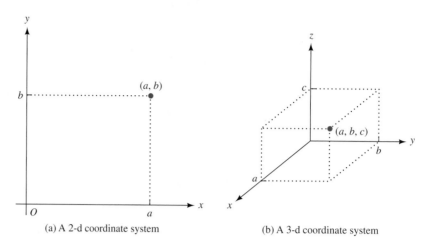

(a) A 2-d coordinate system   (b) A 3-d coordinate system

FIGURE 1
Coordinate systems

Recall the formalities in the xy-plane. A Cartesian coordinate system consists of an origin, labeled O, and horizontal and vertical coordinate axes, labeled x and y, passing through O. On each axis we choose a positive direction (usually "east" and "north") and a unit of measurement (not necessarily the same on both axes).

Given such a coordinate system, every point P in the plane corresponds to one and only one ordered pair (a, b) of real numbers, called the Cartesian coordinates of P. The pair (a, b) can be thought of as P's "Cartesian address." To reach P from the origin, move a units in the positive x-direction and b units in the positive y-direction. ◄

Coordinates in three-dimensional xyz-space work the same way—but with *three* coordinate axes, labeled x, y, and z. Each axis is perpendicular to the other two. ◄ To reach the point $P(a, b, c)$ from the origin, go a units in the positive x-direction, b units in the positive y-direction, and c units in the positive z-direction. As Figure 1 illustrates, the resulting point $P(a, b, c)$ can also be thought of as a corner (the one opposite the origin) of a rectangular solid with dimensions $|a|$, $|b|$, and $|c|$.

If a or b is negative, go the other way.

There's "room" in $\mathbb{R}^3$ for three mutually perpendicular axes. $\mathbb{R}^2$ has room for only two.

Quadrants and octants   The two axes divide the xy-plane into four **quadrants**, defined by the pattern of positive or negative x- and y-coordinates. The analogous regions in xyz-space are called **octants**. The first octant, for instance, consists of all points (x, y, z) with all three coordinates positive. In Figure 1(a), only the first octant is visible. Figure 1(b) gives another view. There are eight octants in all in xyz-space—one for each of the possible patterns of signs of the three coordinates: $(+, +, +)$, $(-, +, +)$, . . . , $(-, -, -)$.

Coordinate planes   One can think of the first octant as a room, with the origin at the lower left corner of the front wall. At this point three walls meet, all at right angles. These "walls" are known as the **coordinate planes**: the yz-plane (the front wall), the xy-plane (the floor), and the xz-plane (the left wall). The coordinate planes correspond to simple equations in the variables x, y, and z. The yz-plane, for example, is the graph of the equation $x = 0$, that is, the set of all points (x, y, z) that satisfy this equation. Similarly, the xy- and xz-planes are graphs of the equations $z = 0$ and $y = 0$, respectively.

Many possible views   The xy-plane, being "flat," is relatively easy to draw. Simulating three-dimensional space on a flat page or computer screen is much harder, and there is always some price to be paid in distortion. For example, in 3-d reality, the x-, y-, and z-axes are all perpendicular to each other, but no flat picture can really show this. The axes in Figure 2, for instance, don't make right angles on the page.

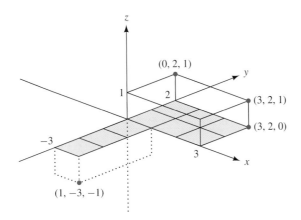

FIGURE 2
Another view of 3-d space

This figure's view of xyz-space is somewhat different from that of the last picture. Observe the following features:

- **Horizontal and vertical**   The xy-plane (in which the shaded "floor tiles" lie) is drawn to appear horizontal. The z-axis is vertical; the positive direction is up. This is a standard convention; we'll follow it consistently.

- **Hidden lines**   The dashed lines in Figure 2 lie "below" the xy-plane. They would be hidden from view if the xy-plane (the "floor") were opaque. How much of xyz-space is considered to be visible is a matter of choice. Sometimes only the first octant is shown.

- **Positive directions**   An arrow on each axis indicates the positive direction. The 3×2 block of shaded squares lies in the first quadrant of the xy-plane. The other shaded squares lie in the plane's fourth quadrant.

- **Plotting points: Positive and negative coordinates**   Any point $P(a, b, c)$ is plotted the same way: From the origin, move a, b, and c units in the positive x-, y-, and z-directions, respectively. Negative coordinates cause no special problem; just move the other way.

- **Where's the viewer?**   The picture is drawn as though the viewer were floating somewhere above the *fourth* quadrant of the xy-plane. In the earlier 3-d picture, by contrast, the viewer floats somewhere above the *first* quadrant. There's nothing sacred about *either* viewing angle; we'll use a variety of viewpoints as we go along. For that matter, so do the various computer plotting packages readers may have at hand.

- **No perspective**   To a human viewer, rectangular boxes like the preceding ones would appear in perspective; the sides would taper toward a vanishing point. The closer the viewer, the more pronounced those effects would be. For the sake of simplicity, we ignore perspective effects in the picture. In effect, the viewer is assumed to be *very* far from the origin, perhaps looking through a telescope.

There is no single "best" picture of a 3-d object; choosing a good or convenient view may depend on properties of the object, what needs emphasis, or even the drawing technology at hand. ➡

Computer, calculator, pencil, sharp stick, . . .

Distance and midpoints

Let $P(x_1, y_1)$ and $Q(x_2, y_2)$ be any two points in the xy-plane. Recall that the distance from P to Q (or from Q to P—it doesn't matter) is given by the familiar Pythagorean formula

$$d(P, Q) = \sqrt{(x_2 - x_1)^2 + (y_2 - y_1)^2}$$

and that the midpoint M of the segment joining P to Q has these "averaged" coordinates:

$$M = \left(\frac{x_1 + x_2}{2}, \frac{y_1 + y_2}{2} \right).$$

The formulas in three dimensions aren't much different.

DEFINITION   The distance between $P(x_1, y_1, z_1)$ and $Q(x_2, y_2, z_2)$ is

$$d(P, Q) = \sqrt{(x_2 - x_1)^2 + (y_2 - y_1)^2 + (z_2 - z_1)^2}.$$

The midpoint of the segment joining P and Q has coordinates

$$M\left(\frac{x_1 + x_2}{2}, \frac{y_1 + y_2}{2}, \frac{z_1 + z_2}{2} \right).$$

In either two or three dimensions, the distance formula reflects the Pythagorean rule. See the exercises for more details.

Both definitions are simply three-dimensional versions of the corresponding formulas in the xy-plane. In both two and three dimensions, for example, distance is computed as the square root of the sum of the squared differences in coordinates. ◄

EXAMPLE 1   Consider the points $P(0, 0, 0)$ and $Q(2, 4, 6)$. Find the distance from P to Q and the midpoint M of the segment joining them. How far is M from P and from Q?

Solution   By the distance formula,

$$d(P, Q) = \sqrt{(2 - 0)^2 + (4 - 0)^2 + (6 - 0)^2} = \sqrt{56} \approx 7.483.$$

According to the formula, the midpoint is $M(1, 2, 3)$; each coordinate of M splits the difference between the corresponding coordinates of P and Q. To see why M deserves the name "midpoint," notice that

$$d(P, M) = \sqrt{(1 - 0)^2 + (2 - 0)^2 + (3 - 0)^2}$$

$$= \sqrt{(2 - 1)^2 + (4 - 2)^2 + (6 - 3)^2} = \sqrt{14} \approx 3.742.$$

Thus, M lies halfway between P and Q, as a midpoint should.   ■

Equations and their graphs

The graph of an equation may or may not be the graph of a function. The unit circle is not a function graph.

The graph of an equation in x and y is the set of all points (x, y) that satisfy the equation. The graph of $x^2 + y^2 = 1$, for instance, is a circle of radius 1 in the xy-plane, centered at the origin. The graph of the equation $x = 0$ is the y-axis. ◄

The same idea applies for three variables: The graph of an equation in x, y, and z is the set of points (x, y, z) in space that satisfy the equation. Figure 3 presents three simple examples (only the first octant is shown).

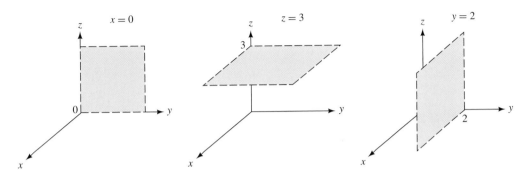

FIGURE 3
Three simple graphs

Solutions of the equation $x = 0$ are points of the form $(0, y, z)$, and so the graph is the yz-plane. Similarly, solutions of $z = 3$ are all points of the form $(x, y, 3)$, and thus the graph is a horizontal plane floating 3 units above the xy-plane. The graph of $y = 2$ is parallel to the xz-plane but moved 2 units in the positive y-direction.

Notice that in each case, the graph of an equation in x, y, and z is a plane—a *two-dimensional* object. In contrast, the graph of one equation in x and y is usually a curve or a line—a *one-dimensional* object. The pattern is the same in both cases: The graph of an equation has dimension one less than the number of variables.

A few special types of graphs in xyz-space deserve special mention.

Planes   A **linear equation** is one of the form $ax + by + cz = d$, where a, b, c, and d are constants and at least one of a, b, and c is nonzero. ➡ All three of the equations just plotted are linear. They illustrate an important general fact:

What goes wrong if $a = b = c = 0$?

> *The graph of any linear equation is a plane.*

(We explain this fact carefully in a later section.) To draw planes in xyz-space, we use the fact that a plane is uniquely determined by three points (unless the points happen to be on a straight line).

EXAMPLE 2   Plot the linear equation $x + 2y + 3z = 3$ in the first octant.

Solution   First we'll find some points (x, y, z) that satisfy $x + 2y + 3z = 3$. If anything, this is too easy—there are infinitely many possibilities. Given *any* values for x and y, the equation determines a corresponding value for z. If, say, $x = 1$ and $y = 1$, then $x + 2y + 3z = 3$ can hold only if $z = 0$. Similarly, setting $y = 2$ and $z = 3$ forces $x = -10$. Among all possible solutions, three of the simplest are

$$P(3, 0, 0); \qquad Q\left(0, \frac{3}{2}, 0\right); \qquad R(0, 0, 1).$$

These solutions are both easy to find (set two coordinates to zero and solve for the third) and easy to plot (they lie on the coordinate axes). Figure 4 shows the plane.

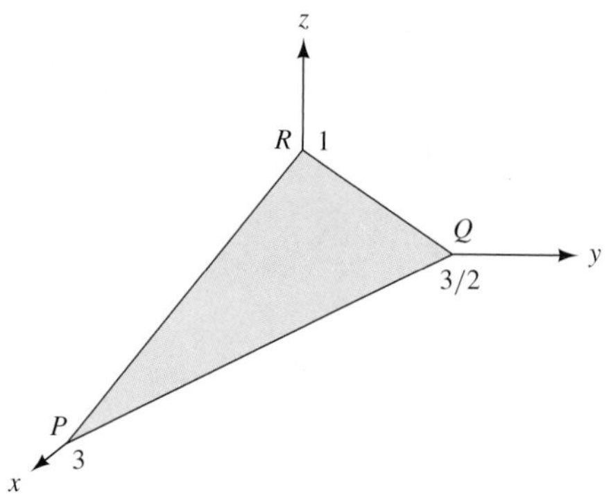

FIGURE 4

A plane in the first octant: $x + 2y + 3z = 3$

Notice the following features:

- **Intercepts** A typical *line* in the xy-plane has x- and y-intercepts where the line intersects the coordinate axes. In a similar sense, a typical *plane* in xyz-space has x-, y-, and z-intercepts. In the figure, the intercepts are P, Q, and R. ◂

Not every line in the xy-plane intercepts both axes; not every plane in space intercepts all three axes. See the exercises at the end of the section for more on this.

- **Traces** If we "slice" a surface in xyz-space with a plane, the intersection of the surface with the plane is called the **trace** of the surface in that plane. The plane p shown in Figure 4 meets each of the three coordinate planes in a straight line. Those three lines are therefore the traces of the surface $x + 2y + 3z = 3$ in the xy-plane, the xz-plane, and the yz-plane, respectively.

    It's easy to find equations for these traces. For example, a point (x, y, z) lies both in the plane p *and* in the xy-plane if and only if it satisfies both $x + 2y + 3z = 3$ and $z = 0$. Setting $z = 0$ in the first equation gives $x + 2y = 3$—as expected, the equation of a line in the xy-plane. This line is therefore the trace of p in the xy-plane. ◂ ∎

Look for this line in Figure 4.

Spheres In the plane, a circle of radius $r > 0$ and center $C(a, b)$ is the set of points $P(x, y)$ at distance r from (a, b). Translating this description into symbolic language produces the familiar formula for a circle in the plane:

$$d(P, C) = \sqrt{(x-a)^2 + (y-b)^2} = r, \qquad \text{or} \qquad (x-a)^2 + (y-b)^2 = r^2.$$

(Squaring both sides does no harm and simplifies the equation's appearance.)

The object in space that is analogous to a circle in the plane is a **sphere** of radius r. Like a circle, a sphere is "hollow," similar to an empty orange skin. Adding the interior (the edible part of the orange) produces a **ball**. Like a circle, a sphere is the set of points at some fixed distance—the radius—from a fixed center point. Given a radius $r > 0$ and a center point $C(a, b, c)$, the sphere of radius r centered at C is the set of points (x, y, z) such that

$$d(P, C) = \sqrt{(x-a)^2 + (y-b)^2 + (z-c)^2} = r,$$

or, equivalently,

$$(x-a)^2 + (y-b)^2 + (z-c)^2 = r^2.$$

The simplest example, the **unit sphere**, has center $(0, 0, 0)$ and radius 1. Its equation reduces to this simple form:

$$x^2 + y^2 + z^2 = 1.$$

For instance, circles may not look circular. Depending on the viewing angle, they may look like ellipses.

Drawing circles in the xy-plane is easy, even by hand. Drawing spheres (or any "curved" objects, for that matter) convincingly by hand is much harder. ◂ Fortunately, rough sketches usually suffice.

Completing the square may reveal an equation's spherical form.

EXAMPLE 3   Is the graph of $x^2 - 2x + y^2 - 4y + z^2 - 6z = 0$ a sphere? Which one?

Solution   Completing the square in each variable separately gives

$$x^2 - 2x + y^2 - 4y + z^2 - 6z = 0 \iff$$
$$(x^2 - 2x + 1) + (y^2 - 4y + 4) + (z^2 - 6z + 9) = 1 + 4 + 9 \iff$$
$$(x - 1)^2 + (y - 2)^2 + (z - 3)^2 = 14.$$

The last form shows that our equation describes the sphere of radius $\sqrt{14}$ centered at $(1, 2, 3)$. As the equation shows, this sphere passes through the origin. Figure 5 shows this, too.

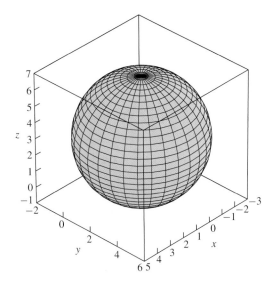

FIGURE 5
Graph of $x^2 - 2x + y^2 - 4y + z^2 - 6z = 0$

■

Cylinders   What is the graph of the equation $y = x^2$? The answer depends on where we're working. In the xy-plane, the graph is the familiar parabola—all points of the form (x, x^2). Here's the graph of the same equation in xyz-space (Figure 6).

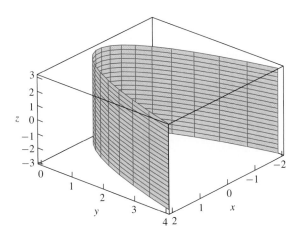

FIGURE 6
Graph of $y = x^2$ in xyz-space

Observe:

In other words, the graph has "vertical walls."

- **A missing variable** The graph is *unrestricted in the z-direction*—it contains all points that lie directly above or below the graph of $y = x^2$ in the xy-plane. ← The graph has this property because z is "missing" in the equation $y = x^2$. This means that if (x, y) satisfies the equation, then so does *every* point (x, y, z), regardless of the value of z.

- **What's a cylinder?** Graphs like this one, in which at least one of the variables is unrestricted, are called **cylinders**. Any equation that omits one or more variables—$y = z$, say—has a cylindrical graph. Plotting cylinders is comparatively simple. If the equation involves only y and z, for instance, we first plot the equation in the yz-plane and then "extend" the graph in the x-direction.

In everyday speech, "cylinder" usually means a circular tube. As the next example illustrates, the mathematical idea of a cylinder is much more general.

EXAMPLE 4 Discuss the graph in xyz-space of the equation $z = 2 + \sin y$. Interpret the result as a cylinder.

The surface, like many graphs, continues forever; a picture shows only part of the graph.

Solution There's no variable x in the equation, and so the graph is unrestricted in the x-direction; that is, it's a cylinder in x. Figure 7 gives a representative view: ←

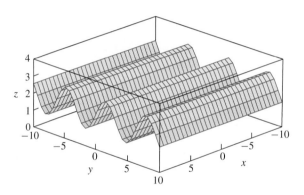

FIGURE 7

Graph of $z = 2 + \sin y$

The graph resembles the surface of an idealized ocean with regular waves moving parallel to the y-axis. Each wave is infinitely long with its trough and crest parallel to the x-axis. Notice especially how the surface meets the yz-plane. The curve of intersection—that is, the trace of the surface in the yz-plane—is the ordinary sine curve $z = 2 + \sin y$. In fact, the entire surface can be thought of as infinitely many identical copies of this curve, one for each value of x. ∎

BASIC EXERCISES

1. We said in this section that the graph of $y = x^2$ is a cylinder in xyz-space unrestricted in the z-direction. (See Figure 6.)

   (a) Plot the equation $z = y^2$ in xyz-space. What is the unrestricted direction?
[HINT: Start in yz-space.]

   (b) Plot the equation $z = x^2$ in xyz-space. What is the unrestricted direction?

2. Plot the equation $x^2 + y^2 = 1$, first in xy-space, then in xyz-space. (The second graph should resemble a vertical pipe—a cylinder in the everyday sense of the word.)

In Exercises 3–6, plot the given equation in xyz-space (a rough sketch is fine); then describe the graph in words.

3. $y^2 - z^2 = 0$ **4.** $y^2 + z^2 = 0$

5. $y^2 + z = -1$

6. $y^2 + z^2 = -1$ [HINT: Does the graph contain any points?]

In Exercises 7–11, find an equation in x, y, and z for the graph in xyz-space of

7. a sphere of radius 2, centered at the origin.

8. a sphere of radius 1, centered at $(1, 1, 1)$.

9. a circular cylinder of radius 1, centered along the y-axis.

10. a circular cylinder of radius 2, centered along the z-axis.

11. a cylindrical surface that resembles an ocean with waves rolling in the x-direction (see Example 4).

12. The equation $z = 3$ omits two variables. Therefore, its graph in xyz-space should be a cylinder in both the x-direction and the y-direction. Is it? What is the trace of the graph in each of the coordinate planes?

13. The graph of $x^2 + y^2 - 6y + z^2 - 4z = 0$ is a sphere. Find the center and radius; then draw the sphere.

14. Consider the unit sphere S with equation $x^2 + y^2 + z^2 = 1$. If we set $z = 0$ in this equation, we get $x^2 + y^2 + z^2 = x^2 + y^2 = 1$. This means, geometrically, that S intersects the xy-plane in the unit circle $x^2 + y^2 = 1$. In mathematical language, the unit circle is the trace of S in the xy-plane. (To put it another way, the unit circle is the "equator" of the unit sphere.)

   (a) Set $x = 0$ in the original equation to find the equation of the intersection of S and the yz-plane.

   (b) What is the intersection of S and the xz-plane? Describe the answer geometrically.

   (c) Use the results in parts (a) and (b) to sketch the part of S that lies in the first octant.
   [HINTS: First draw a set of coordinate axes. Then draw the traces of S in each of the three coordinate planes.]

   (d) Set $z = 1/2$ in the original equation to show that S intersects the plane $z = 1/2$ in the circle of radius $\sqrt{3}/2 \approx 0.87$ centered at $(0, 0)$.

   (e) What is the trace of S in the plane $z = 0.9$? In the plane $z = 1$? In the plane $z = 2$? Explain your answers.

15. The distance formula in xyz-space can be thought of as just another instance of the Pythagorean rule for right triangles. (The square of the hypotenuse is the sum of the squares of the sides.) This exercise illustrates why.

   (a) Plot and label the points $O(0, 0, 0)$, $P(1, 0, 0)$, $Q(1, 2, 0)$, and $R(1, 2, 3)$ in an xyz-coordinate system. Observe that the triangles $\triangle OPQ$ and $\triangle OQR$ are both right triangles. Mark the sides OP, PQ, and QR with their lengths. (The lengths should be obvious from the picture.)

   (b) Use the Pythagorean rule (not the distance formula) on the triangle $\triangle OPQ$ to find the length of OQ.

   (c) Use the Pythagorean rule (not the distance formula) on the triangle $\triangle OQR$ to find the length of OR.

   (d) For comparison, use the distance formula to compute the lengths of OQ and OR.

Any reasonable formula for distance should satisfy some commonsense requirements. For example, the distance $d(P, P)$ from any point P to itself should certainly be zero. So it is. If $P(x, y, z)$ is any point, then the distance formula says $d(P, P) = \sqrt{(x - x)^2 + (y - y)^2 + (z - z)^2} = 0$. In the same spirit, use the distance formula to verify the commonsense properties in Exercises 16–18. Throughout, use the points $P(x, y, z)$ and $Q(a, b, c)$.

16. If $P \neq Q$, then $d(P, Q) > 0$.

17. $d(P, Q) = d(Q, P)$.

18. If M is the midpoint of P and Q, then $d(P, M) = d(M, Q) = d(P, Q)/2$.

19. In this section we said that xyz-space contains eight different octants. List eight points, all with coordinates ± 1, one in each octant. Draw a picture showing all eight points.

In Exercises 20–23, consider the linear equation $Ax + By = C$ and its graph (a line) in the xy-plane. Here A, B, and C are constants such that $A \cdot B \neq 0$.

20. What goes wrong if $A = B = 0$?

21. Find the slope of the line $Ax + By = C$. Which lines have undefined slope?

22. Find the y-intercept of the line $Ax + By = C$. Which lines have no y-intercept?

23. Find the x-intercept of the line $Ax + By = C$. Which lines have no x-intercept?

In Exercises 24–26, consider the linear equation $Ax + By + Cz = D$ and its graph (a plane) in xyz-space. Here A, B, C, and D are constants such that $A \cdot B \cdot C \neq 0$.

24. What goes wrong if $A = B = C = 0$?

25. Find (if possible) an x-intercept of the plane $Ax + By + Cz = D$. (Set $y = 0$ and $z = 0$, then solve for x.) Give an example of a plane with no x-intercept.

26. Find (if possible) a z-intercept of the plane $Ax + By + Cz = D$. Give an example of a plane with no z-intercept.

27. The linear equation $x + 2y + 3z = 3$ defines a plane p in xyz-space. (See Figure 4.)

   (a) Find the equation of the trace of p in the xz-plane. Where does the trace intercept the x- and z-axes?

   (b) Find the equation of the trace of p in the yz-plane. Where does the trace intercept the y- and z-axes?

   (c) Find the equation of the trace of p in the plane $x = 1$.

28. Consider the plane p with equation $4x + 2y + z = 4$.

   (a) Find the x-, y-, and z-intercepts of p. Use them to draw p in the first octant.

   (b) Find the traces of p in each of the three coordinate planes. How do your answers appear in the picture in part (a)?

We said in this section that, as a rule, a line in the xy-plane intercepts both coordinate axes, and a plane in xyz-space intercepts all *three coordinate axes. But exceptions are possible. Some of these are illustrated in Exercises 29–32.*

29. Give an example of a line in the xy-plane that intercepts the x-axis but not the y-axis. Write an equation for your line in the form $ax + by = c$.

30. Consider the plane $x = 1$ in xyz-space. Find all possible intercepts with the three coordinate axes.

31. Consider the plane $x + 2y = 1$ in xyz-space. Find all possible intercepts with the three coordinate axes.

32. Give the equation of a plane in xyz-space that intersects the y-axis and the z-axis but not the x-axis.

33. Suppose that $P_1(x_1, y_1, z_1)$ and $P_2(x_2, y_2, z_2)$ both lie on the plane with equation $Ax + By + Cz = D$. Show that the midpoint of P_1 and P_2 also lies on this plane.

M.2 FUNCTIONS OF SEVERAL VARIABLES

Functions of one variable are the basic objects of single-variable calculus. Functions of two (or more) variables play a similar role in multivariable calculus. In this section we meet such functions and consider some of their rudimentary properties.

Functions of one or more variables

The squaring function, defined for all real numbers x by $f(x) = x^2$, is typical of the functions of beginning calculus: f accepts one number, x, as input and assigns another number, x^2, as output. If, say, $x = 2$, then $f(2) = 4$.

Consider, by contrast, the function g defined by $g(x, y) = x^2 + y^2$. Unlike f, g accepts a *pair* (x, y) of real numbers as inputs. The output is a third real number, $x^2 + y^2$. If, say, $x = 2$ and $y = 1$, then $g(2, 1) = 4 + 1 = 5$.

"One" and "two" count the input variables to f and g.

Naturally enough, f is called a function of one variable and g a function of two variables. ← The difference has to do with domains: The domain of f is the one-dimensional real number line, whereas the domain of g is the two-dimensional xy-plane.

The function f corresponds to the equation $y = x^2$; x is the **independent variable**, and y is the **dependent variable**. The function g corresponds to the equation $z = x^2 + y^2$; here both x and y are independent variables, and z is the dependent variable.

We use f and g to illustrate various similarities and differences between functions of one and two variables. Notice, however, that there's nothing sacred about *two* variables—we can (and do) discuss such functions as

$$h(x, y, z) = x^2 + y^2 + z^2 \quad \text{and} \quad k(x, y, z, w) = x^2 + y^2 + z^2 + w^2,$$

which accept three or more variables. For now, we'll keep things simple by sticking mainly to functions of two variables.

Why bother?   Single-variable calculus is challenging enough. Why complicate things by adding more variables?

It's a fair question. One good practical answer is that functions of several variables are essential for describing and predicting phenomena we care about, both natural and human-made. In economics, for instance, a manufacturer's profit depends on *many* input variables: labor costs, distance to markets, tax rates, and so on. In physics, a satellite's motion through space depends on a variety of forces. In biology, populations rise and fall with variations in climate, food supply, predation, and other factors. The weather varies with both longitude and latitude. In short, our world is multidimensional, and modeling it successfully requires multivariable tools.

See the back page of your newspaper.

The National Weather Service plots multivariable functions every day. ← Here (Figure 1), for instance, are noon surface temperatures on a relatively warm (by Upper Midwest standards) winter day.

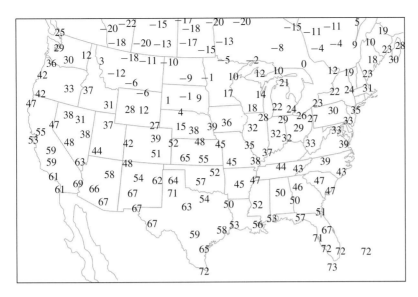

FIGURE 1
Plot of surface temperatures (Fahrenheit)

The map shows how the temperature function varies across its domain, the continental United States.

Functional vocabulary and notation Most of the basic words and notations for functions of several variables are similar to those for functions of one variable. Roughly speaking, a function is a "machine" that accepts inputs and assigns outputs. A bit more formally:

- **Domain** The domain of a function is the set of permissible inputs. The preceding function g, for instance, accepts *any* 2-tuple (x, y) as input, and so the domain is the set $\mathbb{R}^2$.

- **Range** The range of a function is the set of possible outputs. For g, the range is the set $[0, \infty)$ of nonnegative real numbers.

- **Rule** The rule of a function is the method for assigning outputs to inputs. For g, the rule is given by the algebraic formula $g(x, y) = x^2 + y^2$. Not every function has a simple symbolic rule. We often encounter functions given by tables, by graphs, or in other ways.

- **"Arrow" notations** We used the notation $g(x, y) = x^2 + y^2$ to describe a certain function of two variables. Other notations involving arrows are sometimes convenient. The notation

$$g : \mathbb{R}^2 \to \mathbb{R}$$

says that g is a function that accepts *two* real numbers as input and produces *one* real number as output. If we want to emphasize the rule by which g sends inputs to outputs, we can write

$$g : (x, y) \to x^2 + y^2.$$

These notations remind us that a function begins with an input (a 2-tuple in this case) and ends with an output (a single number in this case).

Graphs in one and several variables

To an ant walking along it, the parabola looks like a (one-dimensional) straight line.

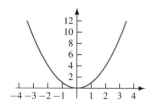

FIGURE 2
Graph of $f(x) = x^2$

The graph of f is the set of all points (x, y) for which $y = f(x) = x^2$. For example, $f(2) = 4$, and so the point $(2, 4)$ lies on the graph. Geometrically, this graph is a curve—a parabola, in this case—in the xy-plane. Like a straight line, the curve $y = x^2$ is a *one*-dimensional object that lives in a *two*-dimensional space—the xy-plane. ◄ Figure 2 shows part of the familiar graph.

The graph of g is the set of all points (x, y, z) for which $z = g(x, y) = x^2 + y^2$. For example, since $g(2, 1) = 5$, the point $(2, 1, 5)$ lies on the g-graph. Figure 3 shows part of the g-graph.

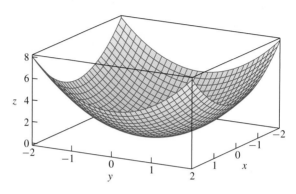

FIGURE 3
Graph of $g(x, y) = x^2 + y^2$

To an ant walking along it, the g-graph looks like a (two-dimensional) flat plane, just as Earth's surface looks flat to a human.

This graph, called a **paraboloid**, is quite different from the parabola shown earlier. Notice especially the dimensions involved: The graph of g is a *two*-dimensional surface hovering in *three*-dimensional xyz-space. ◄ As a rule, the *dimension* of any graph is the number of input variables the function accepts. (This rule of thumb applies to almost every function seen in calculus courses. For the record, however, some very ill-behaved functions do violate this rule.)

In other respects, the graphs of f and g are quite similar to each other. For both functions, the *height* of the graph above a given domain point (a typical input is x_0 for f and (x_0, y_0) for g) tells the corresponding output value. (Typical outputs are $y_0 = f(x_0)$ and $z_0 = g(x_0, y_0)$, respectively.)

Multivariable graphs: Beware   Graphs are at least as important in multivariable calculus as in elementary calculus, but multivariable graphs are usually more complicated; they need extra care in handling. Choosing a "good" viewing window, for example, takes some care even for functions of one variable, and the problem can be stickier still for functions of two variables. The fact that multivariable graphs "live" naturally in three-dimensional (or even higher-dimensional) space—not on a flat page or a computer screen—only adds to the problem. For this and other reasons, we try to look at functions from as many points of view as possible.

Level curves and contour maps   Let $f(x, y)$ be a function of two variables, and let c be a number in the range of f. Then $f(x, y) = c$ is an equation in x and y; its graph is (usually) a curve in the xy-plane. ◄ Such curves have a special name.

Sometimes this "curve" is just one point.

DEFINITION   Let $f(x, y)$ be a function, and let c be a constant. The set of all (x, y) for which $f(x, y) = c$ is called a **level curve** of f. A collection of level curves drawn together is called a **contour map** of f.

Observe:

- **Why "level"?**   The word "level" makes good sense here because $f(x, y)$ has the same value, namely c, at each point along the level curve. In other words, the graph of f is level above the level curve $f = c$.

- **Which level curves to draw?**   Each number c in the range of f has its own level curve. Because the range of a function is usually infinite, we cannot possibly draw *all* the level curves. In practice, we draw some convenient selection of curves, corresponding to *evenly spaced* values of c. The spacing between curves reflects how fast the function increases or decreases.

- **Labels**   We sometimes label level curves with their corresponding output values. Therefore, typical labels might be of the form $z = c$, $f = c$, or even simply c.

EXAMPLE 1   Figure 4 shows a sample of level curves for the function $g(x, y) = x^2 + y^2$. Label each level curve with the appropriate output value. Recall that the graph of g is a paraboloid. How is this shape reflected in the level curves?

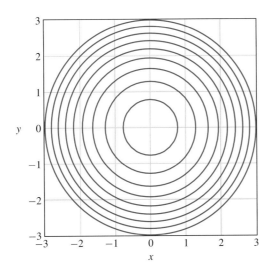

FIGURE 4
Level curves of $g(x, y) = x^2 + y^2$

Solution   Each level curve is a circle about the origin; each is the graph of an equation $x^2 + y^2 = c$ for some c. The circles, from smallest to largest, correspond to the levels $z = 1, z = 2, \ldots, z = 9$.

Notice that level curves of g get closer and closer together as we move outward from the origin. This reflects the fact that the g-graph is a paraboloid—it gets steeper and steeper as we move away from the origin.   ■

EXAMPLE 2   Drawing level curves on a temperature map makes the map easier to read and interpret. (Newspapers usually do this.)

Figure 5 shows another version of the map in Figure 1. The boundaries of shaded regions, called **isotherms**, correspond to the level curves of the temperature function.

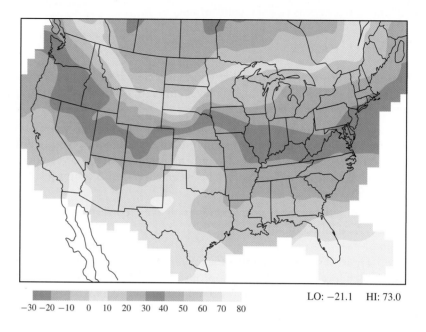

LO: −21.1   HI: 73.0

−30 −20 −10   0   10   20   30   40   50   60   70   80

FIGURE 5
A contour plot of surface temperatures (Fahrenheit)

Linear functions   A linear function of *one* variable is one that can be written in the form $L(x) = a + bx$, where a and b are constants. A linear function of *several* variables has a similar algebraic form.

> **DEFINITION**   A **linear function** of two variables is one that can be written in the form
>
> $$L(x, y) = a + bx + cy,$$
>
> where a, b, and c are constants.

Linear functions of three (or more) variables are similar. A linear function of three variables has the form $L(x, y, z) = a + bx + cy + dz$, where a, b, c, and d are constants.

Why ''linear''?   Why do linear functions deserve that name? In the xy-plane, there's no mystery—the graph of a linear function $y = a + bx$ is a straight *line*. In xyz-space, however, the graph of a linear function $z = a + bx + cy$ is not a line but a plane. (In $xyzw$-space, the graph of a linear function $w = a + bx + cy + dz$ is even less like a line. It's a three-dimensional solid, called a **hypersurface**.) Nevertheless, we call any function that involves only constants and the first power of each variable *linear*, regardless of the number of variables. In the same spirit, we call a function *quadratic* (no matter how many variables are involved) if the variables appear in nonnegative integer powers, none of which exceed 2. ◄

The function $f(x, y) = x^2 + xy + y^2$ is quadratic in this sense.

Readers who have studied linear algebra will recall that the word "linear" (as in "linear transformation") is used in still another way in that subject. Like other very useful English words, "linear" seems to have spawned a whole family of related but not identical meanings.

Why linear functions matter   Linear functions are simple, useful, and easy to work with. Most important for us, linear functions are prototypes, or models, for *all* differentiable functions. Indeed, any differentiable function, in any number of variables, can be called "almost linear" or "locally linear" in much the same sense that an ordinary calculus function $y = f(x)$ looks like a straight line if we zoom in repeatedly on a typical point on its graph.

BASIC EXERCISES

In Exercises 1–5, find the domain and range of the function.

1. $g(x, y) = x^2 + y^2$

2. $h(x, y) = x^2 + y^2 + 3$

3. $j(x, y) = 1/(x^2 + y^2)$

4. $k(x, y) = x^2 - y^2$

5. $m(x, y) = \sqrt{1 - x^2 - y^2}$

6. Let $f(x, y) = y - x^2$, and let $g(x, y) = x - y^2$.

   (a) In the rectangle $[-3, 3] \times [-3, 3]$, draw and label the level curves of f that correspond to $z = -3, z = -2, \ldots,$ $z = 2$, and $z = 3$. What is the shape of each level curve?

   (b) In the rectangle $[-3, 3] \times [-3, 3]$, draw and label the level curves of g that correspond to $z = -3, z = -2, \ldots,$ $z = 2$, and $z = 3$. What is the shape of each level curve?

   (c) How are the results of parts (a) and (b) similar? How are they different?

   (d) Use technology to plot the graphs $z = f(x, y)$ and $z = g(x, y)$ for (x, y) in $[-3, 3] \times [-3, 3]$. Describe briefly, in words, how the two graphs are related to each other.

7. Let $f(x, y) = x^2 + y^2$, and let $g(x, y) = x^2 + y^2 + 1$.

   (a) In the rectangle $[-3, 3] \times [-3, 3]$, draw and label the level curves of f that correspond to $z = 0, z = 2, z = 4,$ $z = 6$, and $z = 8$. What is the shape of each level curve?

   (b) In the rectangle $[-3, 3] \times [-3, 3]$, draw and label the level curves of g that correspond to $z = 1, z = 3, z = 5,$ $z = 7$, and $z = 9$. What is the shape of each level curve?

   (c) How are the results of parts (a) and (b) similar? How are they different?

   (d) Use technology to plot the graphs $z = f(x, y)$ and $z = g(x, y)$ for (x, y) in $[-3, 3] \times [-3, 3]$. Describe briefly, in words, how the two graphs are related to each other.

8. Let f and g be the linear functions $f(x, y) = 2x - 3y$ and $g(x, y) = -2x + 3y$.

   (a) In the rectangle $[-3, 3] \times [-3, 3]$, draw and label the level curves of f that correspond to $z = -5, z = -3,$ $z = -1, z = 1, z = 3$, and $z = 5$. What is the shape of each level curve?

   (b) In the rectangle $[-3, 3] \times [-3, 3]$, draw and label the level curves of g that correspond to $z = -5, z = -3,$ $z = -1, z = 1, z = 3$, and $z = 5$. What is the shape of each level curve?

   (c) How are the results of parts (a) and (b) similar? How are they different?

   (d) What special properties do the level curves of a *linear* function have?

(e) Use technology to plot the graphs $z = f(x, y)$ and $z = g(x, y)$ for (x, y) in $[-3, 3] \times [-3, 3]$. Describe briefly, in words, how the two graphs are related to each other.

9. Let f and g be the functions $f(x, y) = 2 + x^2$ and $g(x, y) = 2 + y^2$.

   (a) In the rectangle $[-3, 3] \times [-3, 3]$, draw and label the level curves of f that correspond to $z = 0, z = 2, z = 4,$ $z = 6$, and $z = 8$. What is the shape of each level curve?

   (b) In the rectangle $[-3, 3] \times [-3, 3]$, draw and label the level curves of g that correspond to $z = 0, z = 2, z = 4,$ $z = 6$, and $z = 8$. What is the shape of each level curve?

   (c) How are the results of parts (a) and (b) similar? How are they different?

   (d) Use technology to plot the graphs $z = f(x, y)$ and $z = g(x, y)$ for (x, y) in $[-3, 3] \times [-3, 3]$. Describe briefly, in words, how the two graphs are related to each other.

   (e) The graphs of f and g are both cylinders (in the sense defined in the text). How do the contour maps of f and g reflect this fact?

In Exercises 10–13, imagine a map of the United States in the usual position. The positive x-direction is east, and the positive y-direction is north. Suppose that the units of x and y are miles and that Los Angeles, California, has coordinates $(0, 0)$. (Several approximations are involved here. The Earth's surface is not flat, and Los Angeles occupies more than a single point.) Let $T(x, y)$ be the temperature, in degrees Celsius, at the location (x, y) at noon, Central Standard Time, on January 1, 2001.

10. What does it mean in weather language to say that $T(0, 0) = 15$?

11. What do the level curves of T mean in weather language? As a rule, would you expect level curves of T to run north and south or east and west? Why?

12. International Falls, Minnesota, is about 1400 miles east and 1100 miles north of Los Angeles. The noon temperature in International Falls on January 1, 2001, was $-15°$ C. What does this mean about $T(x, y)$?

13. Suppose that International Falls was the coldest spot in the country at the time in question. How would you expect the level curves to look near International Falls?

14. For any point (x, y) in the xy-plane, let $f(x, y)$ be the distance from (x, y) to the origin. Then f has the formula $f(x, y) = \sqrt{x^2 + y^2}$.

   (a) Find the domain and range of f.

(b) Plot f. Describe the graph in words.

(c) Draw the level curve of f that passes through $(3, 4)$.

(d) All level curves of f have the same shape. What is it?

15. For any point (x, y) in the xy-plane, let $f(x, y)$ be the distance from (x, y) to the line $x = 1$.

(a) Find a formula for $f(x, y)$.

(b) Plot f. Describe the graph in words.

(c) Draw the level curve that passes through $(3, 4)$.

(d) All level curves of f have the same shape. What is it?

16. Some values of a linear function $L(x, y)$ are tabulated below.

Values of $L(x, y)$							
y＼x	**−3**	**−2**	**−1**	**0**	**1**	**2**	**3**
3	−15	−12	−9	−6	−3	0	3
2	−13	−10	−7	−4	−1	2	5
1	−11	−8	−5	−2	1	4	7
0	−9	−6	−3	0	3	6	9
−1	−7	−4	−1	2	5	8	11
−2	−5	−2	1	4	7	10	13
−3	−3	0	3	6	9	12	15

(a) All the level curves of L are straight lines. Using this fact, draw (in the rectangle $[-3, 3] \times [-3, 3]$) and label the level curves $z = -12$, $z = -8$, $z = -4$, $z = 0$, $z = 4$, $z = 8$, $z = 12$.

(b) Find an equation in x and y for the level line $z = 0$.

(c) Because L is a linear function, its formula has the form $L(x, y) = a + bx + cy$ for some constants a, b, and c. Find numerical values for a, b, and c.

[HINT: The table says that $L(0, 0) = 0$. Therefore, $L(0, 0) = a + b \cdot 0 + c \cdot 0 = 0$, so $a = 0$. Use similar reasoning to find values for b and c.]

(d) Use technology to plot $L(x, y)$ over the rectangle $[-3, 3] \times [-3, 3]$. Is the shape of the graph consistent with the level curves you plotted in part (a)?

M.3   PARTIAL DERIVATIVES

Derivatives in one variable — interpretations

To avoid distractions, we assume that f and f' are continuous functions.

Let f be an ordinary function of one variable. Recall some familiar properties of the derivative function f': ◄

- **Slope**   For any fixed input $x = x_0$, the derivative $f'(x_0)$ tells the *slope* of the f-graph at the point $(x_0, f(x_0))$. The sign of $f'(x_0)$, in particular, tells whether f is increasing or decreasing (rising or falling, in everyday speech) at $x = x_0$.

In a specific example, we would need to specify appropriate units for everything.

- **Rate of change**   The derivative f' can also be interpreted as the *rate function* associated with f, as follows: For any input x_0, $f'(x_0)$ tells the instantaneous rate of change of $f(x)$ with respect to x. If, say, $f(x)$ gives the *position* of a moving object at time x, then $f'(x)$ gives the corresponding *velocity* at time x. ◄

- **Limit**   The derivative $f'(x_0)$ is defined as a limit of difference quotients:

$$f'(x_0) = \lim_{h \to 0} \frac{f(x_0 + h) - f(x_0)}{h}.$$

Similar interpretations hold if $h < 0$, but $h = 0$ is illegal.

For any $h > 0$, the difference quotient can be thought of either as the average rate of change of f over the interval $[x_0, x_0 + h]$ or as the slope of a secant line on the f-graph, over the same interval. ◄ Taking the limit as $h \to 0$ corresponds to finding the instantaneous rate of change of f or, equivalently, the slope of the tangent line to the f-graph at $x = x_0$.

- **Linear approximation**   At a point $(x_0, f(x_0))$ on the curve $y = f(x)$, the tangent line has slope $f'(x_0)$. This tangent line is the graph of the linear function L with equation

$$y = L(x) = f(x_0) + f'(x_0)(x - x_0).$$

The function L is called the **linear approximation** to f at x_0. The name makes sense because the graphs of f and L are close together near x_0. In symbols,

$$L(x) \approx f(x) \quad \text{when} \quad x \approx x_0.$$

Derivatives in several variables

Derivatives are just as important in multivariable calculus as in one-variable calculus, but—not surprisingly—the idea is more complicated for functions of several variables.

Take slope, for instance. The graph of a one-variable function $y = f(x)$ is a curve in the xy-plane. The slope at (x_0, y_0)—a single number—completely describes the graph's direction at (x_0, y_0). The graph of a two-variable function $z = f(x, y)$, by contrast, is a *surface*, which has no single slope at a point. A surface's steepness at a point depends on the direction (uphill, downhill, along the "contour," and so on) taken from the point. ➡ To put it another way, the graph of a one-variable function can be approximated near a given point by a one-dimensional tangent *line*. The graph of a two-variable function can be approximated near a fixed point by a two-dimensional tangent *plane*.

Every hiker knows this. How steep a mountain "feels" depends on the direction of the trail.

Here's the moral: To suit *multivariable* calculus, our notion of derivative must go beyond the simple idea of slope. The idea of linear approximation, not slope, turns out to be the key to extending the idea of the derivative to functions of more than one variable.

In this section we start to extend the derivative idea to functions of several variables. **Partial derivatives** are the simplest multivariable analogues of ordinary derivatives. ➡

"Partial" suggests—correctly—that there's more to the derivative story.

Partial derivatives: The idea   The basic idea of a partial derivative is to differentiate with respect to *one* variable, holding all the others constant. The following easy example illustrates the idea and introduces some useful terminology and notation.

EXAMPLE 1   Let $f(x, y) = x^2 - 3xy + 6$. Find $f_x(x, y)$ and $f_y(x, y)$, the partial derivatives of f with respect to x and y, respectively. Find the numerical values $f_x(2, 1)$ and $f_y(2, 1)$.

Solution   To find f_x, we differentiate $f(x, y) = x^2 - 3xy + 6$ with respect to x, *treating y as a constant*:

$$f_x(x, y) = 2x - 3y.$$

To find f_y, we *treat x as a constant*:

$$f_y(x, y) = -3x.$$

Setting $x = 2$ and $y = 1$ in these formulas gives

$$f_x(2, 1) = 2 \cdot 2 - 3 \cdot 1 = 1; \qquad f_y(2, 1) = -3 \cdot 2 = -6. \qquad \blacksquare$$

The calculations were easy, but what do the results mean? What do they say about how f behaves near $(x, y) = (2, 1)$? Can we interpret the results graphically and numerically? Understanding multivariable derivatives fully is a long-term proposition, but here are some starters.

Holding variables constant   To find the partial derivative f_x of a function f with respect to x, we treat all the other variables as constants. This produces a function of just one variable, x, which we differentiate in the usual way. For example, suppose we fix $y = 3$ in $f(x, y) = x^2 - 3xy + 6$. Then $f(x, y) = f(x, 3) = x^2 - 9x + 6$, and

$$\frac{d}{dx}\bigl(f(x, 3)\bigr) = f_x(x, 3) = 2x - 9.$$

This agrees, as it should, with the general formula $f_x(x, y) = 2x - 3y$ found earlier.

Directional rates of change   Partial derivatives (like ordinary derivatives) can be interpreted as rates or slopes—but with an important proviso about directions. For a function $f(x, y)$ and a point (x_0, y_0) in its domain, the partial derivative $f_x(x_0, y_0)$ tells the rate of change of $f(x, y)$ with respect to x (that's the only variable that's free to move); that is, how fast $f(x, y)$ increases as the input (x, y) moves away from (x_0, y_0) in the *positive* x-direction. The other partial derivative, $f_y(x_0, y_0)$, tells how fast f increases near (x_0, y_0) as y increases. The next example illustrates what this means numerically.

E X A M P L E 2   The function $f(x, y) = x^2 - 3xy + 6$ has partial derivatives $f_x(x, y) = 2x - 3y$ and $f_y(x, y) = -3x$. What does this say about rates of change of f at $(x, y) = (2, 1)$? At $(x, y) = (1, 2)$?

S o l u t i o n   We start at $(2, 1)$. The formulas give $f_x(2, 1) = 1$ and $f_y(2, 1) = -6$. These numbers represent rates of change of f with respect to x and y, respectively, at $(2, 1)$. A table of f-values centered at $(2, 1)$ shows what this means. ◄

The boxed row and column meet at $(2, 1)$, our target point.

Values of $f(x, y) = x^2 - 3xy + 6$ near $(2, 1)$							
y \ x	**1.97**	**1.98**	**1.99**	**2.00**	**2.01**	**2.02**	**2.03**
1.03	3.7936	3.8022	3.8110	3.8200	3.8292	3.8386	3.8482
1.02	3.8527	3.8616	3.8707	3.8800	3.8895	3.8992	3.9091
1.01	3.9118	3.9210	3.9304	3.9400	3.9498	3.9598	3.9700
1.00	3.9709	3.9804	3.9901	4.0000	4.0101	4.0204	4.0309
0.99	4.0300	4.0398	4.0498	4.0600	4.0704	4.0810	4.0918
0.98	4.0891	4.0992	4.1095	4.1200	4.1307	4.1416	4.1527
0.97	4.1482	4.1586	4.1692	4.1800	4.1910	4.2022	4.2136

Reading *up* the boxed column (at each step, y increases by 0.01) shows successive corresponding values of f *decreasing* by about 0.06. Thus, -6 is the rate of change of f with respect to y at $(2, 1)$; equivalently, $f_y(2, 1) = -6$. Similarly, the fact that $f_x(2, 1) = 1$ suggests that values of $f(x, 1)$ should increase at about the same rate as x if $x \approx 2$. Reading *across* the boxed row confirms this expectation.

Similar reasoning applies at $(1, 2)$. The values $f_x(1, 2) = -4$ and $f_y(1, 2) = -3$ mean that f decreases (at different rates) with respect to both x and y near $(1, 2)$. Numerical f-values bear this out:

Values of $f(x, y) = x^2 - 3xy + 6$ near $(1, 2)$					
y \ x	0.98	0.99	1.00	1.01	1.02
2.02	1.0216	0.9807	0.9400	0.8995	0.8592
2.01	1.0510	1.0104	0.9700	0.9298	0.8898
2.00	1.0804	1.0401	1.0000	0.9601	0.9204
1.99	1.1098	1.0698	1.0300	0.9904	0.9510
1.98	1.1392	1.0995	1.0600	1.0207	0.9816

Reading either across or up shows f-values decreasing at rates of about -4 and -3, respectively. ■

Formal definitions Partial derivatives are defined formally as limits, in much the same way as ordinary derivatives are defined.

DEFINITION Let $f(x, y)$ be a function of two variables. The **partial derivative** with respect to x of f at (x_0, y_0), denoted by $f_x(x_0, y_0)$, is defined by

$$f_x(x_0, y_0) = \lim_{h \to 0} \frac{f(x_0 + h, y_0) - f(x_0, y_0)}{h}$$

if the limit exists. The partial derivative with respect to y at (x_0, y_0), denoted by $f_y(x_0, y_0)$, is defined by

$$f_y(x_0, y_0) = \lim_{h \to 0} \frac{f(x_0, y_0 + h) - f(x_0, y_0)}{h}$$

if the limit exists.

The definition says, in effect, that $f_x(x_0, y_0)$ is the ordinary derivative at $x = x_0$ of an ordinary function of one variable—namely, the function given by the rule $x \to f(x, y_0)$. Some further comments and observations follow.

- **Do they exist?** If either limit does not exist, then neither does the corresponding partial derivative. It's possible, for instance, that $f_x(0, 0)$ exists but $f_y(0, 0)$ doesn't. For most functions in this chapter, however, both partial derivatives *do* exist.

- **About domains** For the preceding limits to exist, $f(x, y)$ must be defined at (x_0, y_0) and at nearby points (x, y). In practice, this condition seldom causes trouble. In particular, there's no problem if (x_0, y_0) lies in the interior of the domain of f, that is, if f is defined both at and near (x_0, y_0). (For the record, trouble is likeliest if (x_0, y_0) lies on the edge of the domain of f.)

- **Other notations** As with ordinary derivatives, various notations are used to denote partial derivatives. If, say, $z = f(x, y) = x \sin y$, then all of the expressions

$$f_x, \quad \frac{\partial f}{\partial x}, \quad \frac{\partial z}{\partial x}, \quad \text{and} \quad \frac{\partial}{\partial x}(x \sin y)$$

mean the same thing, as do

$$f_x(x_0, y_0), \quad \frac{\partial f}{\partial x}(x_0, y_0), \quad \frac{\partial z}{\partial x}\bigg|_{(x_0, y_0)}, \quad \text{and} \quad \frac{\partial}{\partial x}(x \sin y)\bigg|_{(x_0, y_0)}.$$

Partial derivatives with respect to y are expressed with similar notations. Notice especially the "curly-d" symbol ∂. It's usually read aloud as "partial."

On a topographic map oriented the usual way, the partial derivatives f_x and f_y describe the steepness of the terrain in the eastward and northward directions, respectively.

Partial derivatives and contour maps   We saw in Example 2, page M-18, how to estimate partial derivatives from a table of function values. Contour maps can be used for the same purpose. Thinking of $f_x(x_0, y_0)$ and $f_y(x_0, y_0)$ as rates suggests how: Use the contour map to measure how fast $z = f(x, y)$ rises or falls near (x_0, y_0) as either x or y increases. ◄

EXAMPLE 3   Figure 1 is a contour map of the function $f(x, y) = x^2 - 3xy + 6$, centered at $(2, 1)$. Use the contour map to estimate the partial derivatives $f_x(2, 1)$ and $f_y(2, 1)$.

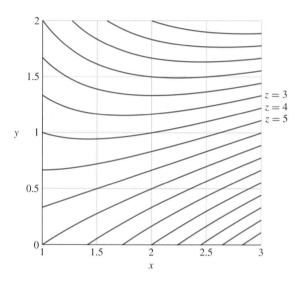

FIGURE 1
Level curves of $f(x, y) = x^2 - 3xy + 6$

Look closely at the contour map to convince yourself of these claims.

Solution   The level curves represent successive integer values of $z = f(x, y)$. In general, values of $f(x, y)$ increase as (x, y) moves toward the lower right. ◄

First we estimate $f_x(2, 1)$. A close look at the figure suggests that $f(2.1, 1) \approx 4.1$. Increasing x by 0.1 increases f by 0.1; this suggests that $f_x(2, 1) \approx 1$. Similarly, $f(2, 1.1) \approx 3.4$, and so increasing y by 0.1 increases f by -0.6; thus, $f_y(2, 1) \approx -6$.   ∎

Partial derivatives and linear approximation

For a differentiable function $y = f(x)$ of one variable, the ordinary derivative can be interpreted in terms of linear approximation. For any point (x_0, y_0) on the f-graph, there's a certain line through this point—the tangent line—that best "fits" the graph near $x = x_0$. The derivative $f'(x_0)$ gives the slope of this tangent line. Because the slope is known, it's easy to find an equation for the tangent line in point–slope form:

$$y - y_0 = f'(x_0)(x - x_0).$$

Equivalently, we can think of the tangent line as the graph of the linear function L defined by

$$y = L(x) = y_0 + f'(x_0)(x - x_0).$$

(Recall that $f(x_0) = y_0$.) The function L is called the **linear approximation** to f at x_0. The name is appropriate for three good reasons:

(i) L is linear;   (ii) $L(x_0) = f(x_0)$;   (iii) $L'(x_0) = f'(x_0)$.

In short, L agrees with f at x_0 as closely as *any* linear function can—both functions have the same value and the same (first) derivative. ➡

The idea of quadratic approximation is similar, except that we'd require agreement in the second derivative too.

Linear approximation in several variables The idea of linear approximation is essentially the same for functions of two (or more) variables: Given a function $f(x, y)$ and a point (x_0, y_0) in the domain of f, we look for a linear function $L(x, y)$ that has the same value and partial derivatives as does f at (x_0, y_0). In other words, we want a linear function L such that

$$L(x_0, y_0) = f(x_0, y_0), \quad L_x(x_0, y_0) = f_x(x_0, y_0), \quad \text{and} \quad L_y(x_0, y_0) = f_y(x_0, y_0).$$

In the one-variable case, the graph of a linear approximation function is called a tangent line. For a function of two variables, the graph of the linear approximation function is called a **tangent plane**. We illustrate and summarize these ideas in the next example.

E X A M P L E 4 Find the linear approximation function L to $f(x, y) = x^2 + y^2$ at $(x_0, y_0) = (2, 1)$. How are graphs of f and L related?

Solution The partial derivatives of f are $f_x(x, y) = 2x$ and $f_y(x, y) = 2y$. Thus, at our base point, $(x_0, y_0) = (2, 1)$:

$$f(2, 1) = 5; \qquad f_x(2, 1) = 4; \qquad f_y(2, 1) = 2.$$

Let's find a linear function L to match these values.

It's easiest to start by writing L in the convenient form

$$L(x, y) = a(x - x_0) + b(y - y_0) + c = a(x - 2) + b(y - 1) + c$$

and then choosing appropriate values for a, b, and c. Because $L(x, y) = a(x - 2) + b(y - 1) + c$, it's easy to see that

$$L(2, 1) = c; \qquad L_x(2, 1) = a; \qquad L_y(2, 1) = b.$$

To match f, we must have $c = 5$, $a = 4$, and $b = 2$; therefore,

$$L(x, y) = a(x - 2) + b(y - 1) + c = 4(x - 2) + 2(y - 1) + 5.$$

Figure 2 shows graphs of f and L together:

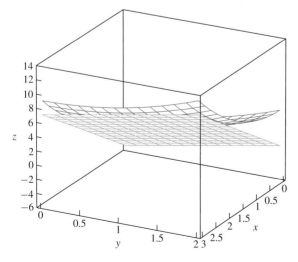

The tangent plane is shown gray.

FIGURE 2
Graphs of f and L together

The figure illustrates the phrases "tangent plane" and "linear approximation": The plane $z = L(x, y)$ touches the surface $z = f(x, y)$ at $(2, 1, 5)$; at this point, moreover, the flat plane fits the curved surface as well as possible. A closer look (Figure 3) at both functions near $(2, 1)$, this time using contour maps, suggests how good the fit really is.

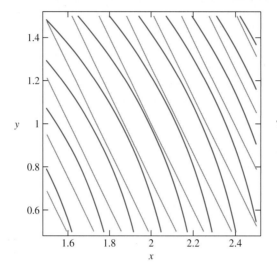

The gray lines are contours of L.

FIGURE 3
Level curves f and L together

The two contour maps are almost identical near $(2, 1)$.   ■

Linear approximation: The general formula   The procedure in Example 4 works the same way for any function of several variables as long as the necessary partial derivatives exist. The following are definitions for two and three variables. (The same idea works for any number of variables.)

> **DEFINITION (Linear approximation)**   Let $f(x, y)$ and $g(x, y, z)$ be functions and suppose that all the following partial derivatives exist. The **linear approximation** to f at (x_0, y_0) is the function
>
> $$L(x, y) = f(x_0, y_0) + f_x(x_0, y_0)(x - x_0) + f_y(x_0, y_0)(y - y_0).$$
>
> The linear approximation to g at (x_0, y_0, z_0) is the function
>
> $$L(x, y, z) = g(x_0, y_0, z_0) + g_x(x_0, y_0, z_0)(x - x_0)$$
> $$+ g_y(x_0, y_0, z_0)(y - y_0) + g_z(x_0, y_0, z_0)(z - z_0).$$

E X A M P L E 5   Find the linear approximation to $g(x, y, z) = x + yz^2$ at $(1, 2, 3)$. Does the answer make numerical or graphical sense?

Solution   Easy calculations give

$$g(1, 2, 3) = 19; \quad g_x(1, 2, 3) = 1; \quad g_y(1, 2, 3) = 9; \quad g_z(1, 2, 3) = 12.$$

The linear approximation function therefore has the form

$$L(x, y, z) = g(1, 2, 3) + g_x(1, 2, 3)(x - 1) + g_y(1, 2, 3)(y - 2) + g_z(1, 2, 3)(z - 3)$$
$$= 19 + 1(x - 1) + 9(y - 2) + 12(z - 3).$$

To see the situation numerically, we tabulate some values of each function.

Values of L and g near $(1, 2, 3)$						
(x, y, z)	$(1, 2, 3)$	$(1.1, 2, 3)$	$(1, 2.1, 3)$	$(1, 2, 3.1)$	$(1.1, 2.1, 3.1)$	$(3, 4, 5)$
$g(x, y, z)$	19	19.1	19.9	20.22	21.281	103
$L(x, y, z)$	19	19.1	19.9	20.2	21.2	63

As the numbers illustrate, $L(x, y, z)$ and $g(x, y, z)$ are close together if, but only if, (x, y, z) is near $(1, 2, 3)$. ➤

To plot ordinary graphs of g and L would require four dimensions. ➤ Instead we plot, for comparison, the level surfaces $L(x, y, z) = 19$ and $g(x, y, z) = 19$, both of which pass through the base point $(1, 2, 3)$. (A **level surface** is like a level curve—it's a set of inputs along which a function has constant output value.) Figure 4 suggests again how similarly $g(x, y, z)$ and $L(x, y, z)$ behave when $(x, y, z) \approx (1, 2, 3)$.

The last column illustrates what happens far from $(1, 2, 3)$.

One for each input variable and one more for the output.

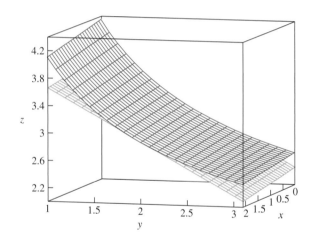

FIGURE 4
A curved surface and a linear approximation

■

BASIC EXERCISES

In Exercises 1–8, find the partial derivative with respect to each variable.

1. $f(x, y) = x^2 - y^2$

2. $f(x, y) = x^2 y^2$

3. $f(x, y) = \dfrac{x^2}{y^2}$

4. $f(x, y) = \cos(xy)$

5. $f(x, y) = \cos x \cos y$

6. $f(x, y) = \dfrac{\cos x}{\cos y}$

7. $f(x, y, z) = xy^2 z^3$

8. $f(x, y, z) = \cos(xyz)$

9. Let $f(x) = x^2$ and $x_0 = 3$.

(a) Let L be the linear approximation to f at x_0. Show that $L(x) = 9 + 6(x - 3)$.

(b) Plot L and f on the same axes. (Choose your own plotting window.) Supposedly, "L linearly approximates f near x_0." How do the graphs illustrate this?

(c) Find an interval $a \le x \le b$ on which $|f(x) - L(x)| < 0.01$. (On this interval, $L(x)$ approximates $f(x)$ within 0.01.) [HINT: This can be done either graphically, by zooming, or symbolically, by solving inequalities.]

10. Redo Exercise 9, but let $f(x) = \sqrt{x}$ and $x_0 = 9$.

11. Let $f(x) = x^2$.

   (a) $f'(3) = 6$. What does this mean about the graph of f? (Use a graph of $f(x)$ for $2.5 \le x \le 3.5$ to illustrate your answer.)

   (b) Plot $f(x)$ over the interval $2.5 \le x \le 3.5$. On your graph, draw the secant lines from $x = 3$ to $x = 3.5$ and from $x = 3$ to $x = 3.1$.

   (c) Find the average rate of change $\Delta y / \Delta x$ of f over the intervals $[3, 3.5]$ and $[3, 3.1]$.

   (d) Find the limit $\displaystyle \lim_{h \to 0} \frac{f(3+h) - f(3)}{h}$. Interpret the answer as a derivative. Does the answer agree with other information already given?

12. Redo Exercise 11, but use $f(x) = x^2 - x$.

Some values of a function $g(x, y)$ are tabulated below. Use the table to answer the questions in Exercises 13–16.

Values of $g(x,y)$				
y ╲ x	**−0.0100**	**0.0000**	**0.0100**	**0.0200**
1.02	2.0603	2.0604	2.0603	2.0600
1.01	2.0300	2.0301	2.0300	2.0297
1.00	1.9999	2.0000	1.9999	1.9996
0.99	1.9700	1.9701	1.9700	1.9697
⋮	⋮	⋮	⋮	⋮
0.02	0.0203	0.0204	0.0203	0.0200
0.01	0.0100	0.0101	0.0100	0.0097
0.00	−0.0001	0.0000	−0.0001	−0.0004
−0.01	−0.0100	−0.0099	−0.0100	−0.0103
y ╲ x	**0.9900**	**1.0000**	**1.0100**	**1.0200**
1.02	1.0803	1.0604	1.0403	1.0200
1.01	1.0500	1.0301	1.0100	0.9897
1.00	1.0199	1.0000	0.9799	0.9596
0.99	0.9900	0.9701	0.9500	0.9297
⋮	⋮	⋮	⋮	⋮
0.02	−0.9597	−0.9796	−0.9997	−1.0200
0.01	−0.9700	−0.9899	−1.0100	−1.0303
0.00	−0.9801	−1.0000	−1.0201	−1.0404
−0.01	−0.9900	−1.0099	−1.0300	−1.0503

13. Use the table to estimate the partial derivatives $g_x(1, 1)$ and $g_y(1, 1)$.

14. It's true that $g_x(0, 0) = 0$ and $g_y(0, 0) = 1$. How do the table entries reflect these facts?

15. (a) Consider the linear function $L(x, y) = 0 + 0x + 1y = y$. Show that $L_x(0, 0) = g_x(0, 0) = 0$, $L_y(0, 0) = g_y(0, 0) = 1$, and $L(0, 0) = g(0, 0) = 0$.

   (b) Fill in the following table of values for the function L from part (a). Compare your results with the tabulated values of g. (The results show how L linearly approximates g near $(0, 0)$.)

Values of $L(x, y)$					
y ╲ x	**−0.02**	**−0.01**	**0.00**	**0.01**	**0.02**
0.02					
0.01					
0.00					
−0.01					
−0.02					

16. Find the linear function $M(x, y)$ such that (i) $M(1, 1) = g(1, 1)$, (ii) $M_x(1, 1) = g_x(1, 1)$, and (iii) $M_y(1, 1) = g_y(1, 1)$. [HINT: One approach is to write $M(x, y) = a + b(x - 1) + c(y - 1)$ and then use the conditions to find values for a, b, and c.]

17. Let $f(x, y) = \sin y + 2$. (The formula is independent of x.)

   (a) Plot f; use the domain $-5 \le x \le 5$, $-5 \le y \le 5$. How does the shape of the graph reflect the fact that f is independent of x? (In Section M.1 we called such graphs cylinders.)

   (b) Find $f_x(x, y)$ and $f_y(x, y)$. How do the answers reflect the fact that f is independent of x?

   (c) Find the linear approximation function L for f at the point $(0, 0)$. How does its form reflect the fact that f is independent of x?

18. Let f be the function in Example 3.

   (a) Use the contour map in Figure 1 to estimate the partial derivatives $f_x(1.5, 1.5)$ and $f_y(1.5, 1.5)$.

   (b) Use the formula $f(x, y) = x^2 - 3xy + 6$ to find $f_x(1.5, 1.5)$ and $f_y(1.5, 1.5)$ exactly.

   (c) Use results of part (b) to find the linear approximation $L(x, y)$ to $f(x, y)$ at $(1.5, 1.5)$.

19. Let $f(x, y) = \sin x$.

   (a) Draw a contour map of f in the rectangle $-\pi \le x \le \pi$, $-2 \le y \le 2$. Show the level curves that correspond to $z = \pm 1$, $z = \pm 0.75$, $z = \pm 0.5$, $z = \pm 0.25$, and $z = 0$.

   (b) Use the level-curve diagram to estimate $f_x(0, 0)$ and $f_y(0, 0)$.

   (c) Use the level-curve diagram to estimate $f_x(\pi/2, 0)$ and $f_y(\pi/2, 0)$.

   (d) The formula shows that $f_y(x, y) = 0$ for all (x, y). How does the contour map reflect this fact?

(e) The formula shows that $f_x(x, y)$ is independent of y. How does the contour map of f reflect this fact?

20. Let $f(x, y) = \cos y$.

(a) Draw a contour map of f in the rectangle $-\pi \le x \le \pi$, $-2 \le y \le 2$. Show the level curves that correspond to $z = \pm 1$, $z = \pm 0.75$, $z = \pm 0.5$, $z = \pm 0.25$, and $z = 0$.

(b) Use the level-curve diagram to estimate $f_x(0, 0)$ and $f_y(0, 0)$.

(c) Use the level-curve diagram to estimate $f_x(\pi/2, 0)$ and $f_y(\pi/2, 0)$.

(d) The formula for f shows that $f_x(x, y) = 0$ for all (x, y). How does the contour map of f reflect this fact?

(e) The formula for f shows that $f_y(x, y)$ is independent of x. How does the contour map of f reflect this fact?

21. Let $f(x, y) = 2x - 3y$.

(a) Draw a contour map of f in the rectangle $[-3, 3] \times [-3, 3]$. Show the level curves that correspond to $z = -5$, $z = -4, z = -3, \ldots, z = 4$, and $z = 5$.

(b) Use your contour map (not the formula) to find $f_x(0, 0)$ and $f_y(0, 0)$.

(c) The formula for f implies that both f_x and f_y are constant functions. How does the contour map of f reflect this fact?

(d) The formula for f implies that for any (x, y), $f_x(x, y) = 2$ and $f_y(x, y) = -3$. How does the contour map reflect the fact that $f_x(x, y)$ is positive but $f_y(x, y)$ is negative?

22. Let $f(x, y) = 2y - x$.

(a) Draw a contour map of f in the rectangle $[-3, 3] \times [-3, 3]$. Show the level curves that correspond to $z = -5$, $z = -4, z = -3, \ldots, z = 4$, and $z = 5$.

(b) Use your contour map (not the formula) to find $f_x(0, 0)$ and $f_y(0, 0)$.

(c) The formula for f implies that $f_x(x, y) = -1$ and $f_y(x, y) = 2$ for all (x, y). How does the contour map reflect these facts? In particular, how does the contour map show that $f_x(x, y)$ is negative but $f_y(x, y)$ is positive?

23. Let $f(x, y) = xy$.

(a) Find the linear approximation function L to f at $(x_0, y_0) = (2, 1)$.

(b) (Do this part by hand.) On one set of xy-axes, draw the level curves $L(x, y) = k$ for $k = 1, 2, 3, 4, 5$. On another set of axes, draw the level curves $f(x, y) = k$ for $k = 1, 2, 3, 4, 5$. (In each case, draw the curves into the square $[0, 3] \times [0, 3]$.)

(c) How do the contour maps in part (b) reflect the fact that L is the linear approximation to f at the point $(2, 1)$? Explain briefly in words.

(d) Use technology to plot contour maps of f and L in the window $[1.8, 2.2] \times [0.8, 1.2]$. Explain what you see.

24. Repeat Exercise 23 using the function $f(x, y) = x^2 - y^2$.

25. Let $f(x, y) = x^2 + y^2$.

(a) Use Figure 4 on page M-13 to estimate the partial derivatives $f_x(1, 2)$ and $f_y(1, 2)$.

(b) Check your answers to part (a) by symbolic differentiation.

(c) Use your answers from part (a) to find the linear approximation $L(x, y)$ to $f(x, y)$ at $(1, 2)$.

(d) On one set of axes, plot the level curves $L(x, y) = k$ and $f(x, y) = k$ for $k = 3, 4, 5, 6, 7$. (Use the window $[0, 3] \times [0, 3]$.) What's special about the point $(1, 2)$?

26. Let $f(x, y) = \sin x + 2y + xy$.

(a) Find the partial derivatives $f_x(x, y)$ and $f_y(x, y)$; then evaluate $f_x(0, 0)$ and $f_y(0, 0)$.

(b) Find a linear function $L(x, y) = a + bx + cy$ such that $L_x(0, 0) = f_x(0, 0)$, $L_y(0, 0) = f_y(0, 0)$, and $L(0, 0) = f(0, 0)$.

(c) Complete the following table (round answers to four decimals). How do the answers reflect the fact that L approximates f closely near $(0, 0)$?

(x, y)	$(0, 0)$	$(0.01, 0.01)$	$(0.1, 0.1)$	$(1, 1)$
$f(x, y)$				
$L(x, y)$				

(d) Use technology to draw contour plots of both f and L in the rectangle $-1 \le x \le 1$, $-1 \le y \le 1$. Label several contours on each. How do the pictures reflect the fact that L approximates f closely near $(0, 0)$?

In Exercises 27–30, find the linear function L that linearly approximates f at the given point (x_0, y_0). (If possible, check your answers graphically by plotting both f and L near (x_0, y_0).)

27. $f(x, y) = x^2 + y^2$; $(x_0, y_0) = (2, 1)$.

28. $f(x, y) = x^2 + y^2$; $(x_0, y_0) = (0, 0)$.

29. $f(x, y) = \sin x + \sin y$; $(x_0, y_0) = (0, 0)$.

30. $f(x, y) = \sin x \sin y$; $(x_0, y_0) = (0, 0)$.

31. Let $f(x, y)$ be a differentiable function of two variables, let (x_0, y_0) be any point in its domain, and let $L(x, y)$ be the linear approximation to f at (x_0, y_0). Show that if f is independent of one of the variables—say, x—then so is L.

32. Suppose that $f(3, 4) = 25$, $f_x(3, 4) = 6$, $f_y(3, 4) = 8$, and $f(4, 5) = 41$.

(a) Find a linear function $L(x, y)$ that approximates f as well as possible near $(3, 4)$.

(b) Use L to estimate $f(2.9, 3.9)$, $f(3.1, 4.1)$, and $f(4, 5)$.

(c) Could f itself be a linear function? Why or why not?

33. Suppose that $g(3, 4) = 5$, $g_x(3, 4) = 3/5$, $g_y(3, 4) = 4/5$, and $g(4, 5) = \sqrt{41}$.

(a) Find a linear function $L(x, y)$ that approximates g as well as possible near $(3, 4)$.

(b) Use L to estimate $g(2.9, 4.1)$ and $g(4, 5)$.

(c) Could g be a linear function? Why or why not?

34. Let $f(x, y) = |y| \cos x$. This exercise explores the fact that the partial derivatives of a function may or may not exist at a given point.

(a) Use technology to plot $z = f(x, y)$ over the rectangle $[-5, 5] \times [-5, 5]$. The graph suggests that there may be trouble with partial derivatives where $y = 0$, that is, along the x-axis. How does the graph suggest this? Which partial derivative (f_x or f_y) seems to be in trouble?

(b) Use the definition to show that $f_y(0, 0)$ does not exist. In other words, explain why the limit $\lim\limits_{h \to 0} \dfrac{f(h, 0) - f(0, 0)}{h}$ does not exist.

(c) Show that $f_x(0, 0)$ does exist; find its value. How does the result appear on the graph?

(d) Use the limit definition to show that $f_y(0, \pi/2)$ does exist; find its value.

(e) How does the graph reflect the result of part (d)? (You may need to do some experimenting with the graph to answer this.)

(f) Find a function $g(x, y)$ for which $g_y(0, 0)$ exists but $g_x(0, 0)$ does not. Use technology to plot its graph.

35. Let $f(x, y) = |x| \, y$.

(a) By experimenting with graphs (use technology!), try to guess where $f_x(x, y)$ and $f_y(x, y)$ do exist and where they don't. (No proofs are needed.)

(b) Find $f_x(0, 0)$.

(c) Explain why $f_x(0, 1)$ does not exist.

M.4 OPTIMIZATION AND PARTIAL DERIVATIVES: A FIRST LOOK

Optimization—finding maximum and minimum values of a function—is as important for multivariable functions as it is for one-variable functions. Functions of several variables are more complicated, but derivatives are still the crucial tool.

A one-variable review

For a one-variable differentiable function $y = f(x)$ on an interval I, finding maximum and minimum values is relatively straightforward. Maxima and minima are found only at stationary points—where $f'(x) = 0$—or at the endpoints (if any) of the interval I. Usually, only a few such "candidate" points exist, and we can check directly which one produces, say, the largest value of f.

"Extremum" (singular) means "either maximum or minimum"; "extrema" is the plural.

A simple idea lies behind all talk of derivatives and extrema: ◄ At a local maximum or minimum point x_0, the graph of a differentiable function f must be "flat," and so $f'(x_0) = 0$. However, a stationary point x_0 might be (i) a local maximum point, (ii) a local minimum point, or (iii) neither. (The function $f(x) = x^3$ at $x = 0$ illustrates the "neither" case.)

For simple one-variable functions, deciding which of (i)–(iii) actually holds is easy. One strategy is to check the sign of $f''(x_0)$. If, say, $f''(x_0) < 0$, then f is concave down at x_0, and so f has a local maximum there. Alternatively, we might just plot f and see directly how it behaves near x_0. ◄

Without technology, plotting f may be hard.

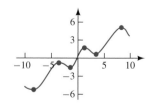

FIGURE 1
Local vs. global extrema

Local talk.   When is a maximum or minimum "local"? When is it "global"? Figure 1 shows the difference:

All six bulleted points (three maximum points and three minimum points) correspond to *local* extrema. At each bulleted point $(x_0, f(x_0))$, $f(x_0)$ is either highest or lowest among *nearby* points $(x, f(x))$. (For present purposes, exactly *how* nearby doesn't matter.) Only the first and last points are *global* extrema because only they represent the largest and smallest values of $f(x)$ among all possibilities shown in the picture. On a larger domain interval—say, $-15 \le x \le 15$—these extrema might not be global.

Local maxima and minima are conceptually simpler than the global variety, are readily recognizable on graphs, and are often convenient to locate symbolically, using derivatives. With all these advantages, local extrema are usually the main tools used to solve optimization problems.

In several variables

If a one-variable function has a local maximum or minimum at x_0, then the ordinary derivative (if it exists) must be zero at x_0. If a two-variable function $f(x, y)$ has a local maximum at (x_0, y_0), then *both* partial derivatives (if they exist) must vanish there:

$$f_x(x_0, y_0) = 0; \qquad f_y(x_0, y_0) = 0.$$

In words, (x_0, y_0) is a stationary point of f. ➡

This is the natural extension of the one-variable idea.

The result sounds plausible, but why is it true? Consider the case of a local maximum. Geometrically, the surface $z = f(x, y)$ has a "peak" above the domain point (x_0, y_0). If we slice the surface with any vertical plane, say the plane $y = y_0$, then the resulting curve—that is, the trace of the surface $z = f(x, y)$ in the plane $y = y_0$—has the equation $z = f(x, y_0)$, and this curve must have a peak at $x = x_0$. We know from one-variable calculus that if the function $x \to f(x, y_0)$ is differentiable, then its derivative must be zero at x_0. In other words,

$$\frac{dz}{dx}(x_0) = f_x(x_0, y_0) = 0.$$

For similar reasons, $f_y(x_0, y_0) = 0$. Here's the general fact:

THEOREM 1 (Extreme points and partial derivatives) Suppose that $f(x, y)$ has a local maximum or a local minimum at (x_0, y_0). If both partial derivatives exist, then

$$f_x(x_0, y_0) = 0 \quad \text{and} \quad f_y(x_0, y_0) = 0.$$

Caution The theorem is useful, but what it *doesn't* say is also important. It does *not* guarantee, in particular, that at a stationary point f *must* assume either a local maximum or a local minimum value. The next example illustrates both sides of this coin.

EXAMPLE 1 Let $f(x, y) = x^2 + y^2$, and let $g(x, y) = xy$. Find all the stationary points of f and g. What happens at each one?

Solution Finding the partial derivatives is easy:

$$f_x(x, y) = 2x; \quad f_y(x, y) = 2y; \quad g_x(x, y) = y; \quad g_y(x, y) = x.$$

Both f and g, therefore, are stationary only at the origin $(0, 0)$. For f the origin is a *minimum* point, because $f(x, y) = x^2 + y^2 \ge 0$ for all (x, y). For g, however, the origin is neither a maximum nor a minimum point, because $g(x, y)$ assumes both positive and negative values near $(0, 0)$. (For example, $g(0.1, -0.1) < 0$, but $g(0.1, 0.1) > 0$.) Contour maps of f and g illustrate their very different behavior near the stationary point. (In

Figures 2 through 5, the level curves are the edges of the shaded regions. Note the key to the right of each contour map.) Figure 2 shows f.

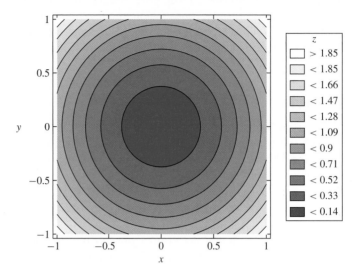

FIGURE 2
A contour map of $f(x, y) = x^2 + y^2$

Level curves of f are circles centered on the "basin" at $(0, 0)$. Thus, the figure suggests a minimum point at $(x, y) = (0, 0)$. The suggestion is correct: $f(0, 0) = 0$, and $f(x, y) \geq 0$ for all (x, y); thus, f assumes a local (and even global) minimum value at $(0, 0)$. ◄ Now look at g near its stationary point (Figure 3):

Look carefully; notice, in particular, that darker regions are "lower."

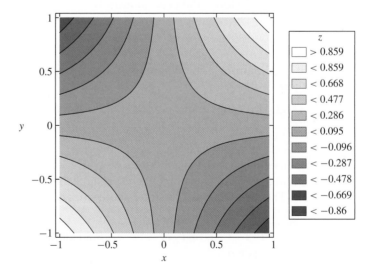

FIGURE 3
A contour map of $g(x, y) = xy$

Think about this carefully. Can you see the saddle in the contour map?

Level curves of g show a "saddle" at $(0, 0)$. (If the surface were literally a saddle, the horse would be walking either "northeast" or "southwest.") ◄ The surface rises above the first and third quadrants and falls below the second and fourth quadrants. ∎

Different types of stationary points Here is one moral of the preceding example: Although the basic strategy for optimizing a function—find the stationary points and analyze them—is exactly the same for functions of one and of several variables, the situation is usually more complicated for functions of several variables. For one thing, finding stationary points may be harder; for another, functions of several variables can behave in more complicated ways near a stationary point. This makes multivariable optimization harder, but also more interesting.

As we did for functions of one variable, we identify three main types of stationary points for a function $f(x, y)$. (The definitions for a three-variable function $g(x, y, z)$ are almost identical.)

- **Local minimum point**   A stationary point (x_0, y_0) is a **local minimum point** for f if $f(x, y) \geq f(x_0, y_0)$ for all (x, y) near (x_0, y_0). (A little more formally, $f(x, y) \geq f(x_0, y_0)$ for all (x, y) in some rectangle surrounding (x_0, y_0).) In this case, we say that f *assumes a local minimum value* at (x_0, y_0). In Example 1, $(0, 0)$ is a local minimum point for $f(x, y) = x^2 + y^2$.

- **Local maximum point**   A stationary point (x_0, y_0) is a **local maximum point** for f if $f(x, y) \leq f(x_0, y_0)$ for all (x, y) near (x_0, y_0). In this case, we say that f *assumes a local maximum value* at (x_0, y_0). Figure 4 illustrates a local maximum point for the function $h(x, y) = 1 - x^2 - y^2$.

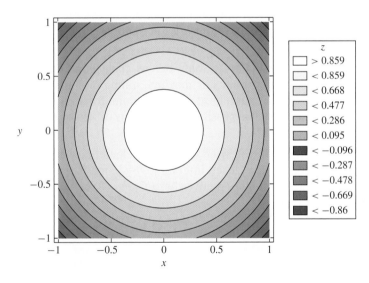

FIGURE 4
Contour plot of $1 - x^2 - y^2$

- **Saddle point**   A stationary point (x_0, y_0) is a **saddle point** for f if f assumes neither a local maximum nor a local minimum at (x_0, y_0). In Example 1, $(0, 0)$ is a saddle point for $g(x, y) = xy$.

As Figure 5 shows, all three possibilities can coexist in close quarters. The function in question is $f(x, y) = \cos(x) \sin(y)$. ➤

Look in the picture for a stationary point of each type just listed.

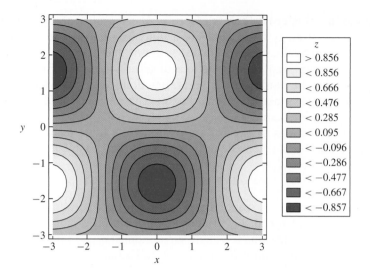

FIGURE 5
Contour plot of cos(x) sin(y)

Optimization: More to the story

There's much more to the story of optimizing functions of several variables; we've taken only a first, short look. Two main questions arise in multivariable calculus courses.

Second derivative tests   In single-variable calculus, the second derivative f'' is sometimes used to classify stationary points as maxima, minima, or neither. For example, if $f'(x_0) = 0$ and $f''(x_0) > 0$, then x_0 is a local minimum point for f. A similar but (necessarily) more sophisticated approach is possible for functions of two or more variables.

Extremes on the boundary   Recall what happens in elementary calculus for a differentiable function $f(x)$ defined on a closed interval $[a, b]$: f may assume its maximum and minimum values either at a stationary point (where $f'(x) = 0$) or at either of the endpoints $x = a$ and $x = b$.

   The situation is similar for a function of two variables defined on a region, such as a rectangle or a circle, that has a definite "edge," or boundary: $f(x, y)$ may assume its maximum and minimum either at a stationary point or somewhere on the boundary of the region. We illustrate the situation, and one way to approach it, with a simple example.

E X A M P L E 2   Where on the rectangle $R = [-1, 1] \times [-1, 1]$ does $g(x, y)$ assume its minimum and maximum values?

Solution   We saw in Example 1, page M-27, that g has only one stationary point, a saddle point, in the interior of R. Therefore, the maximum and minimum values of g must occur somewhere on the boundary of R. A look at the contour plot of g (notice the symmetry) shows that it's enough to look along *any* boundary edge of R, such as the right edge. On this edge we have $x = 1$, and so g behaves like a function of just one variable: $g(x, y) = g(1, y) = y$. Clearly, $g(1, y) = y$ is largest at $y = 1$ and smallest at $y = -1$. We therefore conclude that $g(1, 1) = 1$ and $g(1, -1) = -1$ are, respectively, maximum and minimum values of g on R.   ■

BASIC EXERCISES

1. Let $g(x, y) = xy$. (Its contour map is shown in Figure 3.) To an ant walking along the surface $z = g(x, y)$ from lower left to upper right, the origin seems to be a low spot; another ant walking from upper left to lower right would experience the origin as a high spot.

    (a) An ant walks along the surface from $(0, -1)$ to $(0, 1)$. How does the ant's altitude change along the way?

    (b) Another ant walks along the surface from $(0.5, -1)$ to $(0.5, 1)$. How does the ant's altitude change along the way? Where is the ant highest? How high is the ant there?

2. See Figure 5, the contour map of $f(x, y) = \cos x \sin y$.

    (a) The surface $z = f(x, y)$ resembles an egg carton. Where do the eggs go?

    (b) From the figure alone, estimate the coordinates of a local minimum point, a local maximum point, and a saddle point.

    (c) Use the formula $f(x, y) = \cos x \sin y$ to find (exactly) all the stationary points of f in the rectangle $R = [-3, 3] \times [-3, 3]$.

    (d) Find the maximum and minimum values of f in the rectangle $R = [-3, 3] \times [-3, 3]$.

3. Consider the function $f(x, y) = x(x - 2) \sin y$. Here's a contour map:

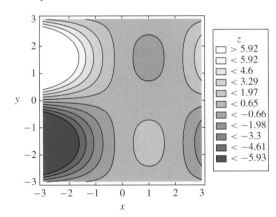

    (a) The function f has two stationary points along the line $x = 1$. Use the figure to estimate their coordinates. What type is each one?

    (b) There are four stationary points inside the rectangle $R = [-3, 3] \times [-3, 3]$. Use the formula for f to find all four.

(c) The contour plot shows that f assumes its maximum and minimum values on $R = [-3, 3] \times [-3, 3]$ somewhere along the left boundary—that is, where $x = -3$. Find these maximum and minimum values.
    [HINT: If $x = -3$, then $f(x, y) = f(-3, y) = 15 \sin y$. This is a function of one variable, defined for $-3 \le y \le 3$.]

In Exercises 4–7, use the formula to find all stationary points of the function. Then use technology (e.g., a properly chosen contour plot or surface plot) to decide what type of stationary point each one is.

4. $f(x, y) = -x^2 - y^2$

5. $f(x, y) = x^2 - y^2$

6. $f(x, y) = 3x^2 + 2y^2$

7. $f(x, y) = xy - y - 2x + 2$

8. Consider the linear function $L(x, y) = 1 + 2x + 3y$. Does L have any stationary points? If so, what type are they? If not, why not?

9. Consider the linear function $L(x, y) = a + bx + cy$, where a, b, and c are constants.

    (a) The graph of L is a plane. Which planes have stationary points? For these planes, where are the stationary points?

    (b) Under what conditions on a, b, and c will L have stationary points? In this case, where are the stationary points? Reconcile your answers with those in part (a).

10. Let $f(x, y) = x^2$. The graph of f is a cylinder unrestricted in the y-direction.

    (a) Use technology to plot the surface $z = f(x, y)$. Where in the xy-plane are the stationary points? What type are they? [HINT: There's a whole line of stationary points.]

    (b) Use partial derivatives of f to find all the stationary points. Reconcile your answer with part (a).

In Exercises 11–13, give an example of a function with the specified properties.
[HINTS: (i) See Exercise 10 for ideas. (ii) Check your answers by plotting.]

11. Every point on the x-axis is a local minimum point.

12. Every point on the line $x = 1$ is a local maximum point.

13. $(3, 4)$ is a local minimum and the function is nonconstant.

M.5 MULTIPLE INTEGRALS AND APPROXIMATING SUMS

The last two sections introduced *derivatives* of functions of several variables and a few of their most basic properties. The next few sections, on *integrals*, continue our flying tour of the basic calculus of functions of two or three variables. This section concerns mainly what multiple integrals are and what they mean. In the following two sections we'll consider more systematically how to calculate multiple integrals.

Integrals and sums

All integrals—single, double, triple, or whatever—are defined to be certain limits of approximating sums (sometimes known as Riemann sums). This important idea is always studied in single-variable calculus, but it may be quickly (and perhaps gratefully) forgotten. Readers whose memories are vague on this score have an excellent excuse: Although integrals are *defined* as limits of approximating sums, they are often *calculated* in an entirely different way, using antiderivatives. Here's a typical calculation: ◄

There's not a Riemann sum in sight.

$$\int_0^1 x^2 \, dx = \frac{x^3}{3} \Bigg]_0^1 = \frac{1}{3}.$$

This method of evaluating an integral—find an antiderivative for the integrand and plug in the endpoints—works just fine, thanks to the fundamental theorem of calculus.

So why bother with approximating sums? Here are two good reasons:

- **Antiderivative trouble** The antiderivative method depends on finding a convenient antiderivative of the integrand. Unfortunately, not every function, even in single-variable calculus, *has* an "elementary" antiderivative, that is, an antiderivative with a symbolic formula built from the usual ingredients. The simple-looking function $f(x) = \sin(x^2)$ is an example. ◄ The best we can do with the integral

Spend a moment looking for an antiderivative formula. Nothing works.

$$I = \int_0^1 \sin(x^2) \, dx,$$

therefore, is to approximate it with some sort of sum using, say, the left rule, the midpoint rule, or the trapezoid rule. (For the record, approximating I with a trapezoid-rule sum with 10 subdivisions gives $I \approx 0.311$.)

- **What integrals mean** The fundamental theorem (when it works) makes *calculating* integrals easy, but approximating sums may illustrate more clearly what the answers mean. Figure 1, for instance, illustrates the sense in which a midpoint-rule sum with four subdivisions approximates the area bounded by the curve $y = x^2$ from $x = 0$ to $x = 1$.

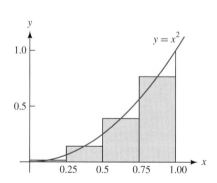

FIGURE 1

A midpoint estimate to $\int_0^1 x^2 \, dx$:
$M_4 = 63/192 \approx 0.328$

The basic idea of an integral as a limit of approximating sums is much the same for functions of two or more variables as for functions of one variable.

Integrals and approximating sums: A review Let's review the main single-variable objects and notations that arise on the way to defining the integral of a function f over an interval $[a, b]$. The first step is to form Riemann sums.

> **DEFINITION** Let $[a, b]$ be partitioned into n subintervals by any $n+1$ points
>
> $$a = x_0 < x_1 < x_2 < \cdots < x_{n-1} < x_n = b;$$
>
> let $\Delta x_i = x_i - x_{i-1}$ denote the width of the ith subinterval. Within each subinterval $[x_{i-1}, x_i]$, choose any point c_i. The sum
>
> $$\sum_{i=1}^{n} f(c_i)\, \Delta x_i = f(c_1)\, \Delta x_1 + f(c_2)\, \Delta x_2 + \cdots + f(c_n)\, \Delta x_n$$
>
> is a **Riemann sum** with n subdivisions for f on $[a, b]$.

Left-rule, right-rule, and midpoint-rule approximating sums all fit this definition. Each of these sums is built from a partition of $[a, b]$ into subintervals of equal length and some consistent scheme for choosing the sampling points c_i. (For left, right, and midpoint sums, respectively, we choose each c_i as the left endpoint, the right endpoint, or the midpoint of the ith subinterval.)

Graphical and numerical intuition suggest that all of these approximating sums (and others) should converge to some fixed number. Geometrically speaking, this number measures the signed area bounded by the f-graph from $x = a$ to $x = b$.

The limit definition of integral makes these ideas precise.

> **DEFINITION** Let the function f be defined on the interval $[a, b]$. The **integral** of f over $[a, b]$, denoted by $\int_a^b f(x)\, dx$, is the number to which all Riemann sums S_n tend as n tends to infinity and as the widths of all subdivisions tend to zero. In symbols,
>
> $$\int_a^b f(x)\, dx = \lim_{n \to \infty} S_n = \lim_{n \to \infty} \sum_{i=1}^{n} f(c_i)\, \Delta x_i$$
>
> if the limit exists.

Honesty dictates a brief admission: The limit in the definition, taken at face value, is a slippery customer. Understanding every ramification of permitting arbitrary partitions and sampling points, for example, can be tricky. Fortunately, these issues need not trouble us for the usual well-behaved functions (e.g., continuous functions) of single-variable and multivariable calculus. For such functions, almost any respectable sort of approximating sum does what we'd expect—approaches the true value of the integral as n tends to infinity.

Two variables: Double integrals

Most of the differences between single-variable integrals and multivariable integrals are technical rather than theoretical. Indeed, the definitions of

$$\iint_R f(x, y)\, dA \quad \text{and} \quad \int_a^b f(x)\, dx$$

are almost identical. Now f is a function of two variables, and R is a region—a rectangle, in the simplest case—in the xy-plane. (The mechanics of evaluating these two types of integrals by antidifferentiation are quite different, on the other hand.)

First, let's list the ingredients that go into defining the **double integral** $\iint_R f(x, y)\, dA$. In each case, look for similarities to, and differences from, the one-variable situation.

- **R, the region of integration** In one variable, the region of integration is always an interval $[a, b]$ in the domain of f; this is implicit in the notation $\int_a^b f(x)\, dx$. In two variables, by contrast, the region of integration, denoted by R, may be almost *any* two-dimensional subset of the plane. In the simplest cases, R is a rectangle $[a, b] \times [c, d]$, and we sometimes write

$$\int_a^b \int_c^d f(x, y) \, dy \, dx, \qquad \text{not} \qquad \iint_R f(x, y) \, dA.$$

(The first notation suggests a fact we'll see in the next section—double integrals can sometimes be calculated by integrating "one variable at a time.")

- **Partitions**   In one variable, we partition an interval $[a, b]$ by cutting it, perhaps unevenly, into smaller intervals with endpoints $a = x_0 < x_1 < x_2 < \cdots < x_n = b$. The "size" of the ith subinterval is simply its length, Δx_i. In two variables, we do much the same thing: We chop the plane region R into m smaller regions $R_1, R_2, R_3, \ldots, R_m$, perhaps of different sizes and shapes. The "size" of a subregion R_i is now taken to be its *area*, denoted by ΔA_i.

    In practice—whatever the number of variables—it's usually convenient to choose the partition in some consistent way. In one variable, using equal-length subintervals is simplest. An analogous procedure in two variables, if R is a rectangle $[a, b] \times [c, d]$, is to cut R by an n-by-n grid in each direction, producing n^2 rectangular subregions in all. (This isn't the only possibility. Another alternative is to cut R into small squares.)

- **Approximating sums**   In one variable, an approximating sum has the form

$$f(c_1) \, \Delta x_1 + f(c_2) \, \Delta x_2 + \cdots + f(c_n) \, \Delta x_n = \sum_{i=1}^n f(c_i) \, \Delta x_i,$$

where each c_i is a sample point chosen from the ith subinterval. A two-variable approximating sum is similar. From each subregion R_i we choose a sampling point $P_i(x_i, y_i)$ and then form the approximating sum

$$S_m = f(P_1) \, \Delta A_1 + f(P_2) \, \Delta A_2 + \cdots + f(P_m) \, \Delta A_m = \sum_{i=1}^m f(P_i) \, \Delta A_i,$$

where ΔA_i is the area of R_i.

Figure 2 illustrates the situation for the function $f(x, y) = x + y$ on the rectangle $R = [0, 4] \times [0, 4]$ with $n^2 = 4^2 = 16$ subdivisions. In each subinterval, sampling is done at the corner closest to the origin.

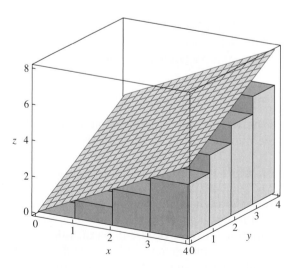

FIGURE 2
The surface $z = x + y$ and a Riemann approximation

The figure shows (among other things) that the approximating sum gives an estimate to the volume bounded above by the surface $z = f(x, y)$ and below by the rectangle $R = [0, 4] \times [0, 4]$. A close look and some back-of-the-envelope calculations show that the approximating sum shown adds up to 48.

It's reasonable to expect these estimates to converge to the "true" volume as n (the number of subdivisions in each direction) increases to infinity. A table of values supports this expectation.

Approximating sums for various m	
Number of subdivisions (m)	Approximating sum (S_m)
2^2	32.00
4^2	48.00
8^2	56.00
12^2	58.67
16^2	60.00
20^2	60.80
24^2	61.33
28^2	61.71
32^2	62.00

Figure 3 shows another approximating sum for the same function over the same interval. This time, however, sampling is done at the *midpoint* of each subrectangle and the approximating sum adds up to 64.

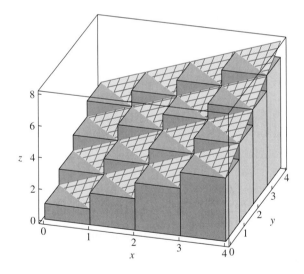

FIGURE 3
The surface $z = x + y$ and a midpoint approximation

- **What's dA?** The symbol "dA" in the double integral resembles the "dx" that appears in single integrals. The "A" reminds us of *area*.

EXAMPLE 1 Consider the double integral $I = \iint_R f(x, y)\,dA$, where $f(x, y) = x + y$ and $R = [0, 4] \times [0, 4]$. Calculate, by hand, an approximating sum S_4 with four equal subdivisions (two in each direction). In each square subregion, evaluate f at the corner nearest to the origin.

Solution   All four subregions are squares with edge length 2 and therefore area 4; all corners have integer coordinates. The sampling points just described are $P_1 = (0, 0)$, $P_2 = (2, 0)$, $P_3 = (0, 2)$, and $P_4 = (2, 2)$. The desired approximating sum is therefore

$$S_4 = \sum_{i=1}^{4} f(P_i)\,\Delta A_i = 0 \cdot 4 + 2 \cdot 4 + 2 \cdot 4 + 4 \cdot 4 = 32. \qquad \blacksquare$$

Some additional subtleties—which we ignore—may arise for very ill-behaved functions.

The integral as a limit   We defined the one-variable integral $\int_a^b f(x)\,dx$ as a limit of approximating sums. The double integral $\iint_R f(x, y)\,dA$ can be defined in a similar way. The following definition is adequate for well-behaved functions $f(x, y)$. ◄

> **DEFINITION**   Let the function $f(x, y)$ be defined on the region R and let S_m be an approximating sum with m subdivisions. Let I be a number such that S_m tends to I whenever m tends to infinity and the diameter of all subdivisions tends to zero. Then I is the double integral of f over R, and we write
>
> $$I = \iint_R f(x, y)\,dA = \lim_{m \to \infty} \sum_{i=1}^{m} f(P_i)\,\Delta A_i.$$

As in the one-variable setting, the limit definition—although crucial to understanding integrals and often useful for approximating them—almost never lends itself to calculating integrals *exactly*. Fortunately, there are methods based on antidifferentiation for this purpose. We'll see some soon.

Triple sums and triple integrals

The idea of integral can be extended to dimension three (and even higher dimensions). We will consider only the simplest case: the **triple integral** of a function $g(x, y, z)$ over a rectangular parallelepiped (a "brick," to put it humbly) $R = [a, b] \times [c, d] \times [e, f]$, denoted by either

$$\iiint_R g(x, y, z)\,dV \qquad \text{or} \qquad \int_a^b \int_c^d \int_e^f g(x, y, z)\,dz\,dy\,dx.$$

As for double integrals, the second notation suggests, correctly, that such integrals can sometimes be calculated one variable at a time.

Triple integrals, just like single and double integrals, are defined formally as limits of approximating sums. An approximating sum in three dimensions is formed by subdividing a rectangular solid region R into m smaller rectangular subregions R_i, each with volume ΔV_i; choosing a sampling point P_i in each subregion; and then evaluating the sum

$$\sum_{i=1}^{m} f(P_i)\,\Delta V_i.$$

The triple integral is then defined as the limit of such sums as the diameter of all subregions tends to zero.

EXAMPLE 2   Consider the triple integral $I = \iiint_R g(x, y, z)\,dV$, where $g(x, y, z) = x + y + z$ and $R = [0, 2] \times [0, 2] \times [0, 2]$. Calculate an approximating sum S_8 with eight equal subdivisions (two in each direction). In each cubical subregion, evaluate g at the corner nearest to the origin.

Solution All eight subregions are cubes with edge length 1 and (therefore) volume 1; all the corners have integer coordinates. The sampling points just described are
$P_1 = (0, 0, 0)$, $P_2 = (1, 0, 0)$, $P_3 = (0, 1, 0)$, $P_4 = (1, 1, 0)$, $P_5 = (0, 0, 1)$, $P_6 = (1, 0, 1)$, $P_7 = (0, 1, 1)$, and $P_8 = (1, 1, 1)$. The approximating sum is therefore

$$S_8 = \sum_{i=1}^{8} g(P_i)\, \Delta V_i = 0 + 1 + 1 + 2 + 1 + 2 + 2 + 3 = 12.$$ ■

Interpreting multiple integrals

Integrals can be interpreted geometrically, physically, or in other ways. A sampler of possibilities follows.

Double integrals and volume The standard geometric interpretation of a single-variable integral $\int_a^b f(x)\, dx$ is in terms of *area*: If $f(x) \geq 0$, then $\int_a^b f(x)\, dx$ is the area of the region bounded above by the curve $y = f(x)$, bounded below by the interval $[a, b]$ in the x-axis, and having vertical sides. In a similar vein, as we've already seen from pictures, if $f(x, y) \geq 0$ for (x, y) in R, then the double integral $\iint_R f(x, y)\, dA$ measures the *volume* of the three-dimensional solid bounded above by the surface $z = f(x, y)$, bounded below by the region R in the xy-plane, and having sides perpendicular to the xy-plane.

Double integrals and area If R is a region in the xy-plane and g is the constant function $g(x, y) = 1$, then (as the preceding paragraph says) the integral $\iint_R 1\, dA$ represents the volume of the solid S bounded below by R and having vertical sides and *constant height* 1. Recall, however, that the volume of any such "cylindrical" solid S is the *area* of the base times the height. Therefore, in this special case the volume of S happens to be the same as its area. Thus, for any plane region R,

$$\iint_R 1\, dA = \text{area of } R.$$

Surprisingly, this fact is often of practical use in calculating areas of plane regions. The next section contains examples.

Triple integrals and volume Triple integrals pose a special problem having to do with dimensions. There isn't "room" in three-dimensional space even to plot a function $w = g(x, y, z)$—*four* variables would be needed. For this reason, interpreting triple integrals geometrically is, as a rule, difficult or impossible.

There's one important exception to this rule. For reasons similar to those explained in the preceding paragraph, integrating the constant function $g(x, y, z) = 1$ over a solid region R in xyz-space gives the volume of the region R. In symbols:

$$\iiint_R 1\, dV = \text{volume of } R.$$

Density, mass, and multiple integrals Both double and triple integrals can often be interpreted physically in the language of density and mass. (This view has the special advantage of making sense for both double and triple integrals.)

For a double integral $\iint_R f(x, y)\, dA$, one thinks of the plane region R as a flat plate with variable density; at any point (x, y), $f(x, y)$ gives the density, measured in appropriate units (grams per square centimeter, say). From this viewpoint, the double integral $\iint_R f(x, y)\, dA$ is the total mass, in grams, of the plate R.

For a triple integral $\iiint_R g(x, y, z)\, dV$, one imagines a solid region R with variable density; at any point (x, y), $g(x, y, z)$ is the solid's density, measured in appropriate units (grams per cubic centimeter, say). From this viewpoint, the triple integral $\iiint_R g(x, y, z)\, dV$ is the total mass, in grams, of the solid R.

BASIC EXERCISES

1. Calculate by hand (without technology) the midpoint sum with four subdivisions for $\int_0^1 x^2\, dx$. Then check that *Maple* agrees. Finally, compare *Maple*'s answer for 100 subdivisions.

2. Calculate by hand (without technology) the double midpoint sum with $n = 3$ (i.e., nine subdivisions in all) for the integral $\iint_R \sin x \sin y\, dA$ over the rectangle $R = [0, 1] \times [0, 1]$. Check your answer using *Maple*. Finally, compare *Maple*'s answer for $n = 10$.

3. Calculate by hand (without technology) the triple midpoint sum with $n = 2$ (i.e., eight subdivisions in all) for the triple integral $\iiint_R xyz\, dV$ over the cube $R = [0, 4] \times [0, 4] \times [0, 4]$. Check your answer using *Maple*. Finally, compare *Maple*'s answer for $n = 4$.

4. Let $f(x, y) = x + y$, let $R = [0, 4] \times [0, 4]$, and let $I = \iint_R f(x, y)\, dA$. Here is a contour plot of f:

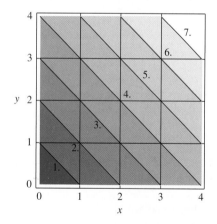

(a) Use the contour plot to evaluate a double midpoint sum for I with $n = 4$ (16 subdivisions in all).

(b) Use *Maple* to check your answer from part (a).

(c) Your answer in part (a) is, in fact, the exact value of the integral I. How does the symmetry of the contour map show this?

5. Let $f(x, y) = x^2 + y^2$, let $R = [0, 4] \times [0, 4]$, and let $I = \iint_R f(x, y)\, dA$. Here is a contour plot of f:

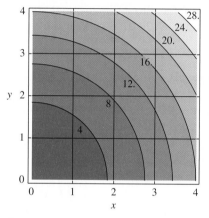

(a) Use the contour plot to evaluate a double midpoint sum for I with $n = 4$ (16 subdivisions in all).

(b) Use *Maple* to check your answer from part (a).

(c) Would you expect your answer from part (a) to overestimate or underestimate the true value of I? How can you tell?

M.6 CALCULATING INTEGRALS BY ITERATION

It works thanks to the fundamental theorem of calculus.

The preceding section was about *defining* multivariable integrals as limits of approximating sums. This section is about *calculating* multivariable integrals, using antidifferentiation. Approximating sums are conceptually simple and (with technology) easy to calculate. But approximating sums are only approximate; to evaluate integrals *exactly*, we'd like an appropriate multivariable version of the single-variable antiderivative method. ◂

How iteration works

The key idea is to integrate a multivariable function *one variable at a time*, treating other variables as constants. The process is called **iterated integration**. ➤ To start, here's an example to show *how* it works; we'll see *why* it works in a moment.

In mathspeak, "iterate" means "repeat."

EXAMPLE 1   Let $f(x, y) = x + y$, and let $R = [0, 4] \times [0, 4]$. Find $\iint_R f(x, y)\, dA$ by iterated integration. (We studied this integral in Section M.5; see the pictures and the table of values starting on page M-35.)

Solution   We integrate first in x, treating y as a constant. Watch each step carefully. ➤

Attaching variable names to the limits of integration is optional, but it can help remind us which variable is involved.

$$\iint_R f(x, y)\, dA = \int_{y=0}^{y=4} \left(\int_{x=0}^{x=4} (x + y)\, dx \right) dy$$

$$= \int_{y=0}^{y=4} \left(\frac{x^2}{2} + xy \Big]_{x=0}^{x=4} \right) dy$$

$$= \int_{y=0}^{y=4} (8 + 4y)\, dy = 64.$$

Observe:

- **The same answer**   The final answer, 64, should be familiar—it's what we estimated in Section M.5, using a midpoint approximation.

- **The answer as volume**   Interpreted as a volume, the answer says that the solid bounded below by $R = [0, 4] \times [0, 4]$, bounded above by the plane $z = x + y$, and having straight vertical walls has volume 64 cubic units. Figure 1 shows the solid in question.

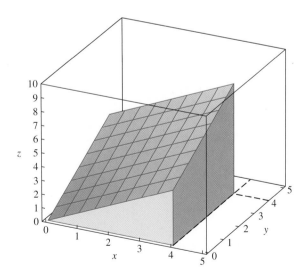

FIGURE 1
A solid bounded above by $z = x + y$

- **Checking answers with technology**   Is the answer 64 geometrically reasonable? Is it symbolically correct? Technology (e.g., *Maple*) can help with both questions. A plot like the preceding one suggests that the answer 64 is at least in the ballpark. As a further check, here is *Maple*'s version of the *symbolic* calculation:

```
> int( int( x+y, x=0..4), y=0..4 );
                    64
```

- **Work from inside out**   Iterated integrals are calculated from the inside out. The "inner" integral (the x-integral in the preceding calculation) is found first.
- **Either order works**   There's nothing sacred about integrating first in x and then in y. For well-behaved functions we can integrate in either order, and both orders give the same answer. We'll return to this question.   ■

Cross-sectional area: An intermediate function   In the preceding example, the inner integral was calculated with respect to x, with y treated as constant. The result was an intermediate function, g, of y alone; the formula is

$$g(y) = \int_0^4 (x + y)\, dx = 8 + 4y.$$

We integrated $g(y)$ with respect to y to get the final answer.

The function g has a nice geometric meaning: For any fixed y_0 in $[0, 4]$, $g(y_0)$ is the area under the curve $z = f(x, y_0)$ and above the xy-plane. In other words, $g(y)$ is the *area* of the cross section of the solid obtained by slicing with the plane $y = y_0$. Figure 2 shows the picture for $y = 2$.

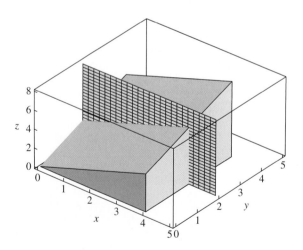

FIGURE 2
The cross section with the plane $y = 2$

Because $g(y) = 8 + 4y$, we get $g(2) = 16$; this is the area of the part of the plane inside the solid. As y runs from $y = 0$ to $y = 4$, $g(y)$ measures the area "swept out" by planes parallel to the one shown. ◄

This area increases with y, as the picture shows.

As a matter of fact, the idea of a cross-sectional-area function is not new; we used the same idea in Chapter 7 when we calculated volumes by integration. Back then, we put it like this:

FACT Suppose that a solid lies with its base on the xy-plane between the vertical planes $x = a$ and $x = b$. For all x in $[a, b]$, let $A(x)$ denote the area of the cross section at x, perpendicular to the x-axis. If $A(x)$ is a continuous function, then

$$\text{Volume} = \int_a^b A(x)\,dx.$$

Why iteration works

Why does the iteration method just demonstrated work? The preceding Fact gives some geometric feeling for the matter, at least when an integral can be thought of as the volume of a three-dimensional solid.

But not all integrals can or should be thought of in this way. A better reason why iteration works—a reason that makes sense in *any* dimension—is based on approximating sums. We describe the idea in two dimensions, but everything transfers readily to three (or even higher) dimensions. The main idea, in a nutshell, is that an approximating sum for a double integral can be grouped either along "rows" or along "columns." We first give an example and then a more general argument.

EXAMPLE 2 Let $f(x, y) = 2 - x^2 y$, let $R = [0, 1] \times [0, 1]$, and let $I = \iint_R f(x, y)\,dA$. Figure 3 is a picture of the solid whose volume is given by I:

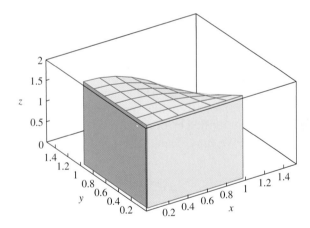

FIGURE 3
The solid under $z = 2 - x^2 y$ and over $[0, 1] \times [0, 1]$

Compute a midpoint approximating sum with 100 equal subdivisions (10 in each direction) for the integral $I = \iint_R f(x, y)\,dA$. Compare the result to the exact value of I found by iteration.

Solution It's easy to calculate I exactly by iteration. This time, for variety, we integrate first in y:

$$I = \int_0^1 \left(\int_0^1 (2 - x^2 y)\,dy \right) dx = \int_0^1 \left(2y - \frac{x^2 y^2}{2} \Big]_0^1 \right) dx$$

$$= \int_0^1 \left(2 - \frac{x^2}{2} \right) dx = \frac{11}{6}.$$

To find an approximating sum, we first tabulate values (rounded to two decimals) of f at the midpoints of all 100 subrectangles of $[0, 1] \times [0, 1]$. For later use, column sums are at the bottom, rounded to one decimal place.

Values of $f(x, y) = 2 - x^2 y$										
y \ x	0.05	0.15	0.25	0.35	0.45	0.55	0.65	0.75	0.85	0.95
0.05	2.00	2.00	2.00	1.99	1.99	1.98	1.98	1.97	1.96	1.95
0.15	2.00	2.00	1.99	1.98	1.97	1.95	1.94	1.92	1.89	1.86
0.25	2.00	1.99	1.98	1.97	1.95	1.92	1.89	1.86	1.82	1.77
0.35	2.00	1.99	1.98	1.96	1.93	1.89	1.85	1.80	1.75	1.68
0.45	2.00	1.99	1.97	1.94	1.91	1.86	1.81	1.75	1.67	1.59
0.55	2.00	1.99	1.97	1.93	1.89	1.83	1.77	1.69	1.60	1.50
0.65	2.00	1.99	1.96	1.92	1.87	1.80	1.73	1.63	1.53	1.41
0.75	2.00	1.98	1.95	1.91	1.85	1.77	1.68	1.58	1.46	1.32
0.85	2.00	1.98	1.95	1.90	1.83	1.74	1.64	1.52	1.39	1.23
0.95	2.00	1.98	1.94	1.88	1.81	1.71	1.60	1.47	1.31	1.14
Sum	20.0	19.9	19.7	19.4	19.0	18.4	17.9	17.2	16.4	15.5

The approximating sum is found by totaling all the function values and multiplying by $1/100$, the area of each subrectangle. (In this case, therefore, the approximating sum is the average of the 100 table entries.) The result is 1.83; it compares nicely with the exact answer, $11/6 \approx 1.8333$.

Now consider any *column* in the table. The last column, for instance, contains numbers of the form $f(0.95, y)$ for 10 equally spaced values of y, with $\Delta y = 0.1$. Therefore, the column sum multiplied by 0.1 (the answer is 1.55) is a Riemann sum for the integral $\int_0^1 f(0.95, y)\,dy$. We can calculate this last integral directly:

$$\int_0^1 f(0.95, y)\,dy = \int_0^1 (2 - 0.95^2 y)\,dy = 1.54875.$$

Integrals are rounded to two decimals.

That's not far from the Riemann sum. Tabulating results for the other columns produces the same pattern. The Riemann sum associated with each column closely approximates the corresponding y-integral. ◄

Column sums and y-integrals										
x	0.05	0.15	0.25	0.35	0.45	0.55	0.65	0.75	0.85	0.95
Column sum	20.0	19.9	19.7	19.4	19.0	18.4	17.9	17.2	16.4	15.5
$\int_0^1 f(x, y)\,dy$	2.00	1.99	1.97	1.94	1.90	1.85	1.79	1.72	1.64	1.55

The calculations are complicated, but the moral is simple: Whether adding up approximating sums or calculating integrals, it's OK to work with one variable at a time. ∎

A general argument Let's summarize the ideas of Example 2 in more general terms. To do so, suppose that we're given a rectangle $[a, b] \times [c, d]$ in the xy-plane and a function f defined on R. We'll see why the integral $I = \iint_R f(x, y)\, dA$ can reasonably be calculated by iteration.

To begin, let's write an approximating sum for I in the sense of the preceding section. First we subdivide both $[a, b]$ and $[c, d]$ into n equal subintervals. Their lengths are $\Delta x = (b - a)/n$ and $\Delta y = (d - c)/n$, respectively. This produces a grid of n^2 subrectangles R_{ij}, where $1 \leq i, j \leq n$; each rectangle has area $\Delta x\, \Delta y$. Let (x_i, y_j) be the midpoint of the (i, j)th rectangle; we'll use these midpoints as sampling points for an approximating sum S_{n^2} for I.

With all ingredients now in place, we can write the approximating sum:

$$S_{n^2} = \sum_{i,j=1}^{n} f(x_i, y_j)\, \Delta x\, \Delta y = f(x_1, y_1)\, \Delta x\, \Delta y + \cdots + f(x_n, y_n)\, \Delta x\, \Delta y.$$

(The subscript on the "Σ" means that we sum over all possible values of *both* i and j from 1 to n.) ➡

There are n^2 summands in all.

We can group and add the summands in any order or pattern we like. Here's one convenient pattern: ➡

It's OK to factor Δy out of each row because it's common to all summands.

$$S_{n^2} = \big(f(x_1, y_1)\, \Delta x + f(x_2, y_1)\, \Delta x + \cdots + f(x_n, y_1)\, \Delta x\big)\Delta y$$

$$+ \big(f(x_1, y_2)\, \Delta x + f(x_2, y_2)\, \Delta x + \cdots + f(x_n, y_2)\, \Delta x\big)\Delta y + \cdots$$

$$+ \big(f(x_1, y_j)\, \Delta x + f(x_2, y_j)\, \Delta x + \cdots + f(x_n, y_j)\, \Delta x\big)\Delta y + \cdots$$

$$+ \big(f(x_1, y_n)\, \Delta x + f(x_2, y_n)\, \Delta x + \cdots + f(x_n, y_n)\, \Delta x\big)\Delta y.$$

Here's the first of two key points:

> *The sum inside parentheses on each line above is a Riemann sum with n subdivisions for a* single-variable *integral in x.*

Specifically, the sum on the first line is a Riemann sum for the integral $\int_a^b f(x, y_1)\, dx$; the sum on the second line approximates $\int_a^b f(x, y_2)\, dx$, and so on. Now if n is large, then all of these sums are close to the integrals they approximate. Thus, for large n,

$$S_{n^2} \approx \int_a^b f(x, y_1)\, dx\ \Delta y + \int_a^b f(x, y_2)\, dx\ \Delta y + \cdots + \int_a^b f(x, y_n)\, dx\ \Delta y.$$

Here's the second key observation:

> *The sum on the right above is a Riemann sum with n subdivisions for the integral $\int_c^d g(y)\, dy$, where $g(y) = \int_a^b f(x, y)\, dx$.*

For large n this sum, too, is near the integral it approximates. ➡ In other words,

This should sound reasonable, and it is indeed true for the well-behaved functions in this chapter. But a rigorous proof is quite subtle; it depends on technical properties of the integrand function.

$$S_{n^2} \approx \sum_{j=1}^{n} g(y)\, \Delta y \approx \int_c^d g(y)\, dy = \int_c^d \left(\int_a^b f(x, y)\, dx \right) dy.$$

This shows (informally) what we hoped to show: For large n,

$$I \approx S_{n^2} \approx \int_c^d \left(\int_a^b f(x, y)\, dx \right) dy.$$

We conclude that we can indeed evaluate a double integral I over a rectangle R by integrating first in x and then in y. Nor does the order matter—we could just as well have reversed the roles of x and y throughout the preceding argument.

Iterated triple integrals   Iteration works exactly the same way for a triple integral defined on a cube in xyz-space.

EXAMPLE 3   Let $f(x, y, z) = x + y + z$, let $R = [0, 2] \times [0, 2] \times [0, 2]$, and let $I = \iiint_R f(x, y, z)\, dV$. Calculate I exactly, by iteration. (Compare Example 2, page M-36, where we calculated a crude approximating sum for I.) What could the answer *mean*?

Solution   We integrate in each of the three variables in turn:

$$\iiint_R f(x, y, z)\, dV = \int_0^2 \left(\int_0^2 \left(\int_0^2 (x + y + z)\, dx \right) dy \right) dz \qquad \text{(integrate in } x\text{)}$$

$$= \int_0^2 \left(\int_0^2 \left(\frac{x^2}{2} + xy + xz \right]_0^2 \right) dy \right) dz$$

$$= \int_0^2 \left(\int_0^2 (2 + 2y + 2z)\, dy \right) dz$$

$$= \int_0^2 \left(2y + y^2 + 2yz \right]_0^2 \right) dz \qquad \text{(integrate in } y\text{)}$$

$$= \int_0^2 (8 + 4z)\, dz = 24 \qquad \text{(integrate in } z\text{)}$$

Maple does the same thing in one fell swoop:

```
> int( int( int( x+y+z, x=0..2), y=0..2), z=0..2);
                        24
```

What the answer means depends on our point of view. If we think of the integrand $f(x, y, z)$ as the density (in, say, grams per cubic centimeter) of the solid R at the point (x, y, z), then the integral tells the *mass* of the solid (in grams).   ■

Integrals over nonrectangular regions

Not every integral of interest is taken over a rectangular domain of integration. It's sometimes useful to integrate over regions with curved boundaries. Much the same process of iteration applies in this case as in that of integrals over rectangles and cubes, but some extra care is needed. We illustrate the process with an example.

E X A M P L E 4 Find $I = \iint_R (2 - x) \, dA$, where R is the plane region between the curves $y = 0$ and $y = x$, and $x = 1$. ➡

It's essential to draw the domain; use an ordinary pair of xy-axes.

Solution The integral gives the volume of the solid figure shown in Figure 4.

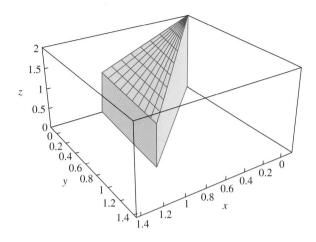

FIGURE 4
A surface with nonrectangular base

We can think of the domain R as bounded by the straight lines $x = 0$ and $x = 1$ on the left and right and by the curves $y = 0$ and $y = x$ on the bottom and top. As in earlier cases, we can integrate by iteration. The *outer* integral, in x, runs from $x = 0$ to $x = 1$. The upper and lower limits of the *inner* integral, however, depend on x, as the following computation shows:

$$
I = \iint_R (2 - x) \, dA = \int_{x=0}^{x=1} \left(\int_{y=0}^{y=x} (2 - x) \, dy \right) dx
$$

$$
= \int_{x=0}^{x=1} \left(2y - xy \big]_{y=0}^{y=x} \right) dx
$$

$$
= \int_{x=0}^{x=1} \left(2x - x^2 \right) dx = \frac{2}{3}.
$$

Again, *Maple* can do the whole thing at once:

```
> int( int( 2-x, y=0..x), x=0..1);
                    2/3
```

In fact, this integral is not much different from the preceding ones, which were over rectangular domains. The variable limits of integration in the inner integral simply reflect the fact that the domain of integration has varying "heights," depending on x. ∎

Changing the order of integration For integrals over rectangles or cubes, we can integrate in any order we like. A similar result holds in the present case, too—as long as the region has an appropriate shape. We illustrate by redoing the preceding example but integrating the variables in the opposite order.

E X A M P L E 5   Redo the integral $I = \iint_R (2-x)\,dA$ of Example 4, but now integrate first in x and then in y.

Solution   This time we think of R as lying between the lines $y = 0$ and $y = 1$. For given y, R starts at the line $x = y$ and ends at the line $x = 1$. Now the calculation is similar to the preceding one:

$$I = \iint_R (2-x)\,dA = \int_{y=0}^{y=1} \left(\int_{x=y}^{x=1} (2-x)\,dx \right) dy$$

$$= \int_{y=0}^{y=1} \left(2x - \frac{x^2}{2} \right]_{x=y}^{x=1} \right) dy$$

$$= \int_{y=0}^{y=1} \left(\frac{3}{2} - 2y + \frac{y^2}{2} \right) dy = \frac{2}{3}.$$

Here's *Maple*'s version:

```
> int( int( 2-x, x=y..1), y=0..1);
```
$$\frac{2}{3}$$
                                                                            ∎

Not always so simple   Things aren't always quite so simple. Some domains of integration lend themselves more naturally to one order of integration than to another; the following exercises give examples.

BASIC EXERCISES

In Exercises 1–5, use iteration to calculate the integral by hand (without technology). Then, check your answer symbolically using Maple. *Finally, plot a 3-D surface over an appropriate domain and check that your answer is reasonable.*

1. $\iint_R \sin x \sin y \, dA;$ $R = [0, 1] \times [0, 1]$

2. $\iint_R \sin(x + y) \, dA;$ $R = [0, 1] \times [0, 1]$

3. $\iint_R (x^2 + y^2) \, dA;$ $R = [0, 4] \times [0, 4]$

4. $\iiint_R x \, dV;$ $V = [0, 1] \times [0, 2] \times [0, 3]$

5. $\iiint_R y \, dV;$ $V = [0, 1] \times [0, 2] \times [0, 3]$

In Exercises 6–8, use iteration to calculate the integral by hand (without technology). In each case, the inner integral should be in y and the outer integral in x. Check your answer symbolically using Maple.

6. $\iint_R (x + y) \, dA;$ R is the region bounded by the curves $y = x$ and $y = x^2$.

7. $\iint_R x \, dA;$ R is the region bounded by the curves $y = x^2$ and $y = \sqrt{x}$.

8. $\iint_R 1 \, dA;$ R is the first quadrant part of the circle $x^2 + y^2 \leq 1$.

9–11. Redo Exercises 6–8, but integrate first in x and then in y.

12. Consider the integral $I = \iint_R (x + y) \, dA$, where R is the region bounded by the curves $y = x^2$ and $y = 1$.

   (a) Calculate I by integrating first in y and then in x.

   (b) Calculate I by integrating first in x and then in y.

13. Let $f(x, y) = x$, and let R be the plane region bounded by the curves $y = e^x$, $y = 0$, $x = 0$, and $x = 1$.

   (a) Calculate $I = \iint_R f(x, y) \, dA$ by integrating first in y and then in x.

   (b) Calculate $I = \iint_R f(x, y) \, dA$ by integrating first in x and then in y. [HINT: First split the region R into two simpler pieces; each simpler piece should be bounded on the left by one curve and on the right by another.]

14. Let $y = f(x)$ be a function, with $f(x) \geq 0$ if $a \leq x \leq b$; let R be the plane region bounded by the curves $y = f(x)$, $y = 0$, $x = a$, and $x = b$.

   (a) What does single-variable calculus say about the area of R?

   (b) The double integral $I = \iint_R 1 \, dA$ gives the area of R. Use an iterated integral to reconcile this formula with the one in part (a).

15. Let $x = g(y)$ be a function with $g(y) \geq 0$ if $c \leq y \leq d$; let R be the plane region bounded by the curves $x = g(y)$, $x = 0$, $y = c$, and $y = d$.

    (a) What does single-variable calculus say about the area of R?

(b) The double integral $I = \iint_R 1 \, dA$ gives the area of R. Use an iterated integral to reconcile this formula with the one in part (a).

M.7 DOUBLE INTEGRALS IN POLAR COORDINATES

What makes a double integral $I = \iint_R f(x, y) \, dA$ hard to calculate? Both f and R can play a role: If either one is complicated or messy to describe, or both, then I may be correspondingly ugly. Let's see examples of both "good" and "bad" integrals.

EXAMPLE 1 Discuss $I_1 = \iint_{R_1} x^2 \, dA$ and $I_2 = \iint_{R_2} \sqrt{x^2 + y^2} \, dA$, where R_1 is the rectangle $[0, 1] \times [0, 2\pi]$, and R_2 is the region inside the unit circle $x^2 + y^2 = 1$. ➡

Draw R_1 and R_2 for yourself.

Solution The first integral is easy:

$$I_1 = \int_{x=0}^{x=1} \left(\int_{y=0}^{y=2\pi} x^2 \, dy \right) dx = \int_0^1 \left(x^2 y \right]_0^{2\pi} \right) dx = \int_0^1 2\pi x^2 \, dx = \frac{2\pi}{3}.$$

The ingredients of I_2 are more complicated to describe. ➡ The circular region R_2 can be thought of as bounded by the curves $y = \sqrt{1 - x^2}$ and $y = -\sqrt{1 - x^2}$ on the top and bottom, and by the lines $x = -1$ and $x = 1$ on the left and right. Now we can write I_2 in iterated form:

In xy-coordinates, at least. Polar coordinates will make things simpler.

$$I_2 = \int_{x=1}^{x=-1} \left(\int_{y=-\sqrt{1-x^2}}^{y=\sqrt{1-x^2}} \sqrt{x^2 + y^2} \, dy \right) dx.$$

The integral looks—and is—complicated. Just to get started on the inner integral, we'd need the ugly antiderivative formula

$$\int \sqrt{x^2 + p^2} \, dx = \frac{1}{2} \left(x\sqrt{x^2 + p^2} + p^2 \ln \left| x + \sqrt{x^2 + p^2} \right| \right).$$

Faced with this prospect, we retreat, but only temporarily—we shall return to I_2. ∎

What went wrong, and what to do The integral I_2 in Example 1 led to an unpleasant calculation in x and y for two reasons:

(i) The integrand, $\sqrt{x^2 + y^2}$, has a complicated antiderivative in x or y.

(ii) The domain of integration, although geometrically simple, is messy to describe in rectangular coordinates.

In polar coordinates, on the other hand, both the integrand and the domain have simple, uncluttered formulas. The integrand is

$$f(x, y) = \sqrt{x^2 + y^2} = r.$$

In polar language, the domain of integration is essentially a rectangle; it's defined by the inequalities

$$0 \le r \le 1 \quad \text{and} \quad 0 \le \theta \le 2\pi.$$

In this case, apparently, both the integrand f and the domain R "deserve" to be described in polar coordinates, not in rectangular coordinates. It seems reasonable, therefore, that the double integral I_2 should also be calculated using polar rather than rectangular coordinates.

That hunch is correct. We demonstrate in this section how to calculate double integrals in polar form, using r and θ, as opposed to Cartesian or rectangular form, using x and y. Integrals such as I_2, in which the integrand, the domain of integration, or both are simplest in polar form, are natural candidates for polar treatment.

Polar "rectangles"   A rectangle in Cartesian coordinates is defined by two inequalities of the form

$$a \le x \le b \quad \text{and} \quad c \le y \le d;$$

each of the coordinates x and y ranges through an interval. A **polar rectangle** is defined by two similar inequalities:

$$a \le r \le b; \qquad \alpha \le \theta \le \beta;$$

again, each of the coordinates r and θ ranges through an interval. Figure 1 shows generic pictures of both types of rectangles.

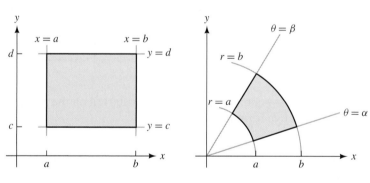

FIGURE 1
Cartesian and polar rectangles

For polar integrals, as for Cartesian ones, rectangles (in the appropriate sense) are the simplest regions over which to integrate.

Polar integration — How it works and what it means

A double integral $I = \iint_R f(x, y)\, dA$ in rectangular coordinates, where $R = [a, b] \times [c, d]$ is an ordinary rectangle, is written in iterated form as follows: ◄

We integrate first in y this time.

$$\iint_R f(x, y)\, dA = \int_a^b \int_c^d f(x, y)\, dy\, dx.$$

(Here and later, we omit some parentheses. It's always understood that the inner integral is done first.)

Now suppose we're given a double integral $I = \iint_R g(r, \theta)\, dA$ in polar coordinates, where R is now a *polar* rectangle defined by inequalities

$$a \leq r \leq b \quad \text{and} \quad \alpha \leq \theta \leq \beta,$$

and $g(r, \theta)$ is a function defined on R. The following Fact gives the appropriate integral formula.

FACT (Double integrals in polar coordinates) Let g and R be as before. Then

$$\iint_R g(r, \theta)\, dA = \int_{\theta=\alpha}^{\theta=\beta} \int_{r=a}^{r=b} g(r, \theta)\, r\, dr\, d\theta.$$

The formula prompts several important observations:

- **Trading x and y for r and θ** Any function $f(x, y)$ can be "traded" for an equivalent function $g(r, \theta)$ by using the relations

$$x = r\cos\theta \quad \text{and} \quad y = r\sin\theta.$$

The same method works for *equations* in x and y. The equation $x = y$, for example, says in polar coordinates that $r\cos\theta = r\sin\theta$ or, equivalently, that $\tan\theta = 1$. This polar equation describes the same line as the original Cartesian equation. We'll use these principles when evaluating polar integrals.

- **A useful mnemonic** Compare the preceding formulas for integrating in rectangular and polar coordinates. An important difference between the two has to do with the "dA" expression. The full mathematical story is much deeper. ➡ But as a quick aid to memory, the following formulas are very handy:

  We won't go into great depth, but we'll give some informal justification soon.

$$dA = dx\, dy \quad \text{for Cartesian coordinates;}$$
$$dA = r\, dr\, d\theta \quad \text{for polar coordinates.}$$

- **That extra factor of r** What's that mysterious extra r doing in the polar formula $dA = r\, dr\, d\theta$? Why not just $dA = dr\, d\theta$? We certainly owe the reader an explanation. We'll honor that debt in a moment when we discuss *why* the formula works. First, however, let's see *that* it works.

EXAMPLE 2 Let R_2 be the region inside the unit circle. Use polar coordinates to calculate that troublesome integral $I_2 = \iint_{R_2} \sqrt{x^2 + y^2}\, dA$ from Example 1, page M-47.

Solution First we write all the data in polar form. For the integrand, we have $f(x, y) = \sqrt{x^2 + y^2} = r = g(r, \theta)$. For the domain of integration, we translate the

Cartesian equation $x^2 + y^2 = 1$ into its (simpler!) polar form, $r = 1$. The rest is easy:

$$
\iint_{R_2} \sqrt{x^2 + y^2}\, dA = \int_{\theta=0}^{\theta=2\pi} \int_{r=0}^{r=1} r\, dA
$$

$$
= \int_{\theta=0}^{\theta=2\pi} \int_{r=0}^{r=1} r^2\, dr\, d\theta \qquad (da = r\, dr\, d\theta)
$$

$$
= \int_{\theta=0}^{\theta=2\pi} \frac{r^3}{3}\Big]_0^1 d\theta \qquad (\text{integrate in } r)
$$

$$
= \int_{\theta=0}^{\theta=2\pi} \frac{1}{3}\, d\theta = \frac{2\pi}{3}. \qquad (\text{integrate in } \theta)
$$

Notice especially the similarity to the integral I_1 of Example 1—I_1 and I_2 turned out to have the same value. This is no accident. After rewriting in polar coordinates, I_2 turned out to be the *same integral* in r and θ as I_1 is in x and y. ∎

We discussed some possible interpretations in Section M.5.

Polar integrals have exactly the same interpretations as any other double integrals. ◄ Depending on the situation and on our point of view, an integral might represent a volume, the area of a plane region, the mass of a thin plate, or many other things.

For example, both the integrals I_1 and I_2 of Example 1, page M-47, can be interpreted as volumes of solids. For I_2, the solid lies above the unit disk and below the surface $z = \sqrt{x^2 + y^2}$, as shown.

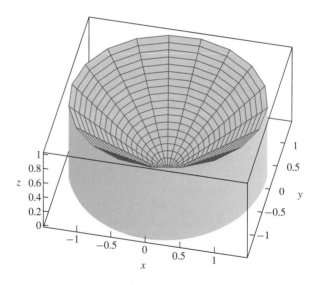

FIGURE 2
A polar solid

Does this seem reasonable from the picture, given the general size of units?

As we just calculated, the solid shown in Figure 2 has volume $2\pi/3 \approx 2.094$ cubic units. ◄

Polar integration — Why it works

Why does the polar integration formula work? Where, especially, does the r in $dA = r\, dr\, d\theta$ come from?

All properties of integrals—whether in Cartesian, polar, or any other form—stem ultimately from properties of the approximating sums that are used to define integrals. For any function f defined on a region R, we have

$$\iint_R f \, dA = \lim_{m \to \infty} \sum_{i=1}^{m} f(P_i) \Delta A_i,$$

where ΔA_i is the area of the ith subregion of R, and P_i is a sampling point chosen inside this subregion.

If $R = [a, b] \times [c, d]$ is a Cartesian rectangle, then it's natural to subdivide R into smaller rectangles, each with sides Δx and Δy. Any such rectangle has area $\Delta A_i = \Delta x \, \Delta y$. In the limit that defines the integral, therefore, $dA = dx \, dy$.

If R is a polar rectangle, the picture is a little different. In this case, a "polar grid" is the natural way to subdivide R. In Figure 3, each black dot represents a "centered" sampling point in its respective subdivision.

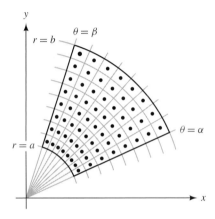

FIGURE 3
A polar grid on a polar rectangle

Notice the following features of the grid:

- **Similar subregions** All subregions correspond to the same $\Delta\theta$ (the angle between any two adjacent dotted radial lines) and the same Δr (the radial distance from one arc to the next).

- **But not identical** Similar as they are, the subregions shown are not identical. Here's the key point:

  In a polar grid, the subregions have different sizes. The area depends not only on $\Delta\theta$ and Δr but also on r: Larger values of r produce larger subregions.

  This fact explains the difference between polar and Cartesian integrals, and hints at why that extra r is needed.

- **The area of one subregion** Each of the preceding subregions is a small polar rectangle with polar dimensions $\Delta\theta$ and Δr, inner radius r, and outer radius $r + \Delta r$. It's an important fact that, for such a polar rectangle,

$$\text{Area} = \Delta A_i = \frac{r + r + \Delta r}{2} \Delta r \, \Delta\theta.$$

(We'll leave verification of this straightforward fact to the exercises.) The first factor on the right is crucial—it represents the *average* radius of the given subregion, that is, the r-coordinate of the ith bulleted midpoint (r_i, θ_i) in the illustration. Therefore,

$$\Delta A_i = r_i \, \Delta r \, \Delta\theta.$$

This is just what we've been waiting for. It shows that, for a polar rectangular region, a midpoint approximating sum has the form

$$\sum_{i=1}^{m} f(r_i, \theta_i) \, \Delta A_i = \sum_{i=1}^{m} f(r_i, \theta_i) \, r_i \, \Delta r \, \Delta \theta.$$

The integral itself therefore has the limiting form $\iint_R f(r, \theta) \, r \, dr \, d\theta$.

Polar integrals over nonrectangular regions   Polar integrals, like Cartesian integrals, can be taken over nonrectangular regions. The method is similar.

> **FACT**   Let R be the region bounded by the radial lines $\theta = \alpha$ and $\theta = \beta$, by an inner curve $r = r_1(\theta)$, and by an outer curve $r = r_2(\theta)$. ("Inner" and "outer" are understood relative to the origin.) Let $g(r, \theta)$ be a function defined on R. Then
>
> $$\iint_R g \, dA = \int_{\theta=\alpha}^{\theta=\beta} \int_{r=r_1(\theta)}^{r=r_2(\theta)} g(r, \theta) \, r \, dr \, d\theta.$$

We illustrate with an example.

EXAMPLE 3   Use a polar integral to find the area of the region R inside the cardioid $r = 1 + \cos\theta$.

It applies equally well in polar coordinates!

Solution   We use a familiar principle: ◄

$$\text{Area of } R = \iint_R 1 \, dA,$$

but we calculate the integral in *polar* form, as follows:

$$\iint_R 1 \, dA = \int_{\theta=0}^{\theta=2\pi} \int_{r=0}^{r=1+\cos\theta} r \, dr \, d\theta$$

$$= \int_{\theta=0}^{\theta=2\pi} \frac{r^2}{2} \Bigg]_0^{1+\cos\theta} d\theta$$

$$= \int_{\theta=0}^{\theta=2\pi} \frac{(1+\cos\theta)^2}{2} \, d\theta.$$

The last integral takes a little doing by hand. *Maple* has no trouble:

```
> int( (1+cos(t))^2, t=0 .. 2*Pi );
              3*Pi/2
```

■

BASIC EXERCISES

1. Let R be the polar rectangle defined by $a \leq r \leq b$ and $\alpha \leq \theta \leq \beta$.

   (a) Show that the area of R is $\dfrac{a+b}{2}(b-a)(\beta-\alpha)$.

   (b) Use part (a) to show that a polar rectangle with dimensions Δr and $\Delta \theta$ and inner radius r has area $\dfrac{r+r+\Delta r}{2} \Delta r \, \Delta \theta = \dfrac{2r+\Delta r}{2} \Delta r \, \Delta \theta.$

2. Let $f(x, y) = y$, let R be the upper half of the region inside the unit circle $x^2 + y^2 = 1$, and let $I = \iint_R f \, dA$.

   (a) Calculate I as an iterated integral in rectangular coordinates with the inner integral in y.

   (b) Calculate I as an iterated integral in rectangular coordinates with the inner integral in x.

   (c) Calculate I as an iterated integral in polar coordinates.

3. Let I_1 be the integral defined in Example 1.

   (a) Draw the solid whose volume is given by I_1.

   (b) Evaluate I_1 again but with the inner integral in x, not y.

In Exercises 4–6, use a polar double integral to evaluate the area of the region. (Draw each region first.)

4. Find the area of the region inside the cardioid $r = 1 + \sin\theta$.

5. Find the area of the region bounded by $y = x$, $y = 0$, and $x = 1$. Could you find the answer another way?
   [HINT: First write the boundary equations in polar form.]

6. Find the area of the region bounded by the circle of radius 1/2 centered at $(0, 1/2)$. Could you find the answer another way? [HINT: One approach is to write a Cartesian equation for the circle first and then change it to polar form.]

7. Use polar coordinates to find $\iint_R \dfrac{1}{\sqrt{x^2 + y^2}}\, dA$, where R is the region inside the cardioid $r = 1 + \sin\theta$ and above the x-axis.

8. Use polar coordinates to find the volume of the solid under the surface $z = 1 - x^2 - y^2$ and above the xy-plane. [HINT: First decide where the surface intersects the xy-plane.]

9. Use polar coordinates to find the volume of the conical solid under the surface $z = 1 - \sqrt{x^2 + y^2}$ and above the xy-plane. [HINT: First decide where the surface intersects the xy-plane.]

SELECTIONS FROM VOLUME 1

3.4 INVERSE FUNCTIONS AND THEIR DERIVATIVES; INVERSE TRIGONOMETRIC FUNCTIONS

We've seen in earlier sections how to find derivatives of new functions built from old. In this section we study how new functions can be produced by *inverting* old ones, and how to find derivatives of the new functions. The arcsine, arccosine, and arctangent functions—inverses of the standard trigonometric functions—are our most important examples.

Identities and inverses

The number 0 is an **identity** for ordinary addition because $a + 0 = a$ for all numbers a. Similarly, the number 1 is an identity for multiplication because $a \cdot 1 = a$ for all numbers a. A similar notion of "identity" applies to function composition, but in this setting the appropriate identity is a *function*:

> **DEFINITION** The **identity function** I is defined by $I(x) = x$.

To see why I deserves the name "identity," let f be *any* real-valued function. Then, for any input x,

$$(f \circ I)(x) = f(I(x)) = f(x) \quad \text{and} \quad (I \circ f)(x) = I(f(x)) = f(x).$$

Thus, composing any function f with I (in either order) gives f.

Inverse functions The word "inverse" has several different (but related) meanings in mathematics. For instance, 3 and -3 are additive inverses because their *sum* is 0 (the additive identity), while 3 and $1/3$ are multiplicative inverses because their *product* is the multiplicative identity. Two *functions* are inverses if their *composition* is the identity function:

> **DEFINITION (Inverse functions)** Let f and g be functions. If
> $$(f \circ g)(x) = x \quad \text{and} \quad (g \circ f)(x) = x$$
> for all x in the domains of g and f, respectively, then f and g are **inverse functions**. In this case, we write $g = f^{-1}$ (and $f = g^{-1}$).

Beware: $f^{-1}(x) \neq 1/f(x)$.

h adds 3; h^{-1} takes 3 away.

In particular, the inverse of a function f is another function f^{-1}, one that "undoes" the effect of f. ◄ Example 1 illustrates the idea.

> **EXAMPLE 1** If $h(x) = x + 3$, then $h^{-1}(x) = x - 3$. ◄ For any input x,
>
> $$\left(h^{-1} \circ h\right)(x) = h^{-1}(x+3) = (x+3) - 3 = x = I(x);$$
> $$\left(h \circ h^{-1}\right)(x) = h(x-3) = (x-3) + 3 = x = I(x).$$
>
> Thus, $h^{-1} \circ h = I = h \circ h^{-1}$, and so h and h^{-1} are indeed inverses. ■

We met these functions in Sections 1.3 and 2.6.

Famous inverses: logs and exponentials We've already studied the most important pair of inverse functions: logarithmic and exponential functions. ◄ Indeed, we *defined* logarithms as inverses of exponentials. In base e, for instance, we've seen that

$$e^{\ln x} = x \quad \text{if } x > 0; \quad \ln(e^x) = x \quad \text{for all real } x;$$

similar identities hold for other bases.

> **EXAMPLE 2** The functions $f(x) = 10^x$ and $g(x) = \log_{10} x$ are inverses. What does this mean when $x = 2$?
>
> **Solution** Because $f(2) = 10^2 = 100$, we must have $g(100) = \log_{10} 100 = 2$. Similarly, $g(2) = \log_{10} 2 \approx 0.30103$; the inverse relationship means $f(0.30103) = 10^{0.30103} \approx 2$. ◄ ■

Your calculator "knows" these numerical facts.

Inverse functions and symmetry If f and f^{-1} are inverse functions, then $f(a) = b \iff f^{-1}(b) = a$ for every a in the domain of f. This condition has a nice graphical meaning:

A point (a, b) lies on the graph of f if and only if (b, a) lies on the graph of f^{-1}.

Note that (b, a) and (a, b) are reflections of each other across the line $y = x$, as are the graphs of f and f^{-1}. Figure 1 illustrates this symmetry.

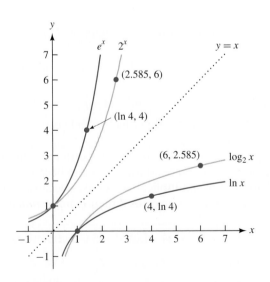

FIGURE 1
Logarithmic and exponential functions: inverses

Notice especially the three marked pairs of points: they reflect (literally!) the facts that 2^x and $\log_2 x$ are inverse functions, as are e^x and $\ln x$.

Formulas for inverse functions If a function f happens to have a simple algebraic formula, then f^{-1} may have one, too, and we may be able to find it algebraically. Here is the key idea:

If f and f^{-1} are inverses, then $y = f^{-1}(x) \iff x = f(y)$.

Therefore, to find a formula for $f^{-1}(x)$ we can try to solve the equation $x = f(y)$ for y in terms of x. Doing so is hard for some functions but easy for others.

EXAMPLE 3 Let $f(x) = 2x + 3$. Find a formula for $f^{-1}(x)$.

Solution We solve $x = f(y)$ for y:

$$x = f(y) = 2y + 3 \implies x - 3 = 2y \implies \frac{x-3}{2} = y = f^{-1}(x).$$

The graphs of $y = 2x + 3$ and $y = (x - 3)/2$ have the expected symmetry across the line $y = x$. ➜ Notice, too, that the inverse of a linear function is linear; reflecting a line produces a new line. ■

Draw them for yourself.

Derivatives of inverse functions

How are derivatives of f and f^{-1} related to each other? Figure 2 suggests the answer.

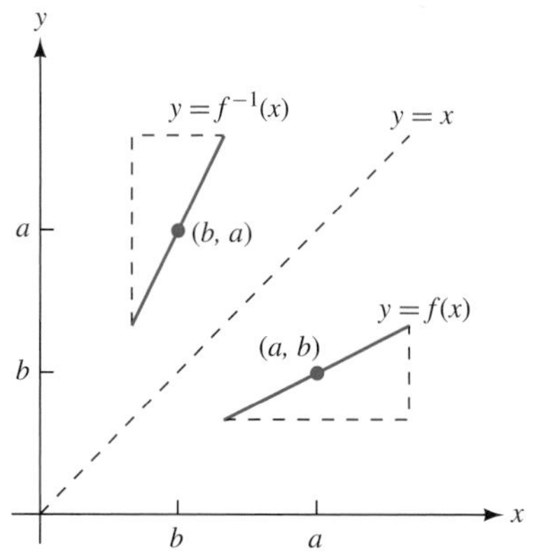

The segments through (a, b) and (b, a) have reciprocal slopes. Compare rises and runs to see why.

FIGURE 2
Comparing derivatives of f and f^{-1}

The picture shows tiny pieces of the graphs of f and f^{-1}. (The pieces look like straight lines because f is differentiable at $x = a$.) The reflective symmetry also shows that the *slopes* $f'(a)$ and $f^{-1'}(b)$ of the two pieces are *reciprocals*—the rise in one is the run in the other, and vice versa. We summarize these observations in a form we'll use later:

> **THEOREM 1 (Derivatives of inverse functions)** Let f and g be inverse functions, and suppose $f(a) = b$. If $f'(a)$ exists and $f'(a) \neq 0$, then $g'(b)$ also exists, and $g'(b) = 1/f'(a)$. More generally,
>
> $$g'(x) = \frac{1}{f'(g(x))}$$
>
> whenever the right side makes sense.

EXAMPLE 4 In Example 3 we found that $f(x) = 2x + 3$ and $g(x) = (x - 3)/2$ are inverses. It's easy to find $g'(x)$ directly, but what does Theorem 1 say?

Solution By Theorem 1, $g'(x) = 1/f'(g(x))$. But $f'(x) = 2$ for *all* inputs x, so $g'(x) = 1/2$. ■

Of course, we could have differentiated $g(x) = (x - 2)/3$ at a glance. We'll apply Theorem 1 more impressively later in this section.

Which functions have inverses?

Does every number have an additive inverse? A multiplicative inverse?

Not every function *has* an inverse. ◄ For instance, no constant function $f(x) = k$ has an inverse because for *any* function g and any input x,

$$(f \circ g)(x) = f(g(x)) = k.$$

Thus $f \circ g$ is *still* a constant function—not the identity function.

So which functions *do* have inverses? One way to decide is to think graphically: If f has an inverse, then reflecting the curve $y = f(x)$ across the line $y = x$ should produce a new *function graph*, not just a new curve.

EXAMPLE 5 Does $f(x) = x^2$ have an inverse? How are square roots involved?

Solution Figure 3 shows the curves $y = x^2$ and $x = y^2$; the second is the result of reflecting the first about $y = x$.

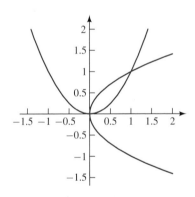

FIGURE 3
The curves $x = y^2$ and $y = x^2$

There's an important difference between the two curves: The rightward-opening parabola $x = y^2$ is *not* the graph of a function because each positive x-value corresponds to *two* different y-values. Indeed, solving $x = y^2$ algebraically for y gives *two* possible inverse functions: $y = \sqrt{x}$ and $y = -\sqrt{x}$.

Is one of these functions the "true" inverse? No—as the figure shows, $y = \sqrt{x}$ (the *upper* half of the curve $x = y^2$) and $y = -\sqrt{x}$ (the *lower* half) are, respectively, the reflections across $y = x$ of the *right* and *left* halves of the original parabola $y = x^2$. Informally speaking, $y = \sqrt{x}$ is an inverse for the "right half" of $y = x^2$, and $y = -\sqrt{x}$ is *A similar description applies* an inverse for the "left half." More formally, $y = \sqrt{x}$ is an inverse for the function *to $y = -\sqrt{x}$.* $f(x) = x^2$ when we "restrict the domain" of f to nonnegative inputs. ◄ ■

One-to-one functions; restricting domains   Example 5 illustrates a more general point: For a function $y = f(x)$ to have an inverse, each output y can come from only *one* input x. This condition has an official name:

> A function f is **one-to-one** if different inputs to f give different outputs. Equivalently, $x_1 \ne x_2 \implies f(x_1) \ne f(x_2)$.

This condition, stated graphically, is the **horizontal line test** of precalculus fame:

> No horizontal line crosses the f-graph more than once.

The *full* curve $y = x^2$ in Example 5 fails this test—which is why we had trouble choosing among possible inverse functions. But the right and left halves of the curve $y = x^2$ (with $x > 0$ and $x < 0$, respectively) pass the test, and so let us choose inverse functions unambiguously. A useful trick, in other words, is to **restrict the domain** of f to an interval I (perhaps infinite) on which f *is* one-to-one. The restricted function then *has* an inverse; the following Fact gives the fine print:

FACT   Suppose that f is one-to-one on an interval I. Then f (with domain restricted to I) has an inverse function f^{-1}, and the graphs of f and f^{-1} are reflections of each other across $y = x$.

Inverse trigonometric functions

Like the squaring function, the sine, cosine, tangent, and other basic trigonometric functions need their domains restricted to become one-to-one. With this taken care of, the trigonometric functions have well-behaved inverses—some rate their own buttons on scientific calculators.

"Arc" and "inverse" notations   Two standard notations are used for inverse trigonometric functions: $\arcsin x$, $\arccos x$, and $\arctan x$ are synonyms for $\sin^{-1} x$, $\cos^{-1} x$, and $\tan^{-1} x$. ➜ We'll use both forms as convenient; the "arc" version reminds us of angles and the unit circle, while "inverse" recalls function inversion.

Don't confuse $\sin^{-1} x$ *with* $1/\sin x$.

Domain surgery   Trigonometric functions are far from one-to-one on their full natural domains. Because they repeat themselves endlessly on intervals of length 2π, their graphs fail the horizontal line test *miserably*. (The x-axis, for instance, intersects the sine curve infinitely often.)

We'll *make* the trigonometric functions one-to-one through the same "surgical" strategy we applied to the squaring function: amputate the offending part of each function's domain. Graphs will show us where to cut.

Inverting sines and cosines   The sine function is one-to-one on the interval $[-\pi/2, \pi/2]$. (We could have chosen other intervals, but this one, centered at 0, seems natural.) Figure 4(a) shows the sine curve with the part corresponding to $[-\pi/2, \pi/2]$ highlighted. Figure 4(b) shows both the (restricted) sine function and its inverse, the **arcsine** function, which is defined formally below.

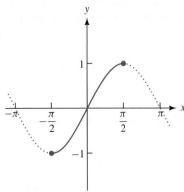
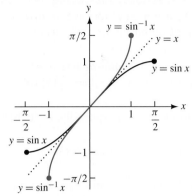

(a) Restricting the sine function   (b) The sine and arcsine functions

FIGURE 4
Constructing the arcsine

> **DEFINITION** For x in $[-1, 1]$, $y = \sin^{-1} x$ (or $\arcsin x$) is defined by the conditions
>
> $$\text{(i)} \quad x = \sin y; \quad \text{(ii)} \quad -\frac{\pi}{2} \le y \le \frac{\pi}{2}.$$
>
> In words: $\arcsin x$ is the angle between $-\pi/2$ and $\pi/2$ whose sine is x.

Restricting $\cos x$ to $[-\pi/2, \pi/2]$ wouldn't work. Do you see why not?

The **arccosine** is found in a similar way. The cosine function is one-to-one if restricted to the interval $[0, \pi]$. ← Reflecting this part of the cosine graph across the line $y = x$ produces the arccosine: graphs (Figure 5) and a definition follow.

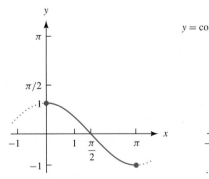
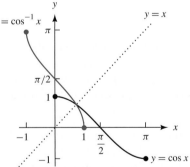

(a) Restricting the cosine function   (b) The cosine and arccosine functions

FIGURE 5
Constructing the arccosine

> **DEFINITION** For x in $[-1, 1]$, $y = \cos^{-1} x$ (or $\arccos x$) is defined by the conditions
>
> $$\text{(i)} \quad x = \cos y; \quad \text{(ii)} \quad 0 \le y \le \pi.$$
>
> In words: $\arccos x$ is the angle between 0 and π whose cosine is x.

Inverting the tangent function The ordinary tangent function becomes one-to-one if restricted to any of its "branches." We'll use the "middle" branch, through the origin; reflecting it across the line $y = x$ produces the **arctangent** function. Graphs (Figure 6) and a formal definition follow.

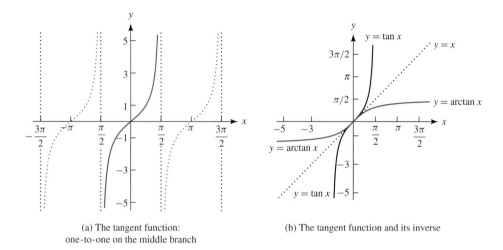

(a) The tangent function: one-to-one on the middle branch

(b) The tangent function and its inverse

FIGURE 6
Constructing the arctangent

DEFINITION Let x be any real number. Then $y = \tan^{-1} x$ (or $y = \arctan x$) means that

$$\text{(i)} \quad x = \tan y; \quad \text{(ii)} \quad -\pi/2 < y < \pi/2.$$

Notice, in particular, that although the (restricted) tangent function has a small domain, its range is the *full* set of real numbers. Therefore, the arctangent function has range $(-\pi/2, \pi/2)$ but accepts *all* real numbers as inputs.

Domains and ranges: take care ... The most obvious property of inverse trigonometric functions is that they "undo" trigonometric functions. A little trickier—but important for using these functions successfully—is keeping careful track of domains and ranges. A table might help (the ordinary trigonometric functions have restricted domains).

Function	sine	arcsine	cosine	arccosine	tangent	arctangent
Domain	$[-\pi/2, \pi/2]$	$[-1, 1]$	$[0, \pi]$	$[-1, 1]$	$(-\pi/2, \pi/2)$	$(-\infty, \infty)$
Range	$[-1, 1]$	$[-\pi/2, \pi/2]$	$[-1, 1]$	$[0, \pi]$	$(-\infty, \infty)$	$(-\pi/2, \pi/2)$

Other inverse trigonometric functions All six trigonometric functions *have* inverses, but $\operatorname{arcsec} x$, $\operatorname{arccsc} x$, and $\operatorname{arccot} x$ are seldom seen in calculus courses. We'll content ourselves here with a few words about the arcsecant function.

The simplest definition of $\operatorname{arcsec} x$ starts with the connection between secants and cosines: $\sec t = 1/\cos t$ (unless $\cos x = 0$). So it's no surprise to find a related connection between the "arc" versions of both functions:

> **DEFINITION** For any x with $|x| \geq 1$, $\operatorname{arcsec} x = \sec^{-1} x = \cos^{-1}(1/x)$.

Note that the definition does what we'd expect of an inverse function—if $y = \operatorname{arcsec} x = \arccos(1/x)$, then

$$\sec y = \frac{1}{\cos y} = \frac{1}{\cos(\arccos(1/x))} = \frac{1}{1/x} = x.$$

Figure 7 shows the secant and arcsecant functions; the dotted lines are asymptotes.

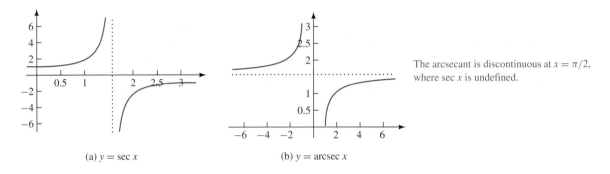

The arcsecant is discontinuous at $x = \pi/2$, where $\sec x$ is undefined.

(a) $y = \sec x$

(b) $y = \operatorname{arcsec} x$

FIGURE 7
The (restricted) secant and arcsecant functions

Working with inverse trigonometric functions

Inverse and ordinary trigonometric functions often tangle together to produce expressions like these:

$$\sin\left(\sin^{-1} x\right); \quad \cos\left(\sin^{-1} x\right); \quad \sin\left(\tan^{-1} x\right).$$

Diagrams like those in Figure 8 can help us untangle such expressions.

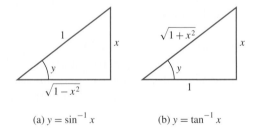

(a) $y = \sin^{-1} x$

(b) $y = \tan^{-1} x$

FIGURE 8
Relating inverse trigonometric functions

In the left-hand figure, the angle labeled y is the arcsine of x (because $\sin y = x$). The figure shows (among other things) that

$$\cos(\sin^{-1} x) = \frac{\text{adjacent}}{\text{hypotenuse}} = \sqrt{1 - x^2}.$$

In the second figure, $y = \tan^{-1} x$ (because $\tan y = x$), so

$$\sin(\tan^{-1} x) = \frac{\text{opposite}}{\text{hypotenuse}} = \frac{x}{\sqrt{1 + x^2}}.$$

Note that these equations hold only for values of x in the domains of the functions in question.

Derivatives   Ordinary trigonometric functions have trigonometric derivatives. It's a little surprising, then, that inverse trigonometric functions have *algebraic* derivatives. To find these derivatives we can use Theorem 1, the chain rule, or implicit differentiation. We'll first list the derivative formulas and then give comments, examples, and proofs.

Function	$\arcsin x$	$\arccos x$	$\arctan x$	$\operatorname{arcsec} x$
Derivative	$\dfrac{1}{\sqrt{1-x^2}}$	$\dfrac{-1}{\sqrt{1-x^2}}$	$\dfrac{1}{1+x^2}$	$\dfrac{1}{\lvert x \rvert \sqrt{x^2-1}}$

Observe:

- **Domains**   The functions $\arcsin x$, $\arccos x$, and $\operatorname{arcsec} x$ are defined on limited domains; the same is true, naturally, of their derivatives. The arcsine function, for instance, has domain $[-1, 1]$; its derivative has the (slightly) smaller domain $(-1, 1)$. ➤

  This section's exercises have more details on domains.

- **Not much difference**   The derivatives of $\arcsin x$ and $\arccos x$ differ only by a sign, so $(\arcsin x + \arccos x)' = 0$ for all legal inputs x. This is no accident—for all x in $[-1, 1]$,

$$\arcsin x + \arccos x = \frac{\pi}{2}.$$

  (The graphs of $\arcsin x$ and $\arccos x$ convey the same information.) ➤

  Plot both graphs on the same axes to see how.

- **Graphical evidence**   We'll show symbolically *why* these formulas hold, but graphs give immediate evidence *that* they hold. For instance, Figure 9 shows graphs of the arctangent function and its claimed derivative.

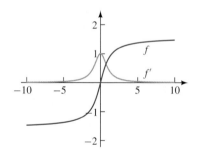

FIGURE 9
$f(x) = \arctan x$ and $f'(x) = \frac{1}{1+x^2}$

Proving the formulas   The derivative formulas can be proved in several different ways. ➤ We illustrate two calculations by example. The remaining derivatives of inverse trigonometric functions can be found in similar ways.

The underlying ideas are similar in each case.

EXAMPLE 6   Use Theorem 1 to differentiate the arcsine function.

Solution   Applying Theorem 1 with $g(x) = \arcsin x$ and $f(x) = \sin x$ gives

$$(\arcsin x)' = \frac{1}{\cos(\arcsin x)} = \frac{1}{\sqrt{1-x^2}};$$

the last equation follows from a calculation related to Figure 8. ∎

EXAMPLE 7   Use implicit differentiation to differentiate the arctangent function.

Solution   If $y = \arctan x$, then $x = \tan y$. Differentiating this equation implicitly with respect to x gives

$$1 = \sec^2 y \frac{dy}{dx}, \quad \text{or} \quad \frac{dy}{dx} = \frac{1}{\sec^2(\arctan x)}.$$

The last expression *can* be simplified using the ideas near Figure 8. For variety, we'll use instead the trigonometric identity $\sec^2 t = 1 + \tan^2 t$:

$$\frac{1}{\sec^2(\arctan x)} = \frac{1}{1 + \tan^2(\arctan x)} = \frac{1}{1 + x^2}.$$

(We used the identity $\tan(\arctan x) = x$ in the last step.) ∎

BASIC EXERCISES

In Exercises 1–6, find the number exactly (in radians).

1. $\arcsin(1)$

2. $\arcsin\left(\sqrt{3}/2\right)$

3. $\arccos(-1)$

4. $\arccos\left(-\sqrt{2}/2\right)$

5. $\arctan(1)$

6. $\arctan\left(\sqrt{3}\right)$

In Exercises 7–12, rewrite the given expression as an algebraic expression. For example, $\cos(\arcsin x) = \sqrt{1 - x^2}$.

7. $\sin(\arccos x)$

8. $\tan(\arcsin x)$

9. $\cos(\arctan x)$

10. $\tan(\text{arcsec } x)$

11. $\cos\left(\arctan(2x)\right)$

12. $\sin\left(\arctan(2x)\right)$

In Exercises 13–24, find $f'(x)$.

13. $f(x) = \arctan(2x)$

14. $f(x) = \arctan(x^2)$

15. $f(x) = \sqrt{\arcsin x}$

16. $f(x) = \arcsin(\sqrt{x})$

17. $f(x) = \arcsin(e^x)$

18. $f(x) = e^{\arctan x}$

19. $f(x) = \arctan(\ln x)$

20. $f(x) = \arccos(2x + 3)$

21. $f(x) = \text{arcsec}(x/2)$

22. $f(x) = \arcsin x / \arccos x$

23. $f(x) = x^2 \arctan(\sqrt{x})$

24. $f(x) = \ln(2 + \arcsin x)$

25. Find four values of x for which $\cos x = 1/2$.

26. Use a calculator to find four values of x (two positive, two negative) for which $\tan x = 2$.

27. Suppose that the point $(-3, 5)$ is on the graph of f. Find a point on the graph of f^{-1}.

28. Suppose that $f^{-1}(2) = 3$. Find a point on the graph of f.

FURTHER EXERCISES

Let f and g be inverse functions, both defined for all x. In Exercises 29–32, explain briefly why the statement is true.

29. If f has a horizontal asymptote, then g has a vertical asymptote.

30. If $f'(x) > 0$ for all x, then $g'(x) > 0$ for all x. [HINT: Use Theorem 1, page S-3.]

31. If $f'(x) < 0$ for all x, then $g'(x) < 0$ for all x.

32. If $f'(1) = 0$ and $g(3) = 1$, then $g'(3)$ is undefined.

In Exercises 33–36, $f(x) = a + bx$, where a and b are any constants, $b \neq 0$.

33. Explain why f has an inverse function g.

34. Does f have an inverse function g if $b = 0$? Justify your answer.

35. Find a formula for the inverse function $g(x) = f^{-1}(x)$.

36. Let $g(x) = f^{-1}(x)$. Find formulas for f' and g'. Are the answers consistent with Theorem 1, page S-3?

The equation $\arcsin(\sin x) = x$ suggests that the graph of $y = \arcsin(\sin x)$, over any interval, might be the straight line $y = x$. It isn't (see below). Exercises 37–39 explore this fact.

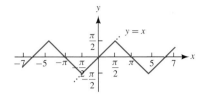

37. Explain why $\arcsin(\sin 5) \neq 5$.

38. For which values of x *does* the equation $\arcsin(\sin x) = x$ hold?

39. Let $f(x) = \arcsin(\sin x)$. Show that $f'(x) = \dfrac{\cos x}{|\cos x|}$.

40. What is the domain of the derivative of the arcsine function? Give a geometric explanation for why it is not the same as the domain of the arcsine function.

41. What is the domain of the derivative of the arctangent function? Is it the same as the domain of the arctangent function?

42. What is the domain of the derivative of the arcsecant function? Is it the same as the domain of the arcsecant function?

43. Let $f(x) = \arctan x$.

   (a) Plot graphs of f and f' on the same axes.

   (b) How is the fact that the tangent function has vertical asymptotes at $x = \pm\pi/2$ reflected in the graph of the arctangent function?

   (c) Is f even, odd, or neither?

   (d) Show that f is increasing on $(-\infty, \infty)$. How does the graph of f' reflect this fact?

   (e) Where is f concave up? Concave down? Does f have any inflection points? If so, where?

44. Let $f(x) = \arcsin x$.

   (a) Plot graphs of f and f' on the same axes.

   (b) Is f even, odd, or neither?

   (c) Show that f is increasing on $(-1, 1)$. How does the graph of f' reflect this fact?

   (d) Where is f concave up? Concave down? Does f have any inflection points? If so, where?

45. Let $f(x) = \arccos x$.

   (a) Plot graphs of f and f' on the same axes.

   (b) Is f even, odd, or neither?

   (c) Show that f is decreasing on $(-1, 1)$. How does the graph of f' reflect this fact?

   (d) Where is f concave up? Concave down? Does f have any inflection points? If so, where are they located?

46. **(a)** The function $\operatorname{arcsec} x$ accepts as inputs only values of x for which $|x| \geq 1$. Why is this restriction necessary? [HINT: What's the domain of $\arccos x$?]

   (b) What's the *range* of $\operatorname{arcsec} x$? [HINT: What's the range of $\arccos x$?]

47. Let $f(x) = \operatorname{arcsec} x$.

   (a) Plot graphs of f and f' using the plotting window $[-5, 5] \times [0, 5]$. [NOTE: If your calculator or computer

doesn't recognize the arcsecant function, trick it by plotting $\arccos(1/x)$ instead.]

   (b) Is f even, odd, or neither?

   (c) Show that f is increasing everywhere in the interior of its domain. How does the graph of f' reflect this fact?

   (d) Where is f concave up? Concave down? Does f have any inflection points? If so, where are they located?

48. **(a)** By drawing an appropriately labeled right triangle, show that $\arctan x + \arctan(1/x) = \pi/2$ if $x > 0$.

   (b) Use calculus to show that $\arctan x + \arctan(1/x) = \pi/2$ for all $x > 0$. [HINT: Start by differentiating both sides of the equation.]

   (c) Find an equation similar to the one in part (a) that is valid if $x < 0$.

Use calculus to prove the identities in Exercises 49–53. [HINT: Start by showing that the derivative of the expression on the left is equal to the derivative of the expression on the right.]

49. $\arcsin(-x) = -\arcsin(x)$

50. $\arccos(-x) = \pi - \arccos(x)$

51. $\arccos x = \pi/2 - \arcsin x$

52. $\arctan\left(x/\sqrt{1-x^2}\right) = \arcsin x$

53. $\arcsin\left(\dfrac{x-1}{x+1}\right) = 2\arctan(\sqrt{x}) - \pi/2$

54. **(a)** Show that $2\arcsin x = \arccos(1 - 2x^2)$ if $0 \leq x \leq 1$.

   (b) Find an identity similar to the one in part (a) that is valid if $-1 \leq x \leq 0$.

55. Let $f(x) = \arctan\left(\dfrac{1+x}{1-x}\right)$.

   (a) Assume that $x < 1$. Show that there is a constant C such that $f(x) = C + \arctan x$.

   (b) Use the result in part (a) to evaluate $\lim_{x \to 1^-} f(x)$.

   (c) Evaluate $\lim_{x \to 1^+} f(x)$.

56. **(a)** Show that $\tan(\operatorname{arcsec} x) = \sqrt{x^2 - 1}$ if $x \geq 1$.

   (b) Find an algebraic expression for $\tan(\operatorname{arcsec} x)$ if $x \leq -1$.

57. **(a)** Use the identity $\tan(x + y) = \dfrac{\tan x + \tan y}{1 - \tan x \tan y}$ to derive the identity

$$\arctan x + \arctan y = \arctan\left(\dfrac{x+y}{1-xy}\right).$$

[HINT: Start by writing $x = \arctan u$ and $y = \arctan v$.]

   (b) What conditions must be satisfied by x and y for the identity derived in part (a) to hold?

58. **(a)** Use the identity in part (a) of the previous exercise to show that $\pi/4 = \arctan(1/2) + \arctan(1/3)$.

   (b) Show that $\pi/4 = 2\arctan(1/4) + \arctan(7/23)$. [HINT: Use the identity in part (a) of the previous exercise twice.]

59. Show that $x/(1 + x^2) \leq \arctan x \leq x$ for all $x \geq 0$.

4.2 MORE ON LIMITS: LIMITS INVOLVING INFINITY AND l'HÔPITAL'S RULE

In Chapter 2 we studied limits and some of their properties and related ideas; we applied limits mainly in defining the derivative. In this section we revisit the limit theme. First we define another variant on the basic limit idea—limits that involve the ∞ symbol—and interpret results graphically in terms of asymptotes. Then we show how to use derivatives and a special result, l'Hôpital's rule, to calculate some challenging limits.

Limits at infinity and infinite limits

Read the → symbol as "approaches."

Limits at infinity and infinite limits (they're different!) describe the behavior of functions when either inputs or outputs increase or decrease "without bound." Informal definitions and examples will convey the idea. ◄

> **DEFINITION (Limit at infinity)** $\lim\limits_{x \to \infty} f(x) = L$ means that $f(x) \to L$ as $x \to \infty$ (that is, as x increases without bound).

> **DEFINITION (Infinite limit)** $\lim\limits_{x \to a} f(x) = \infty$ means that $f(x) \to \infty$ (that is, $f(x)$ "blows up") as $x \to a$.

Other variants exist, including limits as $x \to -\infty$ and one-sided infinite limits. Defining all of them separately would be tedious; what each means is usually clear from context.

EXAMPLE 1 Interpret the following limit statements:

$$\lim_{x \to \infty} \frac{1}{x} = 0; \qquad \lim_{x \to 0} \frac{1}{x^2} = \infty;$$
$$\lim_{x \to 0^-} \frac{1}{x} = -\infty; \qquad \lim_{x \to \infty} \sin x \text{ does not exist.}$$

Solution The limit $\lim\limits_{x \to \infty} \sin x$ does not exist because the sine function oscillates forever, never settling down near any single value. The graphs in Figure 1 help explain the first three limit statements.

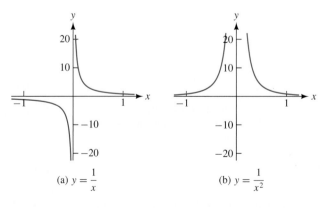

(a) $y = \dfrac{1}{x}$ (b) $y = \dfrac{1}{x^2}$

FIGURE 1

Notice especially the *graphical* meaning of each infinite limit and limit at infinity: each one corresponds to a vertical or horizontal **asymptote**. We'll refine this observation shortly. ∎

New limits from old   When calculating limits of *any* sort it's convenient to reduce complicated limits to combinations of simpler ones. The next example illustrates the process with a limit at infinity.

EXAMPLE 2   Find $\displaystyle\lim_{x\to\infty} \frac{x^2+3}{2x^2-5}$.

Solution   With a graph it would be easy to guess that the limit is 1/2. ➡ Algebra clinches the case:

Plot the function yourself if you like.

$$\lim_{x\to\infty} \frac{x^2+3}{2x^2-5} = \lim_{x\to\infty} \frac{x^2\cdot(1+3/x^2)}{x^2\cdot(2-5/x^2)} \qquad \text{(factor out } x^2\text{)}$$

$$= \lim_{x\to\infty} \frac{1+3/x^2}{2-5/x^2} \qquad \text{(cancel)}$$

$$= \frac{\displaystyle\lim_{x\to\infty} 1 + \lim_{x\to\infty} 3/x^2}{\displaystyle\lim_{x\to\infty} 2 - \lim_{x\to\infty} 5/x^2} \qquad \text{(separate limits)}$$

$$= \frac{1+0}{2-0} = 1/2. \qquad \text{(combine results)}$$

∎

We used several plausible facts in the preceding calculation: that $\displaystyle\lim_{x\to\infty} 3/x^2 = \lim_{x\to\infty} 5/x^2 = 0$, that the limit of a sum or quotient is the sum or quotient of the limits, and so on. Such operations are justified by Theorem 2—a version of which we saw already in Section 2.3. What's new is that the theorem applies (if the hypotheses are satisfied) to infinite limits and limits at infinity.

THEOREM 2 (Algebra with limits)   Suppose that

$$\lim_{x\to a} f(x) = L \qquad \text{and} \qquad \lim_{x\to a} g(x) = M,$$

where L and M are finite numbers; a may be $\pm\infty$. Let k be any constant. Then

(i) $\displaystyle\lim_{x\to a} kf(x) = kL$;

(ii) $\displaystyle\lim_{x\to a} \big(f(x) + g(x)\big) = L + M$;

(iii) $\displaystyle\lim_{x\to a} f(x)\cdot g(x) = L\cdot M$;

(iv) $\displaystyle\lim_{x\to a} \frac{f(x)}{g(x)} = \frac{L}{M}$ (if $M \neq 0$).

Observe:

- **Infinite limits**   The rules also hold for infinite limits (that is, when $L = \pm\infty$, $M = \pm\infty$, or both) but only if appropriate care is taken with uses of the infinity symbol. (We return to this question shortly.)

- **"Obvious" limits**   The theorem is useful only if we *know* some limits to start with. To get such a foothold, we'll take as "obvious" limits such as the following:

$$\lim_{x \to \infty} x^2 = \infty; \quad \lim_{x \to \infty} \frac{1}{x^2} = 0; \quad \lim_{x \to \infty} e^x = \infty; \quad \lim_{x \to -\infty} e^x = 0.$$

Asymptotes and limits

The connections among asymptotes, limits at infinity, and infinite limits are easily summarized:

- **Horizontal asymptotes; limits at infinity**   The horizontal line $y = L$ is an asymptote to the graph of f if f approaches L as $x \to \pm\infty$—that is, if

$$\lim_{x \to \infty} f(x) = L \quad \text{or} \quad \lim_{x \to -\infty} f(x) = L.$$

- **Vertical asymptotes; infinite limits**   The vertical line $x = a$ is an asymptote to the graph of f if f blows up (or blows down) at a—that is, if

$$\lim_{x \to a^-} f(x) = \pm\infty \quad \text{or} \quad \lim_{x \to a^+} f(x) = \pm\infty.$$

Limits and asymptotes of rational functions   Information about asymptotes is useful for *any* function. Horizontal asymptotes describe a function's long-term behavior, while vertical asymptotes reveal sudden spikes and other anomalies. For polynomials and rational functions (quotients of polynomials) the story is especially simple.

Polynomials' asymptotes   Polynomials never have *vertical* asymptotes. A vertical asymptote occurs at a *finite* value of x near which a function's value is unbounded. Because polynomials have the pleasant form

$$p(x) = a_0 + a_1 x + \cdots + a_n x^n ,$$

free of troublesome denominators, polynomials cannot blow up (or down) except as x approaches $\pm\infty$.

Every *constant* polynomial has a horizontal asymptote—the graph itself. If, say, $p(x) = 3$, then the p-graph is the horizontal line $y = 3$. Nonconstant polynomials, on the other hand, never have horizontal asymptotes; the following Fact explains why.

> **FACT**   If $p(x)$ is a nonconstant polynomial, then
> $$\lim_{x \to \infty} p(x) = \pm\infty; \quad \lim_{x \to -\infty} p(x) = \pm\infty.$$

We won't prove this fact formally, but we'll illustrate why it holds.

EXAMPLE 3   Let $p(x) = 2x^3 - 53x^2 - 123x$. How does p behave as $x \to \infty$? What if $x \to -\infty$?

Solution   It's clear at a glance that if $x \to \infty$, then $2x^3 \to \infty$, $-53x^2 \to -\infty$, and $-123x \to -\infty$. But what about the sum? The trick is to *factor out the largest power of x*:

$$\lim_{x \to \infty} \left(2x^3 - 53x^2 - 123x \right) = \lim_{x \to \infty} x^3 \left(2 - \frac{53}{x} - \frac{123}{x^2} \right)$$

$$= \lim_{x \to \infty} x^3 \cdot \lim_{x \to \infty} \left(2 - \frac{53}{x} - \frac{123}{x^2} \right)$$

$$= \infty \cdot (2 - 0 - 0) = \infty .$$

Similarly,

$$\lim_{x \to -\infty} \left(2x^3 - 53x^2 - 123x\right) = \lim_{x \to -\infty} x^3 \cdot \lim_{x \to -\infty} \left(2 - \frac{53}{x} - \frac{123}{x^2}\right)$$
$$= -\infty \cdot (2 - 0 - 0) = -\infty.$$

Plotting $p(x)$ over a large x-interval, say $[-10, 10]$, supports our limit calculations. ∎

Algebra with the infinity symbol In the calculations of Example 3 we breezed past two possibly suspicious claims: $\infty \cdot 2 = \infty$ and $-\infty \cdot 2 = -\infty$. Do they really make sense? Because ∞ and $-\infty$ are not numbers, their appearance in equations should always raise flags of caution. Here, fortunately, we're OK: neither $\infty \cdot 2$ nor $-\infty \cdot 2$ is really ambiguous. The following table collects some rules for proper and improper uses of the infinity symbol.

Good and bad uses of the ∞ symbol	
Expression	**What it means; remarks**
$1/\infty = 0$	Usually OK, but if we really mean, say, $\lim\limits_{x \to \infty} 1/x = 0$, why not say so?
$3 \cdot \infty = \infty$	OK as mathematical shorthand. It means: If any quantity increases without bound, then so does three times that quantity.
$\infty + \infty = \infty$	OK again. It means: If two quantities increase without bound, then so does their sum.
$1/0 = \infty$	Wrong. Division by 0 is not defined. Worse, $1/x \to \infty$ as $x \to 0$ from the right, but $1/x \to -\infty$ as $x \to 0$ from the left.
$\infty - \infty = 0$	Wrong again. For example, as $x \to \infty$, both $x^3 \to \infty$ and $x^2 \to \infty$, but $x^3 - x^2 \to \infty$.
$\infty/\infty = 1$	Wrong. For example, as $x \to \infty$, both $x^3 \to \infty$ and $x^2 \to \infty$, but $x^3/x^2 = x \to \infty$.
$0 \cdot \infty = 0$	Wrong. For example, as $x \to \infty$, we have $x^2 \to \infty$ and $1/x \to 0$, but $x^2 \cdot 1/x = x \to \infty$.

Horizontal asymptotes of rational functions: appeal to higher powers There's a simple algebraic strategy for finding horizontal asymptotes of rational functions:

Factor out the highest powers of x from the numerator and denominator. Then cancel as possible.

EXAMPLE 4 Find the horizontal asymptotes (if any) of the rational functions

$$r(x) = \frac{2x^3 - x}{x^2 - 1}, \qquad s(x) = \frac{2x^3 - x}{x^3 - 1}, \qquad \text{and} \qquad t(x) = \frac{2x^3 - x}{x^4 - 1}.$$

Solution Graphs follow soon, but let's not peek. Instead, we'll factor:

$$r(x) = \frac{2x^3 - x}{x^2 - 1} = \frac{x^3(2 - 1/x^2)}{x^2(1 - 1/x^2)} = x \cdot \frac{2 - 1/x^2}{1 - 1/x^2};$$

$$s(x) = \frac{2x^3 - x}{x^3 - 1} = \frac{x^3(2 - 1/x^2)}{x^3(1 - 1/x^3)} = \frac{2 - 1/x^2}{1 - 1/x^3};$$

$$t(x) = \frac{2x^3 - x}{x^4 - 1} = \frac{x^3(2 - 1/x^2)}{x^4(1 - 1/x^4)} = \frac{1}{x} \cdot \frac{2 - 1/x^2}{1 - 1/x^4}.$$

The last form of each function (and Theorem 2) shows that if $x \to \infty$ or $x \to -\infty$, then

$$r(x) \to \infty, \qquad s(x) \to 2, \qquad \text{and} \qquad t(x) \to 0.$$

Thus, s and t have horizontal asymptotes at $y = 2$ and $y = 0$, respectively, while r has no horizontal asymptotes. ∎

*Recall: r is a root of f if
f(r) = 0.*

Vertical asymptotes and roots of the denominator   If a rational function has vertical asymptotes, they can be located only at roots of the denominator. ◄ But not every root of the denominator needs to have an asymptote. If both the numerator *and* denominator have a root at $x = a$, then they have a common factor $(x - a)$, which can be canceled.

EXAMPLE 5   Let r, s, and t be as in Example 4. Find all vertical asymptotes.

Factoring polynomials can be hard, so we authors usually choose easy-to-factor polynomials, or we factor them for you.

Solution   A rational function has a vertical asymptote at any x-value where the denominator is zero and the numerator isn't. To look for asymptotes, we first factor the denominator as much as possible. ◄ For r, s, and t, we get

$$r(x) = \frac{2x^3 - x}{x^2 - 1} = \frac{2x^3 - x}{(x+1)(x-1)};$$

$$s(x) = \frac{2x^3 - x}{x^3 - 1} = \frac{2x^3 - x}{(x-1)(x^2 + x + 1)};$$

$$t(x) = \frac{2x^3 - x}{x^4 - 1} = \frac{2x^3 - x}{(x^2 - 1)(x^2 + 1)} = \frac{2x^3 - x}{(x-1)(x+1)(x^2 + 1)}.$$

The factored denominators show that, for each function, $x = 1$ and $x = -1$ are the only possible vertical asymptotes. Because the numerator $2x^3 - x$ is *not* zero at $x = \pm 1$, it follows that r and t have vertical asymptotes at $x = 1$ and at $x = -1$, and that s has a vertical asymptote at $x = 1$.

On either side of a vertical asymptote, a function may tend either to ∞ or to $-\infty$. What happens for r, s, and t? Graphs make the answers obvious, but even without graphs the question is usually easy to answer. We need only determine the *sign* of each function just to the right and just to the left of each asymptote. For $r(x)$, just to the left of $x = -1$ we see that

$$2x^3 - x < 0, \qquad x + 1 < 0, \qquad \text{and} \qquad x - 1 < 0.$$

r "blows down" to $-\infty$ as $x \to -1$ from the left.

Thus $r(x) < 0$, and so $\lim_{x \to -1^-} r(x) = -\infty$. ◄ A similar sign check reveals that r blows *up* to ∞ as $x \to -1$ from the *right*.

We've waited long enough: Figure 2 shows the graphs and their vertical asymptotes.

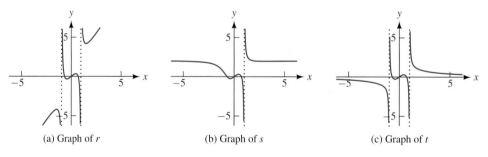

(a) Graph of r (b) Graph of s (c) Graph of t

FIGURE 2
Graphs of r, s, and t

■

Asymptotes of other functions   *Any* function may have asymptotes—with the same connection to infinite limits and limits at infinity as for rational functions. Calculating these limits algebraically may take extra ingenuity.

EXAMPLE 6   Does $f(x) = \sqrt{x^2 + x} - x$ have any horizontal asymptotes?

A graph might help; see for yourself.

Solution   How $f(x)$ behaves as $x \to \infty$ isn't obvious from the formula. ◄ But a table of values suggests a limit of $1/2$:

x	1	2	4	8	16	32	64	128	256
$f(x)$	0.4142	0.4495	0.4721	0.4853	0.4924	0.4962	0.4981	0.4990	0.4995

We'll use algebra to check our guess. Note the trick in the first step—multiplying numerator and denominator by a "conjugate" expression:

$$\lim_{x\to\infty}(\sqrt{x^2+x}-x) = \lim_{x\to\infty}\frac{(\sqrt{x^2+x}-x)(\sqrt{x^2+x}+x)}{\sqrt{x^2+x}+x}$$

$$= \lim_{x\to\infty}\frac{x}{\sqrt{x^2+x}+x} \qquad \text{(multiplying out)}$$

$$= \lim_{x\to\infty}\frac{x}{x\left(\sqrt{1+1/x}+1\right)} \qquad \begin{array}{l}\text{(factor } x \text{ from denominator;}\\ \text{note that } x > 0)\end{array}$$

$$= \lim_{x\to\infty}\frac{1}{\sqrt{1+1/x}+1} = \frac{1}{2}. \qquad ■$$

Composite limits

Limits often involve functions built up by composition; consequently, it's helpful to know how the operations of composition and taking limits interact. First let's discuss some straightforward examples.

EXAMPLE 7 Find and discuss each of the following limits. ➡

Try to guess all three limits before reading on.

$$L_1 = \lim_{x\to 0}\ln(\cos x); \qquad L_2 = \lim_{x\to 0}\ln\left(\frac{\sin x}{x}\right); \qquad L_3 = \lim_{x\to\infty}\exp\left(\frac{x+1}{x+2}\right).$$

Solution The first limit is easiest. As $x \to 0$, $\cos x \to \cos 0$, so

$$\ln(\cos x) \to \ln(\cos 0) = \ln 1 = 0.$$

Why did things work so well? Because both functions involved are continuous at the points that matter: As $x \to 0$, $\cos x \to \cos 0$ because the cosine function is continuous at $x = 0$; as $\cos x \to 1$, $\ln(\cos x) \to \ln 1$ because the natural logarithm function is continuous at $x = 1$.

To find L_2 we reason similarly. Recall first that as $x \to 0$, $(\sin x)/x \to 1$. ➡ Thus, as $x \to 0$,

We've used this fact repeatedly; a formal proof appears in Appendix H.

$$\frac{\sin x}{x} \to 1 \implies \ln\left(\frac{\sin x}{x}\right) \to \ln 1 = 0.$$

(The implication holds because the function $\ln(x)$ is continuous at $x = 1$.)

The limit L_3 is "at infinity," but the same sort of reasoning applies as for L_1 and L_2: ➡

Convince yourself of the first implication.

$$x \to \infty \implies \frac{x+1}{x+2} \to 1 \implies \exp\left(\frac{x+1}{x+2}\right) \to \exp(1) = e.$$

Once again, continuity is key: the second implication holds because the exponential function is continuous at $x = 1$. ■

Composing limits: a general principle One basic principle applies to all three limits in Example 7:

> **FACT (Limit of a composite function)**   Let f and g be functions such that $\lim_{x \to a} g(x) = b$ and f is continuous at $x = b$. (The constant a may be $\pm\infty$.) Then
> $$\lim_{x \to a}(f \circ g)(x) = \lim_{x \to a} f\big(g(x)\big) = f(b).$$

A formal proof, which we omit, appeals to the ϵ–δ definition of limit.

Squeezing limits at infinity

In Section 2.3 we applied the **squeeze principle** (see Theorem 7, page 110) to find some limits for which algebraic techniques don't help. Here's a version that applies to limits at infinity:

> **THEOREM 3 (The squeeze principle)**   Suppose that
> $$f(x) \le g(x) \le h(x)$$
> for all large positive x. If $f(x) \to L$ and $h(x) \to L$ as $x \to \infty$, then $g(x) \to L$ as $x \to \infty$.

EXAMPLE 8   Explain why $\lim\limits_{x \to \infty} \dfrac{\sin x}{x} = 0$.

Solution   Because $-1 \le \sin x \le 1$, the inequality
$$-\frac{1}{x} \le \frac{\sin x}{x} \le \frac{1}{x}$$
holds for all $x > 0$. Because the left and the right sides tend to 0 as $x \to \infty$, so must the middle expression.   ∎

Indeterminate forms

This is mathematicians' jargon for "just by looking."

Many limits are easy to guess "by inspection." ◄ Other limits are less susceptible to intuition. Consider these:
$$\lim_{x \to \infty} \frac{x^2}{2^x}; \qquad \lim_{x \to 0} \frac{\sin(2x)}{x}; \qquad \lim_{x \to \infty} xe^{-x}; \qquad \lim_{x \to \infty} \frac{x^2+1}{2x^2+3}.$$

Such limits are called **indeterminate forms** because, in each case, two conflicting tendencies operate. In the first limit, for instance, both numerator and denominator blow up:
$$x^2 \to \infty \qquad \text{and} \qquad 2^x \to \infty \qquad \text{as } x \to \infty.$$

The limit is therefore called **indeterminate of type** ∞/∞. The key question is how *fast* the numerator and denominator grow compared with each other.

The limit $\lim\limits_{x \to 0} \dfrac{\sin(2x)}{x}$ is ambiguous for another reason: Since both numerator and denominator tend to zero, their ratio is called **indeterminate of type** $0/0$. Again the question concerns the relative *rates* at which numerator and denominator tend to zero.

The limit $\lim\limits_{x \to \infty} xe^{-x}$ involves a product (not a quotient): its two factors behave in conflicting ways:
$$x \to \infty \qquad \text{but} \qquad e^{-x} \to 0.$$

Therefore, the limit is called **indeterminate of type** $\infty \cdot 0$.

The last limit, like the first, is indeterminate of type ∞/∞, but this time no big guns are needed to find the limit. Dividing numerator and denominator by x^2 does the job:

$$\lim_{x\to\infty}\frac{x^2+1}{2x^2+3}=\lim_{x\to\infty}\frac{1+1/x^2}{2+3/x^2}=\frac{1}{2}.$$

Graphical and numerical evidence   The preceding limit calculation worked because the limit happened to involve a rational function; for the other limits above we need more than algebra. Graphical and numerical approaches are a first step. The graphs in Figure 3 strongly suggest that as $x\to\infty$,

$$\frac{x^2}{2^x}\to 0\qquad\text{and}\qquad xe^{-x}\to 0.$$

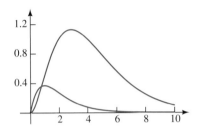

Decide which graph is which—but note that both functions have the same limit at ∞.

FIGURE 3
$$y=\frac{x^2}{2^x}\text{ and }y=xe^{-x}$$

The following table suggests equally convincingly that $\lim\limits_{x\to 0}\dfrac{\sin(2x)}{x}=2$.

x	-0.1	-0.01	-0.0001	$\ldots$	0.0001	0.01	0.1
$\sin(2x)/x$	1.98670	1.99987	1.99999	$\ldots$	1.99999	1.99987	1.98670

l'Hôpital's rule: finding limits by differentiation

Graphics, numerics, and algebra suggested plausible values for all four of the preceding indeterminate forms. l'Hôpital's rule offers another, more powerful approach.

How it works   l'Hôpital's rule says that *under appropriate conditions* (which we'll discuss in a moment) an indeterminate form can be evaluated by *differentiating the numerator and the denominator separately*. In symbols:

$$\lim_{x\to a}\frac{f(x)}{g(x)}=\lim_{x\to a}\frac{f'(x)}{g'(x)}.$$

Deferring legalities for the moment, we show l'Hôpital's rule in action.

EXAMPLE 9   Show that $\lim\limits_{x\to 0}\dfrac{\sin(2x)}{x}=2$.

Solution   By l'Hôpital's rule,

$$\lim_{x\to 0}\frac{\sin(2x)}{x}=\lim_{x\to 0}\frac{2\cos(2x)}{1}.$$

The right-hand limit is no longer indeterminate; simply plugging in $x=0$ gives 2 (as we'd guessed from numerical data).   ■

EXAMPLE 10   Use l'Hôpital's rule to find $\lim\limits_{x \to \infty} \dfrac{x^2}{2^x}$.

Solution   Assuming that the rule applies, we have

$$\lim_{x \to \infty} \frac{x^2}{2^x} = \lim_{x \to \infty} \frac{2x}{2^x \ln 2}.$$

The right-hand limit is *still* indeterminate of type ∞/∞, so we apply the rule *again*:

$$\lim_{x \to \infty} \frac{x^2}{2^x} = \lim_{x \to \infty} \frac{2x}{2^x \ln 2} = \lim_{x \to \infty} \frac{2}{2^x \ln 2 \cdot \ln 2}.$$

The rightmost limit is (at last) *not* indeterminate. Because the numerator remains constant while the denominator blows up, the last limit must be zero. So, therefore, must the first.   ■

EXAMPLE 11   Graphical evidence suggests that $\lim\limits_{x \to \infty} xe^{-x} = 0$. Does l'Hôpital's rule agree?

Solution   Yes—but only after some algebra. To use the rule, we first rewrite the expression as a *quotient* and then differentiate:

$$\lim_{x \to \infty} xe^{-x} = \lim_{x \to \infty} \frac{x}{e^x} = \lim_{x \to \infty} \frac{1}{e^x} = 0,$$

again as expected.   ■

When it works, when it doesn't: careful statements   Using l'Hôpital's rule requires care. The idea that *every* limit of quotients can be handled *à la l'Hôpital* (by differentiating top and bottom) is tempting, but wrong. For instance,

$$\lim_{x \to 0} \frac{x + 42}{x + 1} = 42, \qquad \text{but} \qquad \lim_{x \to 0} \frac{(x + 42)'}{(x + 1)'} = \lim_{x \to 0} \frac{1}{1} = 1.$$

We erred in trying to apply the rule where it doesn't apply—to a limit that wasn't indeterminate to start with.

When *does* the rule work? The following theorem gives full details:

THEOREM 4 (l'Hôpital's rule)   Let f and g be differentiable functions, such that

(a) as $x \to a$, either
  (i) $f(x) \to 0$ and $g(x) \to 0$; or (ii) $f(x) \to \pm\infty$ and $g(x) \to \pm\infty$;
(b) $\lim\limits_{x \to a} \dfrac{f'(x)}{g'(x)}$ exists.

Then

$$\lim_{x \to a} \frac{f(x)}{g(x)} = \lim_{x \to a} \frac{f'(x)}{g'(x)}.$$

Observe:

- **Limits at infinity**   The values $a = \pm\infty$ are allowed in the theorem—luckily, because we've already used the result twice with $a = \infty$. One-sided limits are also permitted.
- **Truly indeterminate**   Hypothesis (a) guarantees that $\lim f/g$ is genuinely indeterminate and is of either type $0/0$ or type ∞/∞.

EXAMPLE 12 Find $\lim\limits_{x\to 0^+} x \ln x$. ➡

The right-hand limit is needed because $\ln x$ is defined only for $x > 0$.

Solution The theorem involves *quotients*, so let's produce one:

$$\lim_{x\to 0^+} x \ln x = \lim_{x\to 0^+} \frac{\ln x}{1/x}.$$

The right-hand form is now indeterminate of type ∞/∞, and

$$\lim_{x\to 0^+} \frac{(\ln x)'}{(1/x)'} = \lim_{x\to 0^+} \frac{1/x}{-1/x^2} = \lim_{x\to 0^+} -x = 0.$$

By l'Hôpital's rule, the original limit is 0. ∎

Why it works: comparing rates The theorem concerns the behavior, as x tends to a, of the ratio $f(x)/g(x)$. If *both* numerator and denominator tend either to zero or to infinity, then the limit—if it exists—depends on the relative *rates* at which f and g tend to their limits. The derivatives $f'(x)$ and $g'(x)$ measure these rates. A formal proof of l'Hôpital's rule makes this general idea rigorous.

We'll omit a formal argument, but we can see the general idea symbolically and graphically. Let's suppose that f and g are differentiable functions, with $f(a) = g(a) = 0$. Then, if $x \approx a$, f and g are close to their respective tangent lines at $x = a$:

$$f(x) \approx f(a) + f'(a)(x - a) = f'(a)(x - a);$$
$$g(x) \approx g(a) + g'(a)(x - a) = g'(a)(x - a).$$

If $g'(a) \neq 0$ and $x \neq a$, then

$$\frac{f(x)}{g(x)} \approx \frac{f'(a)(x - a)}{g'(a)(x - a)} \approx \frac{f'(a)}{g'(a)},$$

so

$$\lim_{x\to a} \frac{f(x)}{g(x)} = \lim_{x\to a} \frac{f'(x)}{g'(x)} = \frac{f'(a)}{g'(a)},$$

which is l'Hôpital's rule in its simplest form.

EXAMPLE 13 By l'Hôpital's rule, $\lim\limits_{x\to 0} \dfrac{\sin(2x)}{\sin x} = \lim\limits_{x\to 0} \dfrac{2\cos(2x)}{\cos x} = 2$. Interpret the result graphically, using tangent lines.

Solution Let $f(x) = \sin(2x)$ and $g(x) = \sin x$. The tangent lines to f and g at $x = 0$ are, respectively, the lines $y = 2x$ and $y = x$. Figure 4 shows graphs of f, g, and these tangent lines.

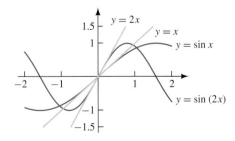

FIGURE 4
$y = \sin(2x)$, $y = \sin x$, and their tangent lines at the origin

The picture shows that, for x near 0, $f(x) \approx 2x$ and $g(x) \approx x$. Therefore,

$$\frac{f(x)}{g(x)} \approx \frac{2x}{x} = 2,$$

as claimed. ∎

BASIC EXERCISES

1. Let $p(x) = (x+1)(2-x)$. Evaluate

  (a) $\lim\limits_{x \to \infty} p(x)$

  (b) $\lim\limits_{x \to -\infty} p(x)$

2. Let $q(x) = x^{17} - 3x^{14} + 39$. Evaluate

  (a) $\lim\limits_{x \to \infty} q(x)$

  (b) $\lim\limits_{x \to -\infty} q(x)$

In Exercises 3–8, use the graphs of f and g shown below to evaluate the limit or to explain why the limit doesn't exist.

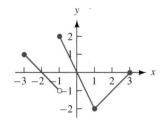
Graph of f

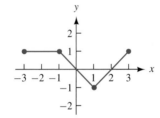
Graph of g

3. $\lim\limits_{x \to 2} \dfrac{f(x)}{g(x)}$

4. $\lim\limits_{x \to -2} \dfrac{g(x)}{f(x)}$

5. $\lim\limits_{x \to -2} \dfrac{x+2}{f(x)}$

6. $\lim\limits_{x \to -2} \dfrac{f(x)}{g(x)-1}$

7. $\lim\limits_{x \to 1} \dfrac{f(x)+2}{g(x)+1}$

8. $\lim\limits_{x \to 2} \dfrac{x^2-4}{g(x)}$

In Exericses 9–24, evaluate the limit.

9. $\lim\limits_{x \to \infty} \dfrac{3}{x^2}$

10. $\lim\limits_{x \to \infty} \dfrac{5}{\sqrt{x}}$

11. $\lim\limits_{t \to \infty} \dfrac{2t+3}{5-4t}$

12. $\lim\limits_{t \to 0} \dfrac{2t+3}{5-4t}$

13. $\lim\limits_{x \to \infty} \dfrac{x^2+1}{x}$

14. $\lim\limits_{x \to \infty} \dfrac{x^2+1}{2x^2+3}$

15. $\lim\limits_{x \to \infty} \dfrac{\sin x}{x}$

16. $\lim\limits_{x \to -\infty} \dfrac{x^3}{2+\cos x}$

17. $\lim\limits_{x \to \infty} \dfrac{2^x}{x^2}$

18. $\lim\limits_{x \to -\infty} \dfrac{2^x}{x^2}$

19. $\lim\limits_{x \to \infty} \dfrac{\ln x}{x^{2/3}}$

20. $\lim\limits_{x \to \infty} \sin\left(\dfrac{\pi}{x}\right)$

21. $\lim\limits_{x \to 0} \sin(\sin x)$

22. $\lim\limits_{x \to \infty} \exp\left((3x^2+x)/2x^3\right)$

23. $\lim\limits_{x \to 0} \cos\left(\dfrac{\tan x}{x}\right)$

24. $\lim\limits_{x \to 0^+} \arctan(x \ln x)$

In Exercises 25–30, use l'Hôpital's rule to evaluate the limit.

25. $\lim\limits_{x \to 1} \dfrac{x^3+x-2}{x^2-3x+2}$

26. $\lim\limits_{x \to 0} \dfrac{5x-\sin x}{x}$

27. $\lim\limits_{x \to 0} \dfrac{1-\cos x}{\sin(2x)}$

28. $\lim\limits_{x \to 0} \dfrac{1-\cos(5x)}{4x+3x^2}$

29. $\lim\limits_{x \to \infty} \dfrac{e^x}{x^2+x}$

30. $\lim\limits_{x \to 0} \dfrac{e^x-x-1}{x^2}$

In Exercises 31–35, use the graph of f shown to estimate the limit.

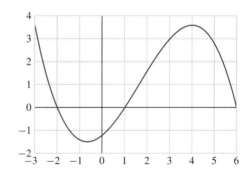

31. $\lim\limits_{x \to 1} \dfrac{f(x)}{x^2-1}$

32. $\lim\limits_{x \to 2} \dfrac{f(x)}{x^2-4}$

33. $\lim\limits_{x \to 4} \dfrac{f(x)}{(x-4)^2}$

34. $\lim\limits_{x \to 1} \dfrac{f(x)}{f(-2x)}$

35. $\lim\limits_{x \to 1} \dfrac{f(x-3)}{f(x+3)}$

36. Explain why l'Hôpital's rule can be used to evaluate $\lim\limits_{\theta \to \pi} \dfrac{\sin\theta}{\theta-\pi}$ but not $\lim\limits_{\theta \to \pi} \dfrac{\cos\theta}{\theta-\pi}$.

37. Can l'Hôpital's rule be used to evaluate $\lim\limits_{x \to \pi/2} \dfrac{\tan x}{x-\pi/2}$? Justify your answer.

38. Can l'Hôpital's rule be used to evaluate $\lim\limits_{\theta \to \infty} \dfrac{\cos\theta}{\theta}$? Justify your answer.

39. Can l'Hôpital's rule be used to evaluate $\lim\limits_{x \to \pi/2^+} \dfrac{\tan x}{x-\pi/2}$? Justify your answer.

FURTHER EXERCISES

40. Let $f(x) = \sqrt[3]{x}$. Explain why $\lim\limits_{x \to 0} f'(x) = \infty$.

41. Suppose that f is a (nonconstant) periodic function. Explain why $\lim\limits_{x \to \infty} f(x)$ does not exist.

42. Sketch the graph of a function f that has *all* of the following properties. (Be sure to label axes and indicate units on each axis.)
   (i) f has domain $(-3, \infty)$
   (ii) f has range $(-\infty, 2)$
   (iii) $f(-1) = 0$; $f(0) = 0$; $f(5) = 0$
   (iv) $\lim\limits_{x \to -3^+} f(x) = -\infty$
   (v) $\lim\limits_{x \to 2} f(x) = -3$
   (vi) $\lim\limits_{x \to \infty} f(x) = 2$

43. Sketch the graph of a function f that has *all* of the following properties:
   (i) f has domain $(-\infty, \infty)$
   (ii) f has range $(-3, \infty)$
   (iii) $\lim\limits_{t \to 2} f(t) = -2$
   (iv) $\lim\limits_{t \to \infty} f(t) = -3$
   (v) $\lim\limits_{t \to -\infty} f(t) = 3$

44. Suppose that $r(x)$ is a polynomial of degree n such that $\lim\limits_{x \to \infty} r(x) = -\infty$ and $\lim\limits_{x \to -\infty} r(x) = -\infty$. Can you tell if n is even or odd? Explain.

45. Suppose that h has a vertical asymptote at $x = 3$, that $\lim\limits_{x \to -\infty} h(x) = \infty$, and that $\lim\limits_{x \to \infty} h(x) = -1$.
   (a) Does h have a horizontal asymptote? Justify your answer.
   (b) Could h be a rational function? Justify your answer.

46. Let $f(x) = \sqrt{x^2 + 1} - x$ and $g(x) = \sqrt{x^2 + x} - x$.
   (a) Use a calculator or a computer to complete the following table.

x	100	1,000	10,000	100,000	1,000,000
$f(x)$					
$g(x)$					

   (b) Use the results you recorded in the table above to guess $\lim\limits_{x \to \infty} f(x)$ and $\lim\limits_{x \to \infty} g(x)$.
   (c) What do your answers to part (b) imply about the graphs of f and g?
   (d) Use algebra to prove your guesses in part (b).

In Exercises 47–50, give examples of polynomials $p(x)$ and $q(x)$ such that $\lim\limits_{x \to \infty} p(x) = \infty$, $\lim\limits_{x \to \infty} q(x) = \infty$, and

47. $\lim\limits_{x \to \infty} \dfrac{p(x)}{q(x)} = \infty$.

48. $\lim\limits_{x \to \infty} \dfrac{p(x)}{q(x)} = 2$.

49. $\lim\limits_{x \to \infty} \dfrac{p(x)}{q(x)} = 0$.

50. $\lim\limits_{x \to \infty} \dfrac{p(x)}{q(x)} = 1$ and $\lim\limits_{x \to \infty} \big(p(x) - q(x)\big) = 3$.

51. Let $f(x) = \dfrac{2x^2 + x + a}{x^2 - 2x - 3}$.
   (a) Find the value of a for which $\lim\limits_{x \to 3} f(x)$ is a finite number.
   (b) For which values of a is $\lim\limits_{x \to \infty} f(x)$ a finite number?

52. Let $f(x) = \begin{cases} \dfrac{\sin x}{x} & \text{if } x \neq 0 \\ 1 & \text{if } x = 0. \end{cases}$
   Explain why f is continuous at $x = 0$.

53. Let $g(x) = \begin{cases} \dfrac{\sin x}{|x|} & \text{if } x \neq 0 \\ 1 & \text{if } x = 0. \end{cases}$
   Is g continuous at $x = 0$? Justify your answer using limits.

54. Let $h(x) = \begin{cases} \dfrac{\sin x}{\sqrt{1 - \cos^2 x}} & \text{if } x \neq 0 \\ 1 & \text{if } x = 0. \end{cases}$
   Is h continuous at $x = 0$? Justify your answer.

In Exercises 55–81, evaluate the limit.

55. $\lim\limits_{x \to \infty} e^{-x} \ln x$

56. $\lim\limits_{x \to 0} x \cot x$

57. $\lim\limits_{x \to 8} \dfrac{x - 8}{\sqrt[3]{x} - 2}$

58. $\lim\limits_{x \to 0^+} \dfrac{\sin x}{x + \sqrt{x}}$

59. $\lim\limits_{x \to 0} \dfrac{\sin x}{x - \sin x}$

60. $\lim\limits_{x \to 0} \dfrac{\tan(3x)}{\ln(1 + x)}$

61. $\lim\limits_{x \to 0} \dfrac{e^x - 1}{x}$

62. $\lim\limits_{x \to 0} \dfrac{\arctan(2x)}{\sin(3x)}$

63. $\lim\limits_{x \to 0} \dfrac{e^x - e^{-x}}{x}$

64. $\lim\limits_{x \to 0} \dfrac{\sin^2 x}{\cos(3x) - 1}$

65. $\lim\limits_{x \to 0} \dfrac{1 - \cos^2 x}{x^2}$

66. $\lim\limits_{x \to 0} \dfrac{1 - x - e^{-x}}{1 - \cos x}$

67. $\lim\limits_{x \to 1} \dfrac{\ln x}{x^2 - x}$

68. $\lim\limits_{x \to \infty} x^2 e^{-x^2/2}$

69. $\lim\limits_{x \to 1} \dfrac{\sin(\pi x)}{x^2 - 1}$

70. $\lim\limits_{x \to 0} \dfrac{1 - x^2 - e^{-x^2}}{x^4}$

71. $\lim\limits_{x \to 1} \dfrac{\cos^3(\pi x/2)}{\sin(\pi x)}$

72. $\lim\limits_{x \to \pi/2} \dfrac{\ln(\sin x)}{(x - \pi/2)^2}$

73. $\lim\limits_{x \to 0^+} x^2 \ln x$

74. $\lim\limits_{x \to \infty} x \sin(1/x)$

75. $\displaystyle\lim_{x\to 0} x^2 \ln(\cos x)$

76. $\displaystyle\lim_{x\to 0^+} \sin x \ln(\sin x)$

77. $\displaystyle\lim_{w\to 0^+} w (\ln w)^2$

78. $\displaystyle\lim_{\theta\to 0} \frac{\cos\theta}{\tan\theta}$

79. $\displaystyle\lim_{x\to\infty} x \left(\frac{\pi}{2} - \arctan x\right)$

80. $\displaystyle\lim_{x\to\pi/2} \left(\frac{\pi}{2} - x\right)\tan x$

81. $\displaystyle\lim_{x\to 0}\left(\frac{1}{\sin x} - \frac{1}{x}\right)$

82. Let $f(k) = \displaystyle\lim_{x\to 2}\frac{x^2 - 5x + 3k}{x^2 - 3x + k}$. Determine $f(k)$ for all k.

83. Suppose that $f(1) = 2$ and $f'(1) = 3$. Evaluate
$\displaystyle\lim_{x\to 1}\frac{(f(x))^2 - 4}{x^2 - 1}$.

84. Suppose that f and f' are continuous functions, that $f(0) = 0$, and that
$$\lim_{x\to 0}\frac{f(x)}{\sin(2x)} = 5.$$
Evaluate $f'(0)$.

85. Suppose that f and g are differentiable functions and that $g(0) \neq 0$. Show that
$$\lim_{x\to 0}\frac{xf(x)}{(e^x - 1)g(x)} = \frac{f(0)}{g(0)}.$$

86. Suppose that
$$\lim_{x\to 0^+}\frac{f(x)}{g(x)} = -2 \quad \text{and} \quad \lim_{x\to 0^-}\frac{f(x)}{g(x)} = 3.$$
Evaluate $\displaystyle\lim_{x\to\infty}\frac{g(1/x)}{f(1/x)}$.

In Exercises 87–92, evaluate the limit.
[HINT: If $\displaystyle\lim_{x\to a}\ln(f(x)) = A$, then $\displaystyle\lim_{x\to a} f(x) = e^A$.]

87. $\displaystyle\lim_{x\to 0^+} x^{2x}$.

88. $\displaystyle\lim_{x\to\infty} (1 + x)^{1/x}$

89. $\displaystyle\lim_{x\to 0}(1 + x)^{1/x}$

90. $\displaystyle\lim_{x\to 0}(e^x + x)^{1/x}$

91. $\displaystyle\lim_{x\to 1}(\ln x)^{\sin x}$

92. $\displaystyle\lim_{x\to(\pi/2)^-}\sin\left(x^{\tan x}\right)$

93. l'Hôpital's rule says that if f is a differentiable function and $f(0) = 0$, then $\displaystyle\lim_{x\to 0}\frac{f(x)}{x} = f'(0)$. Why isn't this a surprise? [HINT: Review the limit definition of the derivative.]

94. Let $f(x) = \begin{cases} 3x - 1 & \text{if } x \neq 1 \\ 3 & \text{if } x = 1. \end{cases}$
Use the definition of the derivative of a function at a point (i.e., as the limit of a difference quotient) to evaluate $f'(1)$ or to explain why this derivative does not exist.

95. Let $f(x) = \begin{cases} 2x + 2 & x < 1 \\ 5 & x = 1 \\ x^2 + 3 & x > 1. \end{cases}$
Does $f'(1)$ exist? Use the limit definition of the derivative of a function at a point to justify your answer.

96. Let $L(t)$ be the length, in centimeters, of a certain species of fish when it is t years old. Suppose that the growth of this species of fish is modeled by the differential equation $L'(t) = 6.48e^{-0.09t}$.

   (a) Assuming $L(0) = 0.0$, find an expression for $L(t)$.

   (b) These fish mature (i.e., begin to reproduce) when they reach a length of 45 cm. At what age does the model predict that the fish mature?

   (c) On the basis of this model, what is the likelihood that a lucky angler will catch one of these fish that is 75 cm long? Justify your answer.

4.4 PARAMETRIC EQUATIONS, PARAMETRIC CURVES

A point P wanders about the xy-plane, tracing its path as it goes. As P travels, its coordinates $x = f(t)$ and $y = g(t)$ are functions of time. In this situation t is called a **parameter**, f and g are **coordinate functions**, and the figure traced out by P is a **parametric curve**. ◄ The first example illustrates the idea and introduces some useful vocabulary.

Here the "parameter" is a variable. In other situations, parameters may be constants.

EXAMPLE 1 At any time t with $0 \leq t \leq 10$, the coordinates of P are given by the **parametric equations**
$$x = t - 2\sin t; \qquad y = 2 - 2\cos t.$$

What curve does P trace out? Where is P at time $t = 1$? In which direction is P moving at $t = 1$?

Solution A simple way to draw a curve is to calculate many points (x, y), plot each one, and "connect the dots." The first step is easy (though tedious if done by hand): For many inputs t, calculate the corresponding x and y. The following table shows a sample of results, rounded to two decimals:

t	0	0.1	0.2	0.3	...	9.8	9.9	10
x	0	−0.10	−0.20	−0.29	...	10.53	10.82	11.09
y	0	0.01	0.04	0.09	...	3.86	3.78	3.68

Figure 1 shows the graphical result, a curve C.

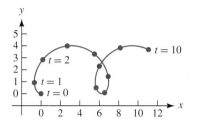

The dots appear at equal *time* intervals
(not *length* intervals).

FIGURE 1
The parametric curve $x = t - 2 \sin t$, $y = 2 - 2 \cos t$

Notice the "dynamic" quality of the figure—the dots indicate both progress in time and a direction of motion. Ordinary function graphs, by contrast, are static objects, lying passively on the page. Let's take a closer look:

- **What the bullets show**   Points corresponding to *integer* values of t are bulleted. At $t = 1$, P has coordinates $(−0.68, 0.92)$; at this instant, P is heading almost due north.
    The bullets on the graph appear at equal *time* ➡ intervals but not at equal *distances* from each other because P speeds up and slows down as it moves. ➡ We'll soon see how to calculate the speed of a parametric curve at a point.

- **Not a function graph**   The full curve C is not the graph of a function $y = f(x)$; some x-values correspond to more than one y-value. ➡ In general, parametric curves may have loops, vertical tangents, and other features not seen in function graphs. Some *pieces* of the curve C, however, do define functions; the part from $t = 2$ to $t = 5$ does so.

- **No t-axis**   The picture shows the x- and y-axes *but no t-axis*: t-values are indicated only by the bulleted points. In most parametric curves t-values don't appear at all. ➡

In this example, t represents time.

When is P moving fastest? Slowest?

x = 6, for instance.

■ *Most graphing calculators don't show t-values graphically, but some calculate t-values numerically.*

A sampler of parametric curves

Parametric curves come in mind-boggling variety. Any choice of two equations $x = f(t)$ and $y = g(t)$ and a t-interval produces a parametric curve. Surprisingly often, the result is beautiful, useful, interesting, or all three. The next several examples hint at some of the possibilities and at connections between parametric curves and ordinary function graphs.

EXAMPLE 2   (Function graphs)   Every ordinary function graph can be written in parametric form. To produce a sine curve, for example, we can use

$$x = t; \qquad y = \sin t; \qquad -2\pi \le t \le 2\pi.$$

Figure 2 shows the curve; the dots appear at equal *time* intervals.

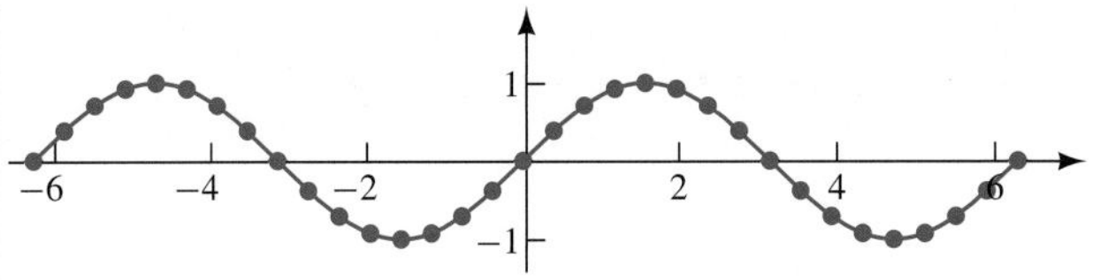

Is the curve traced out at constant speed?
Look at the dots...

FIGURE 2
The parametric curve $x = t$, $y = \sin t$, $-2\pi \leq t \leq 2\pi$ ■

E X A M P L E 3   **(The unit circle, parametrically)**   If a parametric curve has *periodic* coordinate functions, then the curve's shape reflects this fact. The simplest and most important such curve is the **unit circle**, often written in xy form as $x^2 + y^2 = 1$. A simple *parametric* description of the unit circle is as follows:

$$x = \cos t; \qquad y = \sin t; \qquad 0 \leq t \leq 2\pi.$$

Figure 3 shows the curve; dots at integer multiples of $\pi/2$ are bulleted.

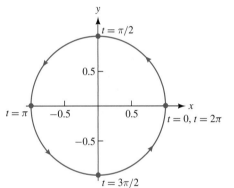

The arrows show the direction of travel.

FIGURE 3
The parametric curve $x = \cos t$, $y = \sin t$, $0 \leq t \leq 2\pi$

This important curve deserves a close look:

- **Interpreting t**   The parameter t can always be thought of as time, but other useful interpretations are sometimes possible. Here, t can be thought of either as the *distance* traveled counterclockwise from $(1, 0)$, or, alternatively, as the radian measure of the *angle* (in radians) determined by the x-axis and the line from the origin to the point $P = (x, y)$.

- **Eliminating t**   Because $x = \cos t$ and $y = \sin t$, we can write

$$x^2 + y^2 = (\cos t)^2 + (\sin t)^2 = 1.$$

  This shows that every point on the parametric curve satisfies the usual Cartesian (xy) equation $x^2 + y^2 = 1$ for the unit circle (as we intended).

      Notice what we just did—we traded *two* equations involving x, y and t for *one* equation involving only x and y. This process is called **eliminating the parameter**; it may or may not be possible for a given parametric curve.

- **Around and around**   The circle is traced *once* as t runs from $t = 0$ to $t = 2\pi$. With a larger t-interval, the same circle would be traced repeatedly; with a smaller t-interval, the circle would be traced only partially. ■

E X A M P L E 4   (Other circles)   The idea in Example 3 works for any circle. If (a, b) is any point in the plane and $r > 0$ is any radius, then the parametric equations

$$x = a + r \cos t; \qquad y = b + r \sin t; \qquad \text{and} \qquad 0 \le t \le 2\pi$$

produce the circle with center (a, b) and radius r. ∎

E X A M P L E 5   (Parametric art)   Curves defined by periodic coordinate functions often have striking shapes. A **Lissajou curve**, for instance, is defined by the equations

$$x = \sin(5t); \qquad y = \sin(6t); \qquad 0 \le t \le 2\pi.$$

(Replacing 5 and 6 with other integers produces other Lissajou figures.) Figure 4 shows our curve.

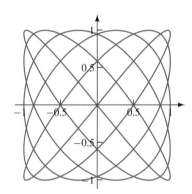

The curve crosses the x-axis 12 times. Can you find these times exactly?

FIGURE 4
The Lissajou curve $x = \sin(5t)$, $y = \sin(6t)$, $0 \le t \le 2\pi$ ∎

E X A M P L E 6   (Same curve, different equations)   Different pairs of parametric equations may produce curves that look exactly the same in the xy-plane. In such cases, labeling t-values can make differences appear. For example, setting

$$x = t; \qquad y = t^2; \qquad -1 \le t \le 1$$

produces a parabolic arc. So does

$$x = t^3; \qquad y = t^6; \qquad -1 \le t \le 1.$$

Figure 5 shows both curves; the dots mark 0.25-second time intervals.

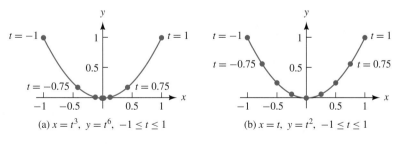

(a) $x = t^3$, $y = t^6$, $-1 \le t \le 1$ (b) $x = t$, $y = t^2$, $-1 \le t \le 1$

FIGURE 5
Two versions of the same curve

The curves themselves are geometrically identical—in each case, eliminating t gives $y = x^2$. Only the different placement of dots suggests the different coordinate functions. ∎

Calculus with parametric curves: speed and slope

Suppose that a parametric curve C has differentiable coordinate functions $x = f(t)$ and $y = g(t)$. What can the derivatives $f'(t)$ and $g'(t)$ say about the situation?

Speed   If t tells time, then $\big(f(t), g(t)\big)$ gives the *position* of P in the xy-plane at time t, and it makes sense to consider the speed of P at time t. ◂ Because the derivatives f' and g' tell, respectively, how fast x and y increase, their appearance is no surprise:

As the bullets on some of the foregoing graphs show, speed varies with time.

> **DEFINITION (Speed of a parametric curve)**   Suppose that the position of a point P at time t, for $a \le t \le b$, is given by differentiable coordinate functions $x = f(t)$ and $y = g(t)$. Then the speed at time t is given by
>
> $$\text{speed of } P \text{ at time } t = \sqrt{f'(t)^2 + g'(t)^2}.$$
>
> Speed is measured in units of distance per unit of time.

Definitions aren't subject to proof, but they *should* appeal to common sense. Does this one?

Speed, properly defined, should tell how fast the distance traveled by P increases with respect to time. Showing carefully that the preceding definition actually does so involves a subtle mathematical approach to **arclength** (distance measured along a curve). We won't study arclength in detail here, but the next example shows that the definition makes good sense for *linear* coordinate functions. ◂

We do define arclength carefully in Chapter 7.

EXAMPLE 7   At time t seconds, a point P has coordinates $x = at + b$ and $y = ct + d$, where $a, b, c,$ and d are constants. How fast is P moving at time t_0?

Solution   The definition says that the speed of P is *constant*:

$$\sqrt{f'(t)^2 + g'(t)^2} = \sqrt{a^2 + c^2}.$$

To see why this result makes sense, notice first that because P has *linear* coordinate functions, P *moves along a straight line*. When $t = t_0$, the point P has coordinates $(x_0, y_0) = (at_0 + b, ct_0 + d)$; by time t_1 the point P has moved to $(at_1 + b, ct_1 + d)$, a distance of

$$\sqrt{(at_1 + b - at_0 - b)^2 + (ct_1 + d - ct_0 - d)^2} = (t_1 - t_0)\sqrt{a^2 + c^2}$$

units. The *average* speed of P from $t = t_0$ to $t = t_1$ is

$$\text{average speed} = \frac{\text{distance}}{\text{time}} = \frac{(t_1 - t_0)\sqrt{a^2 + c^2}}{(t_1 - t_0)} = \sqrt{a^2 + c^2}.$$

Thus, P has the same average speed ($\sqrt{a^2 + c^2}$) over *any* time interval, and so the *instantaneous* speed of P at any time t_0 is also $\sqrt{a^2 + c^2}$, just as the definition says. ■

EXAMPLE 8   Consider again the parametric curve

$$x = f(t) = t - 2\sin t; \qquad y = g(t) = 2 - 2\cos t$$

of Example 1. Find the speed at $t = 3$. When is P moving fastest? Slowest?

Solution The definition says that at any time t,

$$\text{speed} = \sqrt{\left(f'(t)\right)^2 + \left(g'(t)\right)^2} = \sqrt{(1 - 2\cos t)^2 + (2\sin t)^2}$$

$$= \sqrt{1 - 4\cos t + 4\cos^2 t + 4\sin^2 t} \qquad \text{(expanding)}$$

$$= \sqrt{5 - 4\cos t}. \qquad (\cos^2 t + \sin^2 t = 1)$$

The rest is easy. At $t = 3$ the speed is $\sqrt{5 - 4\cos 3} \approx 2.99$. The formula $\sqrt{5 - 4\cos t}$ also shows that the speed is greatest when $\cos t = -1$ (at $t = \pi$, for instance); the speed is least when $\cos t = 1$ (at $t = 0$ and $t = 2\pi$, for instance). Figure 1 "agrees" with these answers, for the speed is related to the distance between dots. ∎

Slope For an ordinary function $y = f(x)$, the derivative $f'(x)$ gives the slope of the f-graph at (x, y). Finding slopes on parametric curves also involves derivatives, but in a different way. Not all functions have derivatives; not all parametric curves have slopes. To avoid needless complications we'll work only with *smooth* curves:

DEFINITION The parametric curve C defined by

$$x = f(t); \qquad y = g(t); \qquad a \le t \le b$$

is **smooth** if f' and g' are continuous functions of t, and f' and g' are not simultaneously zero.

(The second requirement says that the moving point P *never stops*.) For smooth curves, the next theorem tells how to find the slope at a point.

THEOREM 5 (Slope of a parametric curve) Consider a smooth curve with coordinate functions $x = f(t)$ and $y = g(t)$. If $f'(t) \ne 0$, then the slope dy/dx at the point $(x, y) = \left(f(t), g(t)\right)$ is given by

$$\frac{dy}{dx} = \frac{g'(t)}{f'(t)} = \frac{dy/dt}{dx/dt}.$$

Before proving the theorem, let's use it.

EXAMPLE 9 Consider yet again the parametric curve $x = f(t) = t - 2\sin t$, $y = g(t) = 2 - 2\cos t$ of Example 1. Find the slope at $t = 1$. Where is the curve horizontal? Where is it vertical? ➡

Guess first, by looking at the picture.

Solution By Theorem 5, the slope at t is

$$\text{slope at } t = \frac{g'(t)}{f'(t)} = \frac{2\sin t}{1 - 2\cos t},$$

(unless $f'(t) = 0$, where the denominator is zero). Setting $t = 1$ gives

$$\text{slope at time } 1 = \frac{g'(1)}{f'(1)} = \frac{2\sin 1}{1 - 2\cos 1} \approx -20.88.$$

Figure 1 agrees: At the point marked $t = 1$, the curve has large negative slope.

The general slope formula shows that the curve is *horizontal* only when the numerator $2\sin t$ is zero—that is, when t is an integer multiple of π. Again Figure 1 agrees: At $t = \pi$ the curve appears to be horizontal.

Yet again, Figure 1 agrees.

The curve C can be *vertical* only where the denominator $(1 - 2\cos t)$ is zero—that is, when $\cos t = 1/2$. One such value is $t = \pi/3 \approx 1.05$. ◄ ■

Proof of Theorem 5   Suppose that $f'(t_0) \neq 0$; then either $f'(t_0) > 0$ or $f'(t_0) < 0$. If $f'(t_0) > 0$, then $f'(t) > 0$ for t near t_0, and so $x = f(t)$ must be *increasing* near $t = t_0$. Similarly, if $f'(t) < 0$ then $x = f(t)$ must be *decreasing* near $t = t_0$. In either case, $f(t) \neq f(t_0)$ if $t \neq t_0$ and t is near t_0. (This ensures that denominators don't vanish unexpectedly in the following limit computations.)

By definition, the slope in question is

$$\text{slope} = \left.\frac{dy}{dx}\right|_{t=t_0} = \lim_{t \to t_0} \frac{y - y_0}{x - x_0} = \lim_{t \to t_0} \frac{g(t) - g(t_0)}{f(t) - f(t_0)}.$$

We'll be done if we show that the last quantity equals $g'(t_0)/f'(t_0)$.

We start with a little trick—dividing both numerator and denominator by $(t - t_0)$. Then we'll take limits of everything in sight and see what happens. Here goes:

$$\left.\frac{dy}{dx}\right|_{t=t_0} = \lim_{t \to t_0} \frac{g(t) - g(t_0)}{f(t) - f(t_0)} \qquad \text{(by definition)}$$

$$= \lim_{t \to t_0} \frac{\frac{g(t)-g(t_0)}{t-t_0}}{\frac{f(t)-f(t_0)}{t-t_0}} \qquad \text{(by our trick)}$$

$$= \frac{\displaystyle\lim_{t \to t_0} \frac{g(t)-g(t_0)}{t-t_0}}{\displaystyle\lim_{t \to t_0} \frac{f(t)-f(t_0)}{t-t_0}} = \frac{g'(t_0)}{f'(t_0)} \qquad \text{(OK since } f'(t_0) \neq 0\text{)}$$

which is what we wanted to show.   □

Modeling with parametric equations

Parametric equations often help model physical processes that involve quantities that vary in both time and space. **Trajectories**—paths taken by baseballs, missiles, thrown tomatoes, and other missiles—are one such setting.

Major-league batters don't ignore wind resistance.

EXAMPLE 10   A major-league fastball leaves the pitcher's hand traveling horizontally with initial speed 150 ft/sec and initial height 7 feet. Ignoring wind resistance, ◄ describe the ball's trajectory. When and at what speed does the ball cross the plate, 60.5 feet away? Is the pitch a strike? If the batter, catcher, and umpire all miss the ball, where does it land?

Solution   We use an xy-coordinate system with its origin at the pitcher's feet, units measured in feet, and home plate at $(60.5, 0)$; the ball starts at $(0, 7)$. Since we ignore wind resistance, the ball is influenced only by gravity. Thus, the ball's horizontal acceleration is zero, and its vertical (upward) acceleration due to gravity is -32 feet per second per second. ◄

Recall: acceleration is the second derivative of position.

If $x = f(t)$ and $y = g(t)$ are measured in feet and t in seconds since the ball's release, then our information boils down to this:

$$f(0) = 0; \quad g(0) = 0; \quad f'(0) = 150; \quad g'(0) = 0; \quad f''(t) = 0; \quad g''(t) = -32.$$

These facts lead to simple formulas for both f and g. Here is the argument for f:

$$f''(t) = 0 \implies f'(t) = a \implies f(t) = at + b,$$

where a and b are constants, their values as yet unknown. (We found f by antidifferentiating twice.) → To find values for a and b we combine the facts that $f(t) = at + b$, $f(0) = 0$, and $f'(0) = 150$ to conclude that $f(t) = 150t$. A similar argument applies to g; the result is

We did similar "free-fall" calculations in Section 2.5.

$$x = f(t) = 150t; \qquad x = g(t) = 7 - 16t^2.$$

These equations make sense in context, of course, only while the ball remains airborne, that is, until $y = 7 - 16t^2 = 0$, or $t = \sqrt{7}/4 \approx 0.661$ seconds. Figure 6 shows the curve for $0 \le t \le 0.661$.

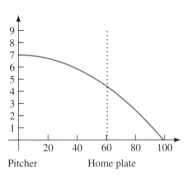

The horizontal and vertical units are different—this isn't a "photo" of the ball's flight.

FIGURE 6
A baseball's trajectory: $x = f(t) = 150t$, $y = g(t) = 7 - 16t^2$, $0 \le t \le 0.661$

Notice especially the trajectory's parabolic shape. In fact, *every* free-falling object either moves vertically or follows a parabolic trajectory.

We can now answer our original questions. The ball crosses the plate when $x = f(t) = 150t = 60.5$, that is, when $t = 121/300 \approx 0.4033$ seconds. A little arithmetic shows that at this time,

$$\text{speed} = \sqrt{f'(0.4033)^2 + g'(0.4033)^2} \approx 150.55$$

feet per second. At the same time, the ball's height is $y = g(0.4033) \approx 4.4$ feet—high but in the strike zone for an average batter. If not hit, the ball lands at $t = 0.661$ seconds; at this time, $x = f(t) = 150 \cdot 0.661 \approx 99.2$ feet—almost 40 feet behind home plate. ■

BASIC EXERCISES

Plot each parametric curve below. (Using a machine is fine, but then draw or copy your own curve on paper.) In each case, mark the direction of travel and label the points corresponding to $t = -1$, $t = 0$, and $t = 1$.

1. $x = t$, $y = \sqrt{1 - t^2}$, $-1 \le t \le 1$

2. $x = t$, $y = -\sqrt{1 - t^2}$, $-1 \le t \le 1$

3. $x = \sqrt{1 - t^2}$, $y = t$, $-1 \le t \le 1$

4. $x = -\sqrt{1 - t^2}$, $y = t$, $-1 \le t \le 1$

5. $x = \sin(\pi t)$, $y = \cos(\pi t)$, $-1 \le t \le 1$

FURTHER EXERCISES

6. Each parametric "curve" below is actually a line segment. In each part, state the beginning point ($t = 0$) and ending point ($t = 1$) of the segment. Then state an equation in x and y for the line each segment determines.

(a) $x = 2 + 3t$, $y = 1 + 2t$, $0 \le t \le 1$

(b) $x = 2 + 3(1 - t)$, $y = 1 + 2(1 - t)$, $0 \le t \le 1$

(c) $x = t$, $y = mt + b$, $0 \le t \le 1$

(d) $x = a + bt$, $y = c + dt$, $0 \le t \le 1$

(e) $x = x_0 + (x_1 - x_0)t$, $y = y_0 + (y_1 - y_0)t$, $0 \le t \le 1$

7. A parametric curve $x = f(t), y = g(t), a \leq t \leq b$ has **constant speed** if its speed is constant in t. Which of the curves in Exercises 1–5 have constant speed?

8. Show that if a curve C has linear coordinate functions $x = at + b$ and $y = ct + d$, then C has constant speed.

9. Consider the curve C shown in Example 1; suppose that t tells time in seconds.

   (a) At which bulleted points would you expect P to be moving quickly? Slowly? Justify your answers.

   (b) Use the curve to estimate the speed of P at $t = 3$. (One approach: Estimate how far P travels—i.e., the length of the curve—over the 1-second interval from $t = 2.5$ to $t = 3.5$. If d is this distance, then d distance units per second is a reasonable speed estimate at $t = 3$.)

   (c) Use the curve to estimate the speed of P at $t = 6$.

10. Plot the parametric curve

$$x = t^3; \qquad y = \sin t^3; \qquad -2 \leq t \leq 2.$$

What familiar curve is produced? Why does the result happen?

11. Let (a, b) be any point in the plane and $r > 0$ any positive number. Consider the parametric equations

$$x = a + r \cos t; \qquad y = b + r \sin t; \quad 0 \leq t \leq 2\pi.$$

   (a) Plot the parametric curve defined above for $(a, b) = (2, 1)$ and $r = 2$. Describe your result in words.

   (b) Show by calculation that if x and y are as above, then $(x - a)^2 + (y - b)^2 = r^2$. Conclude that the curve defined above is the circle with center (a, b) and radius r.

   (c) Write parametric equations for the circle of radius $\sqrt{13}$ centered at $(2, 3)$.

   (d) What "curve" results from the equations above if $r = 0$?

12. Let a and b be any positive numbers, and let a parametric curve C be defined by

$$x = a \cos t; \qquad y = b \sin t; \quad 0 \leq t \leq 2\pi.$$

The resulting curve is an **ellipse**.

   (a) Plot the curve defined above for $a = 2$ and $b = 1$. Describe C in words. Where is the "center" of C? Why do you think the quantities $2a$ and $2b$ are called the **major and minor axes** of C.

   (b) What curve results if $0 \leq t \leq 4\pi$? Why?

   (c) Write parametric equations for an ellipse with major axis 10 and minor axis 6.

   (d) Write parametric equations for another ellipse with major axis 10 and minor axis 6.

   (e) Show that for all t, $\dfrac{x^2}{a^2} + \dfrac{y^2}{b^2} = 1$.

   (f) How does the "ellipse" look if $a = b$? How does its xy-equation look?

   (g) How does an ellipse look if $a = 1000$ and $b = 1$?

   (h) How does an ellipse look if $a = 1$ and $b = 1000$?

13. Consider the Lissajou curve in Example 5.

   (a) Label the points corresponding to $t = 0$, $t = 0.1$, and $t = \pi/2$. Add some arrows to the curve to indicate direction.

   (b) How often in the interval $0 \leq t \leq 2\pi$ does the tracing point P return to $(0, 0)$?

   (c) How would the picture be different if the t-interval $0 \leq t \leq 4\pi$ were used?

14. We stated in this section that if a parametric curve C has linear coefficient functions, then C is a straight line (or part of a line). This exercise explores that fact.

   (a) Plot the parametric curve $x = 2t$, $y = 3t + 4$, $0 \leq t \leq 1$. Where does the curve start? Where does it end? What is its shape?

   (b) Eliminate the variable t in the two equations above to find a single equation in x and y for the line of the previous part.

   (c) Find the slope at $t = t_0$ of the parametric curve $x = 2t$, $y = 3t + 4$, $0 \leq t \leq 1$. Why doesn't the answer depend on t?

15. This exercise pursues the idea that if a parametric curve C has linear coefficient functions, then C is a straight line (or part of a line).

   (a) Consider the parametric curve $x = at + b$, $y = ct + d$, $t_0 \leq t \leq t_1$. Where does the curve start? Where does it end?

   (b) Assume that $a \neq 0$. (Don't assume that $c \neq 0$.) Eliminate the variable t in the two equations above to find one equation in x and y. What line does it describe?

   (c) Assume that $c \neq 0$. (Don't assume that $a \neq 0$.) Eliminate the variable t in the two equations above to find one equation in x and y. What line does it describe?

   (d) What happens if a and c are both zero?

16. This exercise explores the technical assumptions in the definition of smooth curves.

   (a) Consider the "curve" defined for all t by $x = 0$, $y = 0$. Plot C. Does C "deserve" to have a slope at $t = 0$? If so, what slope? If not, why not? Is C smooth in the sense of the definition?

   (b) Consider the curve defined for $-1 \leq t \leq 1$ by $x = t^3$ and $y = t^3$. Plot C. Does C "deserve" to have a slope at $t = 0$? What does Theorem 3 say in this case?

   (c) Consider the curve defined for $-1 \leq t \leq 1$ by $x = t$ and $y = t$. Plot C. Does this C "deserve" to have a slope at $t = 0$? What does Theorem 3 say about slope this time?

17. This exercise concerns Example 10.

   (a) How could the model be made more realistic? What additional information would be needed?

   (b) Use the conditions $g''(t) = -32$, $g'(0) = 0$, and $g(0) = 7$ to show that $g(t) = 7 - 16t^2$, as claimed in the Example.

   (c) Find a formula for $s(t)$, the ball's speed at time t. Plot $s(t)$ over an appropriate interval. How does the graph's shape reflect the physical situation?

18. Use parametric equations to describe the trajectory of a baseball thrown exactly as in Example 10, except that this time the initial velocity is 100 feet per second. Plot the result over an appropriate interval. Is the pitch a strike? At what speed does it cross the plate? (The strike zone is roughly from 1.5 to 4.5 feet above the ground at home plate.)

19. Consider a baseball thrown horizontally, starting from height 7 feet (as in Example 10) but with initial speed s_0.

    (a) Explain briefly why the parametric equations

$$x = s_0 t, \quad y = 7 - 16t^2$$

    describe the ball's trajectory (ignoring wind resistance).

    (b) When does the ball reach home plate?

    (c) Is the trajectory parabolic? Why?

20. This exercise is about the situation in Example 10 but takes wind resistance into account. In practice, wind resistance *does* affect a baseball's trajectory. One possible model of wind resistance (we omit the details) leads to the parametric equations

$$x = \frac{\ln(150 \cdot k \cdot t + 1)}{k}; \quad y = 7 - 16t^2,$$

where the constant k can be thought of as the ball's "drag coefficient"—the smoother the ball, the lower the value of k. For a typical baseball, $k = 0.003$ is reasonable. With this value of k we get

$$x = \frac{1000 \ln(0.45t + 1)}{3}; \quad y = 7 - 16t^2.$$

The resulting trajectory is plotted below; the curve from Example 10 is also shown.

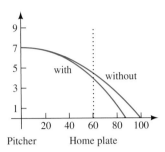

In each part below, use the graphs to give approximate answers; then use appropriate formulas to improve your results.

    (a) At what time does the air-dragged ball cross the plate? How much longer does it take to reach the plate than the dragless ball?

    (b) At what height does the air-dragged ball cross the plate?

    (c) At what speed does the air-dragged ball cross the plate?

    (d) Where does the air-dragged ball land?

21. The situation is as in the previous exercise except that, because the ball is scuffed, its "drag coefficient" is 0.005, not 0.003. Plot the new trajectory. Then answer the same questions as in the previous exercise.

ANSWERS TO SELECTED EXERCISES

Section 5.1

1. 45

3. $A = 12, B = 20$

5. 4

7. $4 + 9\pi/4$

9. 0

11. $4 + 9\pi/4 + 4\pi + 12$

13. -3

15. 6

17. $21/2$

19. 9

21. 2π

23. $-\pi^3/3$

25. 0

27. $\pi^3/3$

29. $A = 6, B = 8$

31. $A = 1/5, B = 1$

33. 74

35. -18

37. Joan

41. 6

43. 3

45. 3

49. 0

51. 0

53. 0

55. $\pi + 2$

57. $\pi^2/8 + 1$

59. 200

67. 7

71. (b) $\int_0^2 \Big(f(x) - g(x) \Big)\, dx$

Section 5.2

3. $A_f(5)$

7. $A_f(-1)$

19. no

21. $F(x) = 3x^2/2 + 2x, G(x) = 3x^2/2 + 2x - 7/2,$
$H(x) = 3x^2/2 + 2x - 2$

23. $F(x) - H(x) = 2$

25. $(-\infty, -2/3); f(x) < 0$

27. no

29. $F(x) = 2x - 3x^2/2, \quad G(x) = 2x - 3x^2/2 - 1/2,$
$H(x) = 2x - 3x^2/2 + 10$

31. -10

33. $(2/3, \infty); f(x) < 0$

35. yes; $(-\infty, \infty)$

37. (a) F is decreasing on $[a, b]$

(b) F is concave up on $[a, b]$

39. (a) $A_f(x) = \begin{cases} 2x + x^2/2, & x \le 0 \\ 2x - x^2/2, & x > 0 \end{cases}$

41. $(-\infty, -2) \cup (2, \infty); f(x) < 0$

43. $(0, \infty); f$ is decreasing

45. $A_f(x) = \begin{cases} 2x + x^2/2 - 3/2, & x \le 0 \\ 2x - x^2/2 - 3/2, & x > 0 \end{cases}$

47. $(-\infty, -2) \cup (2, \infty); f(x) < 0$

49. $(0, \infty); f$ is decreasing

51. $G(x) = H(x) - 2\pi + 1/2$

55. $G(-5) = -3; G(-3) = 0$

57. 1

59. $1 + \sin(x^2)$

61. $F(4)$

63. $F(3) - F(-2)$

69. $\frac{1}{2} x \sqrt{a^2 - x^2} + \frac{a^2}{2} \arcsin(x/a)$

71. (a) $A_f(x) = x^3/3 + 9$ (b) yes

Section 5.3

1. yes

3. yes

5. yes

7. yes

9. $x = -3.5, x = -2, x = 2$

11. $[-5, -3.5) \cup (2, 5]$

13. $F(0) = 0$, $F'(0) = 2$, $F''(0) = 1$

15. $y = x - 0.5$

17. $x = -3, x = -2, x = 1, x = 3$

19. 4

21. 2/3

23. $e^2 - e^{-1}$

25. 1

27. 499/10

29. -1

31. (a) $y = 2x - 2$ (b) no

33. (b) $\frac{1}{2} \sin 9$

35. yes; at $x = 1$, $x = 5$, and $x = 9$

37. $x = 3$

39. yes

41. no

43. $f(2) = 12 + e^2 + \sin 2$

49. $2x \sqrt[3]{1 + x^4}$

51. $-2/3$

53. 2

55. $g'(4) = 2$

57. $(1, 4)$

59. $g'(4) = 4$

61. $x = 5$

63. $-4/3$

65. (a) $D(t) = 800 - 10t$; $P(t) = 900$

      (c) $I(t) = (5 - R/2)t^2 + 100t + 1680$

      (d) $R = 50$

Section 5.4

1. $-\frac{1}{8}(4x + 3)^{-2} + C$

3. $e^{\sin x} + C$

5. $\frac{1}{2}(\arctan x)^2 + C$

7. $-e^{1/x} + C$

9. $\arctan(e^x) + C$

11. $a = 4, b = 1$; $(\pi/4 - \arctan 4)/2$

13. $a = 1, b = 19$; 45/361

17. $2(1 + \sqrt{x}) - 2\ln\left|1 + \sqrt{x}\right|$

19. $-\dfrac{1}{b}\ln\left|a + \dfrac{b}{x}\right| + C$

21. $\pi/3$

23. $\frac{1}{2}\sin(2x + 3) + C$

25. $\frac{1}{15}(3x - 2)^5 + C$

27. $\frac{1}{2}\ln(1 + x^4) + C$

29. $-\frac{1}{2}\ln|1 - 2x| + C$

31. $\frac{1}{10}(3 - 2x)^{5/2} - \frac{1}{2}(3 - 2x)^{3/2} + C$

33. $-\frac{2}{3}\left(1 + x^{-1}\right)^{3/2} + C$

35. $\frac{1}{12}\left(x^4 - 1\right)^3 + C$

37. $\frac{1}{3}\arctan\left(x^3\right) + C$

39. $\frac{1}{18}\left(4x^3 + 5\right)^{3/2} + C$

41. $\frac{1}{2}\ln\left(x^2 + 1\right) + 4\arctan x + C$

43. $-\frac{1}{3}\left(x^2 + 3x + 5\right)^{-3} + C$

45. $\frac{1}{4}\left(3x^2 + 6x + 5\right)^{2/3} + C$

47. $-\frac{1}{2}(2e^x + 3)^{-1} + C$

49. $2\ln|x + 1| - 1/(x + 1) + C$

51. $-\ln|\cos x| + C$

53. $\frac{1}{2}(\arcsin x)^2 + C$

55. $\frac{5}{6}\ln\left(3x^2 + 4\right) + C$

57. $\arcsin(\tan x) + C$

59. $-\frac{1}{2}\ln\left|\cos\left(x^2\right)\right| + C$

61. $-\frac{1}{3}\csc^3 x + C$

63. $\frac{1}{2}\left(1 + \sqrt{x}\right)^4 + C$

65. $\frac{1}{5}(x^2 + 2)^{5/2} - \frac{2}{3}(x^2 + 2)^{3/2} + C$

67. $-(e^x + 1)^{-1} + C$

69. 6/25

71. $1 - \cos 1$

73. 0

77. $2x^{1/2} - 3x^{1/3} + 6x^{1/6} - 6\ln\left(x^{1/6} + 1\right) + C$

Section 5.5

1. $\frac{1}{3}x - \frac{1}{15}\ln\left(3 + 2e^{5x}\right) + C$

3. $\dfrac{1}{9}\ln\left|\dfrac{x}{3 - x}\right| - \dfrac{1}{3x} + C$

5. $\ln\left|\dfrac{\sqrt{2x + 1} - 1}{\sqrt{2x + 1} + 1}\right|$

7. $\frac{1}{13}e^{2x}\left(2\cos(3x) + 3\sin(3x)\right) + C$

9. $\frac{2}{\sqrt{2}}\arctan\sqrt{\frac{3x - 2}{2}} + C$

11. $x - \arctan x + C$

13. $x - 2\ln|x + 1| + C$

15. $\tan x - \sec x + C$

17. $\frac{1}{3}\arctan(3x) + C$

19. $\frac{1}{10}\tan^2(5x) + \frac{1}{5}\ln\left|\cos(5x)\right| + C$

21. $\dfrac{x}{2} + \dfrac{1}{8}\ln|4x + 5| + C$

23. $\dfrac{1}{2(2x + 3)} + \ln|2x + 3| + C$

25. $\dfrac{1}{2}\ln\left(x^2 + 2\right) + \sqrt{2}\arctan\left(\dfrac{x\sqrt{2}}{2}\right) + C$

27. $\dfrac{5}{12}\ln\left|\dfrac{x + 1}{x + 4}\right| + C$

29. $\frac{1}{2}\left(x^2\sin(x^2) + \cos(x^2)\right) + C$

31. $-\dfrac{2x + 3}{x^2 + 3x + 2} - 2\ln\left|\dfrac{x + 1}{x + 2}\right| + C$

33. $\dfrac{1}{2}\arctan\left(\dfrac{e^x - 1}{2}\right) + C$

35. $\frac{1}{2}\left((x + 2)\sqrt{x^2 + 4x + 1} - 3\ln\left|x + 2 + \sqrt{x^2 + 4x + 1}\right|\right) + C$

37. $-\dfrac{4}{13}\ln|\cos x - 4| - \dfrac{1}{39}\ln\left|\cos x + \dfrac{1}{3}\right| + C$

39. $\frac{1}{9}\left(\sin(3x + 4) - (3x + 4)\cos(3x + 4) + 4\cos(3x + 4)\right) + C$

Section 5.6

1. (a) 4

3. (a) 7

5. (a) 3

7. 24

9. 28   **11.** 28

13. 34

15. $L_4 = 6, R_4 = 10, M_4 = 8, T_4 = 8, I = 8$

17. $L_4 = -8, R_4 = 8, M_4 = 0, T_4 = 0, I = 0$

19.

n	L_n	R_n	M_n	T_n
2	−0.75000	3.75000	1.5000	1.50000
4	0.37500	2.62500	1.5000	1.50000
8	0.93750	2.06250	1.5000	1.50000
16	1.21875	1.78125	1.5000	1.50000
32	1.35938	1.64063	1.5000	1.50000
64	1.42969	1.57031	1.5000	1.50000
128	1.46484	1.53516	1.5000	1.50000
256	1.48242	1.51758	1.5000	1.50000

$I = 1.5$

21.

n	L_n	R_n	M_n	T_n
2	0.06250	0.56250	0.21875	0.31250
4	0.14063	0.39063	0.24219	0.26563
8	0.19141	0.31641	0.24805	0.25391
16	0.21973	0.28222	0.24951	0.25098
32	0.23462	0.26587	0.24988	0.25024
64	0.24225	0.25787	0.24997	0.25006
128	0.24611	0.25392	0.24999	0.25002
256	0.24805	0.25196	0.25000	0.25000

$I = 0.25$

23.

n	L_n	R_n	M_n	T_n
2	1.57080	1.57080	2.22144	1.57080
4	1.89612	1.89612	2.05234	1.89612
8	1.97423	1.97423	2.01291	1.97423
16	1.99357	1.99357	2.00322	1.99357
32	1.99839	1.99839	2.00080	1.99838
64	1.99960	1.99960	2.00020	1.99960
128	1.99990	1.99990	2.00005	1.99990
256	1.99997	1.99997	2.00001	1.99998

$I = 2$

25. 0.85   **27.** 1.44

29. (a) $\int_5^{20} f(x)\,dx$ (b) $\int_0^{15} f(x)\,dx$

    (c) $\int_{2.5}^{17.5} f(x)\,dx$

33. no

Section 5.7

1. $\displaystyle\sum_{k=1}^{50} k$ **3.** $\displaystyle\sum_{i=0}^{49}(2i+1)^2$

5. 137/60   **7.** $\sqrt{2}/2$

9. 153   **11.** 385

13. (a) $\displaystyle\sum_{j=1}^{10} j^2$ (b) $2\displaystyle\sum_{k=1}^{5}(2k)^2$

    (c) $\dfrac{1}{2}\displaystyle\sum_{i=1}^{20}(i/2)^2$ (d) $\dfrac{1}{5}\displaystyle\sum_{n=1}^{50}(n/5)^2$

15. (a) $\displaystyle\sum_{i=1}^{10} i^2$ (b) $2\displaystyle\sum_{j=0}^{4}(1+2j)^2$

    (c) $\dfrac{1}{2}\displaystyle\sum_{k=0}^{19}(1+k/2)^2$ (d) $\dfrac{1}{5}\displaystyle\sum_{k=0}^{49}(1+k/5)^2$

17. (a) $\dfrac{1}{10}\displaystyle\sum_{i=0}^{9}\sqrt[3]{i/10}$ (b) $\dfrac{1}{5}\displaystyle\sum_{i=0}^{4}\sqrt[3]{i/5}$

    (c) $\dfrac{1}{20}\displaystyle\sum_{i=0}^{19}\sqrt[3]{i/20}$ (d) $\dfrac{1}{50}\displaystyle\sum_{i=0}^{49}\sqrt[3]{i/50}$

19. (a) $\dfrac{1}{10}\displaystyle\sum_{i=1}^{10}\sqrt[3]{1+i/10}$ (b) $\dfrac{1}{5}\displaystyle\sum_{i=1}^{5}\sqrt[3]{1+i/5}$

    (c) $\dfrac{1}{20}\displaystyle\sum_{i=1}^{20}\sqrt[3]{1+i/20}$ (d) $\dfrac{1}{50}\displaystyle\sum_{i=1}^{50}\sqrt[3]{1+i/50}$

21. (a) $\dfrac{2}{5}\displaystyle\sum_{i=1}^{10}\sin\left((3+2i/5)^2\right)$ (b) $\dfrac{4}{5}\displaystyle\sum_{j=1}^{5}\sin\left((3+4j/5)^2\right)$

    (c) $\dfrac{1}{5}\displaystyle\sum_{k=1}^{20}\sin\left((3+k/5)^2\right)$ (d) $\dfrac{2}{25}\displaystyle\sum_{n=1}^{50}\sin\left((3+2n/25)^2\right)$

23. (a) $\dfrac{1}{2}\displaystyle\sum_{i=0}^{9}\sqrt{1/4+i/2}$ (b) $\displaystyle\sum_{j=0}^{4}\sqrt{1/2+j}$

    (c) $\dfrac{1}{4}\displaystyle\sum_{k=0}^{19}\sqrt{1/8+k/4}$ (d) $\dfrac{1}{10}\displaystyle\sum_{n=0}^{49}\sqrt{1/20+n/10}$

25. 1/2

39. $\displaystyle\int_0^2 \sin x\,dx = -\cos 2 + \cos 0$

41. $\displaystyle\int_0^1 x^3\,dx = 1/4$

45. (b) $T_6 = 38.75; L_6 = 38; M_3 = 50$

Section 5.8

1. (a) -65

7. (b) $y = x$

13. (a) no

(b) yes

27. $\frac{1}{3}\ln|x| + C$

29. $3\arctan x + C$

31. $-\frac{2}{3}\cos(3x) - \frac{4}{5}\sin(5x) + C$

33. $x + \frac{4}{3}x^{3/2} + \frac{1}{2}x^2 + C$

35. $9\ln|x| - 6x + \frac{1}{2}x^2 + C$

37. $e^x - \frac{1}{2}e^{2x} + C$

39. $-\frac{1}{2}\cos(x^2) + C$

41. $\frac{1}{4}e^{x^4} + C$

43. $\frac{1}{4}(\ln x)^4 + C$

45. $-\frac{1}{2}\left(\ln|\cos x|\right)^2 + C$

51. $\approx 74.5\,^\circ\text{C}$

53. $\dfrac{5}{N}\displaystyle\sum_{j=0}^{N-1}\sqrt{3 \cdot j \cdot \dfrac{5}{N}}$; $\dfrac{5}{N}\displaystyle\sum_{j=0}^{N-1}\sqrt{3 \cdot (j+1) \cdot \dfrac{5}{N}}$; $\dfrac{5}{N}\displaystyle\sum_{j=0}^{N-1}\sqrt{3 \cdot (j+0.5) \cdot \dfrac{5}{N}}$

55. $G(3)$

57. $G(2) - G(-2)$

59. (b) 0

61. $\int_5^3 g(t)\,dt < \int_0^2 g(u)\,du < -1 < h(1) < \int_1^1 g(x)\,dx < h(4) < 3 < 5$

63. no

65. 36

67. 2

69. -6

71. (a) $H(-3) > 0$ (b) $H(-2) = 0$ (c) $H(0) < 0$ (d) $H(2) > 0$

73. $H'(1) = e$

77. $G'(2) = 1$

79. yes

81. -19

83. $-0.5 < S(0) < S(1) < 0.5 < S(3) < S(2)$

85. no

87. 1

89. $z = \sqrt{2\pi}$

91. -3

93. cannot be determined

95. 4

97. 1.7

101. yes

103. yes

111. yes

113. no

115. yes

117. 4/3

119. 5/2

CHAPTER 6

Section 6.1

1. (a) $I = 2$

(c) $|I - L_3| = 0.28446$; $|I - R_3| = 0.21554$; $|I - T_3| = 0.034457$; $|I - M_3| = 0.016525$

(d) 0.00046875

5. (a) $L_{10} = 11.810$; $R_{10} = 22.530$; $T_{10} = 17.170$; $M_{10} = 16.098$; $S_{20} = 16.455$

(b) $|I - T_{10}| \le 10.72$ (c) $|I - S_{20}| \le 1.072$

7. (a) $L_4 = 1.1485$; $R_4 = 1.1805$; $T_4 = 1.1645$; $M_2 = 1.1345$; $S_4 = 1.1545$

(b) $1.1345 \le I \le 1.1645$

9. (a) $|I - L_{20}| \approx 0.02113$; $|I - R_{20}| \approx 0.02094$; $|I - T_{20}| \approx 0.00010$; $|I - M_{20}| \approx 0.00005$;

11. $I - M_n < 0$

13. 0.31027

15. $f(x) = 1/x$; decreasing

17. $f(x) = 1 - x^2$; concave down

19. (b) yes

27. $L_{100} \le T_{100} \le S_{200} \le M_{100} \le R_{100}$

29. must

31. must

33. may

35. may

37. may

39. may

41. must

43. cannot

45. must

47. must

49. must

57. (b) $|I - M_n| \le |I - R_n|$

59. no

61. $R_n \le T_n \le I \le M_n \le L_n$

63. yes

65. $R_n \le M_n \le S_{2n} \le T_n \le L_n$

67. $L_n \le I \le R_n$

69. must

71. may

73. may

Section 6.2

5. (a) $|I - R_8| \le 10/n$ (b) $|I - T_8| = 0$

7. (b) no

9. (a) $I = 2$

(b) $|I - L_3| = 0.28446$; $|I - R_3| = 0.21554$; yes

(c) $|I - T_3| = 0.034457$; $|I - M_3| = 0.016525$; yes

(d) $n \ge 13$

11. 772

13. 65

15. 22

17. 5

19. $f(x) = x$

21. $f(x) = x^2$

23. (a) 1/2

(b) 1/4

25. no

27. (a) 8415   (b) 1683

(c) 144

29. (b) 0.1629

31. 5

33. $n \geq 162$

35. cannot

37. cannot

Section 6.3
1. $167.20 - 9.72 \cdot 1 = 157.48$

5. (b) no

7. (a) $m_i = f\left(t_i, Y(t_i)\right)$ (b) nothing

9. (a) 2   (b) 2.4414

(c) yes

11. (a) 2.0778   (b) 2.1074

13. (a) -0.59374 (b) -0.65330

15. (a) 2.6764   **17.** (b) 0.73092

CHAPTER 7

Section 7.1
1. (b) $L_5 = 4/25$ (c) underestimates

(d) 1/6

3. (a) 1/3   (b) 1/3

(c) 2   (d) 8/3

5. (a) 1.4793

7. $I = \int_0^\pi \sqrt{1 + \cos^2 x}\, dx; \; M_{20} \approx 3.820$

9. (a) 105 mi   (b) 105 ft

(c) 105 ft/min

11. (a)

t	1.0	1.1	1.2	1.3	1.4	1.5
$f'(t)$	-0.83	-1.17	-1.46	-1.70	-1.875	-1.965

t	1.6	1.7	1.8	1.9	2.0
$f'(t)$	-1.985	-1.925	-1.78	-1.57	-1.3

(b) $M_5 = -1.666; \; f(2) - f(1) = -1.666$

(c) 0.1194   (d) 1.9494

(e) $-1.5983, 0.11875$

13. 8/5   **15.** 1/12

17. 16/3

19. $65/32 - 9\ln(9)/4 + 9\ln(4)/4$

21. $e - 1$ **23.** 125/6

25. 8/5   **27.** 1/12

29. 16/3

31. $65/32 - 9\ln(9)/4 + 9\ln(4)/4$

33. $e - 1$ **35.** 125/6

37. ratio = 1.4271   **39.** $\pi/2$

41. $\left(4\sqrt{17} + \ln\left(4 + \sqrt{17}\right) - 2\sqrt{5} - \ln\left(2 + \sqrt{5}\right)\right)/4$

43. $\left(80\sqrt{10} - 13\sqrt{13}\right)/27$

45. $40\sqrt{5}/3 - 8\sqrt{2}/15$ **47.** $36 + \ln(3)/4$

49. 53/6

51. 49/9

53. $\sinh(4) - \sinh(1)$

57. $\approx (4.3271, 31.4470)$

Section 7.2
1. $\approx 12.47 \text{ m}^3$ **3.** $140/3\pi \text{ m}^3$

5. (a) $V = s^2 h/3$ **7.** $1 - 1/\sqrt[3]{2}$

9. (b) yes   **11.** $8^7 \pi/7$

13. $1688\pi/21$ **15.** $128\pi/5$

17. $\pi\left(16/\ln 2 - 8 - 6/(\ln 2)^2\right)$

19. 128/3   **21.** $27\sqrt{3}/2$

23. $\pi r^2 h$ **25.** $95\pi/12 \approx 24.87 \text{ in.}^3$

27. (a) $\approx 1.1514 \times 10^{10} \text{ mi}^3$
(b) $\approx 1.1522 \times 10^{11} \text{ mi}^3$

29. $100\sqrt{3} + 800\pi/3 \text{ ft}^3$ **31.** $M_3 = 33.2 \text{ in.}^3$

33. $45\sqrt{3} + 120\pi \text{ ft}^3$

35. (a) $(b-a)h$ (b) $\pi(b-a)h^2$
(c) $\pi(b-a)\left((h+c)^2 - c^2\right)$
(d) $\pi(b-a)\left(c^2 - (c-h)^2\right)$

37. $5\pi/6$ **39.** $625\pi/6$

41. $\pi(\pi/2 - 1)$ **43.** $\pi h(b^2 - a^2)$

45. $2\pi^2$ **47.** $4\pi/3$

49. $\pi(e^2 + 1)/2$

51. $\pi h(r-a)(r^2 + ra - 2a^2)/3$

55. (b) $KR/2$

Section 7.3
1. (a) $k = 5$ (b) 80 in.-lb

3. (a) $50k$ ft-lb   (b) $k(50 + 10a)$ ft-lb

5. 100,000 ft-lb   **7.** The 50-lb bucket.

9. $\pi \rho g r^4/4$ **11.** $171,360\pi$ ft-lb

13. 12 ft-lb

17. (a) volume of the tank

(b) $62.4 \displaystyle\int_0^h y A(y)\,dy$

Section 7.4

1. (a) $200\,°F$; $160\,°F$; $120\,°F$; $70\,°F$; $40\,°F$

3. $y' = -0.1(y - 70)$, $y(0) = 70$

5. (a) $100\,°F$; $70\,°F$; $0\,°F$

7. $y' = -0.1y$; $y(0) = 190$

11. $80 + 40e$

13. (a) $\displaystyle\lim_{t\to\infty} y(t) = T_r$

17. $y = -1/(x - 2)$

19. $y = -2e^{x^3/3}$

21. $y = 2/(3 - x^2)$

23. (a) $f(y) = 1/y$, $g(x) = x$

(b) $F(y) = \ln y$, $G(x) = x^2/2$

25. (a) $2, 1.5, 1, 0.5$ (b) P_1; P_4

27. $y = \tan(x + C)$

29. $y = \dfrac{2Ae^{4x} + 2}{Ae^{4x} - 1}$

31. $\ln\left(\sec y + \tan y\right) = x^3/3 + C$

Section 7.5

1. (a) $\approx \$251{,}579$ (b) $\approx \$155{,}817$

(c) $\approx 10.011\%$

3. (a) $\approx \$631{,}284$ (b) $\approx \$398{,}519$

(c) 7%

5. $\$53{,}317$; $\$36{,}120$

7. (b) 2π; 360; 360

9. yes

CHAPTER 8

Section 8.1

3. $-(1 + x)e^{-x}$

5. (b) $x\arctan(2x) + \frac{1}{4}\ln(1 + 4x^2)$

7. (b) $(2\cos x - \sin x)e^{2x}$

(c) $(2\sin x - \cos x)e^{2x}/5$

9. $\left(\sin(3x) - 3x\cos(3x)\right)/9$

11. $x\tan x + \ln|\cos x|$

13. $x(\ln x)^2 - 2x\ln x + 2x$

15. $1 - 2e^{-1}$; $M_{20} \approx 0.26435$

17. $(e^2 + 1)/4$; $M_{20} \approx 2.0970$

19. $\sqrt{2}\arctan(\sqrt{2}/2)/12 + 1/12 + \ln(3/4)/6 - \pi/12$; $M_{20} \approx -0.15361$

21. (b) $\sin x - x\cos x$

23. $x(\ln x)^3 - 3x(\ln x)^2 + 6x(\ln x - 1)$

29. (a) $x^3\left(\ln(x^3) - 1\right)/3$

(b) $x^3\ln x - x^3/3$

31. $(x^2 - 1)e^{x^2}/2$

35. $\left(2x^2 + 2x\sin(2x) + \cos(2x)\right)/8$

37 $x^2\sin x - 2\sin x + 2x\cos x$

39. $\ln|\csc x - \cot x| - x\csc x$

41. $(6\ln x - 4)x^{3/2}/27$

43. $x\arctan(1/x) + \frac{1}{2}\ln(x^2 + 1)$

45. $2(\sqrt{x} - 1)e^{\sqrt{x}}$

47. $x\left(\sin(\ln x) - \cos(\ln x)\right)/2$

49. $\sin(\sqrt{x}) - \sqrt{x}\cos(\sqrt{x})$

51. $\frac{2}{3}x^{3/2}\arctan(\sqrt{x}) - (1 + x)/3 + \frac{1}{3}\ln(1 + x)$

53. $\sin x\left(\ln|\sin x| - 1\right)$ **55.** $x\cosh x - \sinh x$

57. $x^2\cosh x - 2x\sinh x + 2\cosh x$

59. (d) $120 - 44e$ **61.** $f(x) = x^4/4$

63. -5

Section 8.2

3. $x - x/(1 + x^2)$

5. (b) $2\ln|x + 1| + 3\ln|x + 2|$

7. (c) $1 + 2x/(x - 1)^2$

(d) $x + 2\ln|x - 1| - 2/(x - 1)$

9. yes   **11.** no

13. no

15. (d) $\frac{1}{2}\ln|x| - \ln|x + 1| + \frac{3}{2}\ln|x - 2|$

17. $\frac{4}{3}\ln|x - 1| - 2\ln|x + 1| + \frac{5}{3}\ln|x + 2|$

19. (e) $-\frac{1}{4}\ln|x| + \frac{5}{8}\ln|x^2 + 4|$

21. $q(x) = 4x + 17$

23. no

25. (e) $\ln|x + 1| + (x + 1)/2 - 1/\left(2(x + 1)^2\right)$

27. (a) $1/\left(3(x - 2)\right) - 1/\left(3(x + 1)\right)$

(b) $2/\left(3(x - 2)\right) + 1/\left(3(x + 1)\right)$

(c) $-\left(2\ln|x - 2| + 7\ln|x + 1|\right)/3$

29. (a) $1/\left(2(x - 1)\right) - (x + 1)/\left(2(x^2 + 1)\right)$

(b) $1/\left(2(x - 1)\right) - (x - 1)/\left(2(x^2 + 1)\right)$

(c) $1/\left(2(x-1)\right)+(x+1)/\left(2(x^2+1)\right)$

(d) $\frac{9}{2}\ln|x-1|-\frac{1}{4}\ln(x^2+1)+\frac{5}{2}\arctan x$

31. $-(3\ln 2+\pi)/10$

33. $\ln(x-2)+\ln(x+3)$

35. $x^{-1}+2\ln|x|+3\ln|x+2|$

37. $x+\frac{1}{4}\ln|x-1|-\frac{1}{4}\ln|x+1|-\frac{1}{2}\arctan x$

39. $x^2/2+\frac{1}{2}\ln|x-1|+\frac{1}{2}\ln|x+1|$

41. $\ln\left|\left(\sqrt{x+1}-1\right)/\left(\sqrt{x+1}+1\right)\right|$

Section 8.3

1. (a) $\sin x$ (b) $-\cos x$

   (c) $\tan x$ (d) $\sec x$

3. no

5. (a) $\sin^5 x/5-\sin^7 x/7$

   (b) $\sin x(5\cos^6 x-8\cos^4 x+\cos^2 x+2)/35$

   (c) yes

9. $\ln\left|x+1+\sqrt{x^2+2x+5}\right|$

11. $\left(6x-\sin(6x)\right)/12$

13. $\sin^3 x/3$

15. $-\cos^3 x/3+\cos^5 x/5$

17. $(x+\cos x\sin x-2\cos^3 x\sin x)/8$

19. $\tan^3 x/3-\tan x+x$

21. $\tan^3 x/3$

23. $\left(\tan x\sec x-\ln|\sec x+\tan x|\right)/2$

25. $\sec^3 x/3-\sec x$

27. $-\left(\sqrt{4-x^2}\right)/4x$

29. $\left(x\sqrt{1-x^2}+\arcsin x\right)/2$

31. $\left(\arcsin x+x\sqrt{1-x^2}-2x\left(1-x^2\right)^{3/2}\right)/8$

35. $\sqrt{x^2-4}/4x$

37. $\sqrt{3}-\pi/3$

39. $\cos^2 x/2-\ln|\cos x|$

41. $-2\left(\cos x\right)^{3/2}/3+4\left(\cos x\right)^{7/2}/7-2\left(\cos x\right)^{11/2}/11$

43. $\left(x\sqrt{1+x^2}+\ln\left|\sqrt{1+x^2}+x\right|\right)/2$

45. $\arctan x+2\ln\left|x/\sqrt{1+x^2}\right|$

49. (b) $\left(ax+\cos(ax)\sin(ax)\right)/2a$

51. (b) $\sin^2(ax)/2a$

55. $\dfrac{(n-1)}{n}\cdot\dfrac{(n-3)}{n-2}\cdots\dfrac{3}{4}\cdot\dfrac{\pi}{4}$

61. $\tan(x/2)$ **63.** $\ln(4+\sqrt{2})-\ln(2)$

65. $I_n=\dfrac{x}{2(n-1)\left(1+x^2\right)^{n-1}}+\dfrac{2n-3}{2n-2}I_{n-1}$

Section 8.4

1. $1/(3+\cos x)$ **3.** $\left(3+4x^2\right)^6/48$

5. $3\left(x^2+4\right)^{2/3}/4$ **7.** $\left(\ln|x|\right)^3/3$

9. $\left(\ln|x|\right)^2/2$ **11.** $x/3-2\ln|3x+2|$

13. $\sin^3(3x)/9$ **15.** $e^{3x^2}/6$

17. $-(2-3x)^{11}/33$ **19.** $\ln|3+\tan x|$

21. $\ln\left|(x-1)/(x+2)\right|/3$ **23.** $x/2+\ln|4x+5|/8$

25. $\arcsin\left(e^x\right)$ **27.** $x\left(\ln|x|-1\right)$

29. $x\arcsin x+\sqrt{1-x^2}$ **31.** $\arctan\left((x+1)/\sqrt{2}\right)/\sqrt{2}$

33. $\arcsin(2x)/2$ **35.** $-\ln|\cos x|$

37. $2\left(1+e^x\right)^{5/2}/5-2\left(1+e^x\right)^{3/2}/3$

39. $\arcsin\left((x+1)/2\right)$

41. $-1/\ln|x|+C$

43. $\left(\ln|3x-2|-\ln|3x+2|\right)/12$

45. $-4\sqrt{4-x^2}+\left(4-x^2\right)^{3/2}/3-x^2\sqrt{4-x^2}-2\left(4-x^2\right)^{3/2}/3$

47. $\ln\left|x^2-1\right|/2$

49. $\ln\left|x+\sqrt{9+x^2}\right|$

51. $-x^2/6-x/9-\ln|1-3x|/27+C$

53. $e^{3x}/3$

55. $x\ln(1+x^2)-2x+2\arctan x$

57. $\left(2x^2\arcsin x+x\sqrt{1-x^2}-\arcsin x\right)/4$

59. $\sqrt{2x+3}$ **61.** $-(2x+1)/\left(8(2x+3)^3\right)$

63. $-\cos^2 x/2$ **65.** $\ln\left|1-e^{-x}\right|$

67. $2\sqrt{x}-2\ln(1+\sqrt{x})$ **69.** $x^3\ln(3x)/3-x^3/9$

71. $2x^{3/2}\ln|x|/3-4x^{3/2}/9$

73. $\ln|x+3|+2\arctan x-\ln(x^2+1)/2$

75. $(3x\sin^3 x+\sin^2 x\cos x+2\cos x)/9$

77. $\ln|x|-\ln(x^2+1)/2$

79. $\left((x+1)(x^2+2x+3)^{3/2}+3(x+1)\sqrt{x^2+2x+3}\right.$

    $\left.+6\ln\left|\left(\sqrt{x^2+2x+3}+x+1\right)/\sqrt{2}\right|\right)/4$

81. $2\sqrt{1+e^x} + \ln\left|1 - \sqrt{1+e^x}\right| - \ln\left|1 + \sqrt{1+e^x}\right|$

83. $\left(2\ln|x+1| - \ln|x^2 - x + 1| + 2\sqrt{3}\arctan\left((2x-1)/\sqrt{3}\right)\right)/6$

85. $x\tan x - x^2/2 - \ln|\sec x|$

87. $\sin x/2 - \sin(3x)/6$

CHAPTER 9

Section 9.1

1. (a) $1 + 2x + 44x^2 - 12x^3 + x^4$

    (b) $f(x) = 160 + 50(x-3) - 10(x-3)^2 + (x-3)^4$

3. (a) 9   (b) 24/5

5. (a) 0   (b) $12!/13^4$

7. $1 - x^2/2 + x^4/24 - x^6/720$; $[-2.13, 2.13]$

9. $x - x^2/2 + x^3/3 - x^4/4 + x^5/5 - x^6/6$; $[-0.61, 0.73]$

11. $x + x^3/6 + x^5/120$; $[-1.74, 1.74]$

13. $\sqrt{2}\big(1 + (x-\pi/4) - (x-\pi/4)^2/2 - (x-\pi/4)^3/6$

    $+ (x-\pi/4)^4/24 + (x-\pi/4)^5/120\big)/2$; $[-0.76, 2.22]$

15. $\big(\sqrt{3} + (x-\pi/3) - \sqrt{3}(x-\pi/3)^2/2 - (x-\pi/3)^3/6$

    $+ \sqrt{3}(x-\pi/3)^4/24 + (x-\pi/3)^5/120\big)/2$; $[-0.41, 2.45]$

17. $1 - (x-1) + (x-1)^2 - (x-1)^3 + (x-1)^4 - (x-1)^5$;

    $[0.58, 1.49]$

19. $(x-1) - (x-1)^2/2 + (x-1)^3/3 - (x-1)^4/4 + (x-1)^5/5$;

    $[0.45, 1.67]$

21. $1 + (x-1)/2 - (x-1)^2/8 + (x-1)^3/16 - 5(x-1)^4/128$

    $+ 7(x-1)^5/256$; $[0.25, 1.97]$

23. $1 - (x-1)/2 + 3(x-1)^2/8 - 5(x-1)^3/16$

    $+ 35(x-1)^4/128 - 63(x-1)^5/256$; $[0.47, 1.64]$

25. no   **27.** $3 - x/4 - x^2/5$

31. $(x-1) - (x-1)^2/2 + (x-1)^3/3 - (x-1)^4/4$

33. $1 - x/2 + 3x^2/8 - 5x^3/16 + 35x^4/128$

35. $x/2 - x^3/48 + x^5/3840$

37. $1 - x + x^2/2 - x^3/6 + x^4/24 - x^5/120$

39. $1 + x^2 + x^4/2 + x^6/6$ **41.** $1 - x^2 + x^4 - x^6$

49. $1 - x^2 + x^4/3 - 2x^6/45$

51. $2\left(x + x^3/3 + x^5/5\right)$

53. $x - x^3/3 + x^5/5$

Section 9.2

1. (a) 0.056   (b) yes

3. (a) $\left|\sin x - P_5(x)\right| \le 4/45$

    (b) $\left|\sin x - P_5(x)\right| \le 0.025$

5. (a) 0.35   (b) 0.015

9. (d) 0.00019

11. 0   **13.** 0

15. (a) $1 + x/2 - x^2/8$ (b) $2\pi\sigma\left(a^2/2r - a^4/8r^3\right)$

17. (a) 33/16   (b) 1/64

Section 9.3

1. (c) $\pi/2$

9. (a) $2/\pi$ (b) $2/\pi + 4\sin(3x)$

    (c) $2/\pi + 4\sin(3x)$

11. 3

15. (a) $a_0 = 1/2; a_k = 0$ for $k \ge 1; b_{2m} = 0$ and

    $b_{2m+1} = 2/\left((2m+1)\pi\right)$ for $m = 1, 2, 3, \ldots$

    (b) $q_1(x) = 1/2 + (2/\pi)\sin x$;

    $q_3(x) = q_1(x) + (2/3\pi)\sin(3x)$;

    $q_5(x) = q_3(x) + (2/5\pi)\sin(5x)$;

    $q_7(x) = q_5(x) + (2/7\pi)\sin(7x)$

17. $\pi/2 - (4/\pi)\cos(x) - (4/9\pi)\cos(3x) - (4/25\pi)\cos(5x)$

    $- (4/49\pi)\cos(5x)$

19. $4 + \pi^2/3 - 4\cos x + \cos(2x) - (4/9)\cos(3x) + (1/4)\cos(4x)$

    $- (4/25)\cos(5x) + (1/9)\cos(6x) - (4/49)\cos(7x)$

    $+ 6\sin(x) - 3\sin(2x) + 2\sin(3x) - (3/2)\sin(4x)$

    $+ (6/5)\sin(5x) - \sin(6x) + (6/7)\sin(7x)$

CHAPTER 10

Section 10.1

3. $\lim_{x \to 1^+} 1/(x^2 \ln x) = \infty$ **7.** 1/2

9. $2e^{-1}$ **11.** 1

13. e **15.** 4

17. 1/100   **19.** yes

23. 0   **25.** (b) no

27. (b) no   **29.** no

31. π **33.** diverges

35. $-3\sqrt[3]{2}$ **37.** $\sqrt{5} + \sqrt{7}$

39. 16/3   **41.** diverges

43. diverges

45. $\pi/2$

47. 2

49. $p > 1$

51. $p < 1$

53. none

55. (b) $\lim\limits_{x \to 0^+} \sin x/x = 1$

57. no

59. $C = 4$

61. $\displaystyle\int_0^1 \frac{du}{1+u^3}$

63. $\displaystyle\int_0^1 \frac{u^2}{u^4+1}\, du$

65. $-\displaystyle\int_0^1 \frac{u \ln u}{u^4+1}\, du$

Section 10.2

1. (a) $1/g(x) < 1/f(x) < 1$
  (b) $1/\big(g(x)\big)^r < 1/\big(f(x)\big)^r < 1$
  (c) $1/\big(g(x)\big)^r < 1/\big(f(x)\big)^r < 1$

9. no

11. $a \geq 11.513$

13. $a \geq 100{,}000$

15. $\displaystyle\int_0^{12} \frac{dx}{x^2+e^x}$

17. $\displaystyle\int_0^8 \frac{\arctan x}{(1+x^2)^3}\, dx$

25. none

27. diverges

29. converges

31. yes

35. 1

Section 10.3

7. (a) 1

(b) -0.5

(c) $\dfrac{1}{\sqrt{2\pi}} \displaystyle\int_{-0.5}^{\infty} e^{-x^2/2}\, dx$

(d) $\dfrac{1}{\sqrt{2\pi}} \displaystyle\int_{-\infty}^{1} e^{-x^2/2}\, dx$

(e) $\dfrac{1}{\sqrt{2\pi}} \displaystyle\int_{-0.5}^{1} e^{-x^2/2}\, dx$

9. (b) ≈ 0.1151

11. ≈ 0.4333

13. ≈ 0.1151

15. ≈ 0.4

17. ≈ 0.15

19. ≈ 0.095581

21. $D \approx 2.987$ in.

25. $k = 1/\pi$

27. (b) $1/\lambda$

29. (a) 1

31. $\sqrt{\pi}$

35. $\pi/4$

37. $1/4$

39. $(6 + \ln 2)/9$

CHAPTER 11

Section 11.1

1. no limit

3. ∞

5. 0

7. $\pi/2$

9. 0

11. 1

13. 0

15. 0

17. 1

19. $a_k = (-1)^k/k$

21. $a_k = k$

23. $a_k = e^{-k}$

27. 1

29. 1

31. 0

33. $x \leq 0$

35. $-\sin 1 < x \leq \sin 1$

39. $e^{-1/2}$

41. (c) converges

47. no

49. yes

53. yes

55. $x \geq 0$

Section 11.2

1. (a) $1/5$; $1/25$; $1/3125$; $1/9765625$; $6/5$; $31/25$; $3906/3125$; $12{,}207{,}031/9{,}765{,}625$
  (d) $5/4$
  (e) $1/20$; $1/100$; $1/12{,}500$; $1/39{,}062{,}500$
  (g) 0

7. $\pi^4/90$

9. (a) $S_n = (n+1)a$

11. $1/8$

13. $e/(e-1)$

15. $\pi^2/(16 - 4\pi)$

17. $-1/48$

21. $S_n = \arctan(n+1)$; $S = \pi/2$

23. $S_n = 1 + 1/\sqrt{2} - 1/\sqrt{n+1} - 1/\sqrt{n+2}$; $S = 1 + 1/\sqrt{2}$

25. (a) 4.97

(b) 5

(c) 0

27. $\pi^2/24$

29. $\pi^2/12$

31. $-1 < x < 1$; $1/(1-x)$

33. $-1 < x < 1$; $x^{10}/(1-x^2)$

35. $-2 < x < 0$; $-(1+x)^3/x$

37. $3/2$

39. diverges

41. $3/2$

43. diverges

45. $9/16$

47. diverges

49. $1/5$

51. diverges

53. 8 feet

59. (a) yes; $S = 3$

(b) $\lim\limits_{k \to \infty} a_k = 0$

61. (b) $a_k = (-1)^k$

Section 11.3

1. (c) $S_{10} \approx 1.6963$

(d) no

3. $\displaystyle\sum_{k=2}^{n} a_k < \int_1^n a(x)\, dx < \sum_{k=1}^{n-1} a_k$

9. $1/2 < S < 3/2$

11. $2e^{-1} \le S \le 3e^{-1}$ **13.** (a) no

17. $1/2 < S < 2$ **19.** $1/\sqrt{2} < S < 2$

25. (c) no

27. (a) nothing
　　(b) converges

31. converges; $N \ge 1000$

33. diverges; $N \ge 2999$

35. converges; $N \ge 4$ **41.** yes

45. converges; $\pi/8 < S < \pi/8 + 3\pi^2/32$

47. diverges   **49.** $1/3 < S < 1/2$

51. converges; $1/2 < S < 5/2$

55. $N \ge 6$

Section 11.4

1. conditionally

3. $14.902 < S < 14.918$

5. (a) $S_{50} \approx 0.23794$
　　(c) $0.23774 < S < 0.23814$

7. no   **9.** $S \approx -0.94985$

11. $S \approx -1.625$ **13.** no

15. $p > 1$ **17.** $p > 1$

19. converges absolutely; $3/4 < S < 1$

21. diverges   **23.** diverges

25. converges absolutely; $0 < S < 1/12$

31. $a_k = (-1)^k/\sqrt{k}$

Section 11.5

3. 2   **5.** 1

7. 1; $(1, 3)$ **9.** 1; $[-6, -4]$

15. $\displaystyle\sum_{k=1}^{\infty} \frac{x^k}{k4^k}$ **17.** $\displaystyle\sum_{k=1}^{\infty} \frac{(x-2)^k}{3^k}$

19. $\displaystyle\sum_{k=1}^{\infty} \frac{(12-x)^k}{k4^k}$

25. (a) $R = 14$ (b) $b = 3$

27. cannot   **29.** may

31. may   **33.** may

35. may   **37.** may

41. 0.820   **43.** $(-\infty, \infty)$

45. 0.581   **47.** $[-9, 1]$

49. 1.198

Section 11.6

1. 2   **3.** 2

5. $\displaystyle\sum_{k=0}^{\infty} (-1)^k x^{k+2}$ **7.** $\displaystyle\sum_{k=1}^{\infty} k(-x)^{k-1}$

9. $\displaystyle\sum_{k=0}^{\infty} (-1)^k \frac{(2x)^{2k+1}}{2k+1}; \; R = 1/2$

11. $\displaystyle\sum_{k=0}^{\infty} (-1)^k \frac{x^{2k+3}}{(2k+1)!}; \; R = \infty$

13. 29/48

17. $\displaystyle\frac{1}{2}\sum_{k=0}^{\infty} (-1)^k \left(\frac{1}{2}\right)^k; \; R = 2$

19. $\displaystyle\sum_{k=0}^{\infty} (-1)^k \left(\frac{x^{2k}}{(2k)!} + \frac{x^{2k+1}}{(2k+1)!}\right); \; R = \infty$

21. $\displaystyle\sum_{k=0}^{\infty} (-1)^{k+1} \frac{(4k^2+2k+1)x^{2k+1}}{(2k+1)!}; \; R = \infty$

29. 1   **31.** 1/2

33. $-1/2$ **35.** 2

37. (b) $[-1, 1)$

41. (a) $\displaystyle\sum_{k=0}^{\infty} \frac{(-1)^k}{(2k+1)\cdot k!} x^{2k+1}$ (b) 26/35

43. $2557/7020 \approx 0.364$

45. $x^3 + 2x^4 + 2x^5 + \frac{5}{6}x^6$

47. $1 + 2x + \frac{5}{2}x^2 + \frac{8}{3}x^3$

49. $1 + x + x^2/2 - x^4/8$

51. $(1-x)^{-2}$ **53.** $x/(1+x)$

Section 11.7

1. (b) 9.1×10^{-5} **3.** $2^{100}/100!$

5. (a) $\displaystyle\sum_{k=0}^{\infty} \frac{(-1)^k x^k}{2^{k+1}}$ (b) $-259!/2^{260}$

9. (b) yes

11. (a) $\displaystyle\sum_{k=0}^{\infty} \frac{(-1)^k x^{2k}}{(2k+1)!}$ (b) $(-\infty, \infty)$

　　(c) $f'''(1) \approx \dfrac{37}{210}$

Section 11.8

1. ∞ **3.** ∞

5. 0   **7.** 0

9. converges absolutely; $S < e$

11. diverges

13. converges absolutely; $S < 1/3$

15. converges absolutely; $1 < S < 1 + \sqrt{\pi}/2$

17. diverges

19. converges absolutely; $1/2 < S < 3$

21. converges absolutely; $3/4 < S < 1$

23. converges conditionally; $-1 < S < -1/2$

25. diverges

27. converges absolutely; $S = 583/120$

29. diverges   **31.** diverges

33. diverges

35. (a) $n^n e^{1-n} < n!$ (b) $N > be$

37. $[-1, 1]$ **39.** $(-1/3, 1/3)$

41. $[-1/3, 1/3]$ **43.** $(1, 5)$

45. $[-3, 5)$ **47.** $[-6, -4]$

49. cannot   **51.** must

53. cannot   **55.** 0

57. $1/3$

59. $\displaystyle\sum_{k=0}^{\infty} \frac{(x \ln 2)^k}{k!}; R = \infty$

61. $\displaystyle -\sum_{k=0}^{\infty} \left(\frac{2^{k+2} + (-1)^k}{2^{k+1}} \right) x^k; R = 1$

63. $a_k = (-1)^{k+1}$

65. (a) no   (b) yes

69. $f(x) = 1 + 3x^4 + 3x^8 + x^{12}$

71. $g(x) \approx 1 - 3x^2/2 + 15x^4/8 - 35x^6/16$

73. $252/625$

CHAPTER V

Section V.1

1. $(3, 5)$ **3.** $(-4, -5)$

5. $\sqrt{5}$ **7.** $\sqrt{65}$

11. (a) $(3, 4)$ (b) $w - v = (-1, 2)$

    (c) $v - w = (1, -2)$

21. (a) $L(0) = (1, 2);\quad L(1) = (3, 5);\quad L(2) = (5, 8);$
    $L(-1) = (-1, -1)$

    (b) ray from $P(1, 2)$ in same direction as the vector $(2, 3)$

    (c) line segment from $(-1, -1)$ to $(3, 5)$

23. $\sqrt{65}$

25. (b) $y = -x$ (c) $5000\sqrt{2}$

27. (a) $p(t) = t(50, 50\sqrt{3}) + t^2(0, -g/2);$
    $v(t) = (50, 50\sqrt{3}) - t(0, g)$

    (b) $t = 100\sqrt{3}/g$ seconds

    (d) $t = 50\sqrt{3}/g$ seconds; $3750/g$ meters

    (e) 50 meters per second, $v = (50, 0)$

29. (a) $(13, -1)$ (b) $(-31/3, 1)$

    (c) $\displaystyle\int_0^2 \sqrt{25t^4 + 30t^3 - 30t^2 - 26t + 17}\, dt$

31. (a) $v(t) = (1, -t)$ (b) $p(t) = (t, -t^2/2)$

    (c) $\displaystyle\int_0^5 \sqrt{1 + t^2}\, dt$

33. 2π **35.** $M_{100} \approx 13.365$

37. $\sqrt{10}$

39. 3

41. $\sqrt{1 + t^2}$

43. $\sqrt{2}\,(e^\pi - 1)$

Section V.2

1. $(\pi, 0); (-\pi, \pi); (\pi, 2\pi)$

3. $(\sqrt{2}, \pi/4); (\sqrt{2}, -7\pi/4); (-\sqrt{2}, -3\pi/4); (-\sqrt{2}, 5\pi/4)$

5. $(\sqrt{5}, 1.107); (\sqrt{5}, -5.176); (-\sqrt{5}, 4.249)$

7. $(\sqrt{17}, 1.326); (\sqrt{17}, -4.957); (-\sqrt{17}, 4.467)$

9. $(\sqrt{2}, \sqrt{2})$ **11.** $(\sqrt{3}/2, 1/2)$

13. $(0.5403, 0.8415)$ **15.** $(\sqrt{2}, \sqrt{2})$

17. $x = 2$ **19.** $y = \sqrt{3}x/2$

21. $r = 3$ **23.** $\tan\theta = 2$

27. (a) $2; 1.866; 1.5; 1; 0.5; 0.134; 0; 0.134; 0.5; 1; 1.5; 1.866; 2$

    (c) x-axis

    (d) symmetric about $x = \pi$

49. center $= (a/2, b/2)$, radius $= \sqrt{a^2 + b^2}/2$

51. (b) $(x^2 + y^2)^3 = x^4$

53. $\sqrt{r_1^2 + r_2^2 - 2r_1 r_2 \cos(\theta_1 - \theta_2)}$

Section V.3

3. (b) $\theta = -\tan\theta$ (c) $\theta = \cot\theta$

    (d) $y = \left(\dfrac{\sin 1 + \cos 1}{\cos 1 - \sin 1} \right)(x - \cos 1) + \sin 1$

5. $\pi/2$ **7.** $\pi a^2/2$

9. $2\pi + 3\sqrt{3}/2$ **11.** $3\pi/4 - 2$

13. $\pi + 1$ **15.** $m/2$

17. (a) $\pi/4n$ (b) $\pi/4; 1/4$

19. $(e^{4\pi} - 1)/4$ **21.** $\pi/3 - \sqrt{3}/4$

23. $x = 2\cos t, y = 2\sin t, 0 \le t \le 2\pi$

27. $x = t\cos t, y = t\sin t, 0 \le t \le 2\pi$

25. $x = 1, y = \tan t, -\pi/4 \le t \le \pi/4$

29. $x = \cos(2t)\cos t, y = \cos(2t)\sin t, 0 \le t \le 2\pi$

CHAPTER M

Section M.1

1. (a) x-direction (b) y-direction

7. $x^2 + y^2 + z^2 = 4$ **9.** $x^2 + z^2 = 1$

11. $z = \sin x$

13. center $= (0, 3, 2)$; radius $= \sqrt{13}$

15. (b) $\sqrt{5}$ (c) $\sqrt{14}$

19. $(1, 1, 1)$, $(1, 1, -1)$, $(1, -1, 1)$, $(1, -1, -1)$, $(-1, 1, 1)$, $(-1, 1, -1), (-1, -1, 1), (-1, -1, -1)$

21. $-A/B; B = 0$

23. $C/A; A = 0$

25. $x = D/A; y + z = 1$

27. (a) $x + 3z = 3; (3, 0, 0); (0, 0, 1)$
   (b) $2y + 3z = 3; (0, 3/2, 0); (0, 0, 1)$
   (c) $2y + 3z = 2$

29. $x = 3$

31. $(1, 0, 0), (0, 1/2, 0)$

Section M.2

1. $\mathbb{R}^2; [0, \infty)$

3. $\Big\{ (x, y) \mid (x, y) \ne (0, 0) \Big\}; (0, \infty)$

5. $\Big\{ (x, y) \mid x^2 + y^2 \le 1 \Big\}; [0, 1]$

7. (a) circles with center $(0, 0)$
   (b) circles with center $(0, 0)$

9. (a) vertical lines
   (b) horizontal lines

11. east–west

15. (a) $f(x, y) = |x - 1|$
   (d) lines parallel to the y-axis

Section M.3

1. $f_x(x, y) = 2x; f_y(x, y) = -2y$

3. $f_x(x, y) = \dfrac{2x}{y^2}; f_y(x, y) = \dfrac{-2x^2}{y^3}$

5. $f_x(x, y) = -\sin(x)\cos(y); f_y(x, y) = -\cos(x)\sin(y)$

7. $f_x(x, y, z) = y^2 z^3; f_y(x, y, z) = 2xyz^3; f_z(x, y, z) = 3xy^2z^2$

9. (c) $[2.9, 3.1]$

11. (c) $6.5, 6.1$ (d) 6

13. $g_x(1, 1) \approx -2.01; g_y(1, 1) \approx 3.01;$

15. (a)

y＼x	-0.02	-0.01	0.00	0.01	0.02
0.02	0.02	0.02	0.02	0.02	0.02
0.01	0.01	0.01	0.01	0.01	0.01
0.00	0.00	0.00	0.00	0.00	0.00
-0.01	-0.01	-0.01	-0.01	-0.01	-0.01
-0.02	-0.02	-0.02	-0.02	-0.02	-0.02

17. (c) $L(x, y) = y + 2$

19. (b) $f_x(0, 0) \approx 1; f_y(0, 0) = 0$
   (c) $f_x(\pi/2, 0) \approx 0; f_y(0, 0) = 0$

21. (b) $f_x(0, 0) = 2, f_y(0, 0) = 3$

23. (a) $L(x, y) = 1(x - 2) + 2(y - 1) + 2$

25. (b) $f_x(1, 2) = 2; f_y(1, 2) = 4$
   (c) $L(x, y) = 2(x - 1) + 4(y - 2) + 5$

27. $L(x, y) = -5 + 4x + 2y$

29. $L(x, y) = x + y$

33. (a) $L(x, y) = 3x/5 + 4y/5$
   (b) $g(2.9, 4.1) \approx 5.02; g(4, 5) \approx 6.4$
   (c) no

35. (b) $f_x(x, 0) = 0$

Section M.4

1. (a) remains constant
   (b) rises all the way; $(0.5, 1)$; 0.5

3. (a) $(1, \pi/2)$ is a local minimum; $(1, -\pi/2)$ is a local maximum
   (b) $(0, 0), (2, 0), (1, \pi/2), (1, -\pi/2)$
   (c) $15; -15$

5. $(0, 0)$; saddle **7.** $(1, 2)$; saddle

9. (b) $a = b = 0$ **11.** $g(x, y) = y^2$

13. $k(x, y) = (x - 3)^2 + (y - 4)^2$

Section M.5

1. $21/64; M_{100} \approx 0.333325$

3. $512; 512; 512$

5. (a) ≈ 168
   (b) $M_{16} = 168$
   (c) underestimate

Section M.6

1. $(\cos(1))^2 - 2\cos(1) + 1$

3. $512/3$

5. 6

7. $3/20$

9. $3/20$

11. $\pi/4$

13. (a) 1

(b) 1

15. (a) area $= \displaystyle\int_c^d g(y)\,dy$

Section M.7

3. (b) $2\pi/3$

5. $1/2$

7. $\pi + 2$

9. $\pi/3$

Selections from Volume 1

Section 3.4

1. $\pi/2$

3. π

5. $\pi/4$

7. $\sqrt{1-x^2}$

9. $1/\sqrt{1+x^2}$

11. $\dfrac{1}{\sqrt{1+4x^2}}$

13. $\dfrac{2}{1+4x^2}$

15. $\frac{1}{2}\left((1-x^2)\arcsin x\right)^{-1/2}$

17. $\dfrac{e^x}{\sqrt{1-e^{2x}}}$

19. $\left(x\left(1+(\ln x)^2\right)\right)^{-1}$

21. $f'(x) = \dfrac{2}{\sqrt{x^4 - 4x^2}}$

23. $2x\arctan(\sqrt{x}) + \dfrac{x^{3/2}}{2+2x}$

25. $x = \pm\pi/3,\, x = \pm 7\pi/3$

27. $(5, -3)$

35. $g(x) = (x-a)/b$

41. $(-\infty, \infty)$, yes

43. (c) odd

(d) $f'(x) > 0$

(e) $(-\infty, 0);\ (0, \infty)$; yes, at $x = 0$

45. (b) neither

(c) $f'(x) < 0$

(d) $(-1, 0);\ (0, 1)$; yes, at $x = 0$

47. (b) neither

(c) $f'(x) > 0$

(d) $(-\infty, -1);\ (1, \infty)$; no

55. (a) $C = \pi/4$

(b) $\pi/2$

(c) $-\pi/2$

57. (b) $-\pi/2 < \arctan x + \arctan y < \pi/2$

Section 4.2

1. (a) $-\infty$

(b) $-\infty$

3. does not exist

5. -1

7. does not exist

9. 0

11. $-\frac{1}{2}$

13. ∞

15. 0

17. ∞

19. 0

21. 0

23. $\cos 1$

25. -4

27. 0

29. ∞

31. 0.75

33. ∞

35. 0

37. no

39. no

45. (a) yes

(b) no

47. $p(x) = x^2 - 2x + 3,\ q(x) = x + 1$

49. $p(x) = x + 1,\ q(x) = x^2 + 2$

51. (a) $a = -21$

(b) $-\infty < a < \infty$

53. no

55. 0

57. 12

59. ∞

61. 1

63. 2

65. 1

67. 1

69. $-\pi/2$

71. 0

73. 0

75. 0

77. 0

79. 1

81. 0

83. 6

87. 1

89. e

91. 0

95. no

Section 4.4

7. No. 5

9. (a) $t = 3, t = 4, t = 9$, and $t = 10$; $t = 0, t = 1, t = 6$, and $t = 7$

(b) ≈ 3 units per second

(c) ≈ 1 unit per second

11. (a) circle of radius 2 with center at $(2, 1)$.

(c) $x = 2 + \sqrt{13}\cos t,\ y = 3 + \sqrt{13}\sin t,\ 0 \le t \le 2\pi$

(d) the point $(2, 3)$

13. (b) 3 times

15. (a) $(at_0 + b, ct_0 + d);\ (at_1 + b, ct_1 + d)$

(b) $y = c(x - b)/a + d$

(c) $x = a(y - d)/c + b$

17. (c) $s(t) = \sqrt{22{,}500 - 1024t^2}$

19. (b) at $t = 60.5/s_0$ seconds

(c) Yes (if $s_0 > 0$)

21. (a) at $t = 4\left(e^{121/400} - 1\right)/3$

(b) ≈ 3.4508 feet

(c) ≈ 111.87 feet per second

(d) $x \approx 80.569$ feet

INDEX

Table of Integrals

Basic Forms

1. $\int x^n dx = \dfrac{x^{n+1}}{n+1}, \ n \neq -1$

2. $\int \dfrac{dx}{x} = \ln |x|$

3. $\int e^x dx = e^x$

4. $\int b^x dx = \dfrac{1}{\ln b} b^x$

5. $\int \sin x \, dx = -\cos x$

6. $\int \cos x \, dx = \sin x$

7. $\int \tan x \, dx = \ln |\sec x| = -\ln |\cos x|$

8. $\int \cot x \, dx = \ln |\sin x| = -\ln |\csc x|$

9. $\int \sec^2 x \, dx = \tan x$

10. $\int \csc^2 x \, dx = -\cot x$

11. $\int \sec x \tan x \, dx = \sec x$

12. $\int \csc x \cot x \, dx = -\csc x$

13. $\int \dfrac{dx}{x^2 + a^2} = \dfrac{1}{a} \arctan\left(\dfrac{x}{a}\right), \ a \neq 0$

14. $\int \dfrac{dx}{x^2 - a^2} = \dfrac{1}{2a} \ln \left| \dfrac{x-a}{x+a} \right|$

15. $\int \dfrac{dx}{\sqrt{a^2 - x^2}} = \arcsin\left(\dfrac{x}{a}\right), \ a > 0$

16. $\int \ln x \, dx = x(\ln x - 1)$

17. $\int \sec x \, dx = \ln |\sec x + \tan x| = \ln \left| \tan \left(\dfrac{x}{2} + \dfrac{\pi}{4} \right) \right|$

18. $\int \csc x \, dx = \ln |\csc x - \cot x| = \ln |\tan(x/2)|$

Expressions Containing $ax + b$

19. $\int (ax+b)^n dx = \dfrac{(ax+b)^{n+1}}{a(n+1)}, \ n \neq -1$

20. $\int \dfrac{dx}{ax+b} = \dfrac{1}{a} \ln |ax+b|$

21. $\int \dfrac{x}{ax+b} dx = \dfrac{x}{a} - \dfrac{b}{a^2} \ln |ax+b|$

22. $\int \dfrac{x}{(ax+b)^2} dx = \dfrac{b}{a^2(ax+b)} + \dfrac{1}{a^2} \ln |ax+b|$

23. $\int \dfrac{dx}{x(ax+b)} = \dfrac{1}{b} \ln \left| \dfrac{x}{ax+b} \right|$

24. $\int \dfrac{dx}{x^2(ax+b)} = -\dfrac{1}{bx} + \dfrac{a}{b^2} \ln \left| \dfrac{ax+b}{x} \right|$

25. $\int \sqrt{ax+b} \, dx = \dfrac{2}{3a} \sqrt{(ax+b)^3}$

26. $\int x\sqrt{ax+b} \, dx = \dfrac{2(3ax-2b)}{15a^2} \sqrt{(ax+b)^3}$

27. $\int \dfrac{dx}{\sqrt{ax+b}} = \dfrac{2\sqrt{ax+b}}{a}$

28. $\int \dfrac{dx}{x\sqrt{ax+b}} = \dfrac{1}{\sqrt{b}} \ln \left| \dfrac{\sqrt{ax+b} - \sqrt{b}}{\sqrt{ax+b} + \sqrt{b}} \right|, \ b > 0$

29. $\int \dfrac{dx}{x\sqrt{ax-b}} = \dfrac{2}{\sqrt{b}} \arctan \left(\sqrt{\dfrac{ax-b}{b}} \right), \ b > 0$

Expressions Containing $ax^2 + c, \ x^2 \pm p^2,$ and $p^2 - x^2, \ p > 0$

30. $\int \dfrac{dx}{p^2 - x^2} = \dfrac{1}{2p} \ln \left| \dfrac{p+x}{p-x} \right|$

31. $\int \dfrac{dx}{ax^2 + c} = \dfrac{1}{\sqrt{ac}} \arctan \left(x\sqrt{\dfrac{a}{c}} \right), \ a > 0, c > 0$

32. $\int \dfrac{dx}{ax^2 - c} = \dfrac{1}{2\sqrt{ac}} \ln \left| \dfrac{x\sqrt{a} - \sqrt{c}}{x\sqrt{a} + \sqrt{c}} \right|, \ a > 0, c > 0$

33. $\int \dfrac{dx}{(ax^2 + c)^n} = \dfrac{1}{2(n-1)c} \dfrac{x}{(ax^2+c)^{n-1}} + \dfrac{2n-3}{2(n-1)c} \int \dfrac{dx}{(ax^2+c)^{n-1}}, \ n > 1$

34. $\int x(ax^2 + c)^n \, dx = \dfrac{1}{2a} \dfrac{(ax^2+c)^{n+1}}{n+1}, \ n \neq -1$

35. $\int \dfrac{x}{ax^2 + c} dx = \dfrac{1}{2a} \ln |ax^2 + c|$

36. $\int \sqrt{x^2 \pm p^2} \, dx = \dfrac{1}{2} \left(x\sqrt{x^2 \pm p^2} \pm p^2 \ln \left| x + \sqrt{x^2 \pm p^2} \right| \right)$

37. $\int \sqrt{p^2 - x^2} \, dx = \dfrac{1}{2} \left(x\sqrt{p^2 - x^2} + p^2 \arcsin \left(\dfrac{x}{p} \right) \right), \ p > 0$

38. $\int \dfrac{dx}{\sqrt{x^2 \pm p^2}} = \ln \left| x + \sqrt{x^2 \pm p^2} \right|$

Expressions Containing Trigonometric Functions

39. $\displaystyle \int \sin^2(ax)\,dx = \frac{x}{2} - \frac{\sin(2ax)}{4a}$

40. $\displaystyle \int \sin^n(ax)\,dx = -\frac{\sin^{n-1}(ax)\cos(ax)}{na} + \frac{n-1}{n}\int \sin^{n-2}(ax)\,dx,\ n > 0$

41. $\displaystyle \int \cos^2(ax)\,dx = \frac{x}{2} + \frac{\sin(2ax)}{4a}$

42. $\displaystyle \int \cos^n(ax)\,dx = \frac{\cos^{n-1}(ax)\sin(ax)}{na} + \frac{n-1}{n}\int \cos^{n-2}(ax)\,dx$

43. $\displaystyle \int \sin(ax)\cos(bx)\,dx = -\frac{\cos((a-b)x)}{2(a-b)} - \frac{\cos((a+b)x)}{2(a+b)},\ a^2 \neq b^2$

44. $\displaystyle \int \sin(ax)\sin(bx)\,dx = \frac{\sin((a-b)x)}{2(a-b)} - \frac{\sin((a+b)x)}{2(a+b)},\ a^2 \neq b^2$

45. $\displaystyle \int \cos(ax)\cos(bx)\,dx = \frac{\sin((a-b)x)}{2(a-b)} + \frac{\sin((a+b)x)}{2(a+b)},\ a^2 \neq b^2$

46. $\displaystyle \int x \sin(ax)\,dx = \frac{1}{a^2}\sin(ax) - \frac{x}{a}\cos(ax)$

47. $\displaystyle \int x \cos(ax)\,dx = \frac{1}{a^2}\cos(ax) + \frac{x}{a}\sin(ax)$

48. $\displaystyle \int x^n \sin(ax)\,dx = -\frac{x^n}{a}\cos(ax) + \frac{n}{a}\int x^{n-1}\cos(ax)\,dx,\ n > 0$

49. $\displaystyle \int x^n \cos(ax)\,dx = \frac{x^n}{a}\sin(ax) - \frac{n}{a}\int x^{n-1}\sin(ax)\,dx,\ n > 0$

50. $\displaystyle \int \tan^n(ax)\,dx = \frac{\tan^{n-1}(ax)}{a(n-1)} - \int \tan^{n-2}(ax)\,dx,\ n \neq 1$

51. $\displaystyle \int \sec^n(ax)\,dx = \frac{\sec^{n-2}(ax)\tan(ax)}{a(n-1)} + \frac{n-2}{n-1}\int \sec^{n-2}(ax)\,dx,\ n \neq 1$

Expressions Containing Exponential and Logarithm Functions

52. $\displaystyle \int x e^{ax}\,dx = \frac{e^{ax}}{a^2}(ax - 1)$

53. $\displaystyle \int x^n e^{ax}\,dx = \frac{1}{a}x^n e^{ax} - \frac{n}{a}\int x^{n-1}e^{ax}\,dx,\ n > 0$

54. $\displaystyle \int e^{ax}\sin(bx)\,dx = \frac{e^{ax}}{a^2+b^2}\big(a\sin(bx) - b\cos(bx)\big)$

55. $\displaystyle \int e^{ax}\cos(bx)\,dx = \frac{e^{ax}}{a^2+b^2}\big(a\cos(bx) + b\sin(bx)\big)$

56. $\displaystyle \int x^n \ln(ax)\,dx = x^{n+1}\left(\frac{\ln(ax)}{n+1} - \frac{1}{(n+1)^2}\right),\ n \neq -1$

57. $\displaystyle \int (\ln x)^n\,dx = x(\ln x)^n - n\int (\ln x)^{n-1}\,dx$

58. $\displaystyle \int \frac{dx}{a+be^{px}} = \frac{x}{a} - \frac{1}{ap}\ln|a + be^{px}|$

Expressions Containing Inverse Trigonometric Functions

59. $\displaystyle \int \arcsin(ax)\,dx = x\arcsin(ax) + \frac{1}{a}\sqrt{1 - a^2x^2}$

60. $\displaystyle \int \arccos(ax)\,dx = x\arccos(ax) - \frac{1}{a}\sqrt{1 - a^2x^2}$

61. $\displaystyle \int \arctan(ax)\,dx = x\arctan(ax) - \frac{1}{2a}\ln(1 + a^2x^2)$

62. $\displaystyle \int \text{arccot}(ax)\,dx = x\,\text{arccot}(ax) + \frac{1}{2a}\ln(1 + a^2x^2)$

63. $\displaystyle \int \text{arcsec}(ax)\,dx = x\,\text{arcsec}(ax) - \frac{1}{a}\ln\left|ax + \sqrt{a^2x^2 - 1}\right|$

64. $\displaystyle \int \text{arccsc}(ax)\,dx = x\,\text{arccsc}(ax) + \frac{1}{a}\ln\left|ax + \sqrt{a^2x^2 - 1}\right|$